ORGANIC SYNTHESIS
TODAY AND TOMORROW

Some Other IUPAC Titles of Interest from Pergamon Press

Books

ANANCHENKO: Frontiers of Bioorganic Chemistry and Molecular Biology

BENOIT & REMPP: Macromolecules

BRITTON & GOODWIN: Carotenoid Chemistry and Biochemistry

BROWN & DAVIES: Organ-Directed Toxicity - Chemical Indices and Mechanisms

CIARDELLI & GIUSTI: Structural Order in Polymers

EGAN & WEST: Harmonization of Collaborative Analytical Studies

FRANZOSINI & SANESI: Thermodynamic and Transport Properties of Organic Salts

FREIDLINA & SKOROVA: Organic Sulfur Chemistry

FUWA: Recent Advances in Analytical Spectroscopy

GOETHALS: Polymeric Amines and Ammonium Salts

HOGFELDT: Stability Constants of Metal-Ion Complexes
        Part A: Inorganic Ligands

KORNHAUSER, RAO & WADDINGTON: Chemical Eduation in the Seventies

LAIDLER: Frontiers of Chemistry

MIYAMOTO: 5th International Congress of Pesticide Chemistry

NOZAKI: 4th International Conference on Organic Synthesis

PERRIN: Stability Constants of Metal-Ion Complexes
        Part B: Organic Ligands

RIGAUDY & KLESNEY: Nomenclature of Organic Chemistry

ST-PIERRE & BROWN: Future Sources of Organic Raw Materials

STEC: Phosphorus Chemistry Directed Towards Biology

Journals

CHEMISTRY INTERNATIONAL, the news magazine for chemists in all fields of specialization in all countries of the world.

PURE AND APPLIED CHEMISTRY, the international research journal publishing proceedings of IUPAC conferences, nomenclature rules and technical reports.

INTERNATIONAL UNION OF PURE AND APPLIED CHEMISTRY
(Organic Chemistry Division)

in conjunction with

American Chemical Society (Organic Division)
University of Wisconsin (Department of Chemistry)

ORGANIC SYNTHESIS
TODAY AND TOMORROW

Proceedings of the
3rd IUPAC Symposium on Organic Synthesis
Madison, Wisconsin, USA, 15-20 June 1980

Edited by

BARRY M. TROST

and

C. RICHARD HUTCHINSON

University of Wisconsin
Madison, Wisconsin, USA

PERGAMON PRESS

OXFORD · NEW YORK · TORONTO · SYDNEY · PARIS · FRANKFURT

U.K.	Pergamon Press Ltd., Headington Hill Hall, Oxford OX3 0BW, England
U.S.A.	Pergamon Press Inc., Maxwell House, Fairview Park, Elmsford, New York 10523, U.S.A.
CANADA	Pergamon Press Canada Ltd., Suite 104, 150 Consumers Rd., Willowdale, Ontario M2J 1P9, Canada
AUSTRALIA	Pergamon Press (Aust.) Pty. Ltd., P.O. Box 544, Potts Point, N.S.W. 2011, Australia
FRANCE	Pergamon Press SARL, 24 rue des Ecoles, 75240 Paris, Cedex 05, France
FEDERAL REPUBLIC OF GERMANY	Pergamon Press GmbH, 6242 Kronberg-Taunus, Hammerweg 6, Federal Republic of Germany

QD262
I86
1980

First edition 1981
Reprinted 1982

British Library Cataloguing in Publication Data
IUPAC Symposium on Organic Synthesis
(3rd: 1980: Wisconsin)
Organic synthesis - today and tomorrow.-
(Symposium series/International Union of Pure
and Applied Chemistry)
1. Chemistry, Organic - Synthesis - Congresses
I. Title II. Trost, Barry Martin
III. Hutchinson, C Richard IV. International
Union of Pure and Applied Chemistry. Organic
Chemistry Division V. American Chemical Society.
Organic Division VI. University of Wisconsin.
Department of Chemistry
547'.2   QD262   80-41859
ISBN 0-08-025268-0

In order to make this volume available as economically and as rapidly as possible the author's typescript has been reproduced in its original form. This method unfortunately has its typographical limitations but it is hoped that they in no way distract the reader.

Printed in Great Britain by A. Wheaton & Co. Ltd., Exeter

CONTENTS

vi   Contents

Organizing Committee

Chairman:
B. M. TROST

Executive Secretary-Treasurer:
C. R. HUTCHINSON

Members:
C. P. CASEY
H. J. REICH
E. VEDEJS

INTERNATIONAL UNION OF PURE AND APPLIED CHEMISTRY

IUPAC Secretariat: Bank Court Chambers, 2-3 Pound Way,
Cowley Centre, Oxford OX4 3YF, UK

PREFACE

   Upon being approached to host and organize the Third IUPAC Symposium on Organic
Synthesis, our reflex response was negative. The thoughts of the effort required in
organizing such an international event were discouraging. Nevertheless, we felt such a
meeting was an important forum for transmitting the newest thoughts in the area and for
stimulating new directions. Organic synthesis enjoys an unprecedented degree of activity
and advances. Surely an international meeting devoted to this theme was timely and the
continental United States was an appropriate site considering that the first meeting was
held in Louvain-La-Neuve, Belgium in 1974 and the second in Jerusalem, Israel in 1978.
Unfortunately for us, we believe in translating our ideas into action. The result was the
birth of Synthesis 80.
   At the time we decided to host the Symposium, which eventually resulted in the
production of this monograph, the organizing committee decided that the main theme of the
Symposium should be a thorough presentation of the synthetic methodology that characterizes
modern synthetic organic chemistry. Initial thoughts centered upon research interests
which would reflect useful and swift carriages for traversing the many avenues of approach
to the synthesis of organic compounds, rather than elegant expositions on the many and
clever ways one could direct their course in achieving the total synthesis of intricate
molecules. Such a goal, though admirable for its intended limitations, was unrealistic for
the simple reason that some of the best synthetic methodology is the logical outcome of an
inherent interest in the synthesis of a molecule which has been chosen as the target of a
research program. Consequently we expected that the eventual lectures would encompass many
facets of modern synthetic organic chemistry.
   Our expectations were well-rewarded by the timeliness and comprehensiveness of the
lectures given by the twenty-eight principal participants of the Symposium in Madison during
the week of 15-20 June, 1980. All of these lectures are included in this monograph, which
we have grouped rather arbitrarily into three sub-divisions. The first group represents
lectures whose focus appeared to be useful synthetic methods for general application. This
distinction clearly applies to the presentations of D.H.R. Barton and G. Cainelli.
Similarly, the lectures given by J. Collman, M. Rosenblum, J. Schwartz, M. Semmelhack and
P. Vollhardt represent the excitement associated with the increasing use of organometallic
reagents for the efficient synthesis of molecules. The second group are lectures which
focused on the development of new chemistry and concepts that would be useful for the construction of molecular sub-assemblies. The versatility of organosilicon (I. Fleming,
P. Magnus) and organoboron (H.C. Brown, C. Still) chemistry, and the valuable role that
heteroatom chemistry plays in organic synthesis (L. Ghosez, H. Gschwend, E.C. Taylor)
are represented here. The third and largest group of lectures clearly illustrates the
interplay of synthetic methodology and the total synthesis of organic compounds. Examples
of the elegant synthesis of nearly all kinds of compounds are contained in the lectures by
P. Grieco, S. Masamune (macrolides), S. Danishefsky, M. Julia, H. Nozaki, R. Schlessinger
(terpenes), R. Kelly, R. Noyori (prostaglandins), C. Szantay, M. Uskokovic (alkaloids),
O. Achmatovitz, B. Fraser-Reid (carbohydrates and molecules made therefrom), S. Pines (beta-
lactams), and L. Paquette (unique carbocycles).
   The Symposium was the third in what promises to be a continuing and healthy series
of international meetings on organic synthesis sponsored by the International Union of Pure
and Applied Chemistry. We are pleased to have been able to foster the growth of these
meetings through our efforts, and those of our colleagues and students whose selfless effort
and continued enthusiasm made the Symposium become an outstanding success. Particular
appreciation must be expressed to Bonnie J. Hagness for her diligent and resourceful
assistance with many tasks that had to be done for the Symposium and for this monograph.

August, 1980   B. M. Trost

Madison, Wisconsin USA   C. R. Hutchinson

ACKNOWLEDGEMENTS

We are grateful to the following organizations and companies for their sponsorship and generous financial assistance.

Sponsors: International Union of Pure and Applied Chemistry

Organic Chemistry Division of The American Chemical Society

Department of Chemistry of The University of Wisconsin

Financial Contributors:

McElvain Fund of The University of Wisconsin

National Science Foundation

Petroleum Research Foundation of The American Chemical Society

Allied Chemical Corporation

Ciba-Geigy Corporation

Dow Chemical Company

E.I. DuPont De Nemours & Company

Eli Lilly & Company

G.D. Searle & Company

Hoffmann-La Roche, Inc.

ICI-Americas Inc.

Merck & Company, Inc.

Pfizer Inc.

Monsanto Company

Sandoz, Inc.

Syntex Corporation

Tennessee Eastman Company

3M Company

Union Carbide Corporation

The Upjohn Company

NEW AND SELECTIVE REACTIONS
AND REAGENTS

D. H. R. Barton and W. B. Motherwell

Institut de Chimie des Substances Naturelles, C.N.R.S., 91190 Gif-sur-Yvette, France

Abstract — Radical reactions are not widely used in the synthetic chemistry of natural products. Nevertheless, they have considerable potential utility, because they are little troubled
by steric congestion. Radical deoxygenation of the secondary
hydroxyl group is an established process. It has now been sup-
plemented by radical fission, induced by electron transfer, of
esters. With hindered esters and with various thiocarbonyl esters, excellent yields of hydrocarbon can now be obtained.
Radical deamination of primary, secondary and tertiary primary
amines has been developed to give good yields of deaminated products. These radical reactions find application in the chemistry
of aminoglycoside antibiotics. Other new radical reactions of
synthetic interest will be presented.

As we have argued elsewhere new reactions of importance in chemistry are discovered by conception, by misconception and by accident (1,2). In fact, most
of the important reactions of synthetic chemistry have been discovered by
accident.

The traditional chemistry of carbohydrate molecules is ionic in mechanism.
Faced by the need to deoxygenate secondary hydroxyl groups in aminoglycoside
antibiotics by a non-SN_2 mechanism we conceived (3) that a radical chain reaction would be the ideal solution to this problem. Radical chain reactions
proceed under neutral conditions and are much less subject to steric hindrance
than ordinary ionic reactions. In addition, they are less prone to rearrangement than cationic or anionic reactions. The reaction that we designed (3)
was the reduction of a thiocarbonyl ester, xanthate ester or thiocarbonyl-
imidazolide by tributyl tin hydride or other tin hydride (or germyl hydride)
(4) reagent (Fig. 1).

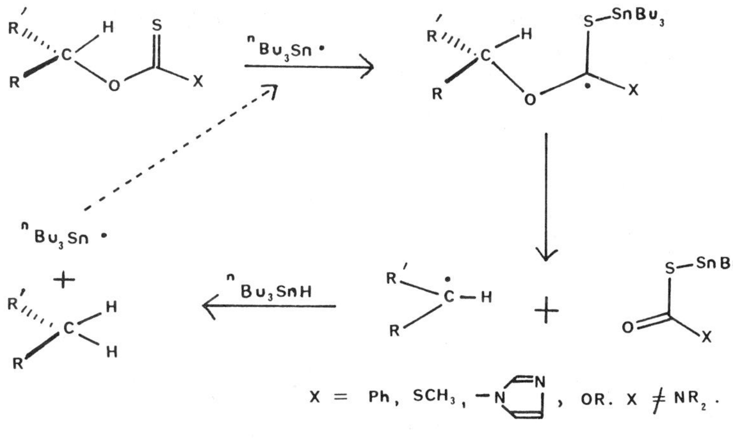

$$X = Ph, \; SCH_3, \; -N\bigcirc^N, \; OR. \; X \neq NR_2.$$

Barton and McCombie, J.C.S. Perk.1 1975, 1574.

FIGURE 1

1

It is not inappropriate to present some examples of the application of this type of reaction, especially in carbohydrate chemistry.

The first example (Fig. 2) shows the synthesis of 3-deoxyglucose and is taken from our original publication (3).

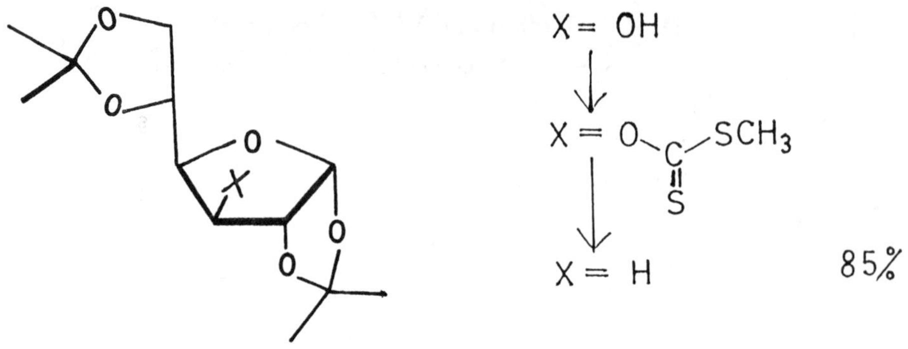

$$X = OH$$
$$\downarrow$$
$$X = O-\underset{\underset{S}{\parallel}}{C}-SCH_3$$
$$\downarrow$$
$$X = H \qquad\qquad 85\%$$

FIGURE 2

It was Bob Stick (5) who applied (Fig. 3) the reaction systematically for the synthesis of several 3-deoxy- sugars and also for the synthesis (using tributyl tin deuteride) of specifically deuterated deoxy-sugars (6). It should be, of course, a good reaction for the synthesis of tritiated molecules.

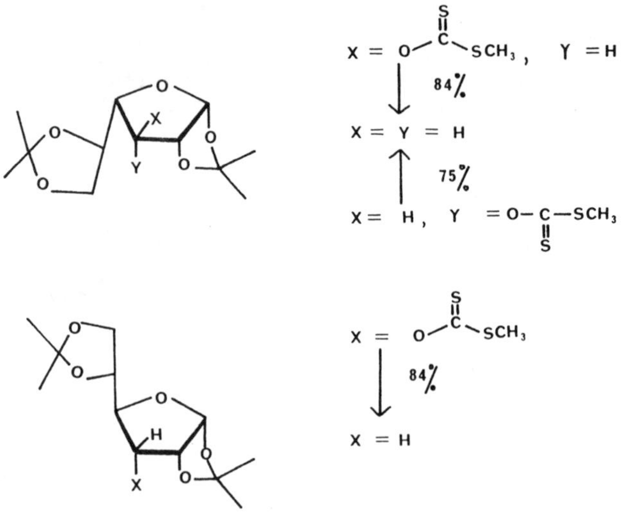

$$X = O-\underset{\underset{S}{\parallel}}{C}-SCH_3, \quad Y = H$$
$$\downarrow 84\%$$
$$X = Y = H$$
$$\uparrow 75\%$$
$$X = H, \quad Y = O-\underset{\underset{S}{\parallel}}{C}-SCH_3$$

$$X = O-\underset{\underset{S}{\parallel}}{C}-SCH_3$$
$$\downarrow 84\%$$
$$X = H$$

Copeland and Stick, Aust. J. Chem., 1977, 30, 1269.

FIGURE 3

Striking examples (Figs. 4, 5 and 6) are provided in the biologically important deoxygenation of aminoglycoside antibiotics by Hayashi et al. (7).

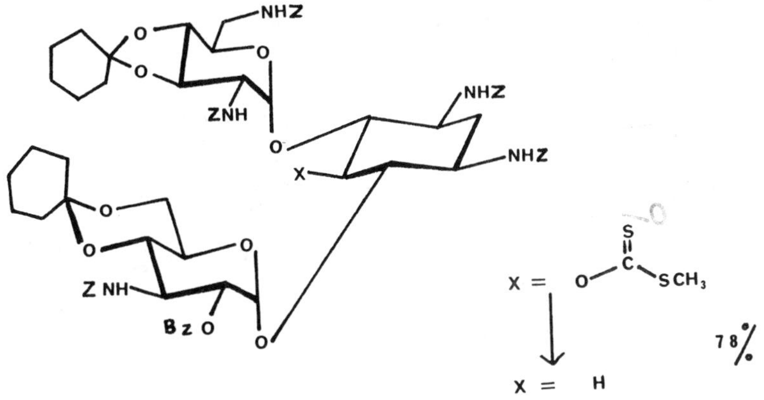

$$X = O-\overset{\overset{\displaystyle S}{\|}}{C}-SCH_3$$

$$X = H \qquad 78\%$$

Hayashi, Iwaoka, Takeda and Ohki,

Chem. Pharm. Bull., 1978, 26, 1786.

FIGURE 4

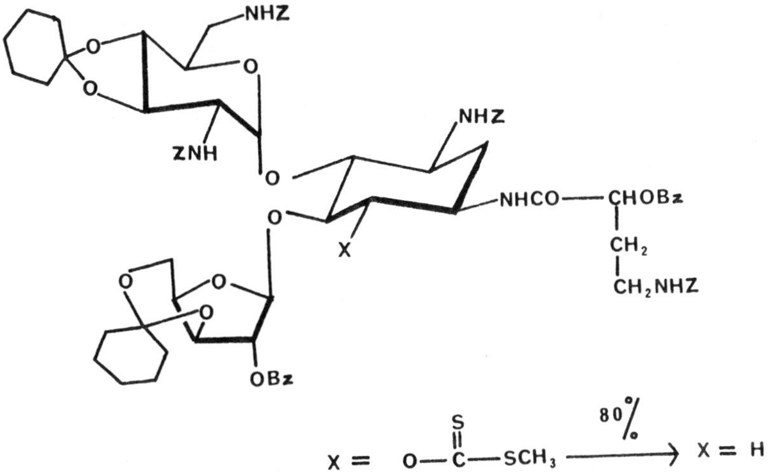

$$X = O-\overset{\overset{\displaystyle S}{\|}}{C}-SCH_3 \xrightarrow{80\%} X = H$$

FIGURE 5

$$X = O-\overset{\overset{\displaystyle S}{\|}}{C}-SCH_3 \longrightarrow X = H \qquad 82\%$$

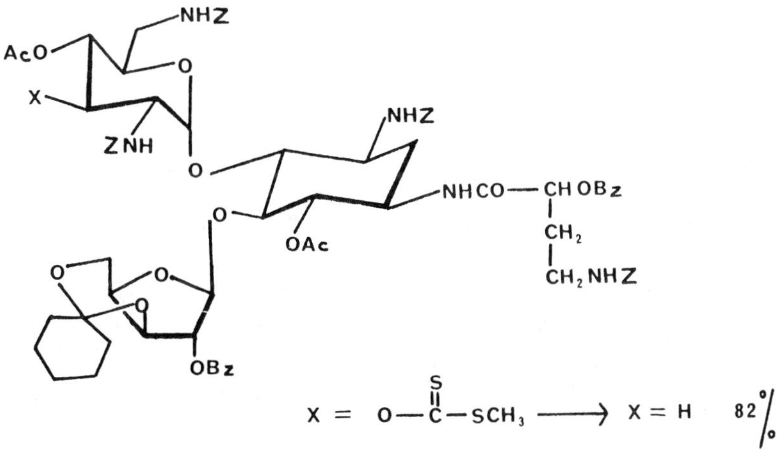

FIGURE 6

The example in Fig. 7 is interesting because the thiocarbonyl radical chain sequence was the only method which permitted the synthesis of the desired deoxy-antibiotic (8). Fig. 8 shows two recent examples that give good yields of 2-deoxygenation of sugars (9).

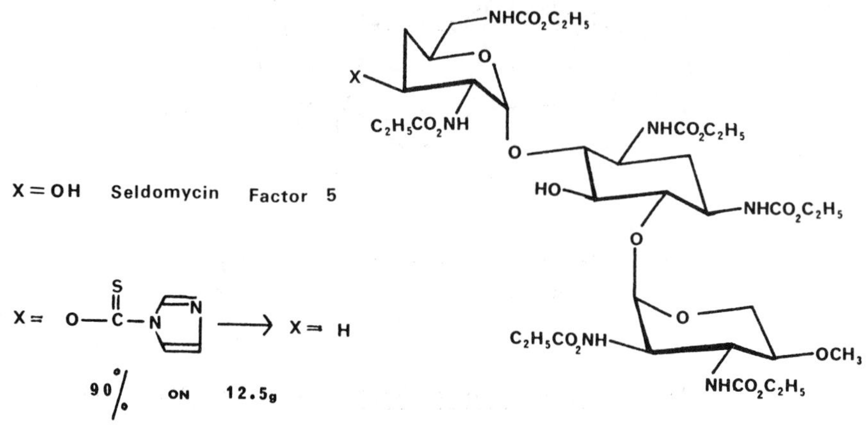

X = OH   Seldomycin   Factor 5

$$X = O-\overset{\overset{\displaystyle S}{\|}}{C}-N\overset{\frown}{\underset{\smile}{N}} \longrightarrow X = H$$

90%   on   12.5g

Carney et al., J. of Antibiotics, 1978, 31, 441.

FIGURE 7

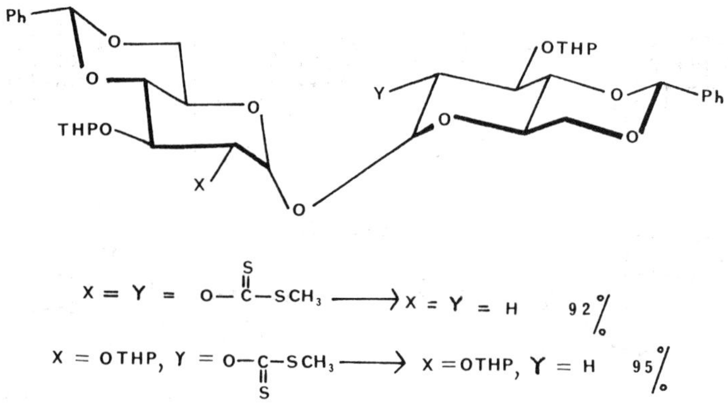

$$X = Y = O-\overset{\overset{\displaystyle S}{\|}}{C}-SCH_3 \longrightarrow X = Y = H \quad 92\%$$

$$X = OTHP, Y = O-\overset{\underset{\displaystyle S}{\|}}{C}-SCH_3 \longrightarrow X = OTHP, Y = H \quad 95\%$$

Defaye et al., Nouveau J. de Chimie, 1980, 4, 59.

FIGURE 8

Finally, the reaction is especially good for the deoxygenation of hindered secondary alcohols. Even the most hindered alcohols give xanthate esters without difficulty. A particularly noteworthy example (10) is shown in Fig. 9.

Primary alcohols are not deoxygenated by thiocarbonyl reduction. Therefore, a thiocarbonate based on a primary alcohol and a secondary alcohol should be reduced only at the secondary position. This has been demonstrated (11) (Fig. 10) for the 5 and 6 positions in sugars. This method of synthesis of 5-deoxy-sugars nicely complements the ionic reactions of 5,6-thiocarbonates which lead (12) to 6-deoxy-sugars. Naturally, when the thiocarbonate is based on two secondary alcohols a mixture of two regioisomeric deoxy-sugars results (Fig. 10).

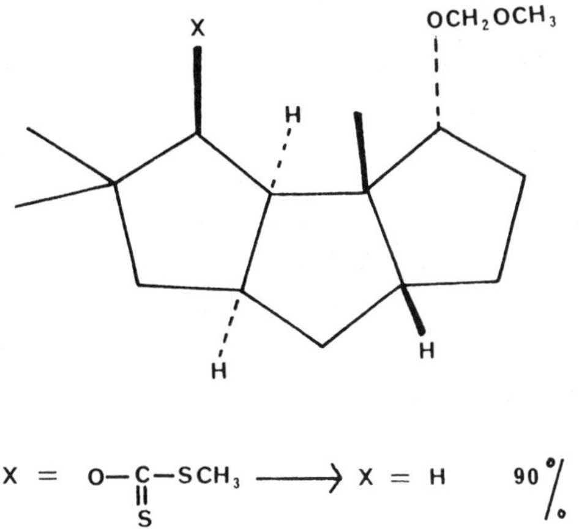

$$X = O-\underset{\underset{S}{\|}}{C}-SCH_3 \longrightarrow X = H \quad 90\%$$

Tatsuta et al., J. Amer. Chem. Soc., 1979, 101, 6116.

FIGURE 9

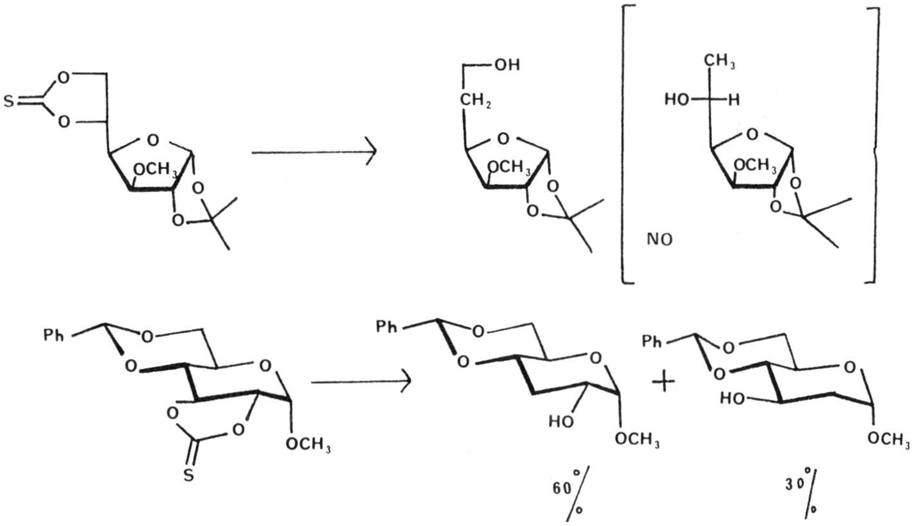

Barton and Subramanian, J. C. S. Perk. 1, 1977, 1718

FIGURE 10

We have, of course, examined the behaviour of α-glycol dithiocarbonyl derivatives. Dithionobenzoates give complex results (3), but 1,2-dixanthate esters
fragment (13) smoothly to give olefins (Fig. 11). Being a radical
elimination reaction this process gives, when aliphatic α-glycols are used,
the more stable trans-olefin (see further below). The 1,2-dixanthate elimination was also discovered independently by Hayashi et al. (7) (Fig. 12).

We now discuss what we consider to be a second, non-chain, radical reaction
for the deoxygenation of alcohols. It was discovered by accident. In a routine reduction by lithium and ethylamine of the diacetate shown in Fig. 13
we confidently expected to obtain the corresponding diol without inversion of
configuration at the (13) α-carbon (14). In fact, Ms Joukhadar, who did the
experiment, obtained a high yield of mono-deoxy-compound (15). After so many
years of routine Bouveault-Blanc reductions to give alcohols this was a surprising accident.

Barton, Barrett, Bielski and McCombie, J.C.S. Chem. Comm., 1977, 866; J.C.S. Perk. I, 1979, 2378.

FIGURE 11

Hayashi, Iwaoka, Takeda and Ohki, Chem. Pharm. Bull., 1978, 26, 1786

FIGURE 12

Ms Laurette Joukhadar.

Boar, McGhie, Misra, Barrett, Barton and Prokopiou, J.C.S. Chem. Comm., 1978, 68.

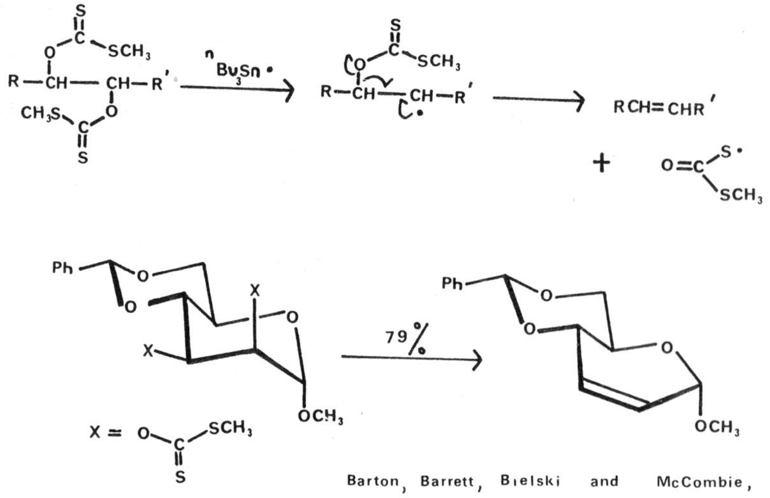

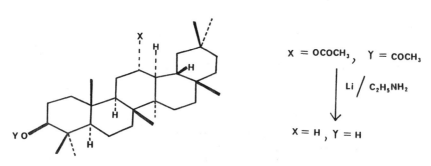

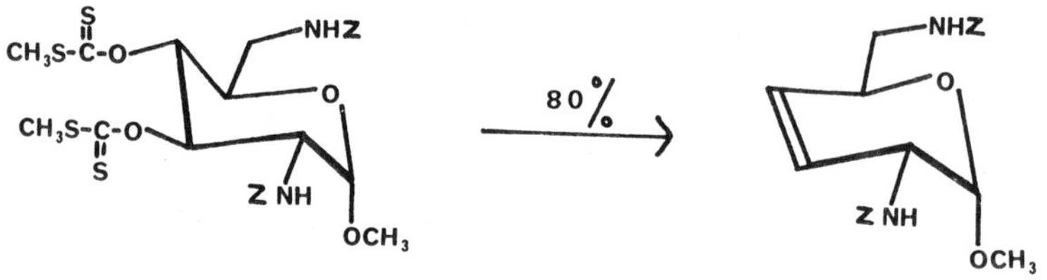

FIGURE 13

The theory of the electron-transfer reduction of esters can be summarised as in Fig. 14. The traditional Bouveault-Blanc reaction involves the cleavage of the carbonyl oxygen bond of the ester and can involve the transfer of 1 or 2 electrons before the cleavage occurs. The alkyl oxygen - carbon cleavage which we accidentally discovered also involves in principle the same transfer of one or two electrons (Fig. 14).

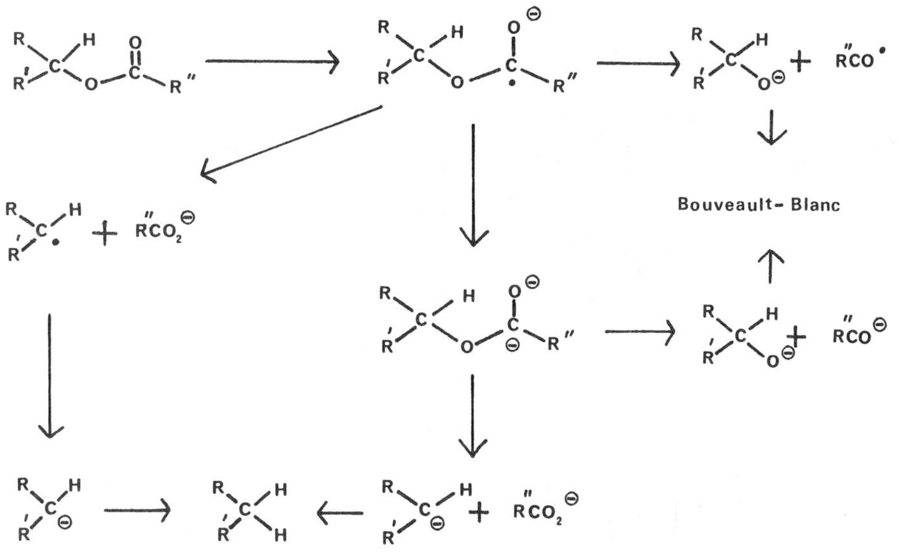

FIGURE 14

The initial example (Fig. 13) of alkyl oxygen-carbon cleavage suggested that a hindered ester was deoxygenated whilst a relatively unhindered ester was not. We investigated systematically this possibility using (Fig. 15) derivatives of cholestane-3β,6β-diol. The 3β-OH (equatorial) is, of course, unhindered whilst the 6β-OH (axial) is moderately hindered. Our respected colleague Dr. A.G.M. Barrett suggested that potassium should be soluble in hindered primary amines like t-butylamine if 18-crown-6 was added. This proved to be a valuable suggestion since one could work conveniently at room temperature, as well, of course, at other temperatures at will. The data reported in this article were obtained under those conditions unless stated to the contrary. Fig. 15 clearly shows that the more hindered is the ester the more completely is it deoxygenated (15).

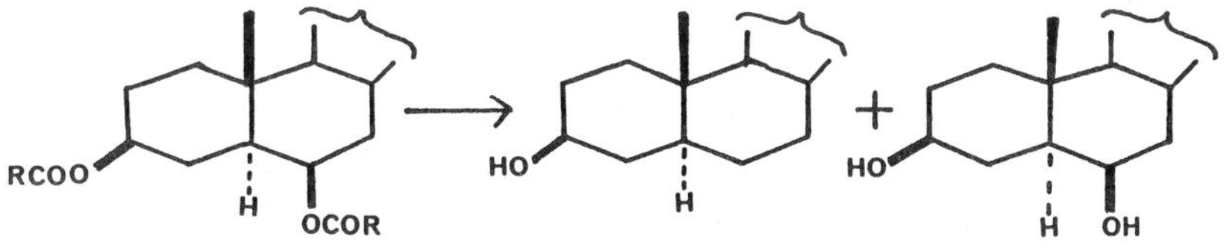

R =	H	$\longrightarrow$	0 %	+	86 %
	Me		56		27
	iso Pr		71		19
	^{t}Bu		79		9

K / 18 – crown – 6 / ^{t}BuNH$_2$ / R.T.

FIGURE 15

This work, and further more extensive investigations (16), has convinced us that the transfer of an electron under non-protonating conditions to an aliphatic or alicyclic ester is <u>normally</u> followed by radical fission to give RR'CH· and R"CO$_2^\ominus$ (Fig. 14). However, this reaction is often not observed because of competitive nucleophilic attack by nucleophiles frequently present in amine or ammonia solutions (see further below).

Thus (Fig. 16), reduction of 3α,5-cyclo-5α-cholestan-6β-yl acetate gave a hydrocarbon fraction (45%) which comprised cholest-5-ene (85%) and 3α,5-cyclo-5α-cholestane (15%). The formation of the former hydrocarbon is indicative (17) of radical opening of the cyclopropane ring. Reduction of tertiary esters always gave excellent yields of hydrocarbon whereas reduction of aromatic esters gave no hydrocarbon (Fig. 16). Tertiary radicals are more stable than secondary so the improved yield is to be expected. The non-formation of hydrocarbon on benzoate reduction is explained by the enhanced stability of the initially formed radical-anion because of the aryl ring.

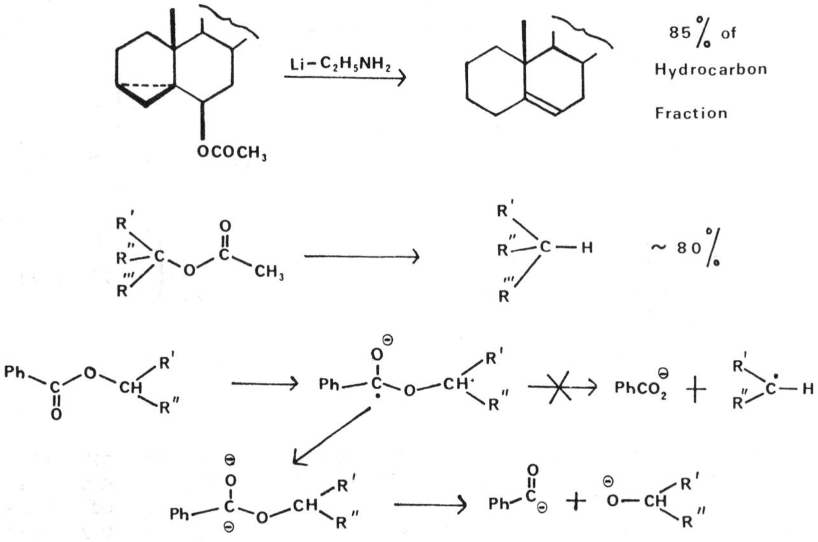

FIGURE 16

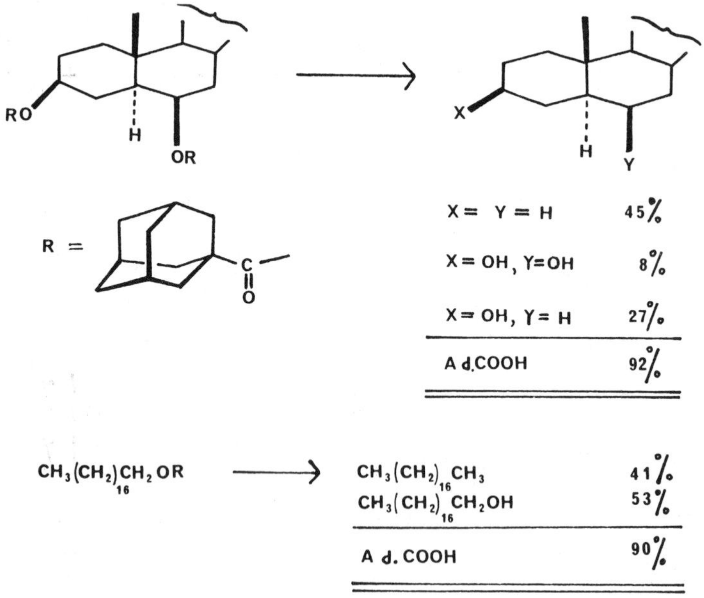

FIGURE 17

An investigation of the fate of the acid fragment of the ester proved revealing. Thus (Fig. 17) reduction of cholestane-3β,6β-diyl diadamantanoate gave extensive deoxygenation, but _exclusive_ recovery of adamantoic acid (92%). The blank experiment showed that the potassium salt of adamantoic acid was stable indefinitely under these reduction conditions. Similarly reduction of the adamantanoate ester of cetyl alcohol gave significant reduction (41%), but exclusive recovery (90%) of the acid fragment.

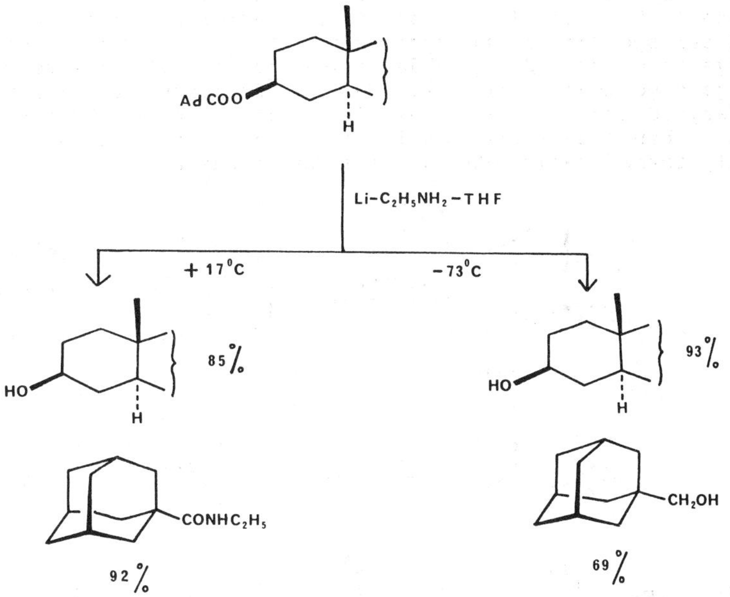

FIGURE 18

Some experiments in ethylamine using lithium as reductant proved interesting (Fig. 18). Attempted reduction of cholestan-3β-yl adamantanoate at 17° gave exclusively unchanged cholestanol (85%) and the ethylamide of adamantanoic acid (92%). Thus the lithium ethylamide, formed from traces of water or catalytically from traces of metal salts, provoked a nucleophilic attack which was faster than the reduction process the speed of which is limited by the

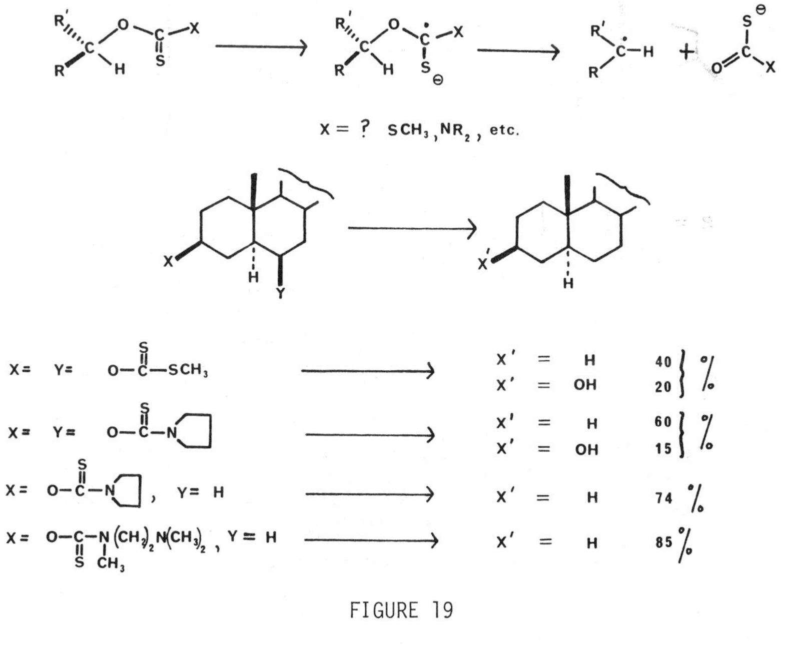

FIGURE 19

dissolution of the metal. In contrast, a reduction at -73° produced normal
Bouveault-Blanc products. We consider that the Bouveault-Blanc reaction takes
place when protonation at the ethereal oxygen of the radical-anion is possible
(Fig. 14), or when the initial radical-anion does not fragment because of
the low temperature and thus is reduced further to the dianion which fragments
to RR'CHO$^\ominus$ and R"CO$^\ominus$. The latter explanation may be responsible for the results obtained at -73°. Incidentally, the protonation of the initial radical
anion on carbonyl oxygen by an alcohol, or by water, is surely excluded by
pKa considerations (18).

Based on those results we were able to conceive of a new family of deoxygenation systems (19). Clearly, if the failure of deoxygenation with primary and
unhindered secondary esters is due to amidolysis, the best way to avoid this
is to take esters which are reformed by the amidolysis reaction. Fig. 19
shows xanthate and, especially, thiocarbonylamide esters which satisfy this
proposition and which give good yields of deoxygenated product. In addition,
they are easy to synthesise by amine displacement on xanthate esters. The
diethylaminothiocarbonyl ester of cetyl alcohol gave 87% of hydrocarbon.

CONCEPTION   (MISCONCEPTION)

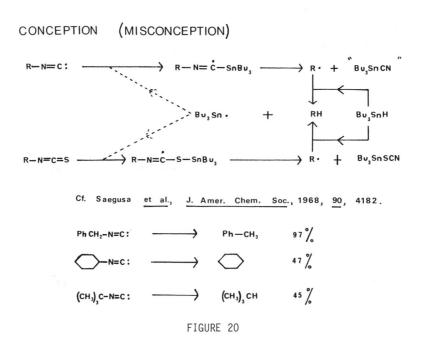

FIGURE 20

In support of our mechanistic proposals we would cite the smooth deoxygenation
of esters when photolysed at 250 mμ in hexamethylphosphoramide (20). This
must be a fragmentation of the radical-anion and it works even for unhindered
esters. Similarly the reduction of esters by sodium in hexamethylphosphoramide affords deoxygenation with varying efficiency (21). Electron transfer
reduction of other sulphur or phosphorus esters has been the subject of recent publications (22).

We now turn to another radical-chain reaction. It would also be interesting
to remove selectively, and by radical processes, the amino-functions present
in aminoglycoside antibiotics. We conceived (Fig. 20) that this should be
possible using isonitriles, isothiocyanates or isoselenocyanates as substrates and a tin hydride as a reducing agent. In fact, we soon discovered that
the reduction of isonitriles by tributyl tin hydride was already known, but
was reported to give poor yields for secondary and tertiary isonitriles (23).
We decided therefore to start with isothiocyanates and isoselenocyanates.
In a suitable model compound (Fig. 21) both gave excellent yields (90%) of
deaminated product. However, the corresponding isonitrile also afforded the
same yield. Indeed, a careful study of the course of the reaction showed
that the isothiocyanate and the isoselenocyanate were first reduced to the
isonitrile before deamination (24).

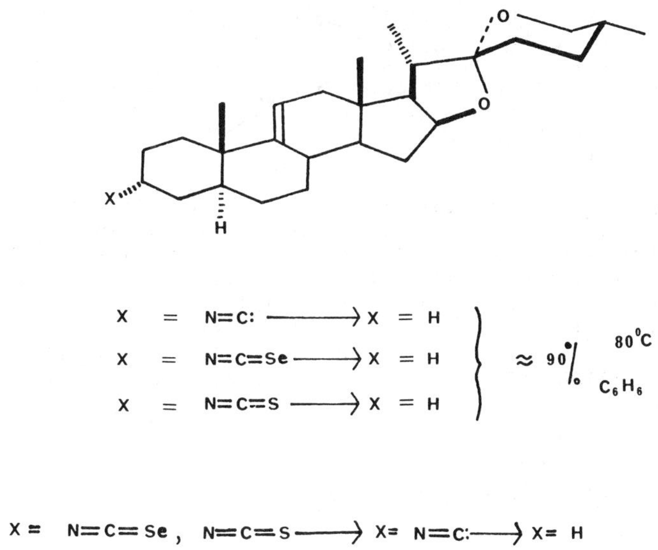

$$X = N\equiv C: \longrightarrow X = H$$
$$X = N\equiv C=Se \longrightarrow X = H$$
$$X = N\equiv C=S \longrightarrow X = H$$

$$\approx 90\% \quad \begin{matrix} 80^{0}C \\ C_6H_6 \end{matrix}$$

$$X = N\equiv C=Se, \quad N\equiv C=S \longrightarrow X= N\equiv C: \longrightarrow X= H$$

FIGURE 21

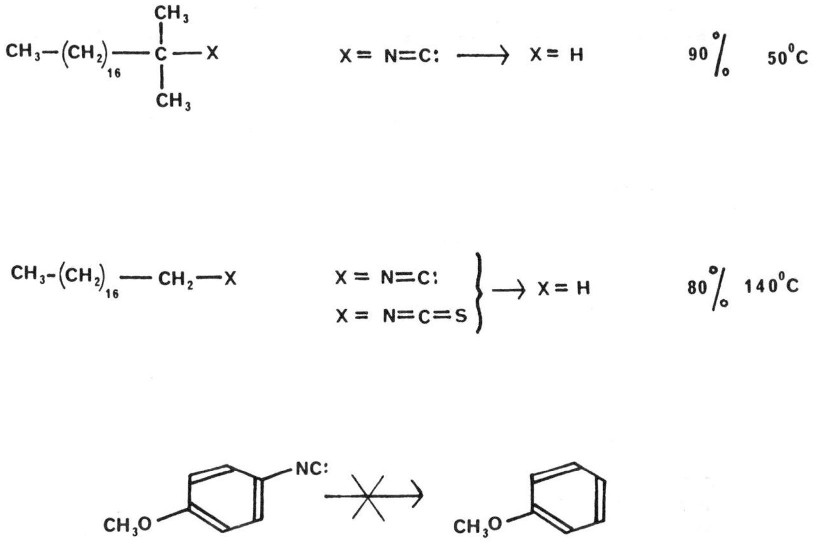

$$X = N\equiv C: \longrightarrow X = H \qquad 90\% \quad 50^{0}C$$

$$X = N\equiv C: \atop X = N\equiv C=S \Big\} \longrightarrow X = H \qquad 80\% \quad 140^{0}C$$

FIGURE 22

The excellent yields for the reduction of secondary isonitriles in benzene or in toluene under reflux could be duplicated (Fig. 22) for tertiary isoni- triles (in benzene) and for primary isonitriles (in xylene). The marked difference in ease of reduction of tertiary, secondary and primary isonitriles makes selective reduction possible. Aromatic isonitriles are not reduced under any of the conditions we have examined.
A simple example of the application of the reaction is given in Fig. 23, which shows two different routes from the readily available glucosamine to 2-deoxyglucose (24).

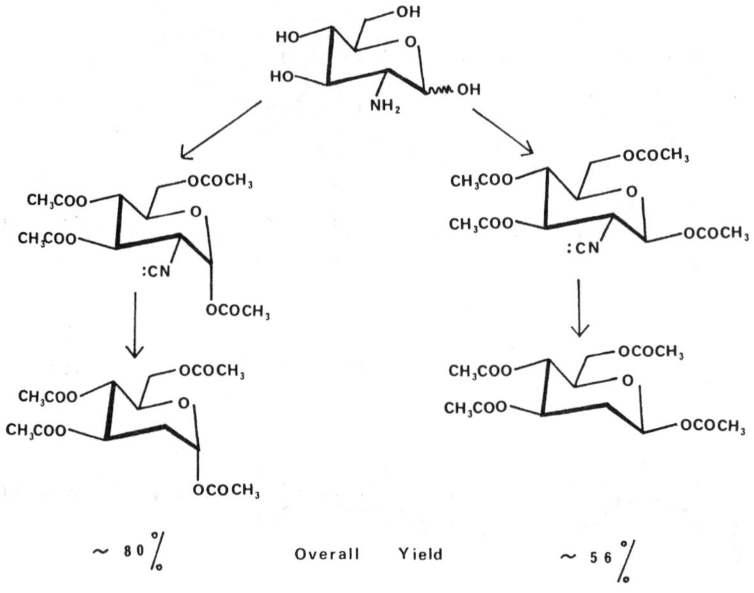

~ 80 %   Overall Yield   ~ 56 %

FIGURE 23

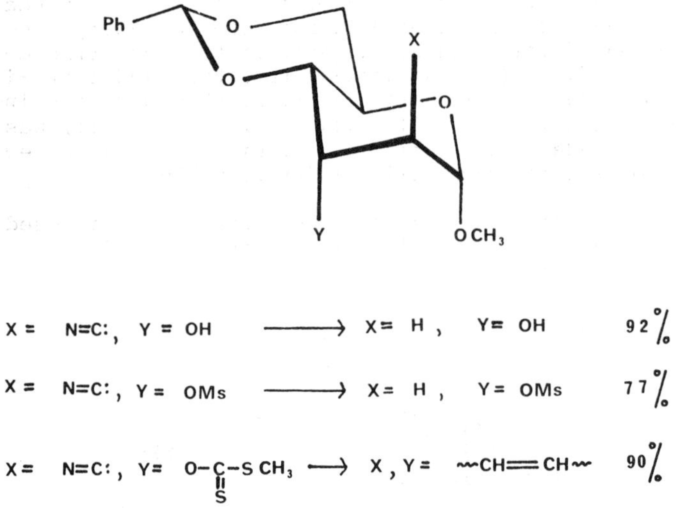

X = N≡C:, Y = OH   ⟶   X= H, Y= OH   92 %

X = N≡C:, Y = OMs   ⟶   X= H, Y= OMs   77 %

X = N≡C:, Y= O-C-S CH₃ ⟶ X, Y = ∼CH═CH∼   90 %
(with S below the C)

FIGURE 24

It was naturally of interest to examine the possibilities of 1,2-elimination
reactions of a radical character based on isonitrile reduction. We used, as
model compounds, the readily available derivatives of glucose shown in
Fig. 24. Neighbouring OH, OAc (Fig. 23) and even O-mesylate were not
effected by the radical reduction of an isonitrile group. In contrast, the
presence of an α-xanthate function afforded an excellent yield of olefin.
We were interested to find out if it was the isonitrile or the xanthate function which was the trigger group for the elimination. Fig. 25 shows that
both (±)- and meso- 1,2-diisonitrilo-bibenzyl gave the saturated bibenzyl on
reduction with minor amounts of toluene. The toluene may come from radical
fragmentation of the starting material. Since the intermediate radical in
this Fig. 25 should lose isonitrile (cyanide) radical even more easily than
in the sugar example in Fig. 24, it is clear that it is the isonitrile function which triggers the elimination of the xanthate residue and not vice-versa.
Olefin forming radical reactions should find a place in Organic Synthesis (25).

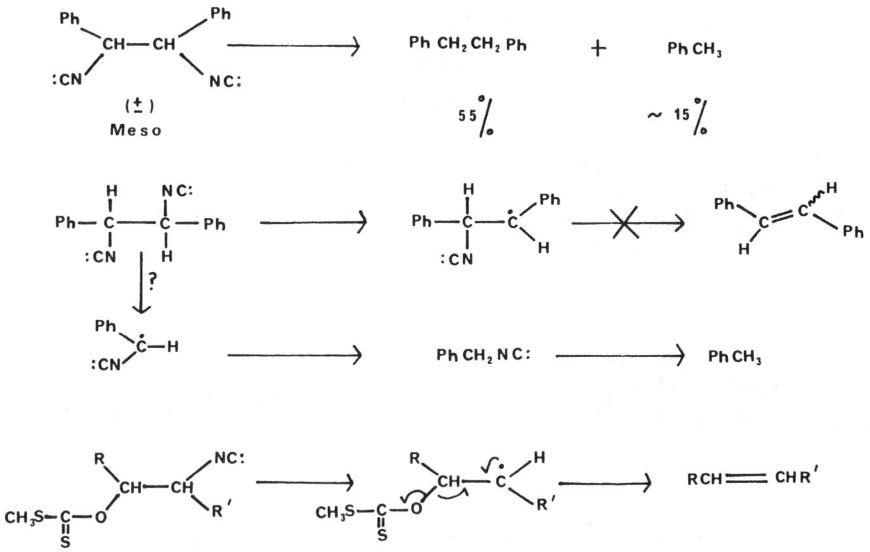

FIGURE 25

We have also made a study of the selective deamination of neamine (Fig. 26) (26). In brief, neamine can be formylated and then acetylated in quantitative yield. This compound can then be dehydrated to isonitrile derivatives in the order indicated ①, ②, ③, ④. The mono-, di-, tri- and tetra-isonitriles thus obtained could all be smoothly deaminated to the corresponding desamino-compounds. The primary isonitrile function, as expected, was less reactive than the other secondary isonitriles. It is clear that selective deamination of aminoglycoside antibiotics will now be possible.

Radical deaminations of the same kind have recently been used in an elegant manner in the manipulation of β-lactam antibiotics (27).

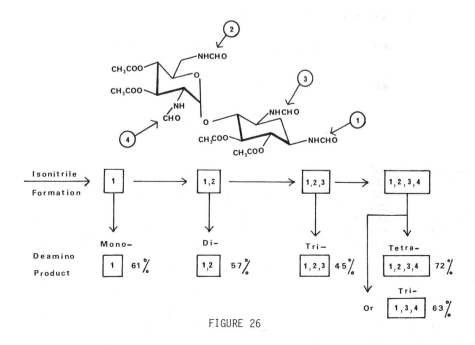

FIGURE 26

CONCEPTION

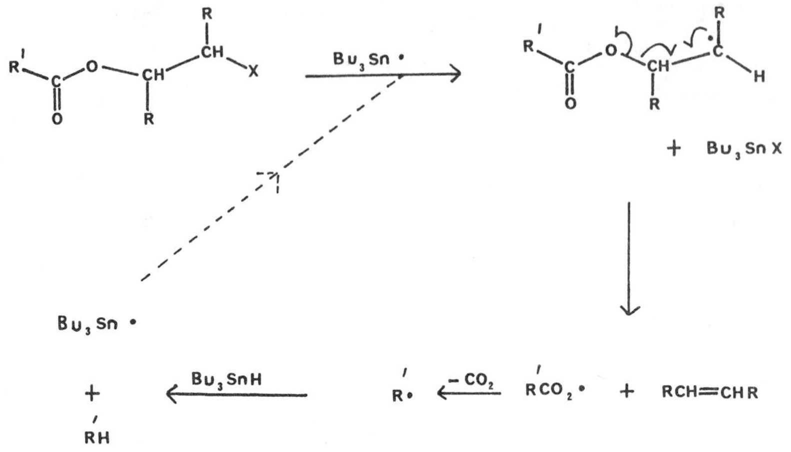

FIGURE 27

Finally, we present a new method for the conversion of a carboxylic acid into the corresponding (nor-) hydrocarbon. At present this operation can be effected by pyrolysis of per-esters in a suitable hydrogen atom donor solvent. However, the temperatures used are relatively high and the yields not satisfactory (28). We conceived (Fig. 27) that an olefin forming β-elimination of the type that we have already discussed several times above would be a novel way of producing cleanly an acyloxy-radical. However, ordinary vicinal halohydrin esters can be reduced smoothly by tributyl tin hydride without any sign of elimination (for example, see Fig. 28). We supposed that a more substantial driving force was needed and therefore concentrated on 9(10)-dihydrophenanthrene derivatives. Our first studies were made using the alcohol obtained by opening phenanthrene-9(10)-epoxide (29) with thiophenoxide anion. Later, we found that the corresponding chlorohydrin (30) esters were even better giving an easier work-up procedure (Fig. 28).

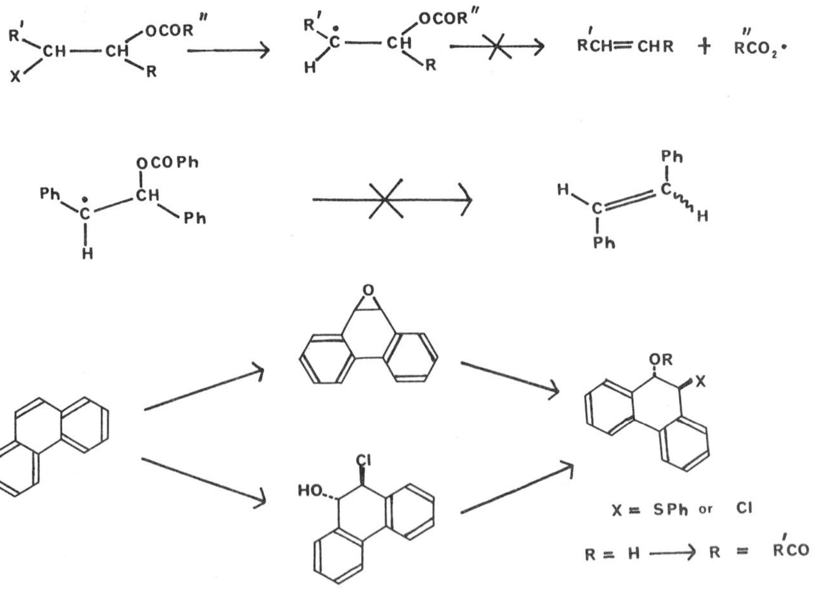

FIGURE 28

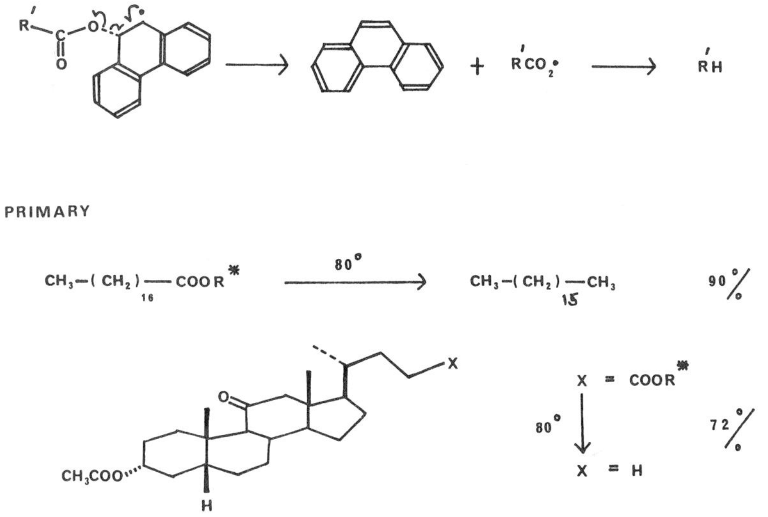

PRIMARY

FIGURE 29

In any case tributyl tin hydride reduction of such esters gave radicals which fragmented (Fig. 29) to give acyloxy-radicals and phenanthrene. A temperature of 80° (boiling benzene) was sufficient for the smooth fragmentation of acyloxy radicals based on primary, secondary and tertiary acids in the aliphatic and alicyclic series (31). Higher temperatures would be needed to obtain good yields in the decarboxylation of certain aromatic acids. Typical examples are shown in Figs. 29, 30 and 31.

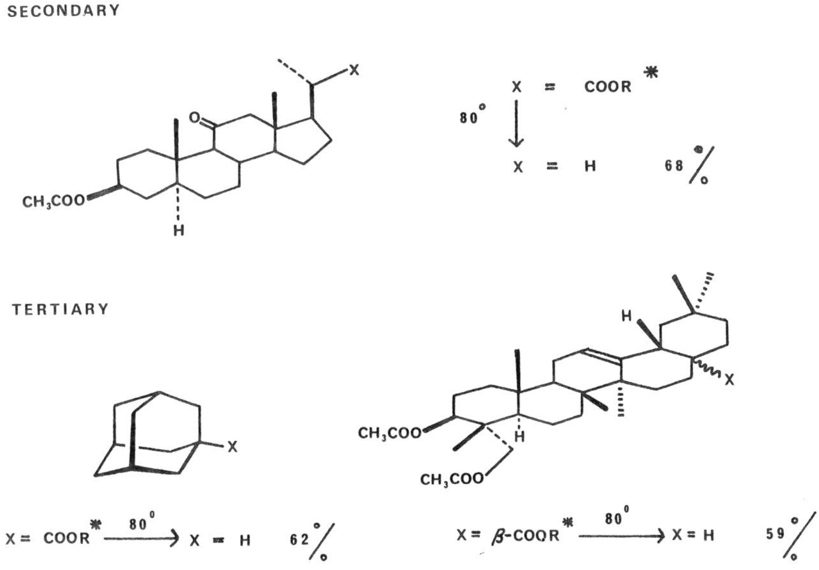

SECONDARY

TERTIARY

FIGURE 30

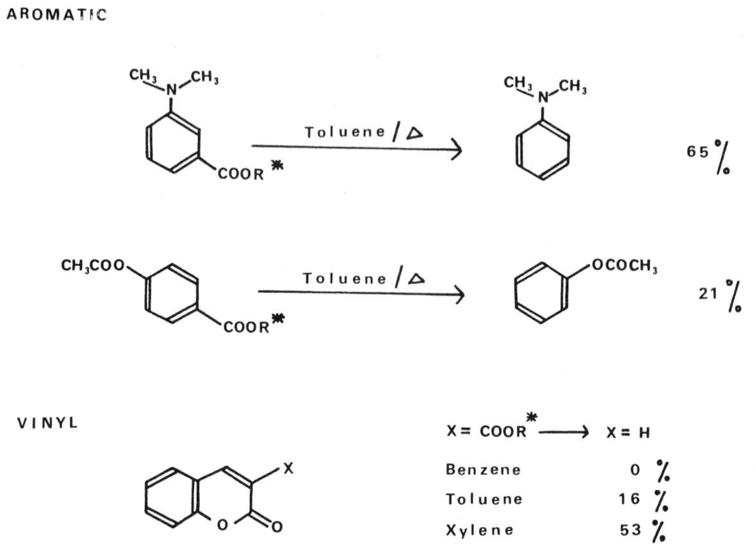

AROMATIC

VINYL

$$X = COOR^* \longrightarrow X = H$$

Benzene	0 %
Toluene	16 %
Xylene	53 %

FIGURE 31

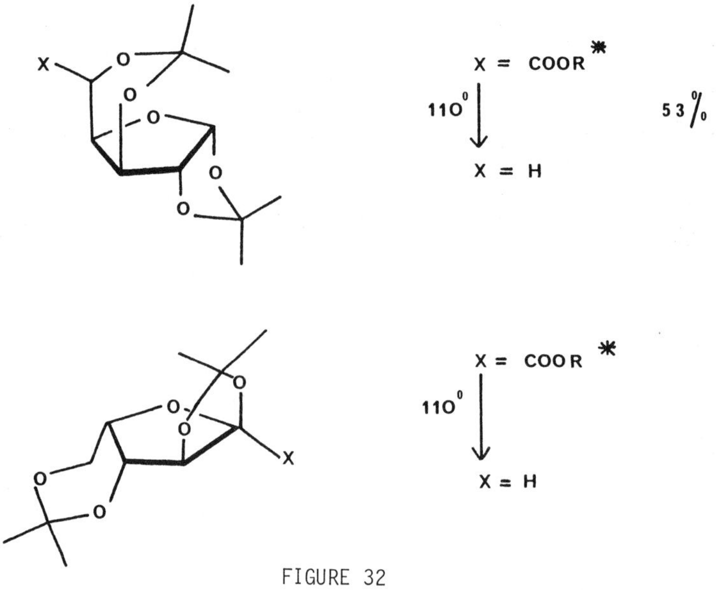

$$X = COOR^*$$
$$110^0 \downarrow \qquad 53 \%$$
$$X = H$$

$$X = COOR^*$$
$$110^0 \downarrow$$
$$X = H$$

FIGURE 32

In Fig. 32 we give two examples of the decarboxylation of sugar acids. No doubt, this reaction can be applied generally to the problem of the reductive decarboxylation of uronic acids.

The above examples will have shown how radical chain reactions can provide useful synthetic procedures especially in carbohydrate chemistry. In addition it is possible in certain cases to discover these reactions by conception.

Acknowledgements — We thank all our collaborators whose names have been cited in the References.

REFERENCES

1. D.H.R. Barton, Pure and Appl. Chem., 49, 1241-1249 (1977).
2. D.H.R. Barton in "Frontiers in Bioorganic Chemistry and Molecular Biology" Ed. Y.U. Ovchinnikov and M.N. Kolosov, Elsevier, Amsterdam, 21-37 (1979).
3. D.H.R. Barton and S.W. McCombie, J. Chem. Soc. Perk. I, 1574 (1975).
4. D.H.R. Barton, R.S. Hay-Motherwell and W.B. Motherwell, unpublished observations.
5. C. Copeland and R.V. Stick, Aust. J. Chem., 30, 1269 (1977) ; J.J. Patroni and R.V. Stick, ibid., 31, 445 (1978).
6. J.J. Patroni and R.V. Stick, J. Chem. Soc. Chem. Comm., 449 (1978) ; Aust. J. Chem., 32, 411 (1979).
7. T. Hayashi, T. Iwaoka, N. Takeda and E. Ohki, Chem. Pharm. Bull., 26, 1786 (1978);see also P.J.L. Daniels and S.W. McCombie, U.S.P.4,053,591 (11/10/1977).
8. R.E. Carney, J.B. McAlpine, M. Jackson, R.S. Stanaszek, W.H. Washburn, M. Cirovic and S.L. Mueller, J. Antibiotics, 31, 441 (1978).
9. J. Defaye, H. Driguez, B. Henrissat and E. Bar-Guilloux, Nouveau J. Chim., 4, 59 (1980).
10. K. Tatsuta, K. Akimoto and M. Kinoshita, J. Amer. Chem. Soc., 101, 6116 (1979).
11. D.H.R. Barton and R. Subramanian, J. Chem. Soc. Chem. Comm., 867 (1976); J. Chem. Soc. Perk. I, 1718 (1977).
12. D.H.R. Barton and R.V. Stick, J. Chem. Soc. Perk. I, 1773 (1975).
13. A.G.M. Barrett, D.H.R. Barton, R. Bielski and S.W. McCombie, J. Chem. Soc. Chem.Comm., 866 (1977) ; A.G.M. Barrett, D.H.R. Barton and R. Bielski, J. Chem. Soc. Perk. I, 2378 (1979).
14. R.B. Boar, L. Joukhadar, M. de Luque, J.F. McGhie, D.H.R. Barton, D. Arigoni, H.G. Brunner and R. Giger, J. Chem. Soc. Perk. I, 2104 (1977).
15. R.B. Boar, L. Joukhadar, J.F. McGhie, S.C. Misra, A.G.M. Barrett, D.H.R. Barton and P.A. Prokopiou, J. Chem. Soc. Chem. Comm., 68 (1978) ; see also A.K. Mallams , H.F. Vernay, D.F. Crowe, G. Detre, M. Tanabe and D.M. Yasuda, J. of Antibiotics, 26, 782 (1973).
16. A.G.M. Barrett, P.A. Prokopiou, D.H.R. Barton, R.B. Boar and J.F. McGhie, J. Chem. Soc. Chem. Comm., 1173 (1979).
17. A.L.J. Beckwith and G. Phillipou, J. Chem. Soc. Chem. Comm., 658 (1971); and references there cited.
18. V. Rautenstrauch, personal communication ; V. Rauteustrauch and M. Geoffrey, J. Amer. Chem. Soc., 98, 5035 (1976) ; E. Hayon and M. Simic, Acc. Chem. Res., 7, 114 (1974).
19. A.G.M. Barrett, P.A. Prokopiou and D.H.R. Barton, J. Chem. Soc. Chem. Comm., 1175 (1979).
20. H. Deshayes, J.P. Pete and C. Portella, Tet. Lett., 2019 (1976) ; J.P. Pete, C. Portella, C. Monneret, J.C. Florent and Q. Khuong-Huu, Synthesis, 774 (1977) ; R. Beugelmans, M.-T. Le Goff, H. Compaignon de Marcheville, Comp. Rend., 269, 1309 (1969).
21. H. Deshayes and J.P. Pete, J. Chem. Soc. Chem. Comm., 567 (1978).
22. O. Oida, H. Sacki, Y. Ohashi and E. Ohki, Chem. Pharm. Bull., 23, 1547 (1975) ; J.A. Marshall and M.E. Lewellyn, J. Org. Chem., 42, 1311 (1977) ; T. Hayashi, N. Takeda, H. Sacki and E. Ohki, Chem. Pharm. Bull., 25, 2134, (1977) ; T. Tsuchiya, F. Nakamura and S. Umezawa, Tet. Lett., 2805 (1979).
23. T. Saegusa, S. Kobayashi, Y. Ito and N. Yasuda, J. Amer. Chem. Soc., 90, 4182 (1968).
24. D.H.R. Barton, G. Bringmann, G. Lamotte, R.S. Hay-Motherwell and W.B. Motherwell, Tet. Lett., 2291 (1979) ; D.H.R. Barton, G. Bringmann, G. Lamotte, R.S. Hay-Motherwell, W.B. Motherwell and A.E.A. Porter, J. Chem. Soc. Perk. I, in press.
25. See also B. Lythgoe and I. Waterhouse, Tet. Lett., 4223 (1977).
26. D.H.R. Barton, G. Bringmann and W.B. Motherwell, J. Chem. Soc. Perk.I, in press.
27. D.I. John, E.J. Thomas and N.D. Tyrrell, J. Chem. Soc. Chem Comm., 345 (1979).
28. Inter alia : K.B. Wiberg, B.R. Lowry and T.H. Colby, J. Amer. Chem. Soc., 83, 3998 (1961) ; P.E. Eaton and T.W. Cole, ibid., 86, 3157 (1964) ; H. Langhals and C. Ruchardt, Chem. Ber., 108, 2156 (1975).
29. S. Krishnan, D.G. Kuhn and G.A. Hamilton, J. Amer. Chem. Soc., 99, 8121 (1977).
30. M.-C. Lasne, S. Masson and A. Thuillier, Bull. Soc. Chim. Fr., 1751 (1973).
31. D.H.R. Barton, H.A. Dowlatshahi, W.B. Motherwell and D. Villemin, J. Chem. Soc. Chem. Comm., in press.

APPLICATION OF SOME POLYMER SUPPORTED REAGENTS TO ORGANIC SYNTHESIS

G. Cainelli, F. Manescalchi and M. Contento

Istituto Chimico "G. Ciamician", Via Selmi 2, Bologna, Italy

Abstract- The preparation and the use of some polymer supported reagents based on commercial readily available anion-exchangers are reported. Preparation of esters form carboxylic acids and of alkyl fluorides from alkyl halides,oxidation of alcohols and allylic halides to carbonyl compounds,conversion of alkyl halides to homologous aldehydes,reduction of aromatic nitro compounds,reductive elimination of 1,2-dibromo derivatives,Wittig-Horner olefination of aldehydes and ketones,α bromination and chlorination of aldehydes and ketones are among the applications studied.Scope and limitations of the reactions and some peculiar features of these polymeric reagents,e.g. their substrate specificity, are breafly discussed.

The advantages of the "solid-phase" method in polypeptyde syntheses,first developed by Merrifield (1),have brought a number of chemists to use polymer supports for different kinds of synthetic operations. An ever increasing number of reviews (Ref.2,3 & 4) on the subject has been published,mainly dealing with the use of functionalised polymers. Ion-exchange resins,historically the earliest examples of polymer supported reagents,have been relatively neglected although they provide the same advantages of more sophisticated polymeric materials,e.g. simple reaction work-up,easy product isolation and mild reaction conditions. Bearing in mind the principles of the phase-transfer catalysis,we have recently carried out in our laboratory a research program with the aim of demonstrating,in this context,the usefulness of anion-exchangers. The latter,indeed,can actually be considered as potential substituents of tetralkylammonium salts,the catalysts usually employed in the phase-transfer technique. It has to be pointed out,however,that ion-exchangers can also be thoroughly dried, for instance by azeotropic distillation,this feature providing two main advantages over traditional phase-transfer catalysis. The high degree of dehydration of the anions linked to the resin,on the one hand avoids undesired hydrolitic side reactions and on the other tremendously increases their nucleophilicity,especially in the case of small anions such as the fluorine one. Furthermore,the ease of removal of the resin from the reaction mixture, permits the use of a substantial excess of reagent,thereby enhancing reaction rates and yields.Finally,as far as our experience is concerned,the choice of the solvent is seldom critical,mostly being determined by the desired reaction temperature:protic solvents such as alcohols and even water have been used in the case of highly reactive species. Thus, commercially available ion-exchange resin can often be used instead of the expensive and time-consuming building of a specific macromolecular backbone. Concerning the structure of the polymeric support,better results were obtained using the so-called macroreticular resins which are characterized by good mechanical properties and by large pores more accessible to the organic molecules. Reactions were generally performed in batch,stirring the functionalized resin with the reagent in a suitable solvent. Sometimes the product can be obtained simply by passing a solution of the substrate down a column packed with the polymeric reagent:this technique is very interesting as it offers the opportunity of a continuous process. Except in only a few cases, the resin can be regenerated by washing with an appropriate reagent solution. In Table 1 are collected some results we have obtained(Ref.5) in the preparation of esters from carboxylic acids and alkyl halides,according to the following scheme

The preparation of the polymeric reagent was accomplished,here and whenever possible,by a neutralization process,starting from the resin in the hydroxide form.

TABLE 1. Esters from resin-bound carboxylate anions and alkyl halides

Acid	Halide	%Yield
C_6H_5COOH	CH_3I	90
	$i-C_3H_7Br$	60
	$C_2H_5O_2CCH_2Br$	99
	$C_6H_5CH_2Cl$	79
	CH_3OTos	91
$c-C_6H_{11}COOH$	CH_3I	76
	$i-C_3H_7Br$	28
$C_6H_5CH=CHCOOH$	CH_3I	97
	$i-C_3H_7Br$	59
$n-C_{11}H_{23}COOH$	CH_3I	93
	$i-C_3H_7Br$	52

All the possible halogen/halogen interchange have been performed (Ref.6) by reaction of alkyl halides with a suitable form of the resin. The most interesting result,from a synthetic point of view,has been,in this context,the preparation of alkyl fluorides:

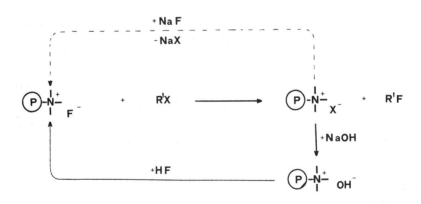

Owing to the effectiveness of this system,extraordinarily mild reaction conditions may be employed and good yields obtained,as shown in Table 2.
Also the basicity of the fluoride ion seems to be strongly enhanced by the polymeric support. Nevertheless,secondary alkyl fluorides are prepared in substantial amounts starting from sulfonates,these substrate being less prone to the elimination reaction than the corresponding alkyl halides.On the other hand,quantitative alkene formation was observed with chloro-cyclohexane.This feature,however,can be turned to advantage,and the 3bromo-2,6,6trimethyl-cyclohex-2-ene-1carbaldehyde was,for instance,prepared from the corresponding 2,3dibromo saturated compound by means of the fluoride form of the resin in refluxing toluene (Ref.7).

TABLE 2. Alkyl fluorides prepared by Hal/F or RSO_2O/F exchange

Substrate	Reaction conditions (solvent/temp./time hrs.)	Alkene %yield	Alkyl fluoride % yield
$n-C_8H_{17}Br$	hexane/reflux/20	12	82
$n-C_8H_{17}Cl$	hexane/reflux/24	13	87
$n-C_8H_{17}OMes$	pentane/reflux/30	2	92
$C_6H_5CH_2Cl$	pentane/reflux/24		99
$C_2H_5OCOCH_2Br$	pentane/r.t./48		65
$C_6H_5COCH_2Br$	pentane/reflux/2		98
$n-C_6H_{13}CH(Br)CH_3$	pentane/reflux/30	73	20
$n-C_6H_{13}CH(OMes)CH_3$	pentane/reflux/30	25	70
$c-C_6H_{11}Cl$	diethylether/reflux/30	50	

Phenyl sulfones were easily obtained (Ref.8) according to the following scheme:

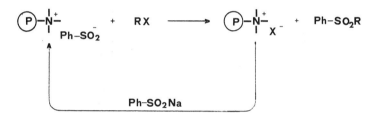

Mild conditions,fast reaction rates and high yields seem to make our method an improvement over other existing procedures.Some results are collected in Table 3.

TABLE 3. Phenyl sulfones from alkyl halides and polymer supported benzensulfinate

Alkyl halide	Reaction time(hrs.)[a]	%Yield
CH_3I	3	95
$n-C_8H_{17}Br$	3	92
$n-C_6H_{13}CH(Br)CH_3$	3	60[b]
$C_6H_5CH_2Cl$	2	93
$(CH_3)_2C=CHCH_2Br$	1.5	95
$C_2H_5OCOCH_2Cl$	2	91
$NCCH_2Cl$	2	95

a - in refluxing benzene
b - in refluxing toluene

Following the same principle,we have polymer-supported the cyanate and thiocyanate ions
(Ref.9) by washing the resin in the chloride form with an aqueous solution of the potassium
salts of the two anions. The cyanate resin results somewhat instable and undergoes decomposition on heating. Removal of water by azeotropic distillation was therefore not possible
and this prevented us from isolation of the expected isocyanates. The products obtained by
treatment of the cyanate resin with alkyl halides in hydrocarbon solvents were identified as
N,N' dialkylureas,arising from a hydrolitic reaction caused on the previously formed isocyanate by the residual water present in the resin. Thoroughly washing the resin in absolute
ethanol and performing the reaction in the same solvent,N-substituted ethyl urethanes were
prepared.

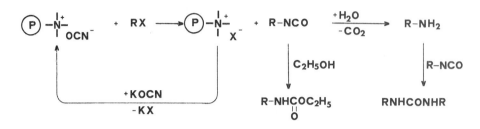

Thiocyanates were similarly obtained in good yields by treating primary or secondary alkyl
bromides in hydrocarbons under reflux. Concerning the regioselectivity of the alkylation of
the bidentate thiocyanate ion,exclusive sulfur attack has been observed in analogy with the
reaction in solution.

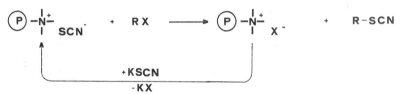

TABLE 4. Reaction of polymer-supported cyanate and thiocyanate with organic halides

Substrate	Product	Reaction conditions[a] (solvent/time hrs.)	%Yield
A: polymer-supported cyanate			
$C_6H_5CH_2Cl$	$(C_6H_5CH_2NH)_2CO$	pentane / 7	90
⟶Br	⟶NH)$_2$CO	benzene / 12	65
$n\text{-}C_6H_{13}CH(Br)CH_3$	$(n\text{-}C_6H_{13}CH(CH_3)NH)_2CO$	benzene / 12	40
$n\text{-}C_8H_{17}Br$	$(n\text{-}C_8H_{17}NH)_2CO$	pentane / 12	93
$C_6H_5CH_2Cl$	$C_6H_5CH_2NHCOOC_2H_5$	ethanol / 6	83
$n\text{-}C_8H_{17}Br$	$n\text{-}C_8H_{17} NHCOOC_2H_5$	ethanol / 6	85
B: polymer-supported thiocyanate			
$n\text{-}C_8H_{17}Br$	$n\text{-}C_8H_{17}SCN$	benzene / 5	90
$C_2H_5OCOCH_2Br$	$C_2H_5OCOCH_2SCN$	pentane / 2	91
$n\text{-}C_6H_{13}CH(Br)CH_3$	$n\text{-}C_6H_{13}CH(SCN)CH_3$	benzene / 12	77

a- under reflux

Till now reactions performed with simple nucleophiles linked to the ion-exchangers have been described. The same kind of support,however,was found to be useful in the preparation of a number of different reagents. As an example,by washing the resin,in the chloride form,with an aqueous solution of chromic acid,a chromate form of the polymer was readily obtained.

$$\text{(P)}-\overset{|}{\underset{|}{N}}{}^{+}\ \bar{C}l\quad +\quad CrO_{3(aq)}\quad \longrightarrow\quad \text{(P)}-\overset{|}{\underset{|}{N}}{}^{+}\ \bar{\ }OCrO_3H\quad +\quad HCl$$

Iodometric titration of the chromate displaced from the resin by reaction with aqueous 1 N potassium hydroxide overnight,proved that a 1:1 exchange between chloride and chromate anions took place. This reagent is remarkably effective in oxidizing primary and secondary alcohols to aldehydes and ketones. Although relatively slow (3-9 hours at 60-80° and 3-4 fold excess of reagent are required for completion) the oxidation is very clean and no traces of acids or other by-products were detected in the reaction mixture.Some significant results are reported in Table 5.

TABLE 5. Oxidation of alcohols to aldehydes and ketones by means of polymer supported chromic acid

Substrate	Reaction conditions[a] (solvent/time hrs.)	%Yield
$n\text{-}C_8H_{17}OH$	hexane / 3	93
$n\text{-}C_{12}H_{23}OH$	benzene / 9	94
$C_6H_5CH_2OH$	THF / 1	98
$C_6H_5CH=CHOH$	$CHCl_3$ / 1	96
$CH_2=C(CH_3)CH_2CH_2OH$	hexane / 3	93
$n\text{-}C_9H_{19}CH(OH)CH_3$	hexane / 3	73
$c\text{-}C_6H_{11}OH$	hexane / 3	77
	benzene / 1	92
	benzene / 3	93
	benzene / 1	90
$C_6H_5CH(OH)C_6H_5$	benzene / 1	77

a - under reflux

The enhanced nucleophilicity of the supported chromate ion provides also an attractive route to carbonyl compounds starting from allylic and benzylic halides (Ref. 11). The latter are probably converted by the resin,in refluxing benzene,to chromate esters which spontaneously decompose to aldehydes and ketones. It has to be pointed out,in this case, that the use of the resin is helpful in circumventing the low solubility of the chromate ion in organic solvents. Table 6 summarizes the results obtained.

TABLE 6. Carbonyl compounds from activated halides by means of polymer-supported chromate ion

Halide	Reaction time (min.)[a]	%Yield
$C_6H_5CH_2Cl$	75	95
$C_6H_5CH(Br)C_6H_5$	60	95
$C_2H_5OCOCH=C(CH_3)CH_2Br$	120	95
	120	95

a - in benzene under reflux

Also the iron tetracarbonylferrate monohydride anion $FeH(CO)_4^-$ described by Collmann (12) may be quickly exchanged,by a column technique,with the chloride ion of the resin,in this way being supported on the polymeric matrix (Ref. 13).

$$Fe(CO)_5 \quad + \quad KOH \quad \xrightarrow[H_2O/C_2H_5OH]{H_2O \ \ or} \quad KHFe(CO)_4 \ + \ K_2CO_3 \ + \ H_2O$$

This reagent,prepared immediately before use,is capable of converting primary alkyl bromides to the homologous aldehydes in tetrahydrofuran under reflux for four hours. The choice of the solvent is critical here as the use of benzene,iso-octane and petroleum ether caused the formation of self-condensation products.Several remarkable advantages are peculiar to our technique.First,the resin can be easily dried by washing,under an inert atmosphere,with anhydrous solvents. The iron-containing by-products,moreover, remain bound to the polymer, allowing one to avoid their separation from the organic products,which is the most difficult step of the usual procedure in solution. Finally,there is no need of added ligand to perform the reaction as is necessary with Cooke's procedure (14),the migratory insertion being probably induced,in the presence of the resin,by the halogen anion formed. The proposed mechanism (see scheme below) accounts for the mentioned features of the reaction.

Results are collected in Table 7.

TABLE 7. Reaction of polymer-supported $FeH(CO)_4^-$ anion with alkyl halides

Alkyl halides	Product	%Yield
$n-C_7H_{15}Br$	$n-C_7H_{15}CHO$	90
$n-C_8H_{17}Br$	$n-C_8H_{17}CHO$	90
$n-C_8H_{17}I$	$n-C_8H_{17}CHO$	95
$C_6H_5CH_2CH_2Br$	$C_6H_5CH_2CH_2CHO$	80
$C_2H_5OCO(CH_2)_3Br$	$C_2H_5OCO(CH_2)_3CHO$	85

The hydride resin behaves also as a practical reducing agent which is useful in converting aromatic nitrocompounds to amines (Ref.15), α-bromocarbonyl compounds to the corresponding dehalogenated compounds and 1,2 dibromoderivatives to olefins. The first two reductions are fast and may be accomplished simply by passing a solution of the substrate through a column of the polymeric reagent at room temperature. The reductive dehalogenation requires, on the contrary, higher temperatures and longer reaction times. Results are collected in Tables 8,9 and 10.

TABLE 8. Reduction of nitroarenes by means of polymer-supported $FeH(CO)_4^-$ anion

Substrate	Product[a]	%Yield
$4-ClC_6H_4NO_2$	$4-ClC_6H_4NH_2$	92
$4-C_6H_5OC_6H_4NO_2$	$4-C_6H_5OC_6H_4NH_2$	80
$4-CH_3OC_6H_4NO_2$	$4-CH_3OC_6H_4NH_2$	70
$4-CHOC_6H_4NO_2$	$4-CHOC_6H_4NH_2$	80
$4-NH_2C_6H_4NO_2$	$4-NH_2C_6H_4NH_2$	70
$3-OHC_6H_4NO_2$	$3-OHC_6H_4NH_2$	99

a - in THF at room temperature by column technique

TABLE 9. Reductive dehalogenation by polymer-supported $FeH(CO)_4^-$ anion

Halide	Product[a]	%Yield
$C_6H_5COCH_2Br$	$C_6H_5COCH_3$	90
(2-bromocyclohexanone)	(cyclohexanone)	92
$CH_3(CH_2)_3CH(Br)COOCH_3$	$CH_3(CH_2)_4COOCH_3$	81
$C_6H_5CH(Br)C_6H_5$	$C_6H_5CH_2C_6H_5$	85
$C_6H_5CH(Br)CH(Br)C_6H_5$	$E-C_6H_5CH=CHC_6H_5$	85

a - in THF at room temperature

TABLE 10. Alkenes from 1,2 dibromoderivatives by reaction with polymer-supported $FeH(CO)_4^-$ anion

Substrate	Product[a]	%Yield
$CH_3(CH_2)_5CH(Br)CH_2Br$	$CH_3(CH_2)_5CH=CH_2$	80
$CH_3(CH_2)_{15}CH(Br)CH_2Br$	$CH_3(CH_2)_{15}CH=CH_2$	77
$CH_3(CH_2)_8CH(Br)CH_2Br$	$CH_3(CH_2)_8CH=CH_2$	75
$CH_3(CH_2)_4CH(Br)CH(Br)CH_3$	$CH_3(CH_2)_4CH=CHCH_3$	80

a - in THF under reflux

Phosphonates substituted with electron withdrawing groups(-CN and COOMe) have been supported (Ref.16),owing to the relatively high acidity of the alpha C-H bonds,on a macroreticular anion exchange resin,simply by a neutralization process,according to the following scheme

$$\text{(P)}-\overset{|}{\underset{|}{N}}^{+}_{\ OH^-} \ + \ (EtO)_2\overset{O}{\overset{\uparrow}{P}}CH_2X \ \longrightarrow \ \text{(P)}-\overset{|}{\underset{|}{N}}^{+}_{\ (EtO)_2\overset{O}{\overset{\uparrow}{P}}\bar{C}HX} \ + \ H_2O$$

$$X= -CN, \ -COOR \qquad\qquad\qquad\qquad (I)$$

The treatment of carbonyl compounds with the polymer-bound Wittig-Horner reagent in various solvents for 1-2 hours gave olefins in high yields,at room temperature.

$$(I) \quad + \quad R-\overset{O}{\underset{R'}{C}} \quad \xrightarrow[\text{r.t.}]{\text{THF}} \quad \overset{R}{\underset{R'}{C}}=CH-X$$

Concerning the stereoselectivity of the reaction we obtained,in some instances,an E/Z ratio of about 2/1,a value similar to that observed in the phase-tranfer catalyzed synthesis(Ref. 17). Results are shown in Table 11.

TABLE 11. Reaction of polymer-supported phosphonates with carbonyl compounds

Substrate	Product	%Yield (E/Z isomers)
A:for X = -CN		
$4\text{-}ClC_6H_4CHO$	$4\text{-}ClC_6H_4CH=CHCN$	83 (E 100%)
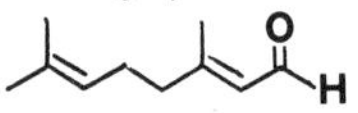 (structure)	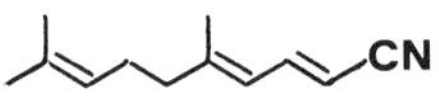 (structure) CN	97 (1/1)
$CH_3(CH_2)_5CHO$	$CH_3(CH_2)_5CH=CHCN$	75 (3/1)
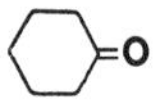 (cyclohexanone) =O	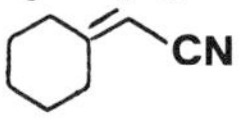 (cyclohexane) CN	97 (E 100%)
$CH_3(CH_2)_8COCH_3$	$CH_3(CH_2)_8C(CH_3)=CHCN$	95 (2/1)
$C_6H_5COCH_3$	$C_6H_5C(CH_3)=CHCN$	90 (E 100%)
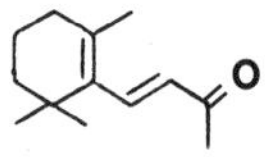 (structure) O	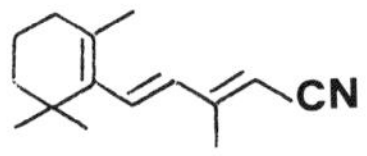 (structure) CN	93 (2/1)
B:for X =-COOCH$_3$		
$4\text{-}ClC_6H_4CHO$	$4\text{-}ClC_6H_4CH=CHCOOCH_3$	97 (E 100%)
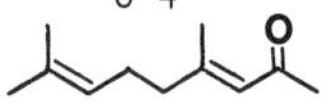 (structure)	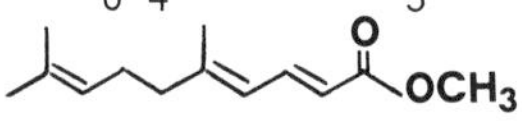 (structure) OCH$_3$	95 (E 100%)
$CH_3(CH_2)_7CHO$	$CH_3(CH_2)_7CH=CHCOOCH_3$	64 (E 100%)

The simultaneous use of the phosphonacetonitrile resin and of an acidic one in tetrahydrofuran/water 9/1,allowed the direct sequential hydrolysis and olefination for dioxolanes of ketones as well as of aromatic and α-β unsaturated aldehydes

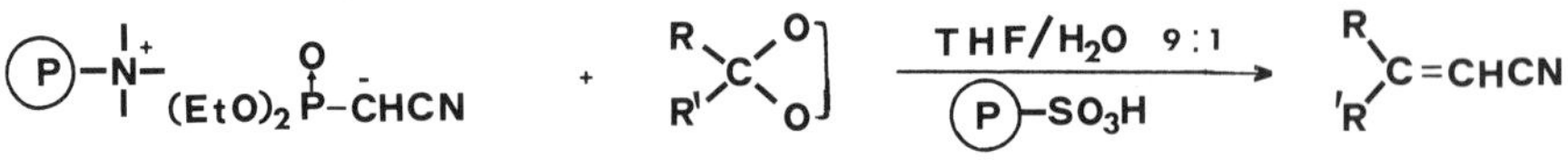

$$\text{(P)}-\overset{|}{\underset{|}{N}}^{+}_{\ (EtO)_2\overset{O}{\overset{\uparrow}{P}}\bar{C}HCN} \ + \ \overset{R}{\underset{R'}{C}}\overset{O}{\underset{O}{]}} \ \xrightarrow[\text{(P)}-SO_3H]{\text{THF/H}_2\text{O} \ 9:1} \ \overset{R}{\underset{R'}{C}}=CHCN$$

TABLE 12. Reaction of polymer-supported phosphonacetonitrile with dioxolanes

Substrate	Product	%Yield
C_6H_5 (dioxolane)	$C_6H_5CH=CHCN$	98
(dioxolane structure)	(CN product)	90
(cyclohexane dioxolane)	(CN product)	90
C_6H_5 (dioxolane)	$C_6H_5C(CH_3)=CHCN$	40

This performance,obviously impossible in solution due to the immediate neutralization of the acidic catalyst by the basic phosphonate reagent is an unique feature of the polymer-supported reagents,an aspect of the resin-bound chemistry which is going to receive more and more attention. Finally we observed that dioxolanes of saturated aldehydes failed to react,owing to the higher stability towards hydrolysis of this type of compound,which could be of potential utility for selective olefination.

Anion-exchange resins have also been found useful for the synthesis of halogenated organic compounds(Ref. 18). The ion-exchangers in the bromide and iodide form were treated with an excess of bromine and chlorine in tetrachloromethane to give respectively the Br_3^-,$BrCl_2^-$ and ICl_2^- anions linked to the polymer.By means of the perbromide reagent we achieved direct bromination of aldehydes,ketones and insaturated compounds,whereas the corresponding chlorination of the same substrates was performed using the ICl_2^- resin. Finally the $BrCl_2^-$ form of the resin allowed us to succeed in the direct chlorobromination of alkenes and alkynes. Results are reported in Tables 13,14 and 15.

TABLE 13. Bromination with polymer-supported perbromide anion

Substrate	Product	Reaction conditions (solvent/temp./time)	%Yield
$n-C_4H_9CH_2CHO$	$n-C_4H_9CH(Br)CHO$	CCl_4/reflux/15min.	95
$n-C_8H_{17}CH_2CHO$	$n-C_8H_{17}CH(Br)CHO$	CH_2Cl_2/reflux/1 hr.	97
$C_6H_5COCH_3$	$C_6H_5COCH_2Br$	THF/r.t./ 12 hrs.	70
$n-C_5H_{11}CH_2COCH_3$	$n-C_5H_{11}CH(Br)COCH_3(65\%)$ $n-C_5H_{11}CH_2COCH_2Br$ (35%)	CH_2Cl_2/reflux/ 1hr.	74
$n-C_5H_{11}CH=CHCH_3$	$n-C_5H_{11}CH(Br)CH(Br)CH_3$	CH_2Cl_2/r.t./15 min.	85
$n-C_4H_9C{\equiv}CH$	$n-C_4H_9C(Br)=CHBr$	CH_2Cl_2/reflux/3hrs.	94

TABLE 14. Chlorination with polymer-supported ICl_2^- anion

Substrate	Product	Reaction conditions (solvent/temp./time)	%Yield
$n-C_4H_9CH_2CHO$	$n-C_4H_9CH(Cl)CHO$	CCl_4/ reflux/ 15 min.	85
$n-C_8H_{17}CH_2CHO$	$n-C_8H_{17}CH(Cl)CHO$	CCl_4/reflux/ 30 min.	88
(cyclohexanone)	(chlorocyclohexanone)	CCl_4/reflux/ 30 min.	45
$C_6H_5COCH_3$	$C_6H_5COCH_2Cl$	CH_2Cl_2/r.t./20 hrs.	66
$n-C_5H_{11}CH=CHCH_3$	$n-C_5H_{11}CH(Cl)CH(Cl)CH_3$	CH_2Cl_2/ r.t./ 2hrs.	83

TABLE 15. Chlorobromination with polymer-supported $BrCl_2^-$ anion

Substrate	Product[a]	%Yield
$n\text{-}C_5H_{11}CH{=}CHCH_3$	$n\text{-}C_5H_{11}CH(Cl)CH(Br)CH_3$ (35%) $n\text{-}C_5H_{11}CH(Br)CH(Cl)CH_3$ (65%)	85
		71
$C_6H_5C{\equiv}CH$	$C_6H_5C(Cl){=}CHBr$	96
$n\text{-}C_8H_{17}C{\equiv}CH$	$n\text{-}C_8H_{17}C(Cl){=}CHBr$	85

a – in dichloromethane at room temperature for 1-12 hours

REFERENCES

1. R.B.Merrifield, J.Am.Chem.Soc. 85,2149-2154 (1963)
2. C.G.Overberger and K.N.Sannes,Angew.Chem.Int.Ed.Engl.13,99-104 (1974)
3. D.C.Neckers,J.Chem.Educ.52,695-702 (1975)
4. C.C.Leznoff, Acc.Chem.Res. 11, 327-333 (1978)
5. G.Cainelli and F.Manescalchi, Synthesis 723-724 (1975)
6. G.Cainelli,F.Manescalchi and M.Panunzio, Synthesis 472-473 (1976)
7. G.Cainelli,G.Cardillo and M.Orena, J.Chem.Soc.Perkin 1 1597-1599 (1979)
8. F.Manescalchi,M.Orena and D.Savoia, Synthesis 445-446 (1979)
9. G.Cainelli,F.Manescalchi and M.Panunzio, Synthesis 141-144 (1979)
10. G.Cainelli,G.Cardillo,M.Orena and S.Sandri, J.Am.Chem.Soc. 98,6737-6738 (1976)
11. G.Cardillo,M.Orena and S.Sandri, Tetrahedron Lett. 3985-3986 (1976)
12. J.P.Collmann,Acc.Chem.Res. 8, 342-347 (1975)
13. G.Cainelli,F.Manescalchi,A.Umani-Ronchi and M.Panunzio,J.Org.Chem. 43,1598-1599 (1978)
14. M.P.Cooke jr.,J.Am.Chem.Soc. 92,6080-6082 (1970)
15. G.P.Boldrini,G.Cainelli and A.Umani-Ronchi unpublished results
16. G.Cainelli,M.Contento,F.Manescalchi and R.Regnoli, J.Chem.Soc.Perkin 1,in press
17. C.Piechucki,Synthesis 869-870 (1974)
18. A.Bongini,G.Cainelli,M.Contento and F.Manescalchi, Synthesis 143-146 (1980)

MOLECULAR ENGINEERING: THE DESIGN AND SYNTHESIS OF CATALYSTS FOR THE RAPID 4-ELECTRON REDUCTION OF MOLECULAR OXYGEN TO WATER

J. P. Collman*, F. C. Anson**, S. Bencosme*, A. Chong*,
T. Collins*, P. Denisevich*, E. Evitt*, T. Geiger**, J. A. Ibers***,
G. Jameson***, Y. Konai*, C. Koval**, K. Meier*, P. Oakley*,
R. Pettman*, E. Schmittou*, and J. Sessler*

*Department of Chemistry, Stanford University, Stanford, California 94305, USA
**Division of Chemistry & Chemical Engineering, The Chemical Laboratories,
California Institute of Technology, Pasadena, California 91125, USA
***Department of Chemistry, Northwestern University, Evanston, Illinois 60201, USA

Abstract - "Molecular engineering" has been used to design and
synthesize a successful catalyst for the 4-electron reduction
of molecular oxygen to water. This catalyst operates at rates
near those of the enzyme cytochrome-c oxidase and two orders
of magnitude greater than the best catalyst currently available, platinum. Knowledge of biochemical and inorganic
reaction mechanisms suggested the catalyst design. This
required construction of binary cyclophane porphyrins with a
cofacial orientation which would permit the simultaneous
interaction of two metal centers with one oxygen molecule.
Early cofacial porphyrins were linked by functionalized aryl
substituents on the meso position of the porphyrin ring and
were ineffective as catalysts. More flexible and sophisticated syntheses led to two series of porphyrins with alkyl
bridges attached at the meso and β positions of the porphyrin
rings. In this way, the length of the bridges could be
varied. The technique of rotating ring-disk voltammetry showed
that the dicobalt β-linked dimer with 4-atom bridges, 17b,
was an excellent catalyst for the 4-electron reduction,
generating <4% hydrogen peroxide. HPLC analysis of the free-
base precursor to 17b showed it was a mixture of three
components, two of which were shown by mass spectroscopic
analysis to bear monochloro substituents on one of the
porphyrin rings. The chlorinated dimer was a better catalyst
than the unfunctionalized parent, and this observation
supports the concept that electronegative substituents may
enhance the properties of these electrode catalysts. The
heterobimetallic cobalt-aluminum and palladium-cobalt
relatives of 17b behave catalytically as monomeric porphyrins.
These and other observations have suggested a mechanism for
this important multi-electron redox reaction which takes place
in aqueous acid at a graphite electrode.

"Molecular engineering", the design and synthesis of catalysts and other
functional molecules, is an important, growing application of organic
synthesis. Initially, a working molecule is designed on the basis of a
mechanistic hypothesis. Once this target molecule has been prepared, it is
examined for catalytic or other activity. On the basis of these results, a
modified target compound is formulated, prepared, and tested. Through this
iterative process, synthesis is closely coupled with physical measurement,
just as drug design has long been linked with physiological testing.

This account describes our successful quest for molecules which catalyze the
4-electron reduction of molecular oxygen to water. The principal application
of such electrode catalysts is in the oxygen cathode of a fuel cell. The

essence of this problem is revealed in Fig. 1, which shows the reversible
standard electrode potentials connecting oxygen and its reduction products,

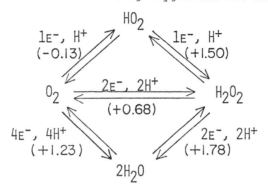

Fig. 1. Standard reduction potentials of oxygen and related
reduction products (1 N acid, V vs. NHE).

hydroperoxyl, hydrogen peroxide, and water. In order to obtain the most
energy from an oxygen electrode, the catalyst should effect rapid 4-electron
reduction of oxygen as close to the thermodynamic potential (+1.23 V) as
possible. Kinetic and thermodynamic arguments which rule out processes
proceeding through the 1- or 2-electron paths are outlined elsewhere (Ref. 1).
Since air is to be used as the oxygen source, the medium must be acidic,
otherwise carbon dioxide would form carbonates. Heat will be released, so
that both the catalyst and the electrode material must withstand boiling acid
over prolonged periods. A highly porous, specially textured, conducting
graphite would be employed as the electrode substrate. Serious mass transport
problems would have to be overcome to achieve the desired current densities
(say, 1.0 amp cm^{-2}). To date, platinum metal is the most effective catalyst
for an acidic oxygen electrode; however, even with platinum, viable current
densities are achieved only at a potential (ca. 0.63 V vs. NHE) which
represents a large, energy-robbing overvoltage.

We looked to biology and to inorganic reaction mechanisms for our initial
catalyst design (Ref. 2). Cytochrome-c oxidase, the enzyme essential for
aerobic metabolism, employs four metals, two at the oxygen-binding site (a
heme and a copper) to reduce oxygen to water. We also knew that iron(II)
porphyrins decompose oxygen through a mechanism involving two porphyrins
acting on one oxygen molecule (Ref. 3). Our scheme was therefore to prepare
binary, cyclophane porphyrins with a cofacial orientation. Metal derivatives
of these "face-to-face" porphyrins would then be attached to graphite electrodes and tested for oxygen reduction in the presence of acidic aqueous
electrolytes. Porphyrins were preferred over other macrocyclic ligands
because the former exhibit good hydrolytic stability. At the start we needed
two things: (1) an immediate source of face-to-face porphyrins (which were
unknown at the time) so that initial physical chemical studies could be
carried out and (2) a more elaborate plan for preparing a series of such
cyclophane porphyrins having a range of interporphyrin separations which might
be suitable for a variety of bridging dioxygen complexes.

The initial goal was reached by coupling α,γ-o-aminophenyl porphyrin, 1, with
phosgene, via isocyanates, to afford a urea-linked, face-to-face porphyrin, 2,
(Fig. 2)(Ref. 2). The monomeric porphyrins were prepared by inelegant
statistical pyrrole coupling, which required tedious chromatography. A range
of physical chemical techniques (^{1}H NMR, ESR of Cu_2 and Co_2 derivatives,
electronic spectra of the free bases, cyclic voltammetry of the Co_2 complex in
benzonitrile) proved diagnostic of the cofacial interaction. However, these
compounds were ineffective as catalysts for the 4-electron reduction of
oxygen--apparently because the interporphyrin separations (∿6 Å) were too
large.

An attempted synthesis of a similar system which should have a smaller interporphyrin separation was unsuccessful. α,γ-Diarylporphyrins having meta
carboxyl and amino functions, 3 and 4, were prepared by total synthesis via
the aryl-substituted dipyrrylmethanes; however, cyclophane coupling to 5
failed (Fig. 3), perhaps because of geometric constraints imposed on this
inflexible system (Ref. 4).

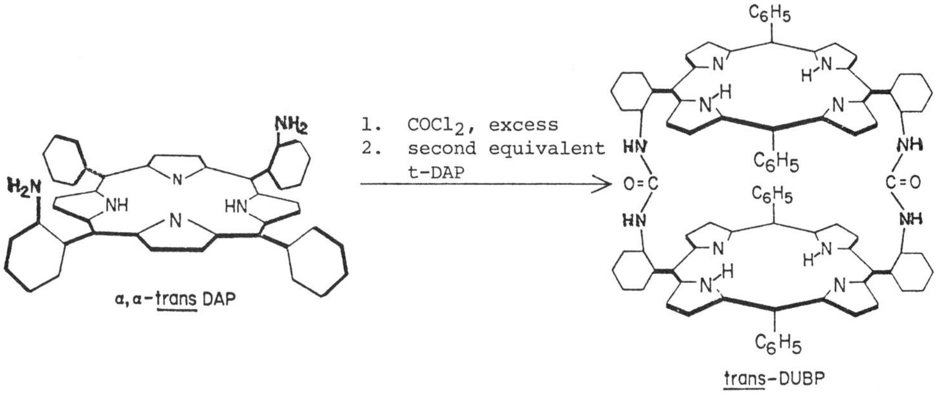

Fig. 2. Synthesis of a diurea-linked binary porphyrin.

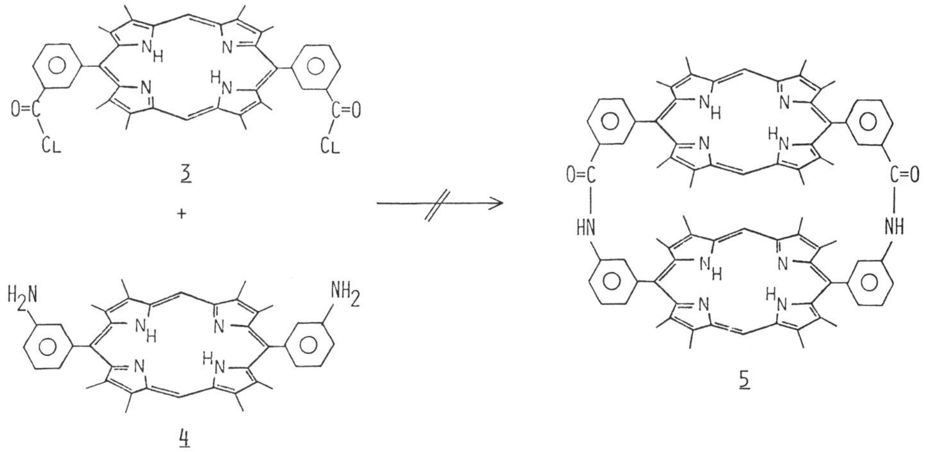

Fig. 3. Attempted coupling of α,γ-diarylporphyrins bearing
meta functionality.

A series of porphyrins bearing functionalized, α,γ-meso alkyl groups were then
prepared by total synthesis (Ref. 4, 5). Coupling of these more flexible
alkyl substituents afforded a series of face-to-face porphyrins with 6, 5, and
4 atoms joining the meso positions, 6, 7, and 8 (Fig. 4)(Ref. 5). Advantage
was taken of this porphyrin synthesis via dipyrrylmethanes such as 9 to
prepare cyclophane porphyrins joined by 4- and 6-atom hydrocarbon links, 10
(Fig. 5). The entire synthesis requries only two steps; however, the yield
of final product is very low (Ref. 5).

For the alkyl meso-linked compounds 6, 7, 8, and 10, the physical chemical
tests which we had previously found to be diagnostic of cofacial porphyrin
interactions indicated that the interporphyrin separations were greater than
values estimated from molecular models. The origin of this effect was
revealed by an X-ray diffraction study of a representative Cu_2 derivative, 11
(Fig. 6). The axes, normal to the center of each porphyrin ring, are offset
by nearly 5 Å, even though the two metals are only 6.3 Å apart and the
distance between the porphyrin planes is 3.95 Å. This unanticipated offset
geometry may explain the fact that in this series cobalt and iron derivatives
fail to exhibit any 4-electron oxygen reduction (vide infra).

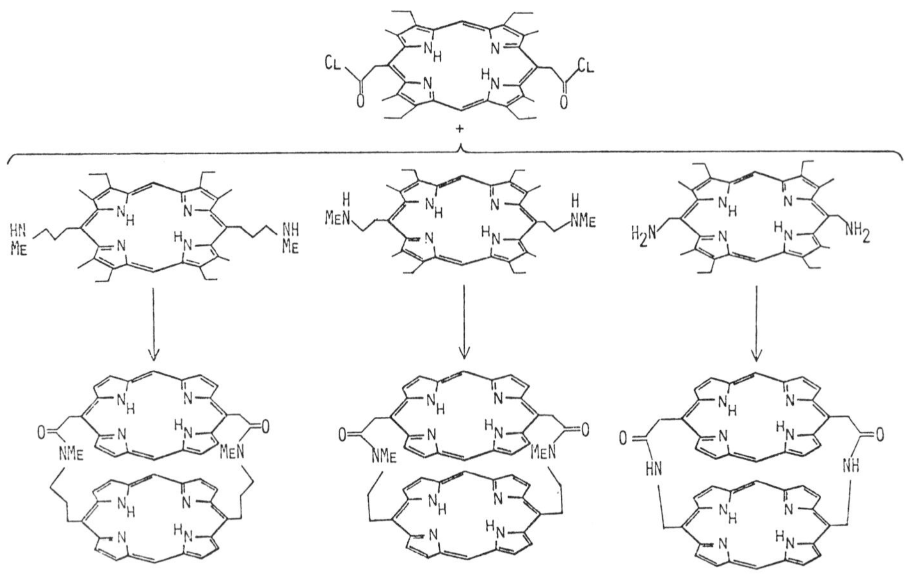

6, 6 ATOMS 6-8 Å

7, 5 ATOMS 5-7 Å

8, 4 ATOMS 4.8-6 Å

Fig. 4. Face-to-face dimer porphyrins derived from monomers bearing α,γ-meso-alkyl substituents. (The alkyl substituents on the porphyrin periphery have been omitted from the dimer structures for clarity.)

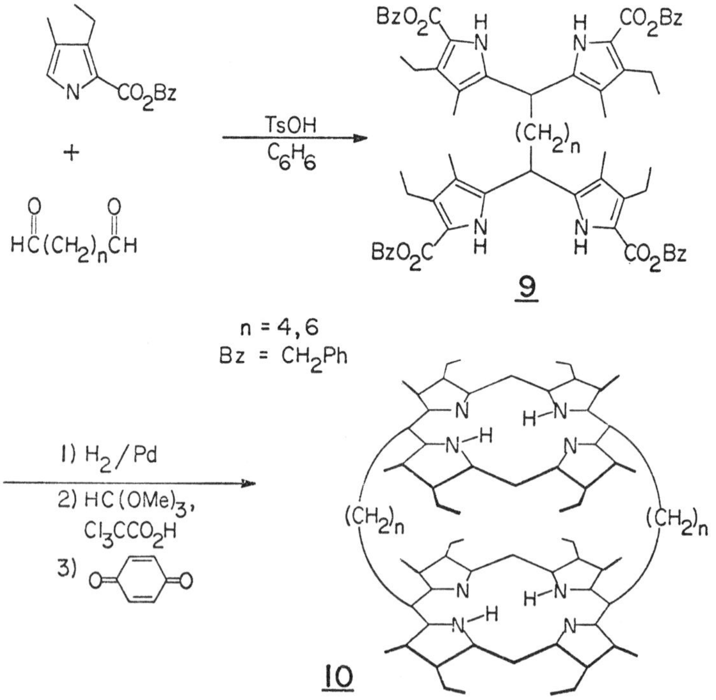

Fig. 5. Synthesis of cyclophane porphryins joined by saturated hydrocarbon linkages.

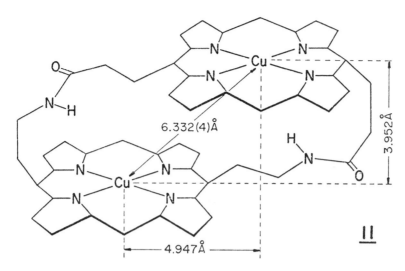

- β-pyrrolic substituents omitted for clarity
- porphyrin planes ruffled

Fig. 6. Simplified representation of the molecular structure
of the Cu$_2$ derivative of the α,γ-_meso_-alkyl-linked, face-to-
face porphyrin dimer with 6-atom bridges.

Fortunately, we were preparing another face-to-face porphyrin series in
parallel. This route depended upon porphyrins which were functionalized in
transverse β-pyrrole positions. The ester-substituted porphyrins 12a and 12b
were prepared by Fischer's route from monopyrroles through the expeditious
modification developed by Battersby (Fig. 7)(Note a). The esters 12a and 12b

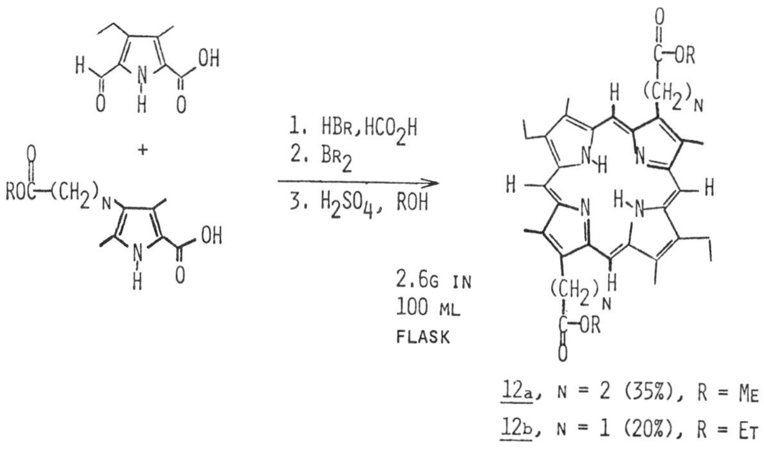

Fig. 7. Synthesis of porphyrins using pyrroles substituted in
the β positions.

Note a: Professor A. Battersby kindly provided us, in advance of publication,
with detailed directions for his synthesis of the bis-proprionic acid ester
porphyrin 12a. We extended this to the synthesis of the bis-acetic acid
porphyrin 12b. In the interim, Battersby independently prepared a related
series of cofacial porphyrin compounds, as did C. Chang (Ref. 6), who also
used a modification of Battersby's synthesis. Other face-to-face porphyrins
have also been reported (Ref. 7).

were converted in good yield to the bis-amines 13a and 13b via Curtius degradation of the corresponding acid hydrazides (Fig. 8). Details are given

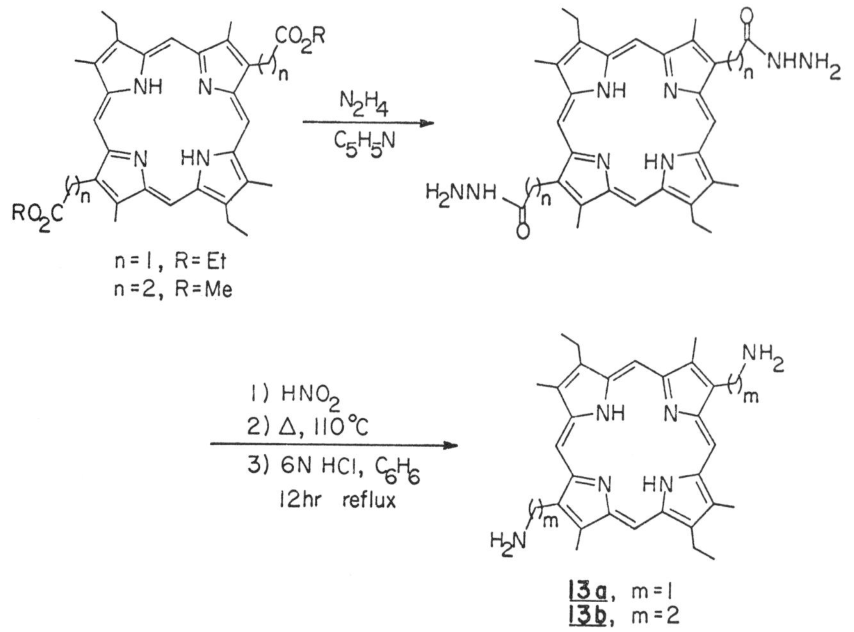
CO2R
NH
N
N
HN
RO2C
n
N2H4
C5H5N
NHNH2
H2NNH
n = 1, R = Et
n = 2, R = Me
1) HNO2
2) Δ, 110°C
3) 6N HCl, C6H6
12hr reflux
NH2
H2N
m
13a, m = 1
13b, m = 2

Fig. 8. Preparation of the bis-amino porphyrins via Curtius degradation of the acid hydrazide.

elsewhere (Ref. 2), but it is important to note here that hydrogen chloride was used to hydrolyze the intermediate isocyanate and that these metal-free porphyrins are very unstable in the presence of both light and atmospheric oxygen. Coupling of the bis-amines 13a and 13b with the p-nitrophenyl esters 14a and 14b in warm (65°C) pyridine afforded a series of β-linked cofacial porphyrins, 15a-c, in good yield (Fig. 9) (Ref. 1).

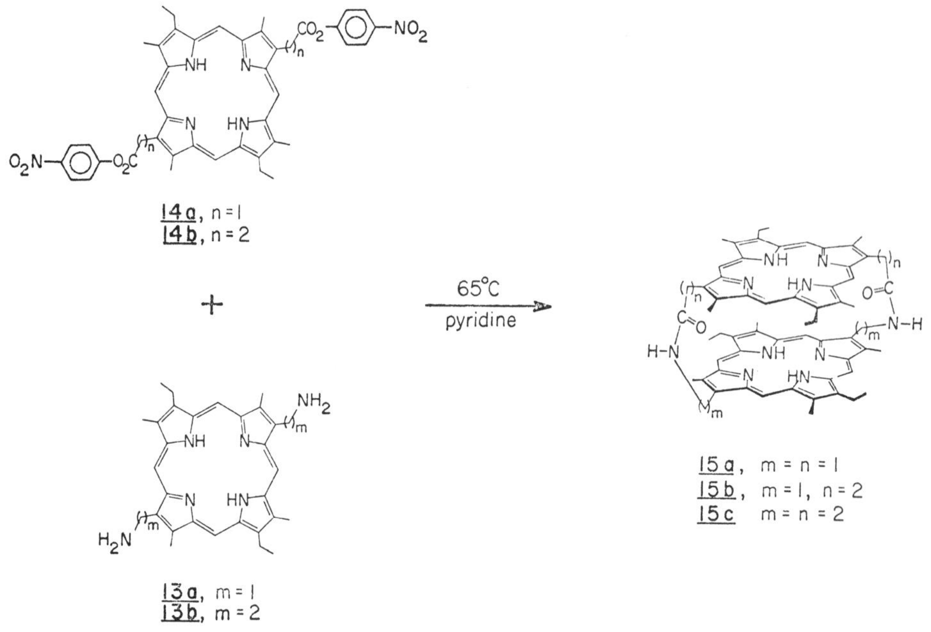
CO2
NO2
O2N
O2C
14a, n = 1
14b, n = 2
+
65°C
pyridine
NH2
H2N
13a, m = 1
13b, m = 2
C=O
O=C
H-N
N-H
15a, m = n = 1
15b, m = 1, n = 2
15c m = n = 2

Fig. 9. Preparation of β-linked face-to-face porphyrin dimers.

The interesting stereochemical possibilities in the cyclophane porphyrins, 15, are shown in Fig. 10 for the 4-atom bridged compounds. Both syn and anti

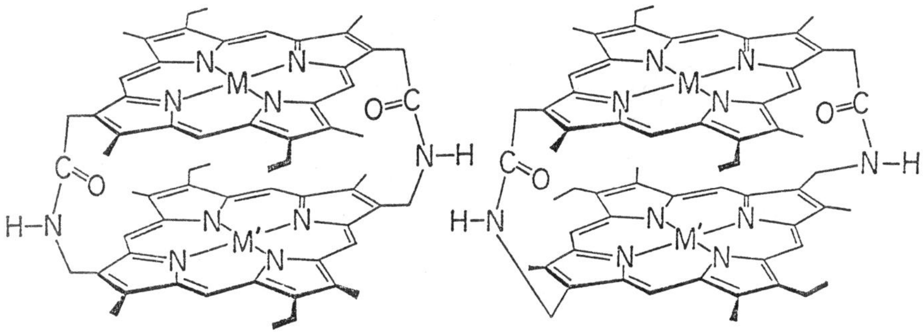

16 syn-   17 anti-

a M = M' = 2H; b M = M' = Co
c M = M' = Fe; d M = M' = Cu
e M = Pd, M' = Co
f M = Co, M' = Al

Fig. 10. Syn and anti diastereoisomers of the β-linked porphyrin dimer containing 4-atom bridges. (Only one member of each enantiomorphic pair is shown.)

diastereoisomers, 16 and 17, are possible, each as an enantiomorphic pair. The rather slow coupling procedure seems to afford a single diastereoisomer, which we believe to be the less-hindered anti isomer, 17. This assumption is based on our observations of four meso hydrogen signals in the 360 MHz ^{1}H NMR spectrum of the free base 17a (Ref. 1). For either syn or anti, each porphyrin should exhibit two different meso hydrogen signals, leading to four signals per diastereoisomer. Furthermore, the metal-free binary porphyrin, 17a, was subsequently found to be homogeneous by analytical HPLC (vide infra). A series of metallated derivatives was prepared, 17b-f, including examples of heterometallic porphyrins. These were examined as catalysts for oxygen reduction.

At this point, it is necessary to explain the method of electrochemical analysis, rotating ring-disk voltammetry (Ref. 8). The basic apparatus is shown in Fig. 11. The electrode assembly comprises a pyrolytic graphite disk with a concentric platinum ring. The porphyrin to be tested as a reduction catalyst is applied to the graphite disk from dilute CH_2Cl_2 solution. The assembly is placed in an oxygen-saturated (1 atm. O_2) aqueous acid electrolyte (0.5 M CH_3CO_2H or 0.5 M $HClO_4$). As the electrode is rotated, the oxygenated electrolyte is drawn to the disk and ejected across the disk and ring (Fig. 11). The disk potential is varied and the disk current signals any oxygen reduction. The ring potential is held constant at +1.4 V vs. NHE so that any hydrogen peroxide reaching the ring is detected as a ring current generated by the oxidation of hydrogen peroxide to oxygen. A platinum ring electrode is useful to catalyze this reaction. By examining the ring and disk current responses as a function of rotation rate, it is possible to measure quantitatively the catalyzed reduction of oxygen at the disk, the number of electrons involved, and the production of unwanted hydrogen peroxide.

A representative current profile for a monomeric cobalt porphyrin, 12b, is shown in Fig. 12. The disk current is plotted along the positive vertical axis. The ring current is plotted along the negative vertical axis. The two parallel lines on the left side show baseline disk and ring currents indicating no current flow and no oxygen reduction or hyrdogen peroxide formation. The clean (catalyst-free) graphite electrode would exhibit such parallel baselines across the entire disk potential range, from +0.9 to +0.1 V, showing that oxygen is not reduced in the absence of a catalyst. Reading from left to right in Fig. 12, as the disk potential reaches about 0.48 V versus the normal hydrogen electrode, the appearance of a disk current indicates some oxygen reduction and, at the same point, the appearance of a ring current detects the formation of hydrogen peroxide from the disk reaction.

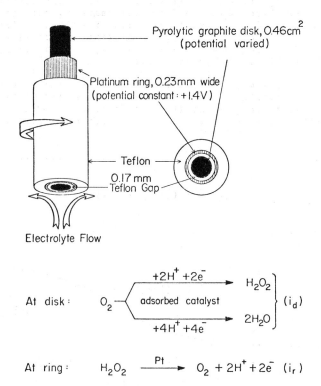

At disk:

$$O_2 \begin{array}{c} \xrightarrow{+2H^+ +2e^-} H_2O_2 \\ \text{adsorbed catalyst} \\ \xrightarrow{+4H^+ +4e^-} 2H_2O \end{array} \left.\right\} (i_d)$$

At ring: $H_2O_2 \xrightarrow{Pt} O_2 + 2H^+ + 2e^-$ (i_r)

Fig. 11. Schematic depiction of a rotating ring-disk electrode assembly and the reactions which take place at the two electrodes.

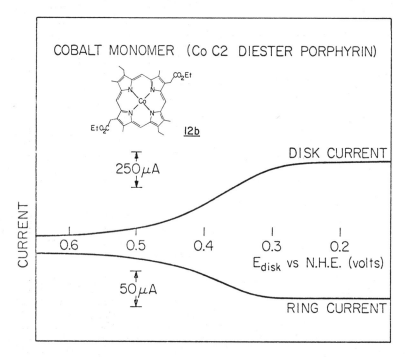

Fig. 12. Disk and ring current vs. the potential of the graphite disk coated with the monomeric cobalt porphyrin illustrated. Experimental conditions: platinum ring +1.4 V vs. NHE; rotation rate 250 rpm; supporting electrolyte 0.5 M CF_3COOH saturated with O_2 at one atmosphere.

The disk and ring currents rise symmetrically as the disk potential is made more reducing (more negative) until a limiting value is reached. A careful analysis of this system indicates that the major reaction at the disk is the 2-electron reduction of oxygen to hyrdogen peroxide, which is not further reduced at the disk (Ref. 1). This 2-electron oxygen reduction, catalyzed by monomeric cobalt porphyrins, is not new (Ref. 9).

Except that hydrogen peroxide production commences at a more positive potential, the current-potential profile of the dicobalt binary porphyrin 18 (Fig. 13) is very similar to that of the simple monomeric cobalt compound 12b. Again, hydrogen peroxide is the major product of oxygen reduction at the disk.

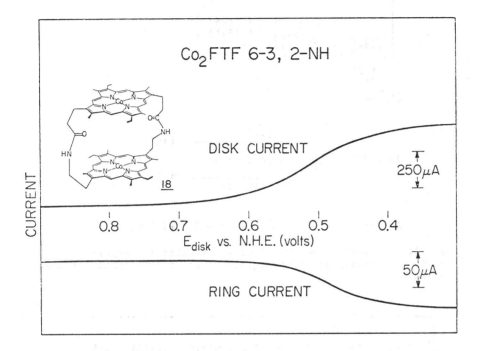

Fig. 13. Disk and ring current vs. the potential of the graphite disk coated with the 6-atom, β-linked, face-to-face dimer. Conditions: same as for Fig. 12,

The elusive goal of 4-electron catalytic oxygen reduction was achieved with the dicobalt face-to-face porphyrin containing four atoms in the linking arms, 17b. The results are revealed in Fig. 14. Oxygen reduction commences at +0.8 V and reaches a maximum by 0.6 V. Over this range the ring current is almost nil, indicating that little or no hydrogen peroxide is produced. Careful analysis of this system shows that a 4-electron reduction of dioxygen is taking place at the disk (Ref. 1).

Once the first successful 4-electron catalyst was in hand, we attempted to improve its stability (vide infra) to make it effective at more positive potentials and to explore the reaction mechanism.

Upon careful determination of the homogeneity of the free base 17a by HPLC, we found minor impurities to be present (Fig. 15). Through repeated, high-resolution HPLC, a few hundred μg of the porphyrins were obtained and analyzed by electronic spectroscopy and mass spectrometry. The fractions all exhibited the blue-shifted Soret bands we have learned to associate with cofacial porphyrins. The mass spectral results showed that the two faster moving, minor components, a and b, were in fact the monochloro derivatives 19a and 19b, whereas the slowest moving component, c, was pure 17a. That the chlorine atoms are in the amine side of the cofacial porphyrin is suggested by HPLC analysis of the monomeric porphyrins from which 17a was prepared. By HPLC the tritylated amine, 20, was found to have three components (Fig. 16), whereas the p-nitrophenyl ester, 14a, was homogeneous. The origin of the meso chlorination can be traced to the hydrogen chloride hydrolysis

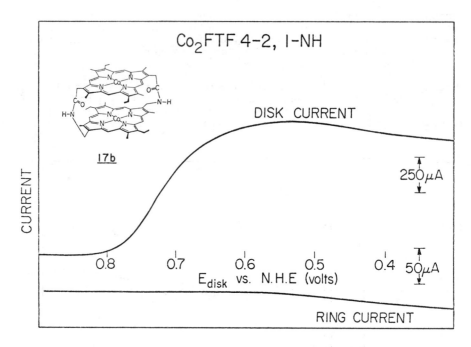

Fig. 14. Disk and ring current vs. the potential of the graphite disk coated with the 4-atom, β-linked, face-to-face dimer. Conditions: same as for Fig. 12.

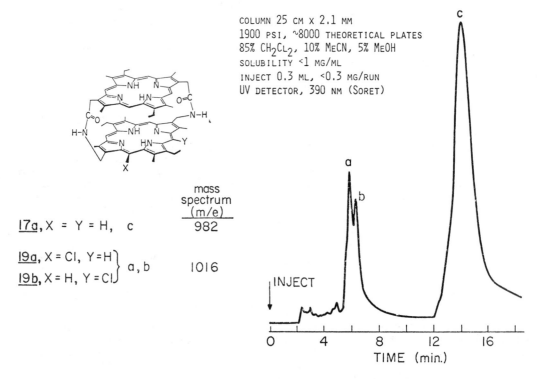

Fig. 15. HPLC chromatographic and mass spectroscopic characterization of the product generated by coupling diester 14a and diamine 13a (Fig. 9).

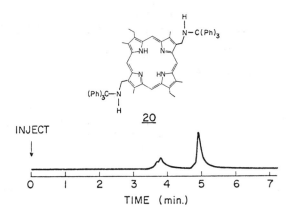

INJECT

Fig. 16. HPLC chromatograph of the trityl derivative of
diamine 13a.

of the isocyanate in the presence of traces of air (Fig. 8). The amine 13a
can be independently chlorinated by a mixture of H_2O_2 and HCl. There are two
different meso positions, but it has not been possible to determine which
regioisomer, 19a or 19b, is peak a or b.

We had anticipated that electronegative substituents might enhance the
properties of these electrode catalysts (Ref. 1). We therefore introduced
cobalt into fractions a, b, and c and examined the electrocatalytic
properties of these dicobalt porphyrins. We were gratified to find that the
complex 21 from fraction b exhibits markedly better activity than pure 17b
from fraction c (Fig. 17). In retrospect, it is clear that the catalytic

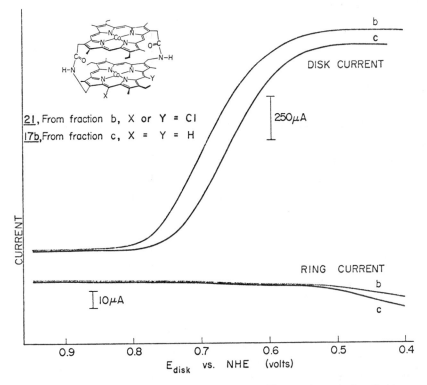

Fig. 17. Disk and ring current vs. the potential of the
graphite disk coated with the dicobalt derivatives of the
HPLC purified components of the mixture shown in Fig. 15.
Conditions: same as for Fig. 12, except rotation rate =
400 rpm.

activity manifest in Fig. 14 derives from a mixture and that the chloro
derivative 21 is playing a major role towards the more positive end of the
current-potential profile. Because of this discovery we have begun to
systematically introduce electronegative substituents into the cofacial
porphyrin catalysts, research which is still in progress.

Now let us turn to the mechanism of this remarkable catalytic 4-electron
reduction. The amount of oxygen reaching the electrode surface is a function
of the rotation rate, ω. This mass transport is described by the Levich
equation, which predicts the limiting disk current, i_{lim}, is proportional to
$\omega^{1/2}$ and the number of electrons taken up for each oxygen molecule which is
reduced. Thus the highest limiting disk currents shown in Fig. 17 for
ω = 400 rpm rise when the rotational rate is increased. As Fig. 18 reveals,

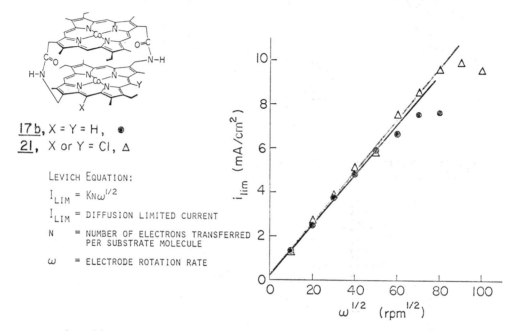

17b, X = Y = H, ◉
21, X or Y = Cl, △

Levich Equation:

$I_{LIM} = K_N \omega^{1/2}$

I_{LIM} = diffusion limited current

N = number of electrons transferred per substrate molecule

ω = electrode rotation rate

Fig. 18. Levich plots for the 4-atom, β-linked, dicobalt
face-to-face porphyrins.

these limiting disk currents increase linearly with the square root of the
rotation rate, showing that the reduction current is limited by the rate of
arrival of oxygen at the disk surface. With the most active catalyst, 21,
the limiting disk current begins to reach a maximum at about 9000 rpm. This
point represents the maximum speed of the reaction. Assuming a monolayer of
catalyst is active on an electrode surface area of 0.46 cm², this limiting
current affords an estimate of the maximum turnover number at ∿100 molecules
of oxygen reduced per dimer per second, a value nearly as great as that of
the enzyme cytochrome-c oxidase and about 200 times greater than platinum
metal centers in a fuel cell operating at +0.63 V. It should be emphasized
that this turnover number is at present an estimate and depends on the actual
active catalyst concentration.

At this point, it is appropriate to consider whether two electroactive metals
are required for the catalytic 4-electron reaction. This point has been
addressed by preparing the cobalt-palladium and cobalt-aluminum cofacial
porphyrins 17e and 17f. The synthesis of 17f is outlined in Fig. 19.
Earlier, we reported that the cobalt-palladium complex 17e behaves as a monomeric cobalt porphyrin, affording only hydrogen peroxide (Ref. 1). The
preliminary characterized cobalt-aluminum compound 17f is of greater interest
because Al(III) has the same formal charge as Co(III), which might be present
during the 4-electron catalytic cycle. The ring-disk analysis of oxygen
reduction (Fig. 20) shows 17f acts like a monomeric cobalt porphyrin; hydrogen
peroxide is the dominant product.

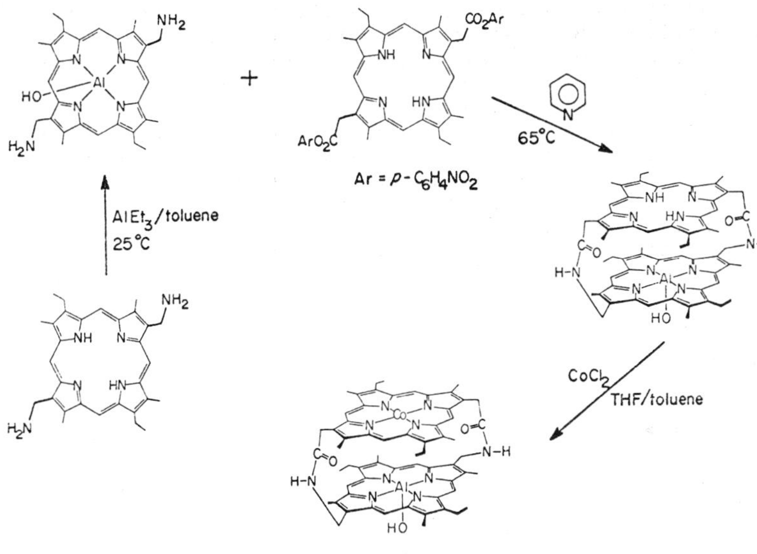

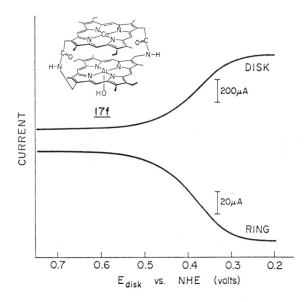

Fig. 19. Synthesis of the cobalt-aluminum, β-linked, cofacial porphyrin.

Fig. 20. Disk and ring current vs. the potential of the graphite disk coated with the cobalt-aluminum cofacial porphyrin. Conditions: same as for Fig. 12, except rotation rate = 400 rpm.

These results and other considerations (Ref. 1) lead us to propose the mechanism shown in Fig. 21. We believe the active form of these catalysts has both cobalt centers reduced to the +II oxidation state. Oxygen binds to both cobalt centers in what may be the rate-determining step. We know that acidic conditions are essential to the 4-electron, but not to the 2-electron process (Ref. 1). Thus we suppose the intermediate dioxygen adduct is protonated as shown for 22 or 23. The latter structure represents an interesting speculation which could account for the exquisite sensitivity of this 4-electron catalytic activity to the geometry of the interporphyrin cavity. A complex such as 23 requires close, coaxial interaction of both cobalt centers. The inactivity of meso-linked compounds such as dicobalt derivatives of 8 and 10 would be explained by the offset geometry of face-to-face porphyrins in the meso series (see Fig. 6). On the other hand, the β-linked compound with a

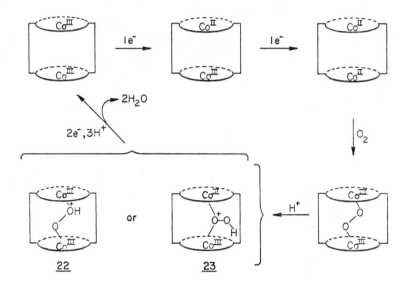

Fig. 21. Proposed mechanism for reduction of O$_2$ by β-linked
cofacial porphyrin **17b**.

5-atom bridge, derived from **15b**, exhibits modest 4-electron activity, whereas
the complex with a 6-atom bridge, **18**, acts like a monomer   (Fig. 13).

There is no direct evidence that during the catalytic cycle dioxygen is bound
to both cobalts within the interporphyrin cavity. There is direct evidence
that dioxygen can bind to both cobalts within the interporphyrin cavity. For
example, in the presence of axial ligands which do not fit in the interporphyrin cavity, such as N-methylimidazole, dioxygen binds to the cobalt(II)
complex **17b**, presumably forming a diamagnetic peroxide-bridged bis-cobalt(III)
complex. Iodine oxidation affords a substance, **24**, whose ESR spectrum
(Fig. 22) is consistent with two cobalt(III) groups collectively binding a

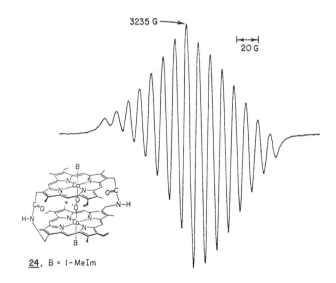

Fig. 22. ESR spectrum of the superoxo derivative of the
cofacial porphyrin **17b**.

superoxide anion. The 15-line ESR spectrum shown in Fig. 22 is consistent
with a hyperfine coupling between the odd electron on superoxide and the two
cobalt nuclei (S = 7/2). Similar results have been reported by Chang with a
related dimer (Ref. 11). Note that <u>24</u> has a nitrogenous axial base and is
not in the presence of a strong acid. Thus the existence of a substance like
<u>24</u> does not afford direct evidence about intermediates which could be avail-
able during the electrocatalytic reaction cycle.

In the context of axial bases and their possible role in the 4-electron
catalytic cycle, it should be mentioned that prior to introduction of the
catalyst the graphite electrode must be polished in order to obtain current-
potential curves such as those shown in Fig. 17. We ascribe this phenomenon
to the production of oxygenated functional groups on the graphite surface.
These groups may be effective ligands for one cobalt(II) center, even at the
low pH that is required for the 4-electron catalysis. Such axial bases would
be expected to increase the oxygen affinity of cobalt(II) porphyrins and
might also influence the redox potentials and facilitate the electron
transfer steps.

A final point concerns the stability of these catalysts under operating
conditions. Catalytic activity diminishes with time. It occurred to us that
the porphyrin might be lost from the electrode surface. We thus prepared an
amino-linked porphyrin, <u>25</u>, by reducing the amide groups in <u>17a</u> (Fig. 23).

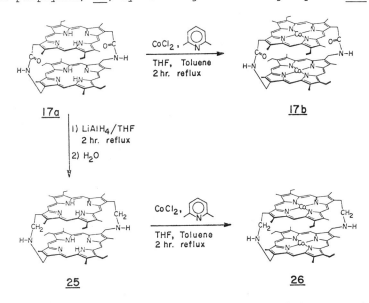

Fig. 23. Preparation of amino-linked cofacial porphyrins.

Care is required to prevent over-reduction. Introduction of cobalt afforded
<u>26</u>, which proved to be a modest 4-electron electrocatalyst. The amine <u>25</u> was
thence elaborated into an amino acid derivative, <u>27</u>, which was subsequently
attached to a polymeric acid chloride that had been precoated on a graphite
electrode (Fig. 24). Alas, the resulting catalyst, <u>28</u>, shows no enhanced
stability. We have subsequently learned that the unwanted peroxide byproduct
of oxygen reduction is the principal cause of catalyst deactivation.

In summary, we note that, although these catalysts are remarkable, they are
still too expensive and fragile to be of technological application. Still,
we may be able to simplify and improve them, using organic synthesis as our
major tool. It will also be important to discover the mechanism of this
reaction. Other systems beckon. Notable goals for future work are the
electrochemical decomposition of water to oxygen (a terminal step in photo-
synthesis) and the electrocatalytic reduction of dinitrogen to ammonia or
hydrazine (at a favorable potential). The first goal will require another
metal, perhaps manganese, as our dicobalt system operates in a potential
range which is thermodynamically uphill for water decomposition. The second
goal will also require other metals, probably a heterometallic, "push-pull"
system.

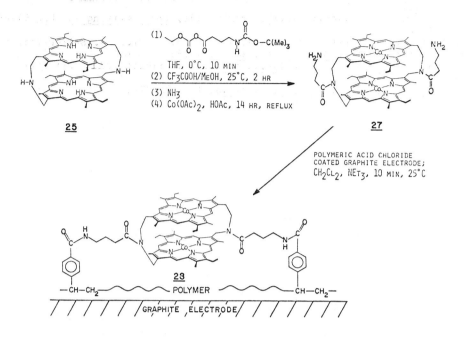

Fig. 24. Derivatization and attachment of the amino-linked porphyrin to a polymer-coated electrode (postulated structure).

Acknowledgement - This work was supported by National Science Foundation grants CHE 78-09443, CHE 77-22722, CHE 78-08716, and GP 23633* (*Magnetic Resonance Laboratory), and by the Institute for Energy Studies, Stanford University. We should like to acknowledge postdoctoral fellowships from the following agencies: Science Research Council, UK (to RP) and the Fond für Stipendien auf dem Gebiete der Chemie, Basel, Switzerland (to KM). This work derives from a collaborative effort which also includes the research groups of Professors M. Boudart and H. Taube (Stanford) and Dr. H. Tennent (Hercules Inc. Research Center). We are indebted to Professor A. Battersby (Cambridge University) for synthetic directions and to Dr. C.J. Wright (Eastman Kodak) for analytical help.

REFERENCES

1. J.P. Collman, P. Denisevich, Y. Konai, M. Marrocco, C. Koval, and F.C. Anson, J. Am. Chem. Soc., in press
2. J.P. Collman, C.M. Elliott, T.R. Halbert, and B.S. Tovrog, Proc. Natl. Acad. Sci. USA 74, 18-22 (1977).
3. J.P. Collman, Acc. Chem. Res. 10, 265-272 (1977).
4. E.R. Schmittou, Ph.D. Dissertation, Stanford University (1979).
5. J.P. Collman, A.O. Chong, G.B. Jameson, R.T. Oakley, E. Rose, E.R. Schmittou, and J.A. Ibers, J. Am. Chem. Soc., in press.
6. (a) C.K. Chang, J. Am. Chem. Soc. 99, 2819-2822 (1977).
    (b) C.K. Chang, Inorganic Compounds with Unusual Properties--II, R. Bruce King, Ed., ACS Adv. Chem. Series No. 173, 162-177 (1979) and references therein.
7. (a) N.E. Kagan, D. Mauzerall, and R.B. Merrifield, J. Am. Chem. Soc. 99, 5484-5486 (1977).
    (b) H. Ogoshi, H. Sugimoto, and Z. Yoshida, Tetrahedron Lett. 169 (1977).
8. (a) W.J. Albery and M.L. Hitchman, Ring Disc Electrodes, Clarendon Press, Oxford (1971).
    (b) H.R. Thirsk and J.A. Harrison, A Guide to the Study of Electrode Kinetics, Academic Press, New York (1972).

9. (a) J.H. Zagal, R.K. Sen, and E. Yeager, J. Electroanal. Chem. 83, 207–213 (1977).
    (b) H. Behret, W. Clauberg, and G. Sandstede, Ber. Bunsenges. 81, 54–60 (1977).
    (c) A review of earlier work is available: H. Jahnke, M. Schönborn, and G. Zimmerman, Top. Curr. Chem. 61, 133–181 (1976).
10. V.G. Levich, Physicochemical Hydrodynamics, Prentice Hall, New York (1962).
11. C.K. Chang, J. Chem. Soc., Chem. Commun. 800 (1977).

SYNTHETIC APPLICATIONS OF THE CHEMISTRY OF DICARBONYL CYCLOPENTADIENYLIRON COMPLEXES

M. Rosenblum, T. C. T. Chang, B. M. Foxman, S. B. Samuels and C. Stockman

Department of Chemistry, Brandeis University, Waltham, MA 02254, USA

Abstract - The use of Fp(vinyl ether) cations [Fp≡C$_5$H$_5$Fe(CO)$_2$] as vinyl cation equivalents is described. Carbon nucleophiles generally add nonregiospecifically to the activated double bond of Fp(olefin) cations. However, nucleophilic addition to vinyl ether complexes (1) is highly regiospecific. The products are (β-alkoxyalkyl)Fp complexes (2) which are in turn readily transformed to Fp(olefin) complexes (3) by brief treatment with acid at low temperatures. Release of the free vinyl group from 3 is simply achieved by brief exposure to iodide, or bromide. The stereochemistry of these reactions and several applications of the above sequence are illustrated.

Coordination of an olefinic center by a transition metal has a profound effect on the chemical reactivity of the double bond. While isolated olefins are generally immune to nucleophilic attack, a complexed olefin is often rendered susceptible to this mode of reaction (Ref. 1). On the other hand, the normal reactivity of the free olefin towards electrophiles is often suppressed in the complexed system. A group of complexes for which this "charge affinity reversal" is well established are the cationic CpFe(CO)$_2$(olefin) complexes:

$$Fp = CpFe(CO)_2$$

These complexes, which are preparable by a number of synthetic routes, add a variety of carbon and heteroatomic nucleophiles, among them enolates, enamines enol silyl ethers, dialkyl cuprates, Grignards reagents (Ref. 2) as well as amines, thiols, phosphines and phosphites (Ref. 3). Both the addition step as well as the reverse process, which is frequently observed in β-heterosubstituted(alkyl) Fp complexes, has been shown to be highly stereospecific and to proceed preferentially trans to the Fp-olefin or Fp-C bond.

In general, Fp(olefin) cations derived from simple alkylated olefins show relatively low regiospecificity in their reactions with nucleophiles, especially with carbon nucleophiles (Ref. 2). This is well illustrated by the
reactions of the Fp(propylene) cation with enolates, enamines and Fp(η^1-
allyl), which yield almost equal amounts of regioisomeric adducts.

Much greater regiospecificity may be achieved with electronically polarized
olefin complexes. Fp complexes of α,β-unsaturated ketones or esters are
particularly good substrates for nucleophilic addition since, in addition to
their high regiospecificity, these cations show a considerably heightened activity as electrophiles. The use of the Fp(methyl vinyl ketone) cation in
ring annelation reactions has earlier been reported (Ref. 4).

A second general class of synthetically useful olefin complexes encompasses Fp complexes of vinyl ethers (Ref. 5). These cations, which also exhibit a high degree of regiospecificity in their reactions with nucleophiles, have been shown to function as vinyl cation equivalents.

Fp(vinyl ether) complexes are readily available by metallation of α-bromo-acetals or ketals by NaFp, and subsequent low temperature acid treatment of the resulting metallated product. The cations, obtained as BF_4 or PF_6 salts, are yellow crystalline solids, which may be stored indefinitely at $0°$. They are, however, very sensitive to hydrolysis, and on treatment with water are readily converted to the corresponding aldehyde or ketone complex. Protonation of these latter complexes generates the free enol complexes, which are relatively strong acids. The simplest of these, Fp(vinyl alcohol)BF_4 (pK_a = -0.75), undergoes self etherification on dissolution in methanol or ethanol.

The formulation of Fp(vinyl ether) complexes as simple olefin complexes is un-doubtedly an imprecise description of the bonding in these substances, since the heteroatom supports part of the positive charge residing on the ligand, with the consequence that metal complexation to the olefin bond is highly un-symmetrical. The vinyl N,N-dimethylamine complex, which shows this effect in its extreme, might easily be described as an (N,N-dimethyliminiumethyl)Fp complex, on the basis its x-ray determined crystallographic structure (Fig. 1). The iron to vinyl carbon distances in this cation differ by almost 0.7 Å and the C_4C_3Fe bond angle is much closer to a tetrahedral angle than to that cal-culated for an iron atom π-complexed to a double bond. Finally, strong res-onance interaction of nitrogen with the adjacent cationic center is seen in the contraction of the $N-C_4$ bond distance to one typically observed in imin-ium salts and amides.

The corresponding methyl vinyl ether complex, for which less precise x-ray data is at present available, suggests a similar but less pronounced in-equality in the iron to vinyl carbon distances. Accordingly, heteroatom sub-stituted Fp(olefin) cations might perhaps best be formulated as resonance hybrids with the extent of contributions from form B largely dependent on the nature of the heteroatom at the β-carbon atom.

A   B

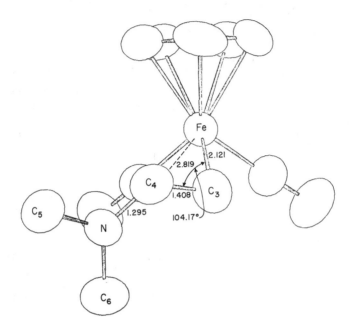

Fig. 1. Molecular structure of the $C_5H_5Fe(CO)_2(CH_2CHNMe_2)$ cation.

Contributions from form B should lead to a decrease in rotational barrier
about the "olefin" bond, and indeed these effects are manifest in the nmr
spectral behavior of a number of vinyl ether and vinyl amine complexes which
we have examined. The methyl vinyl ether cation shows no change in its ABX
resonance pattern in nitromethane solution up to its point of decomposition
at 80°, while the AB absorption pattern of the methyl i-propenyl complex col-
lapses to a singlet above 0° and that of the vinyl N,N-dimethylamine complex
shows coalescence above -60°. For the latter complex the two methyl proton
resonances fail to coalesce even at 100°, reflecting the high barrier to
rotation about the vinyl carbon-nitrogen bond. Approximate rotational barriers derived from these nmr studies are summarized in Table 1.

In view of the above, it is perhaps not surprising that the preparative sequence shown below for the Fp(propenyl ethyl ether) complex leads to the exclusive formation of the cis isomer. Both the cis and trans isomers should
exchange readily at room temperature, but the greater thermodynamic stability
expected for a cis compared with a trans Fp(olefin) complex is well precedented in a number of transition metal-olefin complexes.

TABLE 1. Rotational Barriers in Fp-vinyl ethers and -vinyl amines

$\alpha \equiv\!\!\!\!= \beta$ Z over Fp^+	OMe over Fp^+	OMe over Fp^+	NMe_2 over Fp^+
$\Delta G^{\ddagger}$ (Kcal/mol) $C \overset{\alpha}{=\!\!\!/\!\!\!=} C$	> 18	14.8	11
$C \underset{\beta}{\overset{\curvearrowright}{-\!\!\!/}} Z$			> 19

The sequence, which illustrates the use these vinyl ether complexes as vinyl cation equivalents is summarized in general form below, and makes use of the fact that ligand displacement as shown in the final step may readily be achieved by brief exposure of the derived olefin complex to halide ion.

The use of this sequence for the vinylation of enolates has been examined with a number of cyclohexanone derived enolates. These have been generated from the corresponding trimethylsilyl ethers with butyllithium so as to avoid the presence of any base, aside from the enolate, which might compete for the very reactive complex cation.

Cyclohexanone lithium enolate (1a) reacts rapidly in THF solution -78° with Fp(ethyl vinyl ether)BF$_4$ (2) to give the neutral adduct (3a, 90%) as a single diastereomer (by proton and ^{13}C nmr spectra). The adduct is then transformed by protonation with HBF$_4$·Et$_2$O at -78° to the cation (4a, 89%) and thence by brief exposure to NaI (acetone, 25°, 0.5 h) to the vinyl ketone (5a, 68%).

The vinyl ketone complex (4a) may be further transformed through deprotonation with Et$_3$N (25°) to the conjugated enone (6a) and thence with Me$_3$O$^+$BF$_4^-$ to the methoxydiene complex (7, 24%).

The formation of a single diastereomeric product may be accounted for in terms of preferential orientation of the chiral acceptor component and the prochiral enolate in a manner which minimizes steric interactions in the transition state.

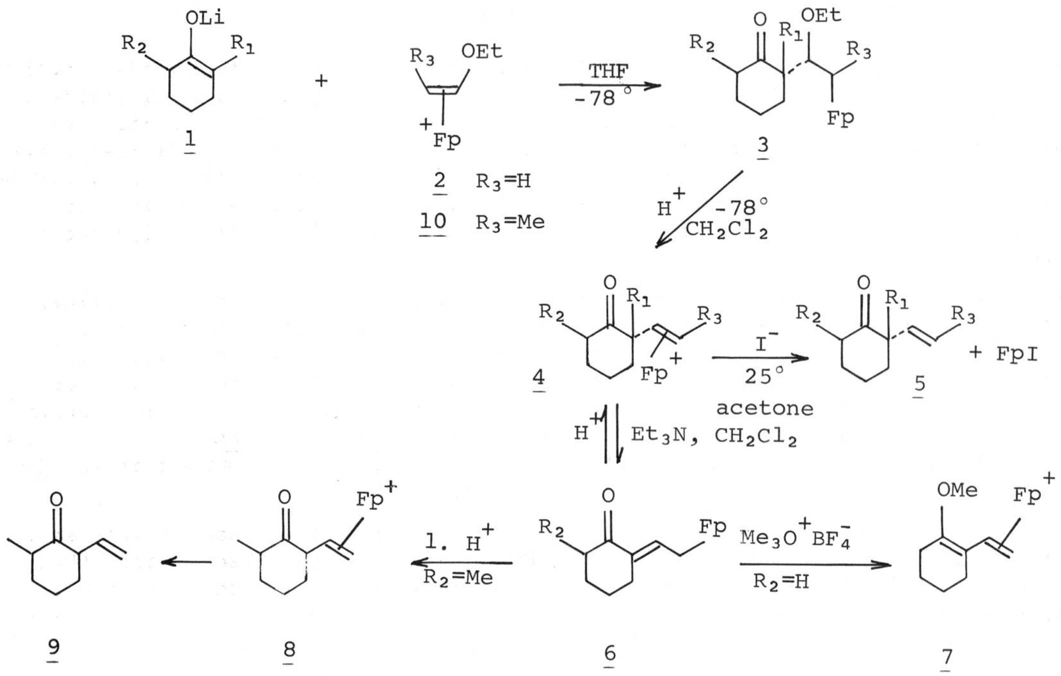

a. $R_1=R_2=R_3=H$

b. $R_1=R_3=H$, $R_2=Me$

c. $R_1=Me$, $R_2=R_3=H$

d. $R_1=R_3=Me$, $R_2=H$

The high reactivity of Fp(vinyl ether) complexes assures alkylation of a kinetically generated enolate free of complications due to equilibration of the enolate or of proton exchange, which are often associated with less re-active electrophiles. Thus 6-methylcyclohexanone enolate (1b) gave the adduct 3b (90%) as a single stereoisomer on reaction with 2. This substance has been assigned trans C-2,6 stereochemistry, assuming preferential axial attack of the vinyl cation on the enolate. Again, the preferential forma-tion of one diastereomer, associated with the relative configurations at C-2 and C-7, may be accounted for in terms of the above orientation of reacting components in the transition state. As before, treatment of 3b with HBF$_4$·Et$_2$O, followed by decomposition of the resulting salt (4b) with NaI, gave trans-2-vinyl-6-methylcyclohexane (5b) in 79% overall yield from 1b.

An attempt to isomerize the trans-2,6-disubstituted cyclohexanone complex (4b) to the more stable cis-isomer (8), through deprotonation with triethyl-amine (25°, 2 h) followed by reprotonation with HBF$_4$·Et$_2$O (-78°, 0.5 h) met with only partial success. The product, obtained in 78% yield after demet-allation with NaI, was found to be a 2:1 mixture of trans- and cis-2-vinyl-6-methylcyclohexanones (5b,9). It is possible that the stereochemistry of protonation of the enone complex (6b) at C-2 is controlled by the spatial orientation of the Fp group which is known to direct electrophillic attack in (η^1-propargyl)Fp complexes trans to the Fp-C bond.

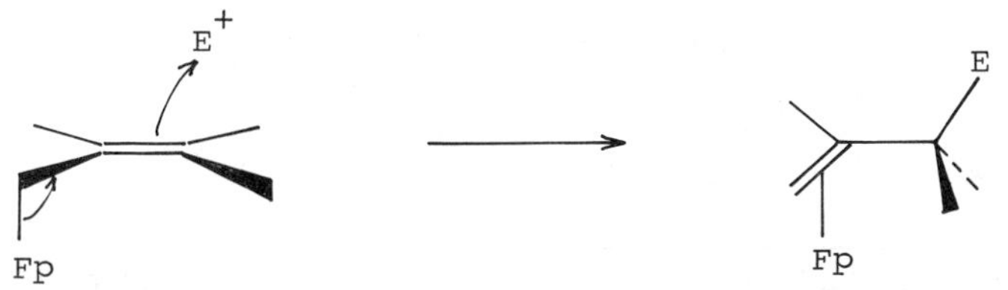

Even the fully substituted 2-methylcyclohexanone enolate (1c) reacts smoothly
with 2 to give the adduct 3c in 90% yield as a 2:3 mixture of diastereomers.
The formation of a mixture of adducts here, in contrast to the experience
with 1a and 1b, is not unanticipated, since the steric preference associated
with the orientation of reacting components for the latter reactants would be
expected to be largely lost when C-2 is methylated. Protonation of product
3c followed by decomposition of the resulting cation (4c) with NaI, gave the
free enone (88%).

The use of the cis-propenyl ethyl ether complex, described earlier, allows
for the introduction of a propenyl substituent at C-2 in 2-methylcyclohex-
anone enolate. The adduct (3d), obtained from the reaction of these components was converted with acid to the cation (4d) and thence to the free
enone with NaI in 78% overall yield. An nmr spectrum of the product, under
spin decoupling conditions, shows it to be entirely the trans isomer (5d) as
would be anticipated for trans addition of the enolate to complex cation 10,
followed by trans elimination of ethanol from adduct (4d).

A further elaboration of this chemistry provides a new and facile synthesis
of α-methylene lactones, (Ref. 6) employing the functionalized vinyl ether
complex (11). This is prepared from pyruvic ester diethyl ketal by the sequence shown below.

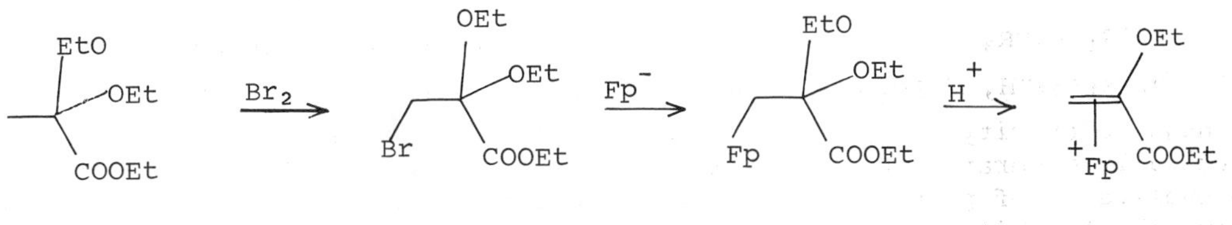

11

The cation complex (11) is less stable than the simple vinyl ether complexes,
but considerably more reactive. Reaction with cyclohexanone lithium enolate
takes place rapidly at -78° in THF solution, from which the adduct 12 may be
isolated in 81% yield, as a mixture of diastereomers. Reduction of 12 with
sodium borohydride in ethanol gave the trans-lactone (13, 72%) and this on
treatment with HPF₆·Et₂O at -78° gave the unstable olefin complex (14, 93%).
The olefin is readily decomplexed by brief exposure to Et₄NBr, liberating
the trans-lactone (15, 80%). In a similar manner, reduction of 12 with L-
selectride in THF at -78° gave the cis- lactone (16, 94%) and this was converted by acid treatment to 17 (91%) and thence by decomplexation to the
free cis-lactone (18, 90%). These transformations are summarized below.

In summary, Fp(vinyl ether) complexes have been shown to function as vinyl
cation equivalents. Their use in the vinylation of cyclohexanone enolates
and in the construction of α-methylene-γ-lactones has been demonstrated. It
seems likely that these complexes will prove to be of value in a number of
synthetic contexts, and that other, more highly functionalized members of
this class of complex can be elaborated and employed in synthesis.

Acknowledgement - This research has been supported by grants from the National Institutes of Health (GM-16395) and from the National Science Foundation (CHE-78-16863).

REFERENCES

1. For leading references see A. Rosan, M. Rosenblum and J. Tancrede, J. Amer. Chem. Soc., 95, 3062(1973). A. J. Birch and I. O. Jenkins in "Transition Metal Organometallics in Organic Synthesis", H. Alper, Ed., Academic Press New York, N.Y., 1976. B. W. Roberts and J. Wong, J. Chem. Soc., Chem. Comm., 1129 (1979), A. Panunzi, A. DeRenzi and G. Paiaro, J. Amer. Chem. Soc., 92, 3488 (1970). B. M. Trost and T. J. Fullerton, J. Amer. Chem. Soc., 95, 292 (1973). M. F. Semmelhack, H. T. Hall, M. Yoskifuji and G. Clark, J. Amer. Chem. Soc., 97, 1247 (1975).
2. P. Lennon, A. M. Rosan and M. Rosenblum, J. Amer. Chem. Soc., 99, 8426 (1977).
3. P. Lennon, M. Madhavarao, A. Rosan and M. Rosenblum, J. Organometal. Chem. 108, 93 (1976).
4. A. Rosan and M. Rosenblum, J. Org. Chem., 40, 3621 (1975).
5. A. Cutler, S. Raghu and M. Rosenblum, J. Organometal. Chem., 77, 381 (1974).
6. P. A. Grieco, Synthesis, 68 (1975).

ZIRCONIUM REAGENTS IN ORGANIC SYNTHESIS

J. Schwartz, F. T. Dayrit and J. S. Temple

Department of Chemistry, Princeton University, Princeton, New Jersey 08544, USA

Abstract - Carbon-carbon bond-forming procedures are described which involve organozirconium complexes in tandem with other metallic species. Mechanistic considerations for these reactions are described.

INTRODUCTION

The synthetic organic chemist today is aware of many types of transition metal-based organometallic reactions which are available for his consideration. The course of many of these reactions has been elucidated from the point of view of stereochemistry, selectivity, and functional group tolerance, for example, and many detailed procedures exist in the literature which enable him to utilize these reactions in his work. In the future, in addition to continuing to discover new types of synthetically useful reactions, the organometallic chemist will increasingly face the challenge of learning to *control* reactivity in organometallic systems so that *selectivity* of product formation can be accomplished. Indeed, as target organic substrates become more complex, the need to know in detail the mode of operation of the organometallic system (which has at its heart the transition metal) will grow in importance. Fortunately, the organometallic chemist, through careful consideration of basic concepts of inorganic chemistry, can develop leverage in manipulating reagent systems of interest through availing himself of techniques known in that field. We describe below two types of synthesis procedures in which organozirconium compounds are used in conjunction with other metallic species in carbon-carbon bond-forming processes. We describe in one case how control of organometallic reactivity can be accomplished by judicious ligand manipulation techniques. In another, we note how the carbon-carbon bond-forming effectiveness of catalyst species can be predicted based on the results of a surveying technique involving simple physical measurements.

GENERAL CONSIDERATIONS FOR TANDEM METAL SYNTHESIS SEQUENCES

Organozirconium compounds can be obtained from Cp_2ZrHCl and olefins or acetylenes (1). In general, these species do not participate directly in carbon-carbon bond-forming reactions, apparently because of the mechanistic requirements for cleavage of the carbon-zirconium bond (2). Through the use of electrophilic metal complexes, cleavage of this bond can be effected to generate a second organometallic compound plus the zirconium salt (3). Because of the carbon-carbon bond-forming propensity of these second metallic species, tandem metal sequences can be employed successfully in which the zirconium hydride functionalizes the organic starting material, and the organic derivative of the second metal is used to generate the carbon-carbon bond. Transmetalation reactions involving palladium and nickel species are described below in which reductive elimination from complexes of these metals provides the means for obtaining the desired carbon-carbon coupled product.

REGIO-CONTROLLED COUPLING OF (π-ALLYLIC)PALLADIUM
COMPLEXES WITH ORGANOZIRCONIUM SPECIES

(π-Allylic)palladium compounds can be obtained, for example, from olefins (4) and have been used in the stereoselective synthesis of a variety of target molecules. In these processes the carbon-carbon bond-forming step involves direct attack by a carbon nucleophile upon the π-allylic ligand. The stereochemical consequences of this coupling process are such that the incoming nucleophile attacks the π-allylic complex on the face *opposite* the one to which the palladium is bound. This synthetic methodology has been extensively developed with regard to problems in steroid synthesis, for example. In these cases the attacking nucleophile is derived from a stabilized carbanion and is the precursor of the steroid side chain. However, because of stereochemical requirements for forming the (π-allylic)Pd complex and of the trans attack of the nucleophile upon it, steroidal compounds thus obtained have the epi configuration at C-20 (5). Transmetalation from zirconium to palladium, though, should provide the means to transfer an organic group to the palladium complex on the *same* side of the allylic unit to which the metal is bound; reductive elimination would give, in the steroidal case, product with the *natural* configuration at C-20.

(π-Allylic)palladium complex $\underline{2}$ could be obtained easily according to modifications of known literature procedures. When this species was treated with alkenylzirconium reagents, the coupled 1,4 diene was obtained in almost quantitative yield. However, when this reaction was applied to palladium

100 (86)

complexes derived from steroidal olefin precursors (in which both termini of the allylic unit are substituted with alkyl groups), the yield of coupled reaction fell. A complex mixture was obtained consisting of the starting olefin and two regioisomers corresponding to coupling at either C-16 or C-20.

To develop procedures such as the one described above into synthetically useful ones requires not only the ability to increase the yield of coupled product but, more importantly, the means to control the regiochemistry of coupling. We describe below our approach to this question of regio control of coupling in these (π-allylic)palladium complexes (6).

Our approach to the question of control derives from the fact that the palladium chloride moiety cannot bind symmetrically to an unsymmetrically substituted π-allylic unit. We propose that, in response to steric differences between the substituents on the termini of the allylic unit, the palladium chloride moiety is displaced slightly toward the less bulky (side chain) terminus. Unsymmetrical binding of palladium chloride units in unsymmetrical π-allylic ligands is well-known (7), and a simple representation can be used to describe an extreme for this unsymmetrical binding.* It is important to note that consideration of this structure suggests that the greater *donor* interaction between the allylic unit and the metal occurs at that terminus which is most closely bound to the metal. In the case of the steroidal π-allylic complex, this steric displacement would translate into donor character being strongest at the side chain terminus.

Rupture of the di-μ-chloro bridge of the π-allylic dimer by treatment with external ligand can occur to give rise to two complexes, one in which the added ligand is bound preferentially trans to the stronger (side chain) donor terminus; the other one involves the binding of the chloride trans to this terminus. Because of bonding considerations known for square planar Pd(II)

---

* This structure is not meant to suggest "σ,π" bonding in this complex. Rather, it is intended to show an extreme perturbation of the unsymmetrically substituted (η^3-allylic) metal unit (8).

complexes, we propose that a donor ligand should preferentially adopt this
latter arrangement and that an acceptor ligand should preferentially adopt
the former (barring major steric repulsions in both instances). If one
assumes that replacement of chloride in these ligand-substituted π-allylic
complexes occurs stereospecifically about the metal, and that reductive
elimination of the carbon-carbon bond occurs with a cis geometry about the
metal, then one can see that the donor-acceptor properties of the added
ligand can be used to control the regiochemistry of the coupling reaction:
Fostering coupling at the side chain terminus should be effected by acceptor
ligands and at the C-16 position by donors.

The donor/acceptor properties of added ligands should also have an effect
upon the activation energy for carbon-carbon bond formation. Reductive
elimination of the carbon-carbon bond is associated with concomitant buildup of charge on the metal [it is reduced from Pd(II) to Pd(0)]. Thus donor
ligands bound to palladium in the transition state for coupling should
retard that process; acceptor ligands should facilitate it.

When these (π-allylic)palladium chloride dimers are treated first with a
bridge-breaking ligand and then with the organozirconium compound, a coupling reaction occurs. The ratio of products coupled at C-20 and C-16 is
consistent with the arguments made above. As well, one finds that when
donor ligands are used, coupling proceeds only slowly at room temperature;
when an acceptor ligand is employed, coupling is rapid even at -78°C (9).

EFFECT OF ADDED LIGANDS ON REGIOSELECTIVITY (4:6)

(All reactions performed at room temperature unless otherwise noted)

Ligand	Equiv Added [a]	4:6 Ratio in Coupled Product
None	—	0.67
PPh$_3$	4.1	1.0
PPh$_2$(o-anisyl)	4.6	1.0
(tri-o-tolyl)phosphine	2.7	0.9
	14.2	0.3
Maleic anhydride	2.1	1.5
	4.1	1.5
	2.9	1.8 [b]
	2.8	6.0 [c]
	3.0	7.0 [d]

[a] per Pd,   [b] 0°C,   [c] -40°C,   [d] -78°C

For example, when maleic anhydride is employed, coupling product is obtained (96% combined yield; <u>4</u>:<u>6</u> ≥ 7:1) from which <u>4</u> can be crystallized by the addition of ethanol. It is important to note that this diene, as predicted by the carbon-carbon bond-forming mechanism, yields cholestanone possessing the *natural* configuration at C-20, when hydrogenated using platinum oxide and then deblocked with aqueous acid.

We believe that this type of ligand control of the regiochemistry of coupling will be of general importance in processes in which unsymmetrically substituted π-allylic complexes are involved.

NICKEL-CATALYZED CONJUGATE ADDITION REACTIONS

We have reported (10) that Ni(AcAc)$_2$, when treated with 1 equiv DiBAH, efficiently catalyzes conjugate addition of alkenylzirconium species to α,β-unsaturated ketones. In elucidating the scope of this conjugate addition reaction, we observed that addition proceeded to dienones with selectivity for product formation that essentially duplicated that noted for a comparable cuprate. It seemed unusual that a nickel-based zirconium conjugate addition

reaction and one involving a cuprate should indeed proceed with the same selectivity unless common intermediates were involved. Cuprate-based conjugate addition reactions have been suggested to proceed by an electron transfer mechanism (11). If Ni(AcAc)$_2$/DiBAH were a one-electron reducing agent, and if conjugate addition occurred by this pathway, then the parallel in selectivities between the nickel and copper-based routes would be understandable: Both would involve the same ketyl. We describe below our studies (12) which show that Ni(AcAc)$_2$/DiBAH can serve as a one-electron reducing agent. Based on this electron transfer concept, we have predicted other types of Ni(AcAc)$_2$/DiBAH-catalyzed coupling reactions and describe these results as well.

The catalytic efficiency of Ni(AcAc)$_2$/DiBAH systems was found to vary as a function of time as follows. A 1:1 mixture of Ni(AcAc)$_2$ and DiBAH was

stirred in THF and aliquots were withdrawn periodically. The aliquots were
then used to catalyze the conjugate addition reaction between an organozirconium substrate and an enone for a fixed time period. The yield of conjugate adduct in this fixed time period was measured and relative yields *vs.*
time were plotted. In this way it was found that catalytic efficiency as a
function of digestion time increased to a maximum after approximately 4-6
hours and then decreased to very low activity after approximately 24 hours of
digestion. If the same aliquots containing Ni(AcAc)$_2$/DiBAH were examined by
cyclic voltammetry (CV), several distinct anodic waves were observed. The
most prominent one occurred at -1.30 volts (*vs.* SCE); when measured as a
function of aliquot, it was found that the peak size of this wave rose and
fell in parallel with catalytic activity of the system as described above.
The reaction between Ni(AcAc)$_2$ and DiBAH was also monitored as a function
of time by noting evolution of isobutane: The formation of 1 equiv isobutane implies that 1 equiv Ni(II) has been reduced to 1 equiv Ni(O). It
was observed that upon mixing Ni(AcAc)$_2$ and DiBAH, a burst of approximately
one-half equiv isobutane occurred almost immediately. Nearly a second half
equiv isobutane was evolved slowly in the subsequent 24-hour period. In
other words, the falloff of catalytic activity paralleled the formation of
Ni(O). Apparently Ni(O) complexes ultimately formed are not effective catalysts for conjugate addition. The fact that the formation of one-half equiv
isobutane occurs concomitantly with generation of the active catalyst species
suggests strongly that Ni(I) is the catalytically active species in these
reactions. Conjugate addition is proposed to occur, therefore, by an electron transfer mechanism in which Ni(I) (likely as aggregate species) delivers
an electron to the enone to yield a ketyl and a Ni(II) species. The ketyl
would then be trapped by the Ni(II) to give an organo Ni(III) compound. The
Ni(III) species would then undergo transmetalation with the zirconium compound to give a Ni(III) complex containing two carbon ligands which then, on
reductive elimination of the carbon-carbon bond, would yield the conjugate
adduct which is the zirconium O-enolate.

It was found that these Zr enolates would undergo trapping with electrophiles
such as formaldehyde (10) or phenylselenenyl bromide (13). By this route the
intermediate first reported by Stork (14) for prostaglandin synthesis could
be obtained simply and in high yield.

PREDICTION OF CARBON-CARBON COUPLING PROCESSES

It had been shown that nickel complexes would effect coupling reactions
between aryl halides and, for example, alkenylzirconium reagents (15). In
this sequence aryl halide is activated as an arylnickel species by oxidative
addition to a low-valent nickel complex [here *in situ* generated $(PPh_3)_4Ni$],
followed by alkenyl group transmetalation from Zr to Ni and reductive elimination of the styrene. Since no buildup of arylnickel compounds is observed,
it is likely that the oxidative addition step of aryl halide determines the
overall rate of the coupling sequence. The oxidative addition reaction
between aryl halides and reduced nickel species has been found (16) to proceed by an electron transfer mechanism. We find that cyclic voltammetric
determinations can be used to predict the efficiency of catalysis of different species in this type of coupling reaction: The ease of activation
of the aryl halide should depend on the one-electron reducing ability of the
nickel catalyst species present in solution. As shown in the table, the
higher catalytic efficiency of $Ni(AcAc)_2$/DiBAH *vs.* L_4Ni, predicted by these
CV determinations, is borne out. (Noteworthy is the observation that L_4Ni
does not catalyze conjugate addition of alkenylzirconium compounds to cyclohexanone, an observation consistent with the notion of an electron transfer
mechanism in this reaction too.)

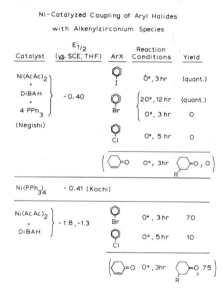

The utility of CV determinations can now be developed as a means to rapidly
ascertain optimal conditions for effecting coupling reactions based on comparable catalyst systems. For example, we find that ligand or solvent modifications give rise to changes in catalytic efficiency of these reduced
nickel catalysts and that these effects correlate with CV determinations
made on the modified systems. Clearly then, it should be possible to
rapidly survey by CV hosts of variously ligated Ni species in a variety of
solvents and use these measurements to determine best conditions for carrying out the catalysis procedure.

CONCLUSIONS

To gain an understanding of organometallic reactivity patterns makes it possible to develop synthesis procedures logically. We have described two different types of synthesis procedures in which modifications of originally observed reactions were performed to optimize desired results. In one case, basic concepts of square-planar (d^8) complex reactivity were exploited. In the other, it was found possible to take advantage of a physical technique previously of interest primarily to analytically oriented chemists to predict and develop efficient catalysts for effecting a desired transformation. Doubtlessly the future will provide us with new examples illustrating how, by resorting to concepts once held to be the exclusive province of the inorganic chemist, the reactions-oriented organometallic chemist will be able to make real headway in the design and control of new procedures for organic synthesis.

Acknowledgment - The authors gratefully acknowledge support for this research provided by the National Science Foundation, Grant No. CHE-790096.

REFERENCES

1. Schwartz, J.; Labinger, J. A. *Angew. Chem., Int. Ed. Engl.* <u>1976</u>, *15*, 333.
2. Labinger, J. A.; Hart, D. W.; Seibert, W. E., III; Schwartz, J. *J. Am. Chem. Soc.* <u>1975</u>, *97*, 3851.
3. Carr, D. B.; Schwartz, J. *J. Am. Chem. Soc.* <u>1979</u>, *101*, 3521.
4. Trost, B. M.; Strege, P. E.; Weber, L.; Fullerton, T. J.; Dietsche, T. J. *J. Am. Chem. Soc.* <u>1978</u>, *100*, 3407.
5. Trost, B. M.; Verhoeven, T. R. *J. Am. Chem. Soc.*, <u>1978</u>, *100*, 3435.
6. Schwartz, J.; Temple, J. S., unpublished results.
7. Mason, R.; Russell, D. R. *Chem. Commun.* <u>1966</u>, 26.
8. Cotton, F. A.; Faller, J. W.; Musco, A. *Inorg. Chem.* <u>1967</u>, *6*, 179.
9. Numata, S.; Kurosawa, H. *J. Organomet. Chem.* <u>1977</u>, *131*, 301.
10. Schwartz, J.; Loots, M. J. *J. Am. Chem. Soc.* <u>1980</u>, *102*, 1333.
11. House, H. O. *Acc. Chem. Res.* <u>1976</u>, *9*, 59.
12. Dayrit, F. M.; Gladkowski, D. E.; Schwartz, J. *J. Am. Chem. Soc.* <u>1980</u>, *102*, 0000.
13. Schwartz, J.; Hayasi, Y. *Tetrahedron Lett.* <u>1980</u>, 1497.
14. Stork, G.; Isobe, M. *J. Am. Chem. Soc.* <u>1975</u>, *97*, 4745, 6260.
15. Negishi, E.; Van Horn, D. *J. Am. Chem. Soc.* <u>1977</u>, *99*, 3168, and references cited therein.
16. Kochi, J. K.; Tsou, T. T. *J. Am. Chem. Soc.* <u>1979</u>, *101*, 6319.

ARENE – METAL COMPLEXES IN ORGANIC CHEMISTRY

M. F. Semmelhack

Department of Chemistry, Princeton University, Princeton, NJ 08544, USA

<u>Abstract</u>- Coordination of an arene with the tricarbonyl chromium
unit induces a strong electron deficiency in the arene ring.
This effect allows metalation of the ring positions, addition of
nucleophiles to the arene π-system, and conjugate addition to
coordinated styrene derivatives. These basic reactions have
been developed into methods for synthesis.

INTRODUCTION

Coordination of an arene ring with a transition metal via the arene pi orbitals often produces a significant electron deficiency in the arene ring. The
effect was first demonstrated by Nicholls and Whiting, who observed that h^6-
(chlorobenzene)chromium tricarbonyl was converted to h^6-(anisole)chromium
tricarbonyl with sodium methoxide in methanol, at a rate similar to p-nitro-
chlorobenzene.[1,2] Subsequently, the effect of a $Cr(CO)_3$ group was found to
enhance the kinetic acidity of arene ring hydrogens,[3] and to stabilize negative charge at a benzylic carbon.[4]

The addition of nucleophiles to coordinated arene ligands produces h^5-cyclo-
hexadienyl complexes (eq. 1), which from h^6-benzenechromium tricarbonyl is the
anionic complex <u>1</u>. An x-ray structure determination of <u>1</u> showed that the
nucleophile bonded at the side opposite $Cr(CO)_3$ (exo)[5], as is commonly ob-

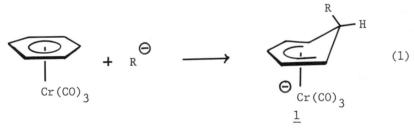

$$(1)$$

served for nucleophilic addition to coordinatively saturated organo-metallic
complexes. We have been concerned with techniques for manipulating the anionic intermediates (analogs of <u>1</u>) into useful organic products and have uncovered two general procedures which have now been applied in the total synthesis of acorenone and acorenone B (below). At the same time, we have been
studying the selectivity in addition of nucleophiles to substituted arene-
$Cr(CO)_3$ complexes,[6] and the addition of nucleophiles to styrene-type ligands.[7]
The latter effort is based on the question of whether a nucleophile will add
to the beta styrene position or directly to the arene ring. We report here
preliminary results concerning selectivity of addition to indole-$Cr(CO)_3$ complexes and nucleophilic addition to a series of styrene-$Cr(CO)_3$ complexes.

ADDITION TO h^6-(STYRENE)CHROMIUM TRICARBONYL COMPLEXES

Addition of 2-lithio-2-methylpropionitrile to h^6-(styrene)chromium tricarbonyl proceeded slowly at -30°. Anion <u>2</u> was postulated but has not yet been observed directly. The intermediate anion is stable at 0° for several hours

and can be trapped with a variety of electrophiles (scheme 1). Oxidation de-

Scheme 1. Addition to h^6-Styrene-$Cr(CO)_3$.

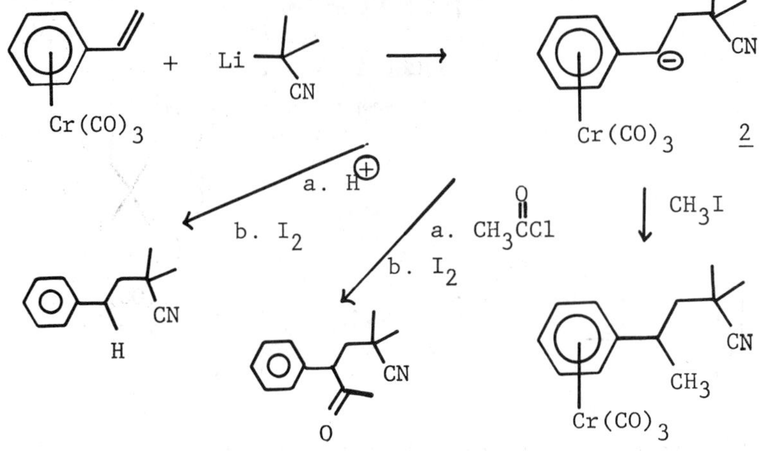

taches the metal and provides the free ligand, generally in good yield. A
forthcoming publication will describe the scope of this reaction, including
other styrene derivatives and a variety of nucleophiles. A particularly interesting case is h^6-(1,2-dihydronaphthalene)chromium tricarbonyl (3). Addition of anion 4 proceeded as before, and the intermediate anion (5) was trapped with protons and with methyl iodide (scheme 2). The product from reaction
with methyl iodide was oxidatively decomposed and the arene (6) was obtained
as a single isomer (homogeneous by tlc, ^{13}C NMR; 65% yield). Spectral evidence and analogy[8] supports the cis stereochemistry for compound 6.

Scheme 2. Addition to h^6-(1,2-Dihydronaphthalene)$Cr(CO)_3$.

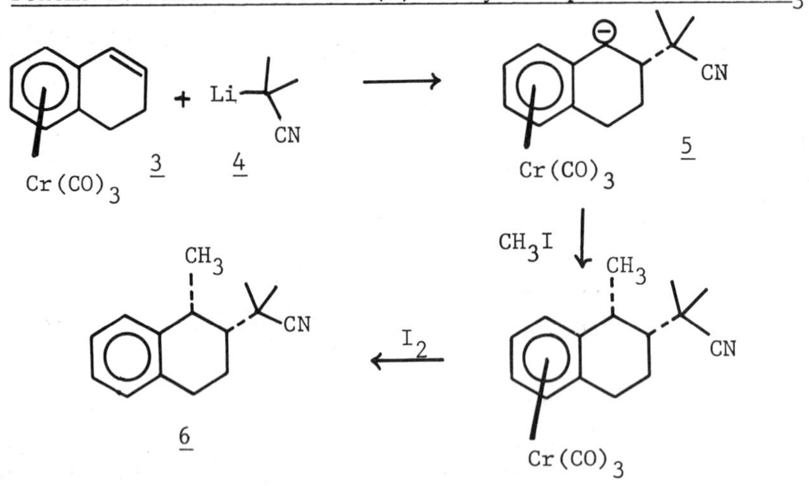

NEW SELECTIVITIES IN ADDITIONS TO SUBSTITUTED ARENE-$Cr(CO)_3$ COMPLEXES

A series of examples of the substituent effect on site of attack in substituted arene-$Cr(CO)_3$ complexes has appeared.[9,10] The process in Scheme 3 illustrates the strong meta selectivity in arene ligands bearing a powerful resonance donor[9], and the general reaction of the intermediate cyclohexadienyl
complexes (ie, 7) with acid, to form substituted cyclohexa-1,3-dienes. From
anisole, 5-substituted-1-methoxycyclohexa-1,3-dienes (ie, 8) are obtained
which can be hydrolyzed to 5-substituted cyclohex-2-en-1-ones.[11]

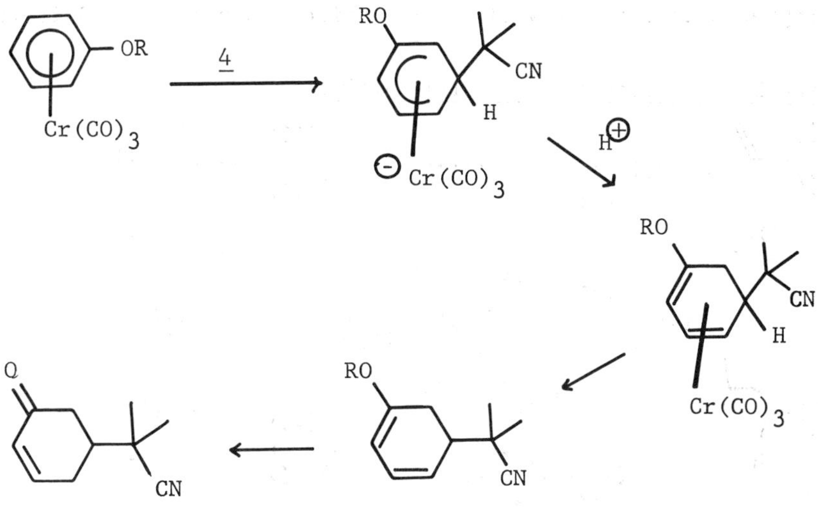

Scheme 3. Conversion of Anisole to Cyclohexenones.

Using the addition/oxidation technique, formal substitution for hydride is observed, and strong donors (eq. 2 and 3) lead to <u>meta</u> substitution, while a trialkylsilyl group (eq. 4) directs the anion to the <u>para</u> position.[9,10] These and other substituent effects have been interpreted in terms of a frontier orbital picture.[6,10]

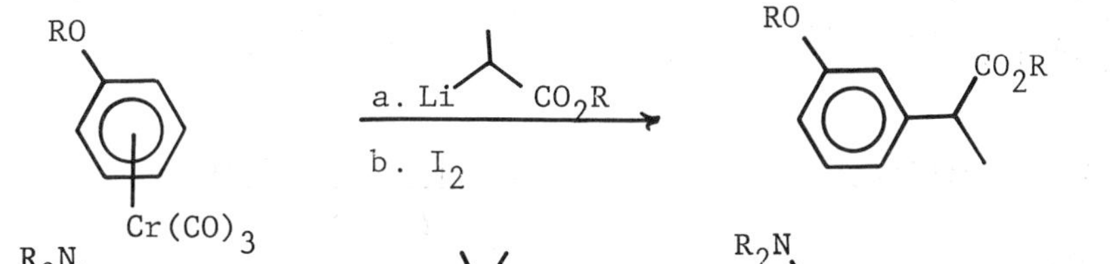

$$(2)$$

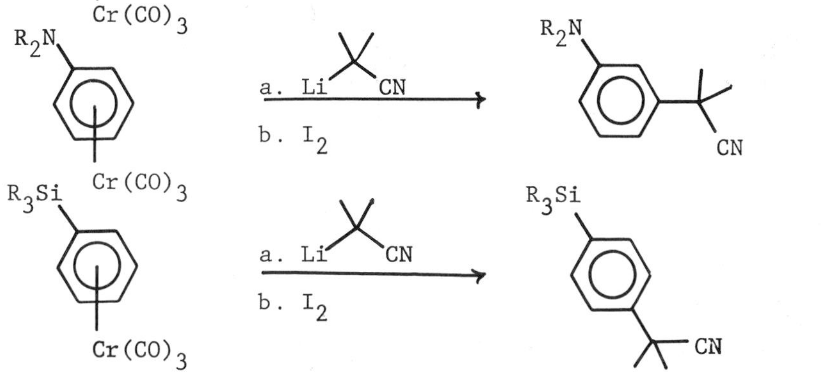

$$(3)$$

$$(4)$$

More recently, indole and other heterocycles have been studied in the nucleophilic substitution reactions. Reaction of h[6]-indole-Cr(CO)$_3$ with anion 4 followed by iodine oxidation produced the 4-substituted indole, 9 in 82% yield (eq. 5a). Similar selectivity was observed with the enolate anion of methyl isobutyrate and lithioacetonitrile. However, with 2-lithio-1,3-dithiane, reaction occurred selectively at carbon 7 (eq. 5b), also in good yield. On the other hand, benzofuran complex 10 gives predominantly 4-substitution with a series of anions including those mentioned above (eq. 6). These elements of selectivity do not fit readily into the frontier orbital picture and are under active study.

The powerful meta-directing effect of a methoxy also operates effectively in intramolecular cases. For example, the simple cyclization of 11 produced only ring-fused product (12, eq. 7)[12], while the 3-methoxy analog 13 gave only spiro-fused product (eq. 8, 14, mixture of diastereomers, 76% yield)[11]

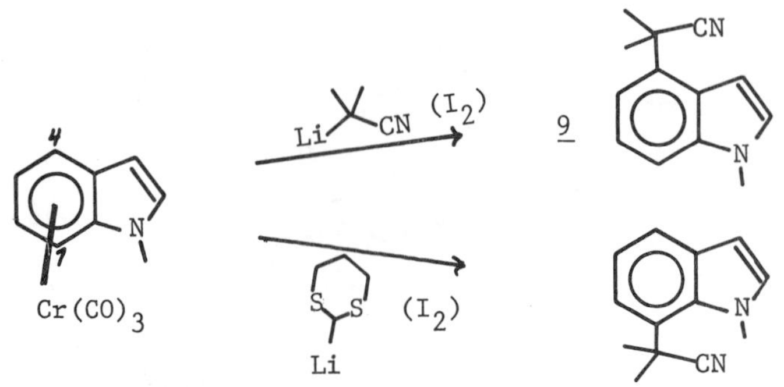

$$(5a)$$

$$(5b)$$

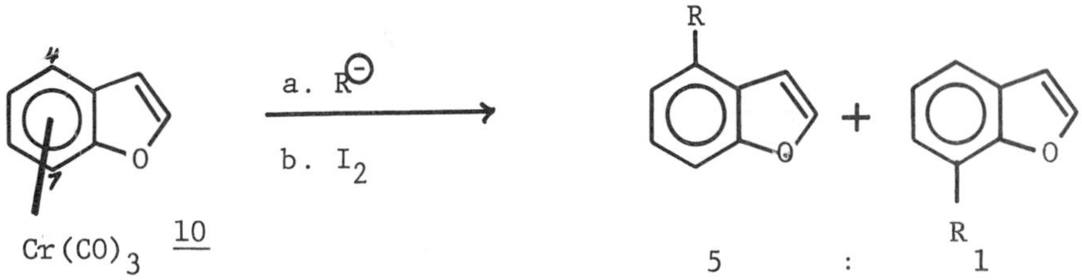

$$(6)$$

latter reaction provided a strategy for a successful approach to the spiro sesquiterpenes, acorenone and acorenone B.

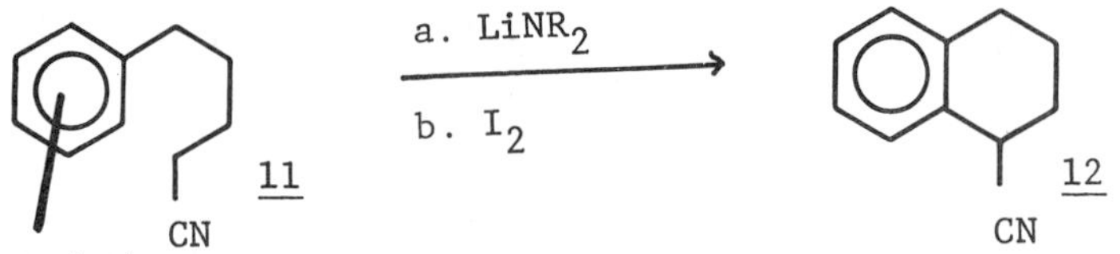

$$(7)$$

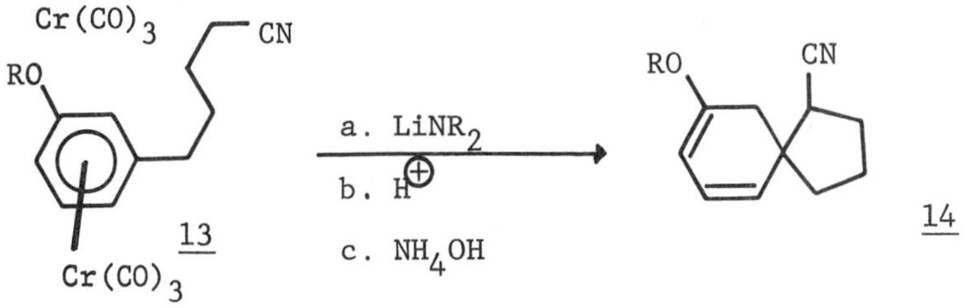

$$(8)$$

Acorenone and acorenone B have the common structural feature of a spiro ring formed from a 2-substituted cyclohexenone and 1,3-dialkylcyclopentane.[13] There are three centers of chirality, the carbons bearing the methyl and isopropyl groups and the spiro carbon. Our strategy (Scheme 4) employs the specific exo addition to coordinated arenes[5] to control the configuration at the spiro carbon. Hydrogenation of an exo-methylene unit is proposed to introduce the cis relationship of the two alkyl groups. The execution of this strategy is outlined in Schemes 5-7, showing all isolated intermediates.

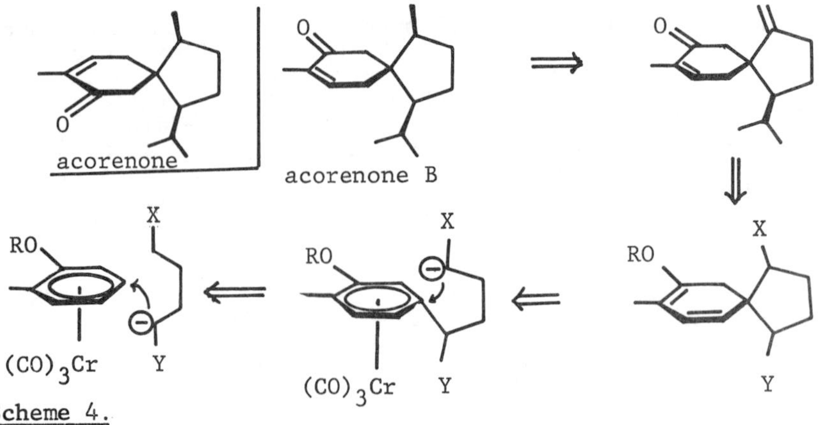

Scheme 4.

As outlined in Scheme 5, the chromium complex 15 was prepared in 95% yield by heating o-methylanisole and chromium hexacarbonyl in dioxane at reflux using an air condenser. Complex 15 was easily crystallized (mp 91-92°) and has been prepared on a 70-g scale. The first crucial bond is formed by reaction of 15 with the cyanohydrin acetal anion 16 followed by oxidation with excess iodine. Treatment with aqueous acid and then base converted the cyanohydrin acetal to a ketone unit, effecting formal nucleophilic substitution for hydrogen by an acyl group. The reaction produced isomer (17) (90-95% yield) contaminated by a trace (<3%) of a positional isomer. The 4-carbon side chain required for cyclization (in 19) was obtained by addition of allylmagnesium bromide to 17 followed by reduction of the resulting tertiary alcohol using "ionic hydrogenation". Anti-Markovnikoff addition of hydrogen bromide followed by displacement by cyanide afforded the nitrile, 19, in 77% yield overall from 17.

Reaction of 19 with excess chromium hexacarbonyl was complete within 12-15 hr at reflux in dioxane, to give a mixture of 20 and the corresponding complex with a $Cr(CO)_5$ unit attached to the nitrile group. Treatment of the mixture with carbon monoxide (350 psi, 25°, 15 hr, THF) followed by chromatography on Florisil gave 20 as a mixture of diastereomers in 84% yield. The ratio of diastereomers was about 60:40 (HPLC and ^{13}C NMR spectra). Complete separation by preparative HPLC gave an easily crystallized solid (20a), mp 90.8-92°, and a low mp solid, 20b, each homogenous by HPLC and ^{13}C NMR.

Scheme 5.

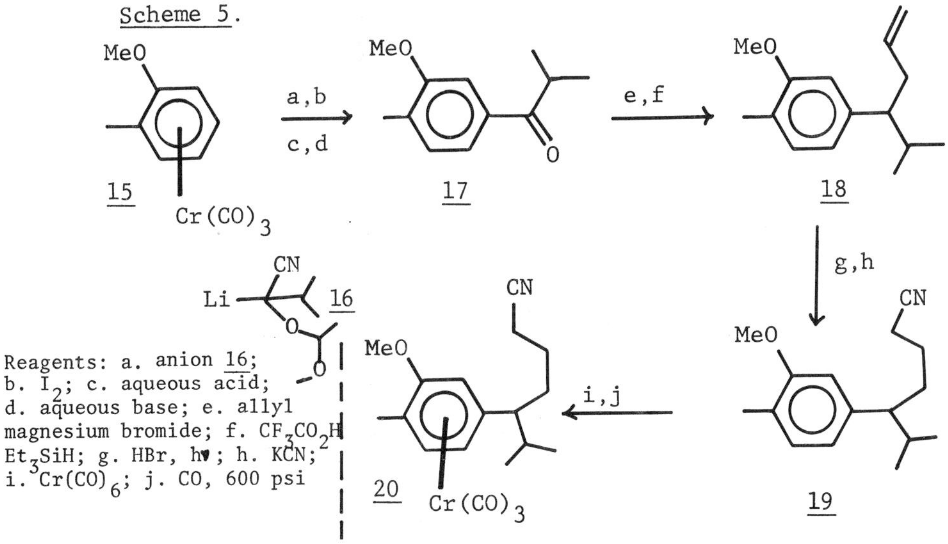

Reagents: a. anion 16; b. I_2; c. aqueous acid; d. aqueous base; e. allyl magnesium bromide; f. CF_3CO_2H Et_3SiH; g. HBr, hν; h. KCN; i. $Cr(CO)_6$; j. CO, 600 psi

Diastereomer 20a was treated with lithium diisopropylamide in THF at -78° (Scheme 6). Then hexamethylphosphoric triamide (HMPA) was added and the mixture was stirred at -78° for 4 hr. Dropwise addition of trifluoromethanesulfonic acid (5-fold molar excess) produced a deep red solution which was poured into a mixture of conc. aqueous ammonium hydroxide and ether (equivolume) cooled to -30°. From the organic layer was isolated a crude product which was treated with a 5% solution of hydrochloric acid (H_2O-CH_3OH, 80°, 18 hr). The resulting mixture of cyclohexenones (spiro and fused ring systems) was separated carefully by chromatography. It consisted of recovered arene 19 (30%), fused ring isomers (20%), and two spirocyclohexenones 21a (40%) and 21b (5%).

Spiroketone 21a was converted to the ethylene ketal 22a and then to ketoketal 23 following the oxidative-decyanation procedure of Watt (Scheme 7)[14]. Spiroketone 21b was converted by the same sequence to 23, verifying that 21a and 21b differ in the orientation of the cyano group. Wittig olefination followed by removal of the ketal unit produced ketone 24. Hydrogenation of the exomethylene group was accomplished stereospecifically (Wilkenson's catalyst) to give racemic acorenoneB, identical in pmr and chromatographic properties with a sample of natural (-)-acorenone B.[15] The overall yield from 21a was ca 45%.

The lower melting complex (20b) was treated in a precisely parallel way. A single spirocyclohexenone (25) was obtained, and none of the other diastereomeric series (ie, 21) was detected by analytical HPLC. The yield of 25 was only 15%, and it was accompanied by both unreacted 19 and fused cyclohexenones. Conversion of 25 to acorenone (26) was accomplished in 40% yield overall in a sequence exactly parallel with scheme 7. The racemic acorenone was shown to

Scheme 6.

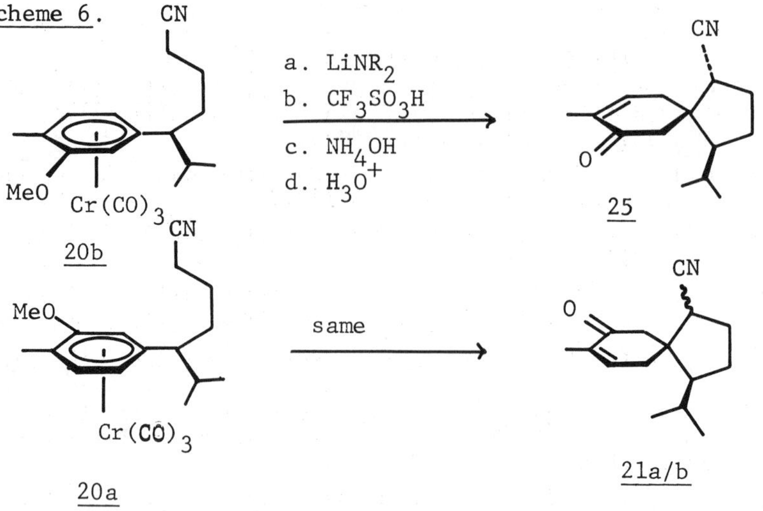

Scheme 7.

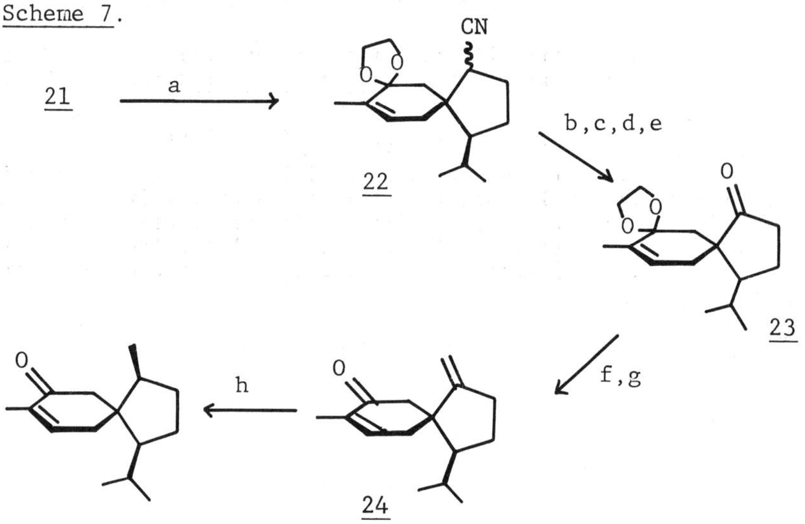

be identical by pmr and chromatographic properties with a sample of natural (-)-acorenone.[15]

The lower efficiency of these spirocyclizations compared to simpler models, especially for diastereomer 20b, was unexpected and efforts are underway to improve the reaction.

Acknowledgement- This work results from the intellectual and experimental contributions of an energetic group of collaborators at Cornell and at Princeton. The specific details mentioned here and not yet published are due to Dr. Ayako Yamashita, Dr. Leonard Keller, Dr. Walter Seufert, Dr. Charles Shuey, Dr. William Wulff, and Mr. John Garcia. I am very grateful for their contributions.

REFERENCES

1. B. Nicholls and M. C. Whiting, J. Chem. Soc. (London), 551 (1959).

2. For reviews, see: (a) M. F. Semmelhack, Ann. NY. Acad. Sci., 295, 36 (1977) and (b) G. Jaouen in "Transition Metals Organometallics in Organic Synthesis," H. Alper, ed., Academic Press, 1978.

3. For examples and leading references, see: M. F. Semmelhack, J. Bisaha, and M. Czarny, J. Amer. Chem. Soc., 101, 768 (1979).

4. For examples see reference 2 and: W. S. Trayhanovsky and R. J. Card, J. Amer. Chem. Soc., 98, 2897 (1972).

5. M. F. Semmelhack, H. T. Hall, Jr., R. Farina, M. Yoshifuji, G. Clark, T. Bargar, K. Hirotsu, and J. Clardy, J. Amer. Chem. Soc., 101, 3535 (1979).

6. M. F. Semmelhack, G. Clark, R. Farina, and M. Saeman, J. Amer. Chem. Soc., 101, 217 (1979).

7. Addition of alkyl-lithium reagents to h^6-styrenechromium tricarbonyl in low yield has been published: G. R. Knox, D. G. Leppard, P. L. Pauson, and W. E. Watts, J. Organomet. Chem., 34, 347 (1972).

8. For examples, see: M. A. Boudeville and H. des Abbayes, Tet. Lett., 2727 (1975).

9. M. F. Semmelhack and G. Clark, J. Amer. Chem. Soc., 99, 1675 (1977).

10. G. Clark, PhD Thesis, Cornell University, 1977.

11. M. F. Semmelhack, J. J. Harrison, and Y. Thebtaranonth, J. Org. Chem., 44, 3275 (1979).

12. M. F. Semmelhack, L. Keller, and Y. Thebtaranonth, J. Amer. Chem. Soc., 99, 959 (1977).

13. For a general review of spirocyclic sesquiterpenes, see: J. A. Marshall, S. F. Brady, and N. H. Andersen, Fortsch. Chem. Org. Naturst., 31, 283 (1974).

14. S. J. Selikson and A. S. Watt, J. Org. Chem., 40, 268 (1975).

15. We are grateful to L. Romero Fonseca (National University of Argentina), Gordon Lange (University of Guelph, Canada), M. Resaro (Givauden, Dubendorf, Switzerland) and James D. White (Oregon State University) for samples of acorenone and acorenone B. We also thank Barry Trost (University of Wisconsin) and J. Wolfe (Technical University, Braunschweig, West Germany) for spectral data.

TRANSITION METAL MEDIATED CARBON-CARBON BOND FORMATIONS: A GENERAL, PARTIALLY CHEMO-, REGIO-, AND STEREOSPECIFIC SYNTHESIS OF ANNELATED CYCLOHEXADIENES FROM ACYCLIC STARTING MATERIALS

Chu-An Chang, C. G. Francisco, T. R. Gadek, J. A. King, Jr.,
E. D. Sternberg, and K. P. C. Vollhardt*

Department of Chemistry, University of California, Berkeley, and the Materials and
Molecular Research Division, Lawrence Berkeley Laboratory, Berkeley,
California 94720, USA

Abstract - The cobalt complex η^5-cyclopentadienyldicarbonylcobalt effects the stoichiometric and chemospecific cyclization of α,δ,ω-diyneenes to complexed tricyclic cyclohexadienes, and of α,ω-enynes with monoacetylenes to complexed bicyclic cyclohexadienes. In several cases stereoselectivity is observed of the newly formed chiral center relative to the cobalt atom. The stereochemistry of the double bond is retained in the product, and in the case of unsymmetrical monoacetylenes the cyclization proceeds regioselectively. The free ligands may be prepared by treatment of the complexes with cupric chloride, or by running the reaction with catalytic amounts of bis(triphenylphosphine)dicarbonyl nickel. The complexed as well as the free ligands may be further functionalized broadening the scope of the synthetic method.

INTRODUCTION

The Diels-Alder reaction (1) in which a conjugated diene reacts with a monoene to form two new bonds generating a six-membered ring has provided the synthetic chemist with a powerful tool in carbo- and heterocycle construction. This is particularly due to its three features of frequently observed selectivity: chemo-, regio-, and stereoselectivity. Its intramolecular version (2) constitutes a rapid synthetic entry into polycyclic molecules of synthetic interest (3). A potentially more useful reaction would be a [2+2+2] cycloaddition of three unsaturated moieties to furnish cyclohexane derivatives by simultaneous formation of three new carbon-carbon or carbon-heteroatom bonds. Such thermal reactions involving alkenes are very infrequent and proceed only in cases in which at least two of the reacting double bonds are lying close enough to allow for some electronic interaction (4). This appears necessary to generate the appropriately matched HOMO-LUMO gaps between the respective reactants in transformations which may be more appropriately described as "homo-Diels-Alder reactions". An example is the addition of alkenes to norbornadiene (1→2) (4):

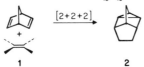

Obviously, entropy problems must play a large role in the lack of success in effecting intermolecular cyclic trimerization of alkenes. One may anticipate that transition metals might provide a suitable matrix on which cyclizations of this type might occur. Whereas there is a plethora of literature reports dealing with transition metal catalyzed alkyne cyclizations (5) to furnish benzene derivatives, similar reports using only alkenes to produce cyclohexanes are exceedingly rare (6).

On the other hand there is some organometallic literature pertaining to the cocyclization of alkynes with alkenes, particularly when employed in a 2:1 ratio resulting in cyclohexadiene derivatives. This transformation may be carried out in various ways. One approach utilizes a metallacyclopentadiene, preformed by stoichiometric reaction of a transition metal complex with two

71

equivalents of an alkyne, and reacts it with an alkene to provide a complexed
cyclohexadiene, as demonstrated for cobalt (Scheme 1) (7). It is not clear
whether the final step in this procedure occurs through a Diels-Alder type

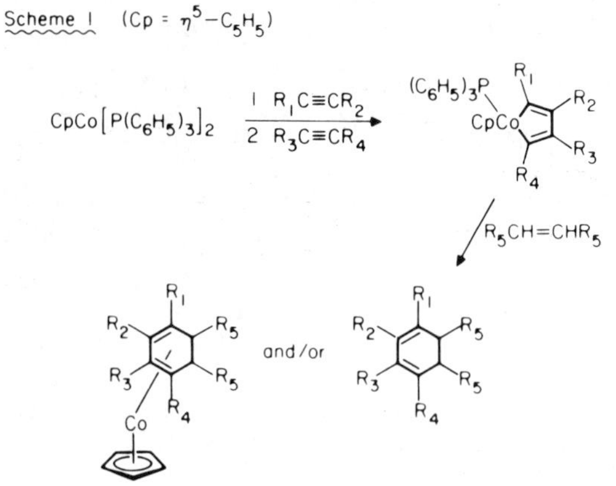

Scheme 1   (Cp = $\eta^5-C_5H_5$)

addition or via insertion of the alkene to furnish an intermediate cobalta-
cycloheptadiene capable of rearrangement to give product. However, the reaction is phosphine inhibited, suggesting precomplexation of alkene followed by
an intramolecular process (7), although some substrates may react directly
(8). On the other hand, the cycloaddition is not stereospecific as is typical for the Diels-Alder reaction giving both exo- and endo-products, although
the alkene stereochemistry appears to remain intact.

Cyclohexadienes may also be formed catalytically. Here metallacyclopentadienes have been invoked as likely intermediates in selective catalytic cocyclizations leading to cyclohexadienes using maleimides (9), 1,3-butadiene
(10), and norbornene (11). However, it is possible that these reactions proceed through metallacyclopentene intermediates by initial coordination of one
equivalent of alkyne and alkene, respectively. Stoichiometric and stepwise
transformations of this type have been observed for cobalt (Scheme 2) (12)
and rhodium (Scheme 3) systems (13). The reason for the particular stereo-

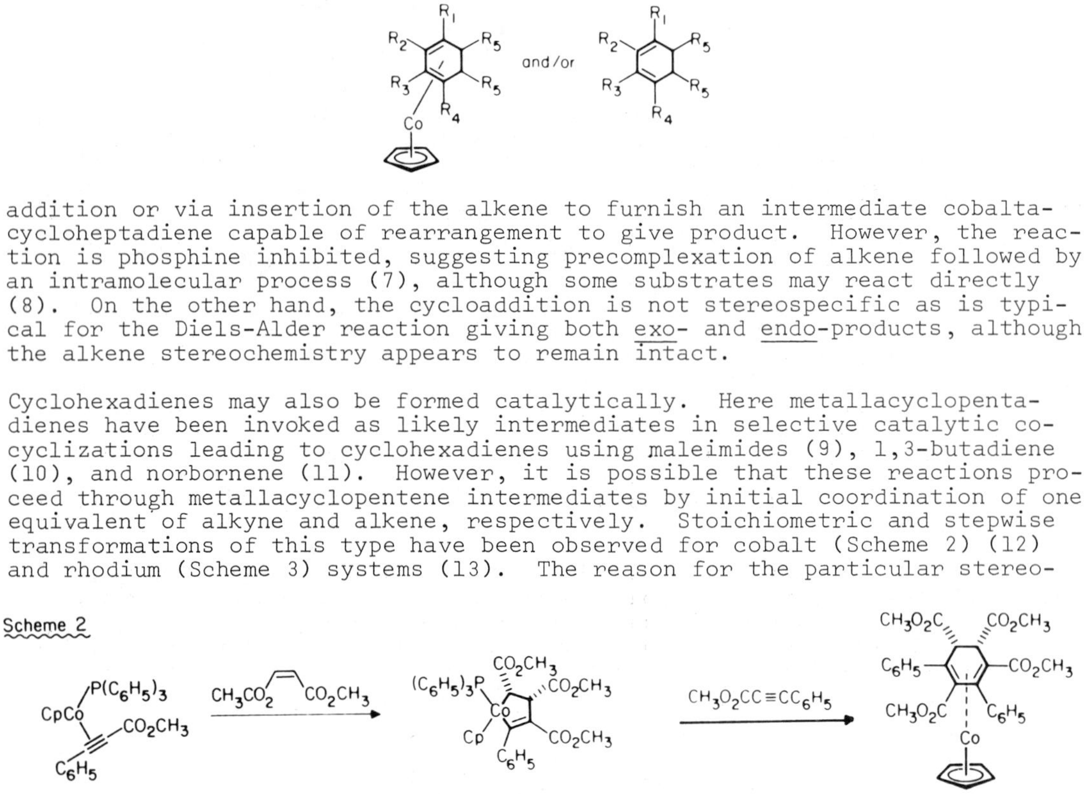

Scheme 2

chemical outcome of these reactions (when observed) is obscure.

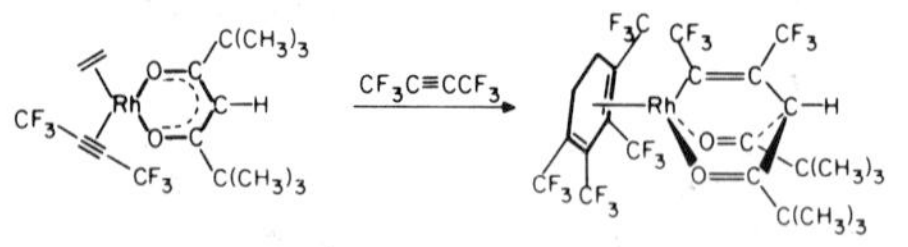

We have been interested in utilizing transition metals, particularly cobalt
(I) complexes, in the chemo-, regio-, and stereoselective construction of
complex molecules of theoretical (14) and synthetic (15) interest. It occurred to us that cocyclizations of α,ω-enynes with monoalkynes and internal
[2+2+2] cycloadditions of α,δ,ω-diyneenes might provide an extremely facile
synthetic entry into polycyclic annelated cyclohexadienes. If successful,
such synthetic methodology might be amenable to further elaboration in natural product and (in the case of short bridges linking the unsaturated moieties in starting material) strained ring synthesis. The following report
is an account of our preliminary work in the area.

INTRAMOLECULAR DIYNEENE CYCLIZATIONS

The starting materials 3-7 required for intramolecular cyclization were prepared by alkylation of the appropriate α,ω-diyne with 5-bromo-1-pentene or
6-bromo-1-hexene [20% DMSO-NaNH$_2$ (two equiv.)-NH$_{3\,liq}$] in adequate yields (ca.
50%). Trimethylsilylation was performed in high yield by direct reaction
of the crude reaction mixtures with trimethylchlorosilane after removal of

ammonia and addition of ether. When compound 3 was subjected to a slight excess of CpCo(CO)$_2$ in refluxing isooctane under N$_2$ over a period of 4-5 days, chromatography on alumina furnished a single crystalline product in 85% yield, the CpCo-diene complex 8 (16). In this reaction three rings are formed simultaneously from an acyclic and achiral precursor by establishing three new carbon-carbon bonds leading stereospecifically to a chiral ligand system. The structure of 8 was assigned based on its analytical, physical, and chemical data. Its stereochemistry was elucidated by high field ^{1}H-NMR

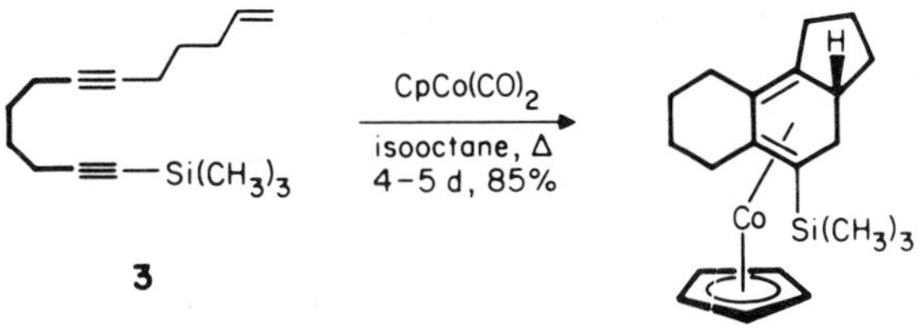

spectroscopy. As observed in many other CpCo-diene complexes the anisotropy of the cobalt leads to a comparatively much stronger shielding of the exo-protons in the saturated part of the cyclohexadiene ligand, when compared with the corresponding endo-protons (17). In the case of 8 the tertiary hydrogen located on the chiral center showed a multiplet resonance at $\delta 0.53$, whereas the adjacent methylene group in the six membered ring exhibited two absorptions in the form of double doublets: one at 1.01 assigned to H$_{exo}$, and one at 1.31 (H$_{endo}$). Appropriate decoupling experiments and deuteration at the endo tertiary position (vide infra) confirmed these assignments. In addition, proton and carbon NMR data on a series of similar compounds with both exo- and endo-configuration at the tertiary carbon correlate well within the series and with the results of an X-ray structural determination of 11a (18). In 8 and the compounds 9-12 to be described shortly, the ^{13}C NMR absorptions are consistently assignable to the following ranges (δ ppm from internal C$_6$D$_6$): η^5-cyclopentadienyls: 78-84, internal complexed diene carbons: 86.7-95.8; terminal diene carbon at ring juncture: 66.0-76.4; other terminal diene carbon (regardless of whether silylated or protonated): 42.1-46.3.

It is difficult to assess what is the origin of the stereospecificity of the cyclization leading to 8. If one assumes initial metallacyclopentadiene formation (Scheme 4) followed by an intramolecular Diels-Alder reaction, the latter would have to proceed in the exo-fashion as shown. On the other hand, the intermediacy of an initial metallacyclopentene intermediate allows

Scheme 4

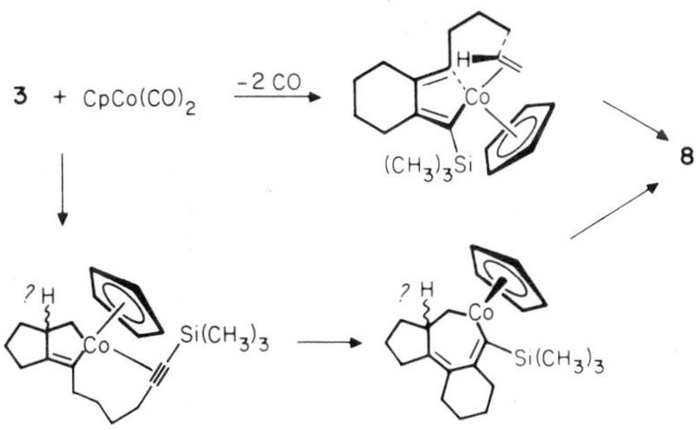

stereochemical fixation and/or equilibration at several stages of the scheme (Scheme 4) making definite mechanistic interpretations impossible. The problem is compounded in difficulty by the absence of any knowledge of the dynamics of the structure of coordinatively unsaturated η^5-cyclopentadienylcobalt ligand systems (19).

Similar seemingly complete stereoselectivity is observed when 4 is exposed to cyclization conditions, although the yield of product 9 is only moderate (35%) and a fair amount of insoluble polymer is formed, possibly due to the instability of starting material and/or another product. Compound 9 appears stable to the reaction conditions. Due to the anisotropy of cobalt and the

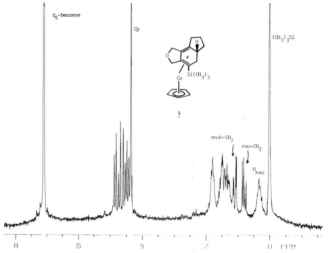

4   **9**

electronegativity of oxygen the high-field ^{1}H-NMR spectrum of 9 (Figure 1) is
particularly easy to interpret, with doublets for each of the diastereotopic
dihydrofuran protons, and the characteristic high field double doublet pattern and multiplet for the hydrogens on the central ring.

In order to get some feeling for the scope of the reaction several other examples were run, all of which resulted in mixtures of isomers. For example,

Fig.1 180 MHz ^{1}H NMR spectrum of 9 in C_6D_6.

cyclization of 5 yielded a 1:1 mixture of <u>exo-endo</u>-isomers 10a,b, separable
by fractional crystallization (which gave pure 10a), or high pressure liquid

5   **10a**   **10b**
                                          1:1

chromatography using a reverse phase nitrogen or argon saturated column.
The proton NMR spectrum of 10b is characterized by the absence of the high
field absorption found in 10a for the tertiary proton, and the presence of a
new absorption at low field assigned by decoupling experiments to the analogous proton in 10b, now in the <u>endo</u>-position. The disappearance of any
stereoselectivity on simply removing a trimethylsilyl group when going from 3
and 4 to 5 is surprising and not understood.

Similarly unexpected is the lack of selectivity in the cyclization of 6. The

6   **11a**   **11b**
                                          1:1

equimolar mixture of the products 11a,b yields one crystalline isomer, 11a,
the structure of which was unambiguously established by single crystal X-ray
diffraction (18). The result of the X-ray investigation was in accord with
the structural assignment derived on the basis of spectral considerations.

Finally, conversion of 7 gave not only the anticipated exo- and endo-products 12a and 12b but also a third isomer (ratio 3:1:2), which appears to be related to the products by a hydrogen shift. As in all of the previous cases, the exo-isomer 12a is the most readily crystallizable.

In order to test whether the above cyclization method could also be successfully applied to cases where the alkene unit is located internally with respect to the two alkyne functions the two isomeric cis- and trans- 1,12-tridecadiyne-6-enes 15 and 16 were prepared (Scheme 5). Pure 15 can be obtained by Wittig reaction of 5-hexyn-1-al with the appropriate phosphonium salt (bu-

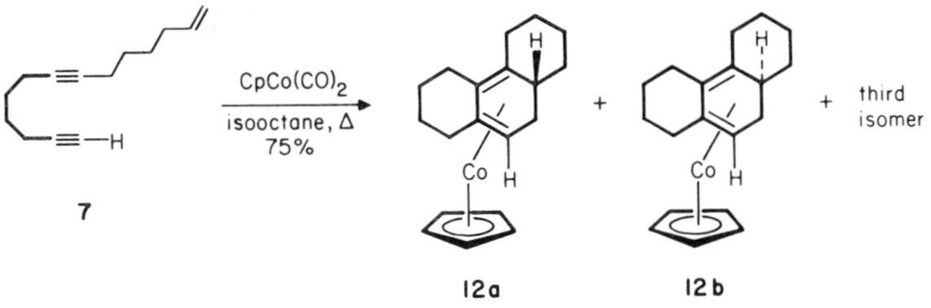

tyllithium, DMSO, 48%). Similarly, 6-heptynal 13 may be converted to the

Scheme 5

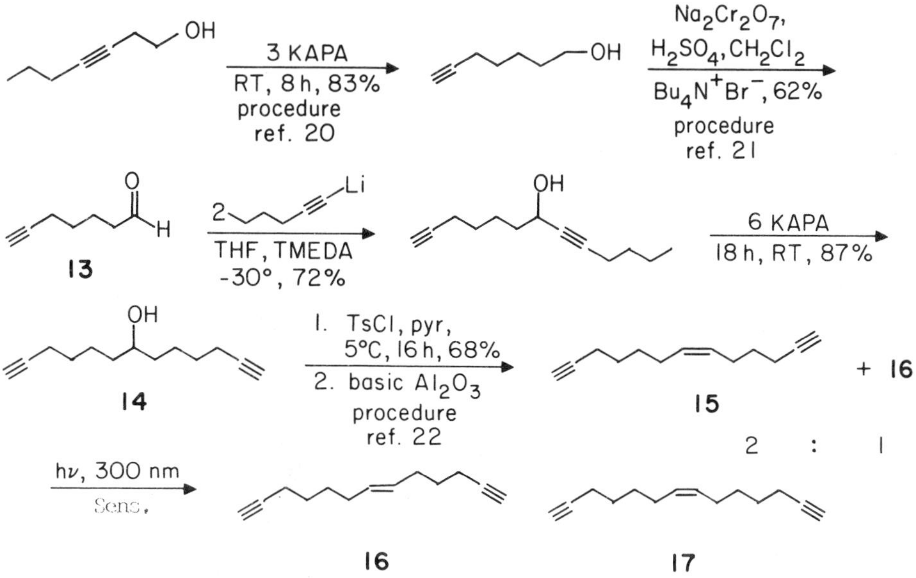

symmetrical cis-diyneene 17. Cyclization of these molecules proceeded smoothly under the usual conditions [CpCo(CO)$_2$, isooctane, 4-7 d] giving the results shown below:

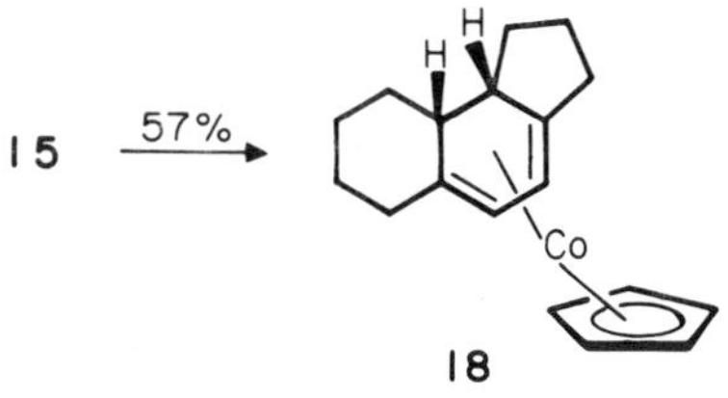

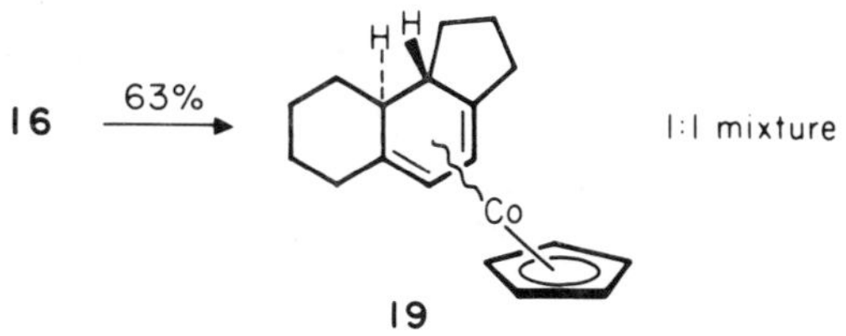

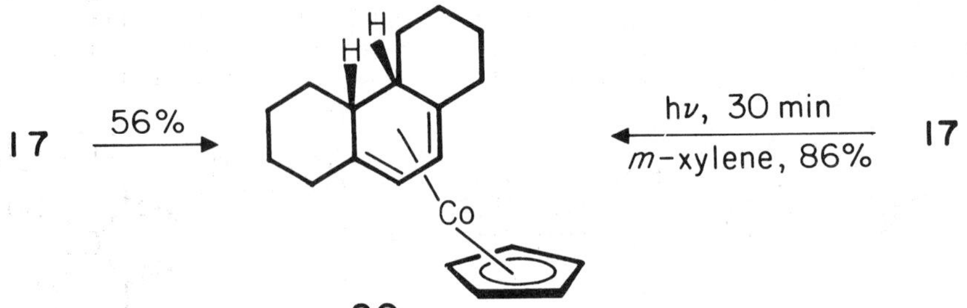

The stereoselectivity in the reaction of the cis-isomers 15 and 17 giving exclusively 18 and 20 is again remarkable. No other isomer was detectable by
^{13}C NMR investigation of the crude reaction mixture. On the other hand, it
is not surprising that 19 yields the two possible syn and anti-isomers since,
whatever their mechanism of formation, one would expect very little difference in the steric bias of a five membered versus six membered ring chain
with respect to the CpCo-unit.

Another fascinating aspect of the above reactions is the recently discovered
fact that exposure of the starting material (e.g. 17) to visible light tremendously accelerates the cyclization with concomitant improvement in yield.
Although this effect has not yet been generalized, this change promises to
significantly facilitate the application of the described methodology. It is
not yet clear whether the photochemical rate acceleration is associated with
a change in mechanism. It could well be that light simply improves carbon
monoxide extrusion to generate open coordination sites on cobalt (23). However, whether this indeed occurs and is necessary even in thermal cyclizations catalyzed by CpCo(CO)$_2$ has been questioned (24). We have made some

rather peculiar observations in this connection which need further investigation but will be reported here in preliminary form. Since most of this work
was carried out with 3 we will restrict our remarks to this compound. In a
kinetic analysis of the cyclization of 3 to 8 we have noted pronounced induction periods of up to 36h. Moreover, we have noted that once reaction has
started, it accelerates and is usually complete within several hours. Interestingly, this induction period virtually disappears on addition of triphenylphosphine! However, in these cases the number of "fast" turnovers depends on
the amount of phosphine added, one full equivalent being necessary for complete "fast" conversion of starting material. Spurts of activity are also
noted on addition of benzoyl peroxide or azodiisobutyronitrile, although in
these cases the catalytically active species appears to deactivate after a
short while. Improved catalytic activity is also seen with preformed CpCo-
(CO)[P(C$_6$H$_5$)$_3$] (25), but no asymmetric induction is observed in the presence
of chiral and optically active phosphine [(+)-neomenthyldiphenylphosphine].
These data suggest the possibility of a radical chain mechanism being responsible for catalytic action (26).

All of the above cyclizations are not catalytic but stoichiometric in cobalt.
Although, as will be seen, the attached CpCo-group may serve a useful function
in the further chemistry of the ligand, and although the latter may be readily and efficiently removed by cupric ion oxidation, a catalytic process
would be eminently desirable from a synthetic organic point of view. We have
recently discovered that this may be achieved by using a different transition
metal complex: bis(triphenylphosphine)dicarbonyl nickel. This species catalytically (0.08 molar equivalents) converts 3 to the free ligand present in
8 in good yield. Other systems are under investigation.

INTERMOLECULAR [2+2+2] CYCLIZATIONS

Cocyclizations of α,ω-enynes with monoalkynes can also be effected by using
CpCo(CO)$_2$ although we have as yet not been able to find conditions which give
good yields. One of the dilemmas we appear to face is the ready competing
rearrangement of intermediate metallacyclopentadienes (c.f. 22) to stable cyclobutadiene complexes (c.f. 23), rather than alkene incorporation (c.f. 24).

The starting enyne 21 was prepared from 6-bromo-1-hexene with excess sodium acetylide in liquid ammonia (70%) and then added to neat bis(trimethylsilyl)-acetylene (BTMSA) over a period of 2d. The major product was the cyclobutadiene complex 23 and only small amounts of a 1:1 mixture of the two isomers of 24 were obtained.

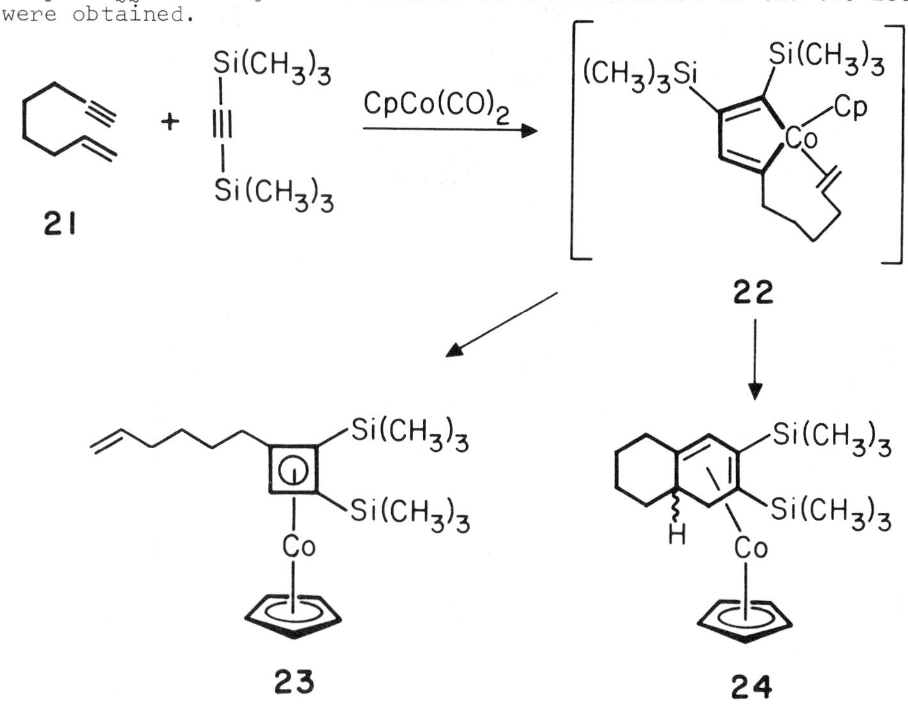

Since both 23 and 24 are presumably derived from the same intermediate 22, it became of interest to determine whether the kinetically formed 23 could be induced to thermally revert to 22 in order to ultimately give 24. Unfortunately, this proved not to be the case. Compound 23 is photolytically and thermally quite inert. In the gas phase flash pyrolysis at very low pressures resulted in decomposition with regeneration of the alkyne components of the four ring (27).

We had previously observed that cyclobutadiene formation is much less pronounced with unsilylated alkynes (15). This turned out to be advantageous when trimethylsilylheptyne 25 was cocylized with 21 to give the exo- and endo-isomers of 26 (1:1 ratio), and none of the cyclobutadiene isomer analogous to 23, but a complexed dimer of 25, namely 27. Both products are formed regioselectively, and the structure of 26 is tentatively assigned as shown

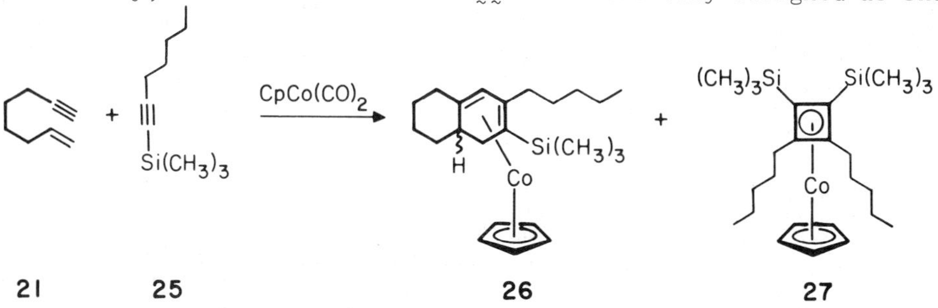

based on the expectation that intermediate metallacycles prefer the silyl group α to the metal for both steric and electronic reasons (15). The substitution pattern of 27 is distinguishable from its alternative in the mass spectrum by its fragmentation pattern (28). This type of regioselective dimerization has been observed by us previously (15). Although yields (ca. 25%) require improvement, the observed regioselectivity could become synthetically useful.

A fascinating extension of the above reaction leads stereospecifically to a strained ring system incorporating a transition metal stabilized bridgehead double bond. Thus, cyclization of 1,5-hexeneyne 28 with BTMSA furnished 29 (33%), in addition to 30 (22%). So far, 29 constitutes the only example of

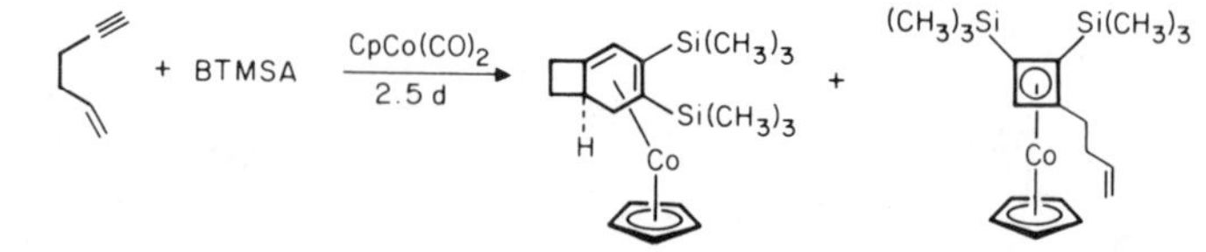

28   29   30

cyclohexadiene formation in which the tertiary hydrogen comes to lie in the syn-position with respect to the metal. Again the origin of this selectivity is not clear. Compound 29 is thermally stable although similar but noncomplexed bicyclic alkene systems containing bridgehead double bonds are unstable particularly with respect to dimerization (29). An unsuccessful attempt was made to construct an even more strained tricycle 31, but a (presumably kinetic) mixture of cyclobutadiene isomers was obtained instead, all separated by high pressure liquid chromatography and fully characterized (Scheme 6).

Scheme 6

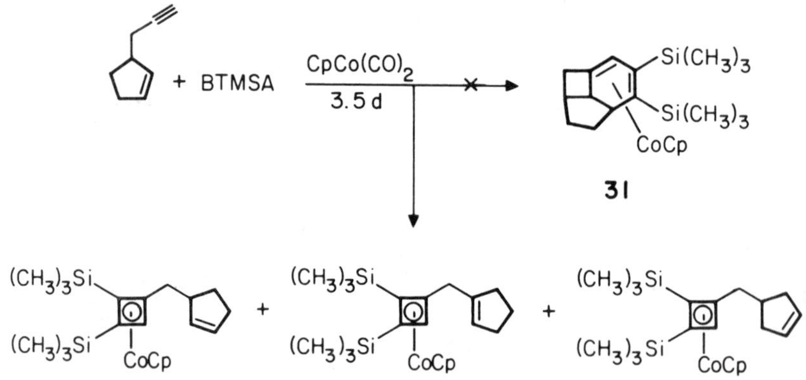

31

2 diastereomers
6%   32%   62%

DECOMPLEXATIONS AND LIGAND REACTIONS

Cobalt diene complexes of the type described above may ordinarily be readily demetallated by oxidative degradations. We have, however, found some of the (particularly silylated) polycyclic dienes to be very sensitive to acidic conditions and several of the stronger oxidizing agents commonly used (e.g. Ce^{4+} etc.). Frequent polymerization, desilylation, and aromatization was the outcome of reactions carried out under such conditions. With bromine the latter two occur simultaneously (e.g. 8→32, and 11a,b→33). The ligand may, however, be removed intact in excellent yields by using cupric chloride in acetonitrile with added triethylamine.

8   $\xrightarrow[80\%]{Br_2, CH_2Cl_2}$

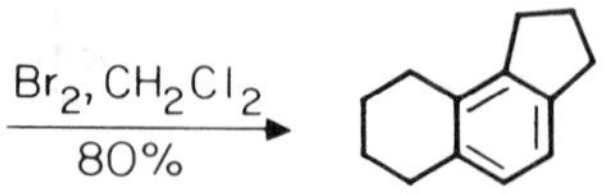

32

11a,b $\xrightarrow[80\%]{Br_2, CH_2Cl_2}$
(1:1 mixture)

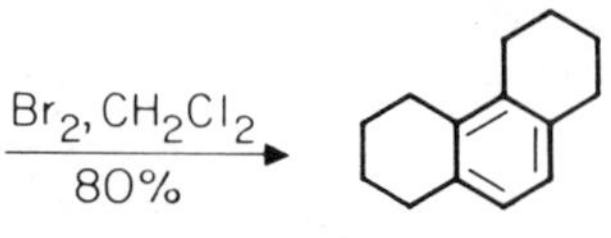

33

The free silylated dienes derived from 8, 9, and 11 appeared potentially useful intermediates in synthesis, although the chemistry of such systems (as opposed to that of vinylsilanes) is relatively unexplored (30). In this connection it was of interest to determine which one of the two double bonds was more susceptible to attack by electrophilic oxygen. Treatment of 34 with m-chloroperbenzoic acid (CH$_2$Cl$_2$, 0°C) gave none of the expected epoxides but the unusual products 35 (20%) and 36 (55%), each of which as only one respective stereoisomer, as shown by ^{13}C-NMR. The alcohol 35 was crystalline, a further reflection of its isomeric purity. Whereas 35 decomposed on treatment with acid, 36 gave the α,β-unsaturated ketone 37. A plausible (31) mechanism that rationalizes the outcome of the oxidation of 34 is shown in

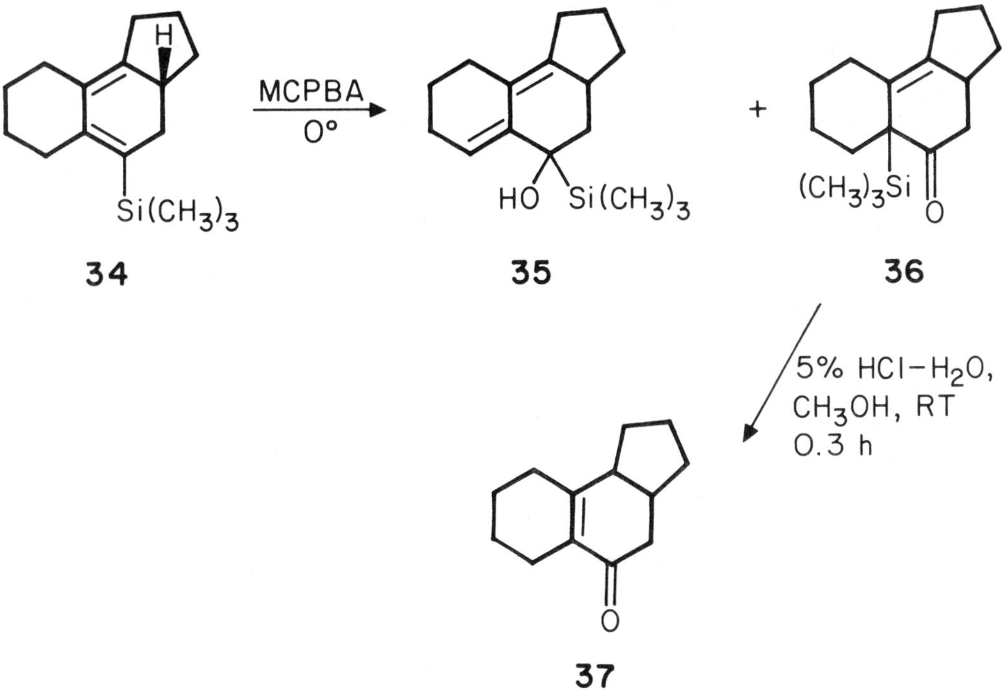

Scheme 7. Initial epoxidation appears to occur exclusively at the silylated double bond. However, this epoxide opens on protonation to furnish an inter-

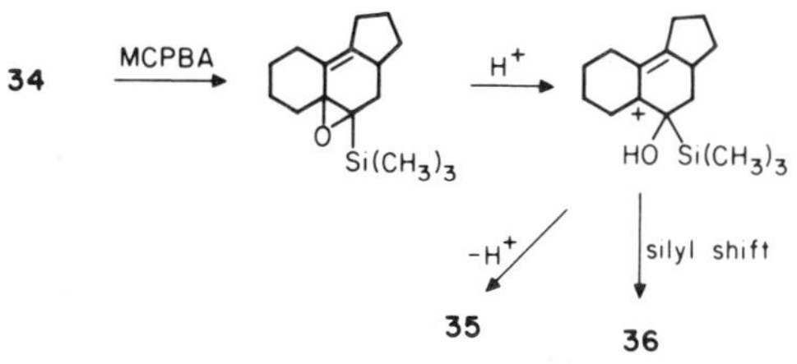

mediate highly (tertiary, allylic, β-silylic) stabilized carbonium ion from which 35 may be derived by proton loss, and 36 by a silyl shift.

Other ligand reactions may be envisioned (30) on 34 and similar molecules and
these are under investigation.

REACTIONS OF COMPLEXED DIENES

The noncatalytic formation of cyclopentadienylcyclohexadiene cobalt complexes
from acyclic diyneenes may be turned to synthetic advantage when the metal is
used as an electronically stabilizing and sterically blocking group. Drawing
on the extensively developed chemistry of pentadienylcation iron tricarbonyl
systems (32) one might envisage further synthetic elaboration of the tricyclic ligand in complexes of the type 8. Indeed, 8 may be subjected to hydride abstraction by trityl cation to furnish the air stable 38 in good
yield. When 38 was treated with NaBD₄ it regenerated 8 with a deuterium atom
at the tertiary carbon, clearly recognizable by the simplified pattern (two
doublets) for the methylene absorptions of the central ring in the ¹H-NMR
spectrum of 8-d. In addition, deprotonation of 38 appeared to occur leading
to the labile benzene complex 39. The latter was generated pure by reaction
38 with a mild base, such as K_2CO_3 (85%). This appears to be the first reported benzene complex of neutral CpCo. It seems to be bound in the tetra-
hapto-mode as shown, a suggestion which is corroborated by the 13C-NMR spectrum which resembles that of 8 closely,except for the lack of two saturated
carbon absorptions and the presence of two olefinic peaks at δ111.2 and 145.0,
clearly indicative of the presence of two uncomplexed vinylic carbons. The
complex 39 liberates the ligand on standing, particularly in the presence of
potentially coordinating species. Thus, not surprisingly, 38 gave the aromatic ligand directly when treated with KCN in a two-phase system (CH_3OH-
petroleum ether, RT, 87%). The reaction appeared to proceed through initial
carbon-carbon bond formation to the cyanide anion since the yellow color char-

acteristic of cation 38 dissolved in the methanol was quickly replaced by the
deep red color typical of CpCo-diene systems now appearing in the hydrocarbon
layer. Within seconds discoloration occurred and only the free ligand in 39
could be isolated. This finding indicated that other but irreversibly binding carbon nucleophiles might also find 38 a useful substrate. Indeed, t-
butyllithium, despite its considerable bulky character, converted 38 to 40 in
good yield. Surprisingly, however, reaction with methyllithium under the
same conditions gave mainly 41, in addition to some 39 (15%)! Butyllithium
added under similar conditions gave the same result. Although the structure

of 41 is tentative, it is heavily supported by mass spectral and NMR data.
In the former rapid aromatization, loss of a trimethylsilyl group and the
methylcyclopentadienyl ligand constitute major fragmentation pathways. In
the latter a relatively high field trimethylsilyl absorption is seen when
compared with 8 and its analogs, the central methylene hydrogens appear as
double doublets, and there is no vinyl proton nor a silicon bound terminal
vinyl carbon signal detectable. In addition, the four different Cp-hydrogens
and the attached methyl group are all clearly resolved. Compound 41 might be
thought of as being derived through methyllithium attack at the Cp ring and
a possibly cobalt mediated migration of the Cp-hydrogen positioned endo with
respect to the metal to the upper ring to regenerate what appears to be the
thermodynamically more advantageous η^5-CpCo-η^4-diene electronic arrangement.
Should the initial attack at Cp be indeed directly nucleophilic this result
would constitute a violation of the Davis-Green-Mingos rules which require
reaction at the open π-ligand to precede that at the closed one (33). This
result warrants further investigation.

CONCLUSION

The one step construction of bicyclic and tricyclic ring systems from acyclic
unsaturated precursors should provide a viable simplification of the preparation of molecules of synthetic interest containing such carbon skeletons.
This applies particularly to steroids, terpenes, and selected alkaloids. We
have shown previously (15) that CpCo(CO)$_2$ is compatible with extensive functionality, and other potential transition metal catalysts have been shown to
have similar characteristics (5), indicating areas of immediate application.

One of the interesting aspects of the reported methodology is the fact that
chiral molecules are constructed from achiral precursors in a transition
metal mediated step. Obviously, the potential for asymmetric induction by
using optically active catalysts presents itself. This will be particularly
applicable in cases where the outcome of the metal induced cyclization is
stereospecific (e.g. 3→8, 28→29). Further uses of silicon to mask functionality and the synthetic exploitation of cations of the type 38 should provide
additional flexibility. These aspects of the reported chemistry are under
active investigation.

Acknowledgements This work was supported by NIH (GM 22479) and NSF
(CHE 79-03954). C.G.F. is grateful for a postdoctoral fellowship
from the Spanish-North-American Scientific Committee. K.P.C.V. is an
Alfred P. Sloan Foundation Fellow (1976-80) and a Camille and Henry
Dreyfus Teacher-Scholar (1978-83).

REFERENCES

1. R. Huisgen, R. Grashey, and J. Sauer in S. Patai, ed., The Chemistry of
    Alkenes, p. 739, Interscience, New York (1964); A. Wasserman, The
    Diels-Alder Reaction, Elsevier, Amsterdam, (1965); A. S. Onishenko,
    Diene Synthesis, Israel Program for Scientific Translation, Jerusalem,
    (1964).
2. W. Oppolzer, Synthesis, 793 (1978); Angew. Chem., 89, 10 (1977); Angew.
    Chem., Int. Ed. Engl., 16, 23 (1977); R. G. Carlson, Ann. Rep. Med.
    Chem., 9, 270 (1974).
3. For some recent work see: D. F. Taber and B. P. Gunn, J. Am. Chem. Soc.,
    101, 3993 (1979); F. Näf, R. Decorzant, and W. Thommer, Helv. Chim.
    Acta, 62, 114 (1979); S. M. Weinreb, N. A. Khatri, and V. Shringare-
    pure, J. Am. Chem. Soc., 101, 5073 (1979); S. Weinreb, R. W. Franck,
    and B. Nader, ibid., 102, 1154 (1980); E. J. Corey, R. L. Danheiser,
    S. Chandrasekaran, G. E. Keck, B. Gopalan, S. D. Larsen, P. Siret, and
    J.-L. Gras, ibid., 100, 8034 (1978); G. Stork and D. J. Morgans, ibid.,
    101, 7110 (1979); L.-F. Tietze, G. V. Kiedrowski, K. Harms, W. Clegg,
    and G. Sheldrick, Angew. Chem., 92, 130 (1980); Angew. Chem., Int. Ed.
    Engl., 19, 134 (1980); T. Kametani, Pure Appl. Chem., 51, 747 (1979);
    R. L. Funk and K.P.C. Vollhardt, Chem. Soc. Rev., in press; K.P.C.
    Vollhardt, Ann. N.Y. Acad. Sci., 333, 241 (1980).
4. R. B. Woodward and R. Hoffmann, Angew. Chem., 81, 797 (1969); Angew.
    Chem., Int. Ed. Engl., 8, 781 (1969).
5. C. W. Bird, Transition Metal Intermediates in Organic Synthesis, Chapter
    1, Academic Press, New York, (1967); F. L. Bowden and A.B.P. Lever,

Organomet. Chem. Rev., 3, 227 (1968); W. Hübel in I. Wender and P.
Pino, eds., Organic Synthesis via Metal Carbonyls, Vol. I, Chapter 2,
Wiley, New York, (1968); P. M. Maitlis, Pure Appl. Chem., 30, 427
(1972); L. P. Yur'eva, Russ. Chem. Rev., 43, 48 (1974); S. Otsuka and
A. Nakamura, Adv. Organometal. Chem., 14, 245 (1976); R. S. Dickson
and P. J. Fraser, ibid., 12, 323 (1974); E. Müller, Synthesis, 761
(1974); K.P.C. Vollhardt, Acc. Chem. Res., 10, 1 (1977); H. Bönnemann,
Angew. Chem., 90, 517 (1978); Angew. Chem., Int. Ed. Engl., 17, 505 (1978).

6. See, for example, P. Binger, A. Brinkman, and J. McMeeking, Liebigs Ann.
    Chem., 1065 (1977); F. Tureček, V. Hanuš, P. Sedmera, H. Antropiusová,
    and K. Mach, Tetrahedron, 35, 1463 (1979); J.T.M. Evers and A. Mackor,
    Tetrahedron Lett., 415 (1980) and the references therein. For tran-
    sition metal catalyzed homo-Diels-Alder reactions, see, for example,
    J. E. Lyons, H. K. Myers, and A. Schneider, Ann. N.Y. Acad. Sci., 333,
    273 (1980).

7. Y. Wakatsuki, T. Kuramitsu, and H. Yamazaki, Tetrahedron Lett., 4549
    (1974); Y. Wakatsuki and H. Yamazaki, J. Organometal. Chem., 139, 169
    (1977).

8. η^5-Cyclopentadienyltrimethylphosphine cobaltatetramethylcyclopentadiene
    reacts with dimethylacetylene dicarboxylic ester directly without pri-
    or phosphine dissociation: D. R. McAlister, J. E. Bercaw, and R. G.
    Bergman, J. Am. Chem. Soc., 99, 1666 (1977).

9. A. J. Chalk, J. Am. Chem. Soc., 94, 5928 (1979).

10. P. W. Jolly and G. Wilke, The Organic Chemistry of Nickel, Vol. II, Aca-
    demic Press, New York, (1975).

11. L. D. Brown, K. Itoh, H. Suzuki, K. Hirai, and J. A. Ibers, J. Am. Chem.
    Soc., 100, 8232 (1978); H. Suzuki, K. Itoh, Y. Ishii, K. Simon, and
    J. A. Ibers, ibid., 98, 8494 (1976).

12. Y. Wakatsuki, K. Aoki, and Y. Yamazaki, J. Am. Chem. Soc., 101, 1123
    (1979).

13. J. H. Barlow, G. R. Clark, M. G. Curl, M. E. Howden, R.D.W. Kemmitt, and
    D. R. Russell, J. Organometal. Chem., 144, C47 (1978); see also: D. M.
    Barlex, A. C. Jarvis, R.D.W. Kemmitt, and B. Y. Kimura, J. Chem. Soc.,
    Dalton Trans., 2549 (1972); A. C. Jarvis and R.D.W. Kemmitt, J. Orga-
    nometal. Chem., 136, 121 (1977); P. Caddy, M. Green, E. O'Brien, L. E.
    Smart, and P. Woodward, Angew. Chem., 89, 671 (1977); Angew. Chem.,
    Int. Ed. Engl., 16, 648 (1977); C. E. Dean, R.D.W. Kemmitt, D. R. Rus-
    sell, and M. D. Schilling, J. Organometal. Chem., 187, C1 (1980).

14. C. J. Saward and K.P.C. Vollhardt, Tetrahedron Lett., 4539 (1975); R. L.
    Funk and K.P.C. Vollhardt, Angew. Chem., 88, 63 (1976); Angew. Chem.,
    Int. Ed. Engl., 15, 53 (1976); R. L. Hillard III and K.P.C. Vollhardt,
    J. Am. Chem. Soc., 98, 3579 (1976); R. L. Hillard III and K.P.C. Voll-
    hardt, Angew. Chem., 89, 413 (1979); Angew. Chem., Int. Ed. Engl., 16,
    399 (1977); J. R. Fritch and K.P.C. Vollhardt, J. Am. Chem. Soc., 100,
    3643 (1978); P. Perkins and K.P.C. Vollhardt, Angew. Chem., 90, 648
    (1978); Angew. Chem., Int. Ed. Engl., 17, 615 (1978); W.G.L. Aalbers-
    berg and K.P.C. Vollhardt, Tetrahedron Lett., 1939 (1979); R. I. Du-
    clos, K.P.C. Vollhardt, and L. S. Yee, J. Organometal. Chem., 174,
    109 (1979); K.P.C. Vollhardt, Pure Appl. Chem., 52, 1645 (1980).

15. K.P.C. Vollhardt, Chem. Ztg., 103, 309 (1979); E.R.F. Gesing, J. A. Sin-
    clair, and K.P.C. Vollhardt, J. Chem. Soc., Chem. Commun., 286 (1980);
    see also ref. 3 and 5.

16. E. D. Sternberg and K.P.C. Vollhardt, J. Am. Chem. Soc., 102, 0000 (1980).

17. R. B. King, P. M. Treichel, and F.G.A. Stone, J. Am. Chem. Soc., 83,
    3593 (1961); G. E. Herbereich and J. Schwarzer, Chem. Ber., 103, 2016
    (1970); G. E. Herberich and R. Michelbrink, ibid., 103, 3615 (1970);
    see also: Gmelins Handbuch der anorganischen Chemie, Supplement to
    the 8th edition, Vol. 5, part I, p. 340, (1973).

18. S. D. Darling, private communication.

19. P. Hofmann, private communication. For the theoretical treatment of a
    related problem, see: P. Hofmann, Angew. Chem., 89, 551 (1977);
    Angew. Chem., Int. Ed. Engl., 16, 536 (1977).

20. J. C. Lindhoudt, G. L. van Mourik, and H.J.J. Pabon, Tetrahedron Lett.,
    2565 (1976); C. A. Brown and A. Yamashita, J. Chem. Soc., Chem. Commun.
    959 (1976).

21. D. Pletcher and S.J.D. Tait, J. Chem. Soc., Perkin Trans. II, 789 (1979).

22. G. H. Posner, G. M. Gurria, and K. A. Babiak, J. Org. Chem., 42, 3173
    (1977).

23. K.P.C. Vollhardt, J. E. Bercaw, and R. G. Bergman, J. Organometal. Chem.,
    97, 283 (1975); J. Am. Chem. Soc., 96, 4998 (1974); W.-S. Lee and H.
    H. Brintzinger, J. Organometal. Chem., 127, 87 (1977).

24. O. Crichton, A. J. Rest, and D. J. Taylor, J. Chem. Soc., Dalton Trans.,

167 (1980).
25. R. B. King, Inorg. Chem., 5, 82 (1966).
26. See: T. L. Brown, Ann. N.Y. Acad. Sci., 333, 80 (1980); J. Halpern, Pure Appl. Chem., 51, 2171 (1979).
27. J. R. Fritch and K.P.C. Vollhardt, Angew. Chem., 91, 439 (1979); Angew. Chem., Int. Ed. Engl., 18, 409 (1979).
28. J. R. Fritch and K.P.C. Vollhardt, J. Am. Chem. Soc., 100, 3643 (1978).
29. A. Greenberg and F. J. Liebman, Strained Organic Molecules, Academic Press, New York, (1978).
30. T. H. Chan and I. Fleming, Synthesis, 761 (1979); I. Fleming and A. Percival, J. Chem. Soc., Chem. Commun., 681 (1976); J.-P. Pillot, J. Dunogues, and R. Calas, J. Chem. Res. (S) 268 (1977); L. L. Koshutina and V. I. Koshutin, Zh. Obshch. Khim., 48, 932 (1978); V. I. Koshutin, ibid., 48, 1665 (1978); M. E. Jung and B. Gaede, Tetrahedron, 35, 621 (1979); K. Yamamoto, M. Ohta, and J. Tsuji, Chem. Lett., 713 (1979).
31. E. W. Colvin, Chem. Soc. Rev., 7, 15 (1978); C. M. Robbins and G. H. Whitham, J. Chem. Soc., 697 (1976); M. Obayashi, K. Utimoto, and H. Nozaki, Tetrahedron Lett., 1807 (1977); 1383 (1978); id., Bull. Chem. Soc. Jap., 52, 2646 (1979).
32. For recent reviews, see: A. J. Birch, Ann. N.Y. Acad. Sci., 333, 107 (1980); A. J. Birch and I. D. Jenkins, in H. Alper, ed., Transition Metal Organometallics in Organic Synthesis, Academic Press, New York, chapter 1, (1976).
33. S. G. Davies, M.L.H. Green, and D.M.P. Mingos, Tetrahedron, 34, 3047 (1978).

SOME USES OF SILICON COMPOUNDS IN ORGANIC SYNTHESIS

I. Fleming

University Chemical Laboratory, Lensfield Road, Cambridge CB2 1EW, England

Abstract—1. The amount of γ-phenylthioalkylation of silyl dienol ethers
is increased when the triphenylsilyl ether is used in place of the trimethylsilyl ether. Phenylthiomethylation is the least γ-selective carbon electrophile of several tried so far. 2. The acid-catalysed reactions of a range
of γ-silyl tertiary alcohols cleanly give rearrangement, in which the silyl
group controls the outcome. The reactions are similar in several respects
to the rearrangements of the corresponding pinacols, except that the silicon
controlled reactions are usually cleaner and give higher yields. The reaction is particularly useful for setting up quaternary carbon atoms.

1. SILYL ENOL ETHERS

Phenylthioalkylation of silyl enol ethers followed by desulphurisation (Scheme 1) is a high-
yielding method for the alkylation of aldehydes, ketones, and esters (1-3). This method

Scheme 1

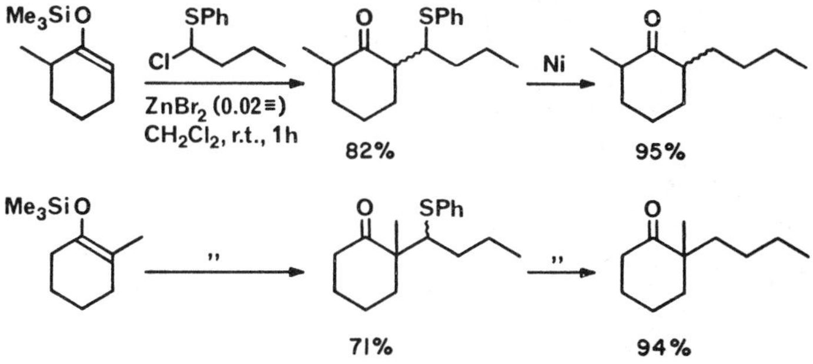

largely solves the well-known problems of polyalkylation and loss of regioselectivity inherent in the use of enolates. The phenylthioalkyl chlorides are easily made in high yield

Scheme 2

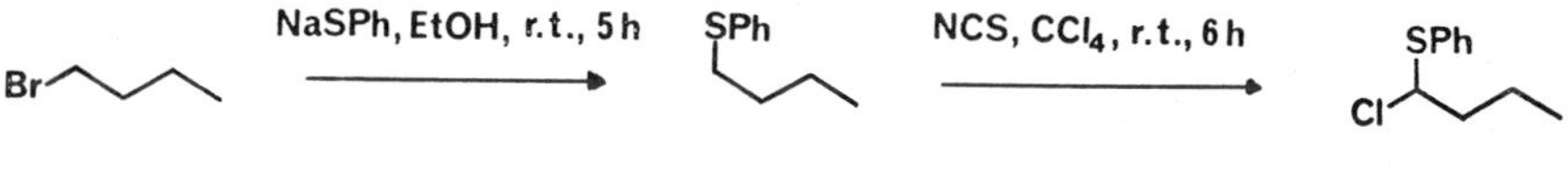

from the corresponding alkyl bromides (Scheme 2). Early experiments (4) established that
phenylthioalkylation of silyl dienol ethers gave good yields of a mixture of α- and γ-phenyl-
thioalkyl enones (Scheme 3), with the latter a substantial and often the major product.

Scheme 3

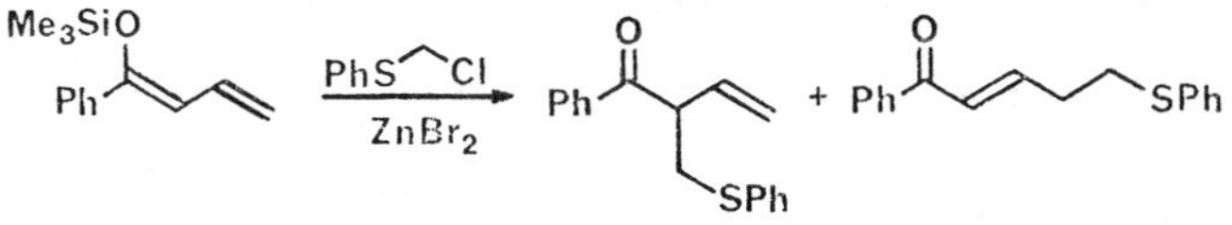

85

Quite small changes in the structure of the silyl dienol ether and the phenylthioalkyl halide gave an example (Scheme 4) in which the reaction took place completely at the γ position, and

Scheme 4

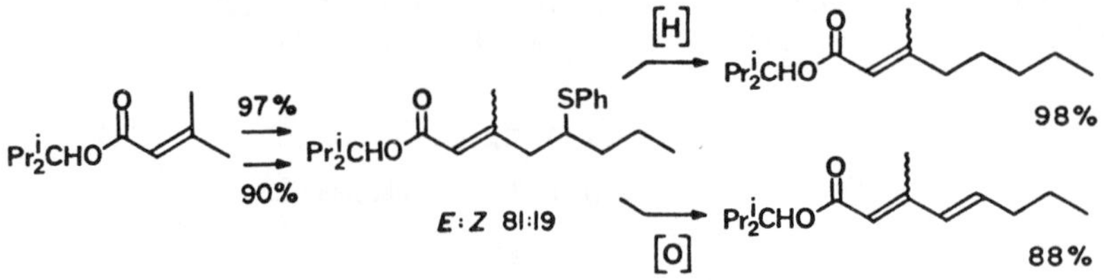

the silyl dienol ether involved in that reaction gave largely, or only, γ-alkylation or acylation with a variety of other carbon electrophiles (Scheme 5).

Scheme 5

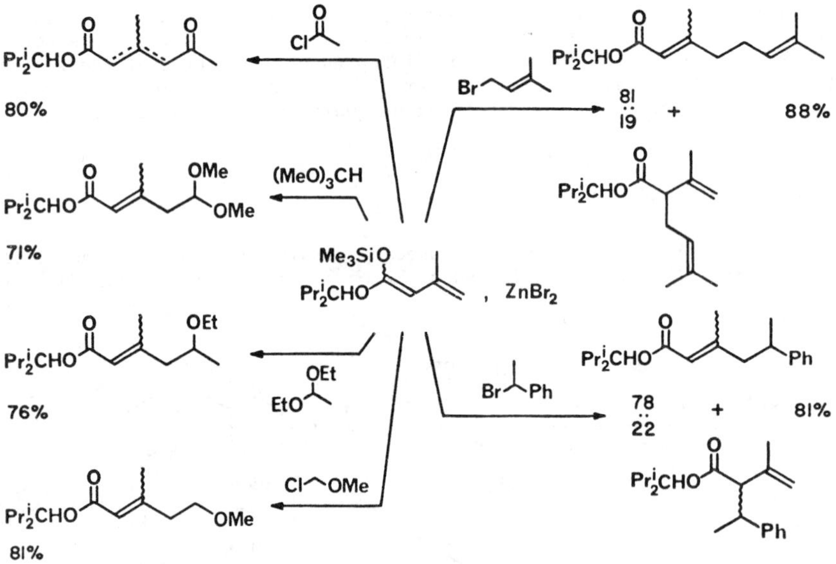

This work, and Mukaiyama's (5) on which it was based, showed that silyl dienol ethers were very promising substrates for solving the long-standing problem of getting d^4 reactivity (6) from enones, but, except in special cases, the proportion of α-attack was still enough to be troublesome. However, before we could look for a general solution to this problem, we needed to know what was the least γ-selective electrophile (E^+ in Scheme 6). If we could improve the proportion of γ-attack in that case, whatever it turned out to be, we might reasonably

Scheme 6

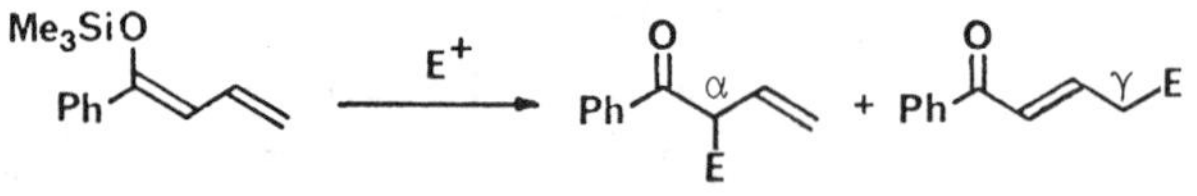

claim to have solved the problem. Before our work, nothing much was known about the α:γ selectivity of electrophiles, mainly because γ-attack at all has been so rare.

We started with phenylthiomethylation of the silyl enol ether of crotonophenone (Scheme 3). This was one of the few cases in which we had already got more α than γ-attack (α:γ 55:45). As it happens, all other electrophiles we have tried so far give, to a greater or lesser extent, a higher proportion of γ-alkylation (Table 1).

What makes some electrophiles more γ-selective than others? We have far too few results to answer a question like this with any conviction, but, if we assume, that the active electrophile in each case is the cation in column 2 of Table 1, we can see that, on the whole, the better stabilised the cation, the more γ-attack it shows. The more stabilised the cation, the more selective it will be, and also, since its LUMO-energy will be lower and its charge more dispersed, the softer it will be. This fits in well with the expectation that the γ-position is the softer site of a dienol derivative (7). Recently, Danishefsky has found that

Eschenmoser's salt (the Mannich intermediate) is also a highly γ-selective electrophile for silyl dienol ethers (8). This fits in well with our analysis, since it too is a highly stabilised cation, more so probably than most of those listed in Table 1.

TABLE 1. α:γ Ratios for the reaction of various electrophiles as in Scheme 6

Reagent	Electrophile	α:γ	Yield
PhS⌒Cl, ZnBr$_2$	PhS —⁺	55 : 45	65%
MeO⌒Cl "	MeO —⁺	47 : 53	53%
(EtO)$_2$CHMe "	EtO —⁺/	40 : 60	78%
PhS⌒$^{Pr^n}$Cl "	PhS —⁺/$^{Pr^n}$	34 : 66	85%
(EtO)$_2$CMe$_2$, TiCl$_4$	EtO —⁺⟨	(0 : 100)	78%
(MeO)$_3$CH "	MeO —⁺⟨ OMe	(0 : 100)	52%

Having established that phenylthiomethylation was as little γ-selective a reaction as we were likely to find, we turned to the problem of increasing the proportion of γ-attack. In our published work (4), we had tested the effect of changing the substituents on C-1 and C-3 of the dienol ether, and now we have looked at changes in the structure of the electrophile. There remains one highly significant place where we have been conservative: the silyl group. Since this group does not appear in the product, we can freely make changes in it, and then, if the idea works, we should have a general solution to the problem. Our first thought was that a larger silyl group might hinder attack at the α more than at the γ-position. This hope was rapidly dashed: the triethylsilyl and t-butyldimethylsilyl ethers both gave more α-attack than the trimethylsilyl ether had (Table 2). However, this result gave us two useful leads. In the first place, it established that the silyl group was still attached at the time of reaction; before this result, it had always been possible, though unlikely, that the role of the Lewis acid was to replace the silicon. Moreover, it was gratifying to see that changing the silyl group at least had some effect, even if it was the wrong one. In the second place, this result established that electronic effects as far away as the substituent on the silicon atom could influence the reaction. We turned therefore to substituents which

TABLE 2. α:γ Ratios in the phenylthiomethylation:

as a function of the structure of the silyl group

R$_3$	α:γ	Yield
Me, Me, Me	55 : 45	65%
Et, Et, Et	62 : 38	75%
Me, Me, Bu^t	77 : 23	62%
Me, Me, Ph	39 : 61	91%
Me Ph, Ph	32 : 68	74%
Ph, Ph, Ph	14 : 86	93%

would be effectively electron-withdrawing. We find that replacement of the methyl groups by phenyl does almost all that we could have asked for (Table 2, bottom line). With three phenyl groups, and our least γ-selective electrophile, we get an α:γ ratio of 14:86 and a 93% yield.

2. γ-SILYL ALCOHOLS

The greater part of our interest in the uses of silicon compounds in organic synthesis might be defined loosely as the control of carbonium ion chemistry. The presence of a silyl group controls the reactions of vinyl- and allylsilanes with electrophiles, because the cationic intermediates lose the silyl group to give alkenes in which the position of the double bond has been determined by where the silyl group was placed. Concurrent with our work in this area (9-13), we were also investigating the synthesis of the oxindole alkaloid gelsemine (1) (14, 15). We planned to synthesise this molecule by way of an intermediate, which we can

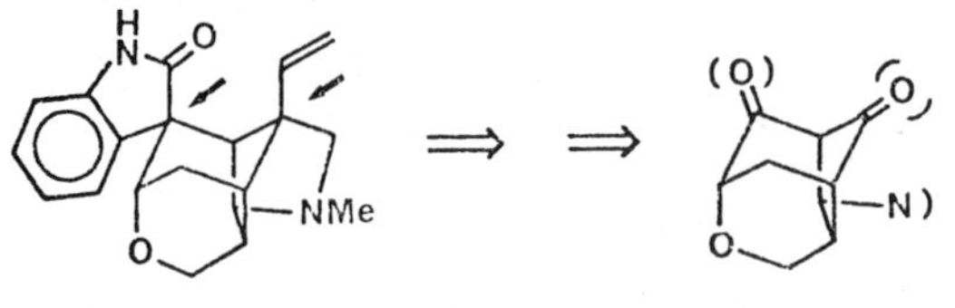

(1)   (2)

represent by the formula (2). Assuming that we can make this intermediate, and we have, so far (15), got as far as the keto diester (3), we shall still need to convert each of the bracketed ketone groups into fully substituted quaternary carbons (16). This need has, in

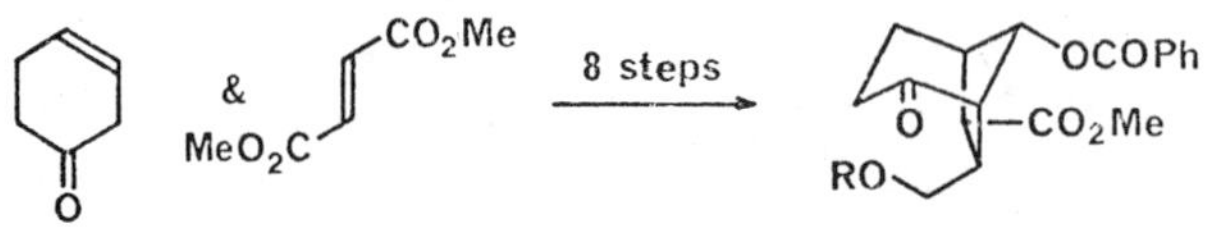

(3)

the meantime, influenced the direction in which our organosilicon work has developed. Thus, we developed (17) a two-step solution to this problem (Scheme 7), which was designed to deal

Scheme 7

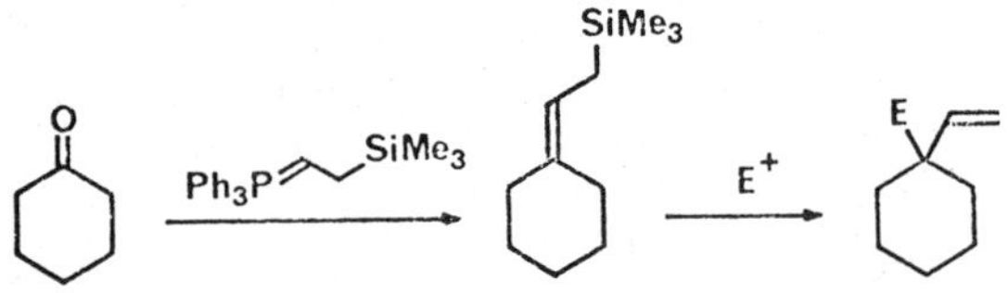

with the elaboration of the ketone group on the right-hand side of (2). However, we were aware that the Wittig reaction, the first step of the sequence (18), might not work too well, as indeed it does not (19) with cyclopentanone in place of cyclohexanone (Scheme 8), when we

Scheme 8

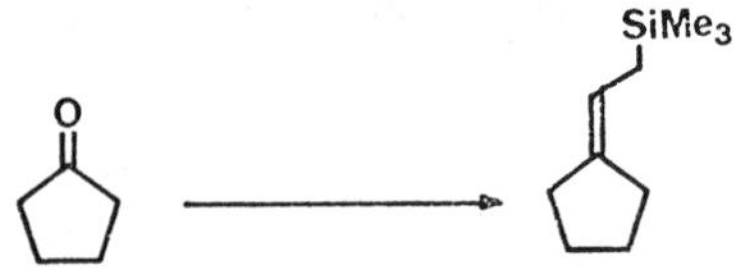

get a yield of only 15%. In anticipation of this problem, we have now found that the allylsilane (5) can be prepared by the reaction of our silyl-copper reagent (20,21) on the allyl acetate (4), and that, at least with a proton, the electrophilic substitution of this allylsilane is uneventful (5 → 6) (22).

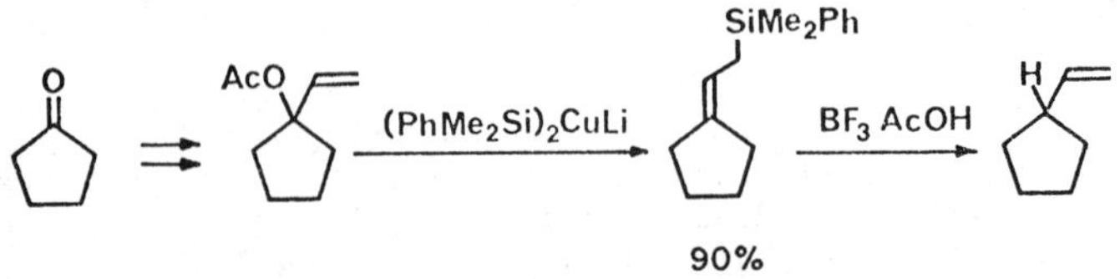

(4)   (5)   (6)

Before we can see how we have approached the problem associated with the development of the other (left-hand) carbonyl group of 2, we must return to another theme of our silicon work, which is the control of carbonium ion rearrangements. Knowing that a silyl group is usually lost from a β-silyl cation, we became interested in what would happen to the homologous cations of general structure (7) (Scheme 9). Two plausible pathways exist, both of which have

Scheme 9

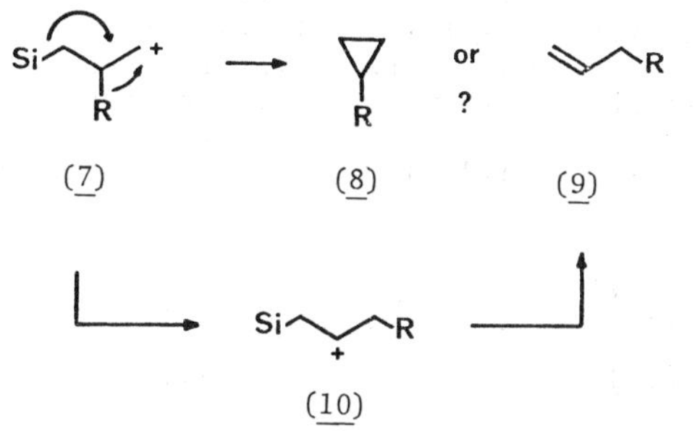

precedent. The silyl group could be lost directly (7 → 8), with formation of a cyclopropane. This has precedent in the reaction of the bromide (11) with aluminium chloride (23), and in

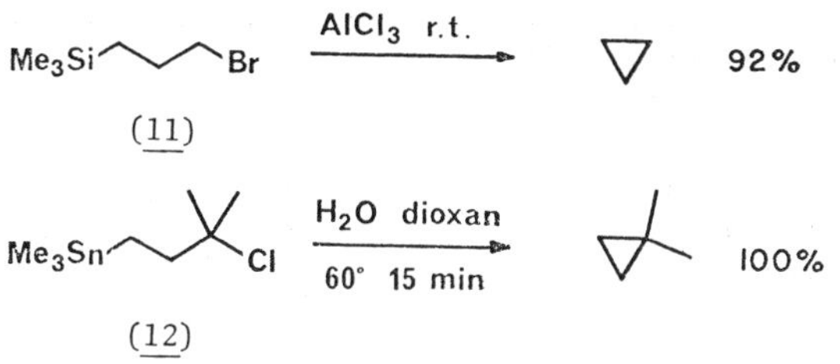

the solvolysis of the tin analogue (12) (24). Alternatively, the silyl group might encourage rearrangement to produce a cation (10), in which the silyl group is a stabilising influence, and from which the silyl group will be displaced with ease (10 → 9). We observed this type of reaction in our work (25) on diphenylphosphinoyl migration (Scheme 10) and phenylthio migration (26) (Scheme 11). Rearrangement of a carbon migrating group was part of our work (27)

Scheme 10

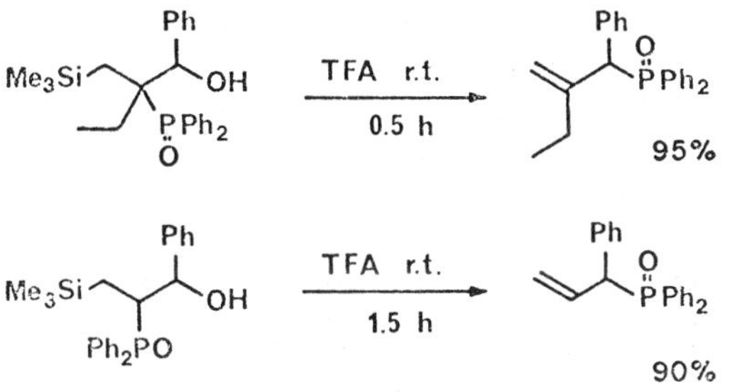

Scheme 11

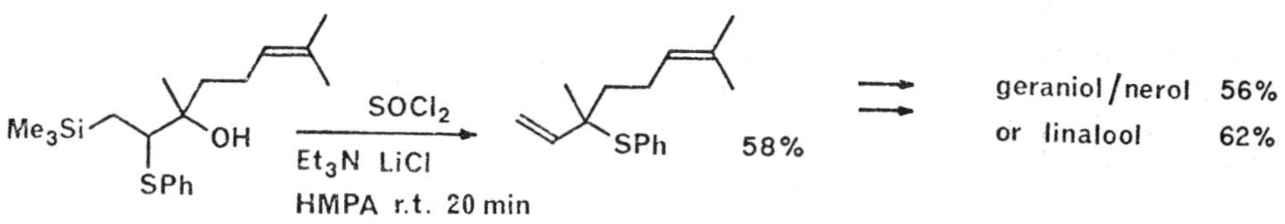

on a synthesis of 7-functionalised norbornenes (Scheme 12); this work is also one of our approaches to the synthesis of gelsemine. A related rearrangement in the norbornyl series has also been observed in the tin series by Traylor (28). Finally, both pathways have been observed by Sakurai (29) in the acylations of homoallyl silanes shown in Scheme 13. When the intermediate cation is secondary (the top and bottom lines) cyclopropane formation is the major reaction, but when the intermediate cation is tertiary (the middle line) rearrangement

of hydride took place. We decided to look at this problem in a general way, and have confin-
ed ourselves for the time being to the reactions of tertiary alcohols of general structure
(13) (Scheme 14) (30). These react under very mild conditions (one equivalent of boron tri-

Scheme 12

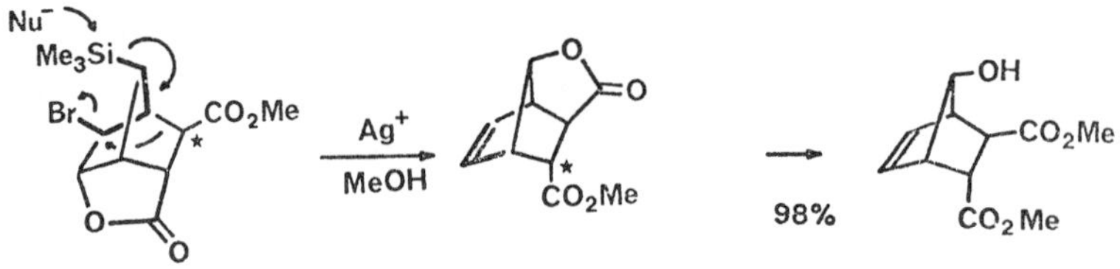

Scheme 13

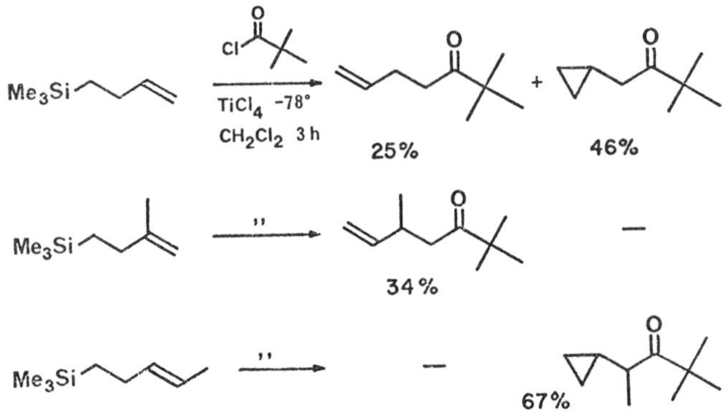

Scheme 14

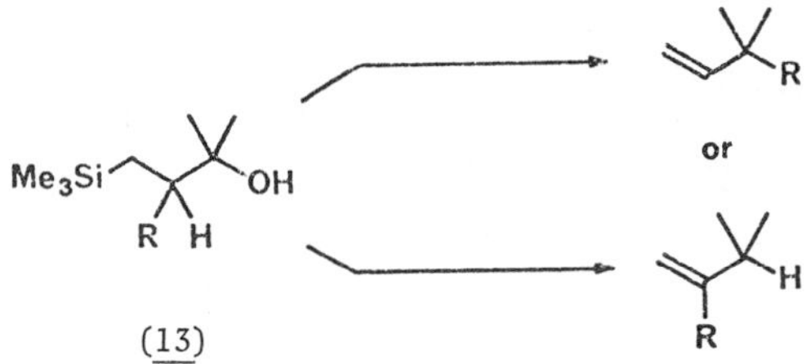

(13)

fluoride-acetic acid in dichloromethane at 0°C for 5 minutes is fairly typical), and, so far,
in every case we observe only the rearrangement pathways, and no cyclopropane formation. Thus
the alcohol (14) cleanly gave 3-methylbutene (15) by hydride shift. Although the cations
(16 and 17) may not be discrete intermediates, the rearrangement is effectively that of a

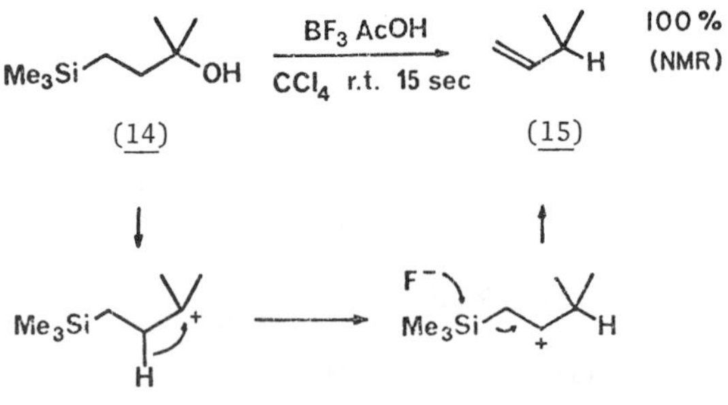

tertiary cation to a secondary cation, driven by the stabilisation provided by the silyl
group and/or by the loss of the silyl group (17 → 15). A useful analogy to this superficial-
ly 'uphill' rearrangement is the pinacol rearrangement of the diol (18), which gives isobut-

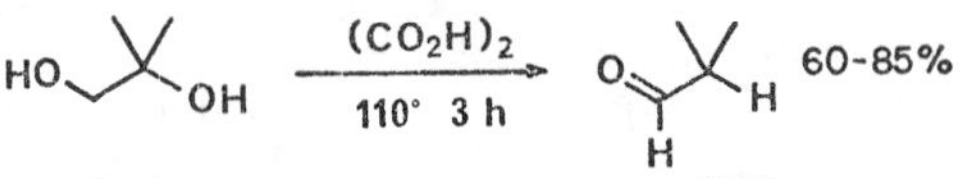

yraldehyde (19) under conditions of kinetic control (31). Two other examples of silicon controlled hydride shift are shown in Scheme 15, where the yields are now those of isolated and
distilled products. Phenyl groups can also be induced to rearrange (Scheme 16), and again

Scheme 15

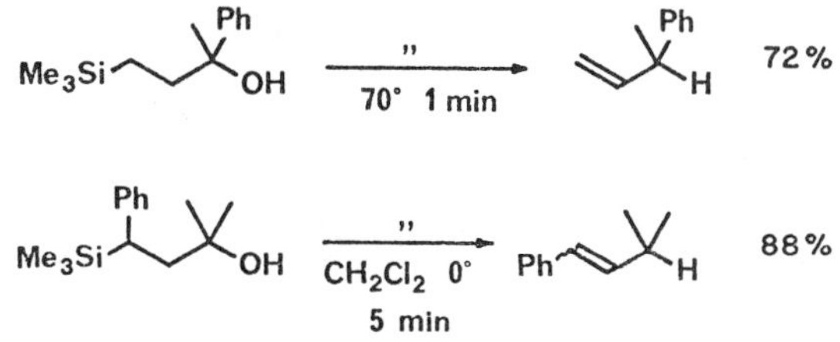

Scheme 16

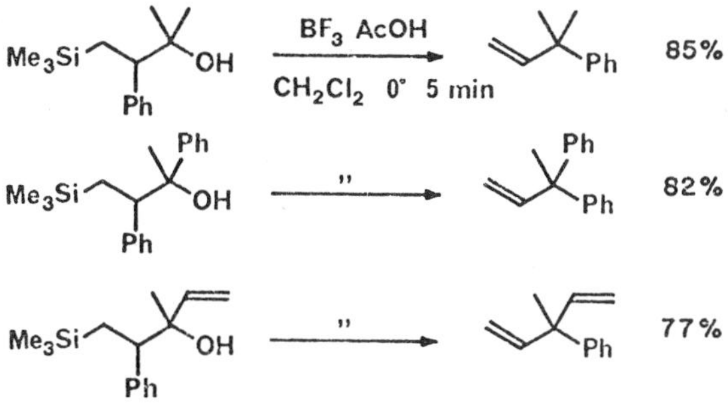

there is analogy in the corresponding pinacol rearrangement (20 → 21)(32); however, the latter reaction suffers from the ease with which the major kinetic product (21) itself rearranges to the isomeric ketones (22 and 23). This did not seem to be a problem in the silicon

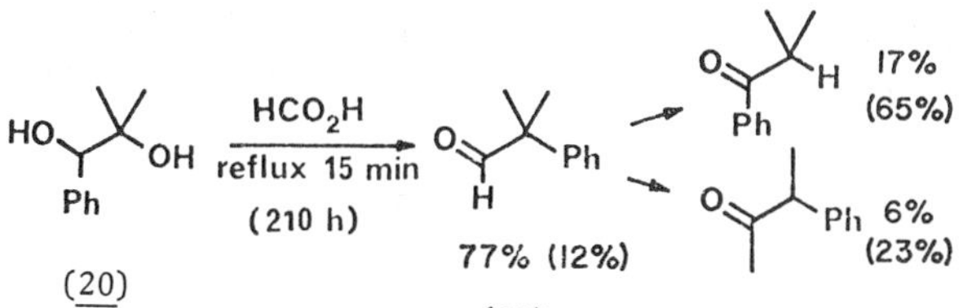

series. Silicon also induced alkyl migration in the ring-expansion (24 → 25), which has its
counterpart in the pinacol rearrangement itself (27 → 28)(33). This time, however, it was in
the silicon-controlled rearrangement that the kinetic product (presumably 25) was unstable to

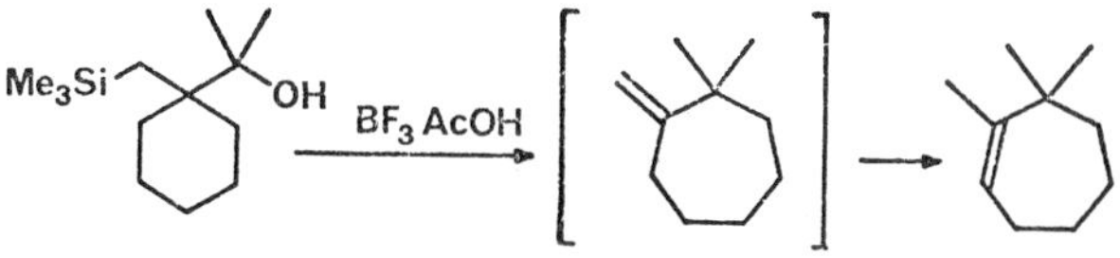

the reaction conditions, and, not surprisingly, gave the cycloheptene (26).

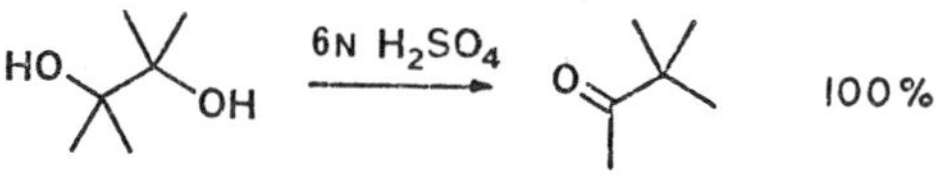

The most interesting cases are those in which there is a choice of alkyl or hydrogen migration (Scheme 14, R = alkyl). We have seen both, and in each case our observations have paral-
lels in the corresponding pinacol rearrangements. Thus hydrogen migrates when there are two
phenyl groups at the migration terminus (Scheme 17), and this is the major type of reaction
in the corresponding epoxide and pinacol rearrangements (34,35). Hydrogen also migrates in
preference to methyl (Scheme 18, top line) but ethyl and isopropyl migrate in preference to
hydrogen (Scheme 18, second and third lines). This pattern too is similar to pinacol and related chemistry: hydrogen migrates in preference to methyl (Scheme 19)(36,37), but ethyl mig-

Scheme 17

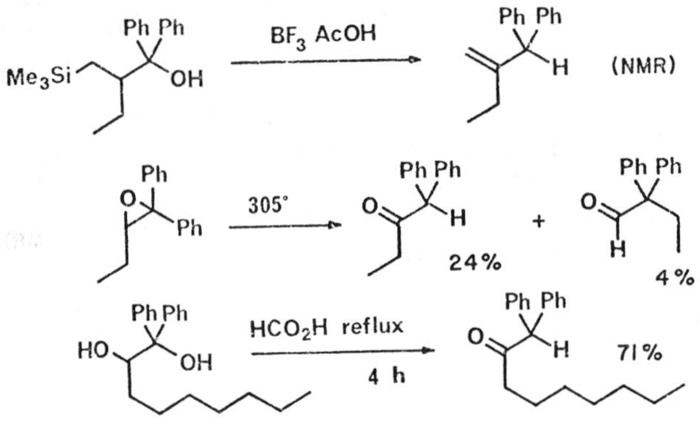

Scheme 18

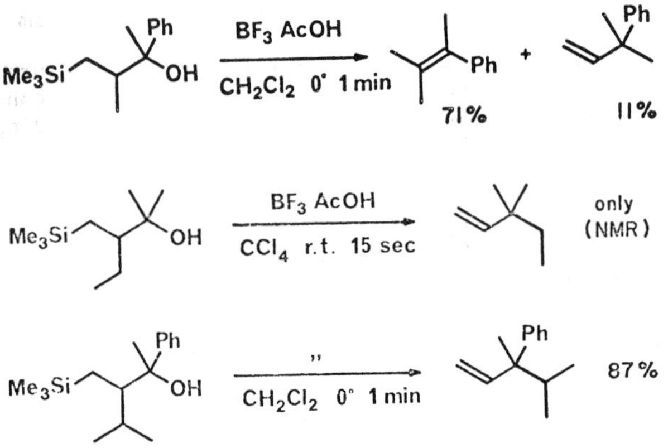

Scheme 19

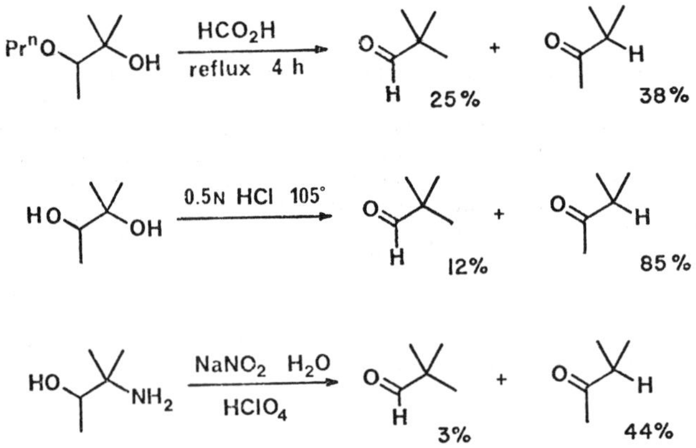

Scheme 20

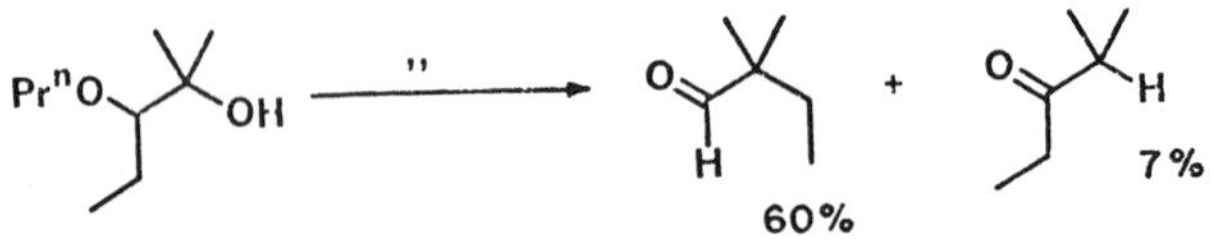

rates in preference to hydrogen (Scheme 20)(36). In a cyclic case, (29), both stereoisomers gave hydrogen migration followed by a prototropic shift of the double bond (Scheme 21), and this is the major pathway with the corresponding pinacols (30)(38). The rearrangement of the

Scheme 21

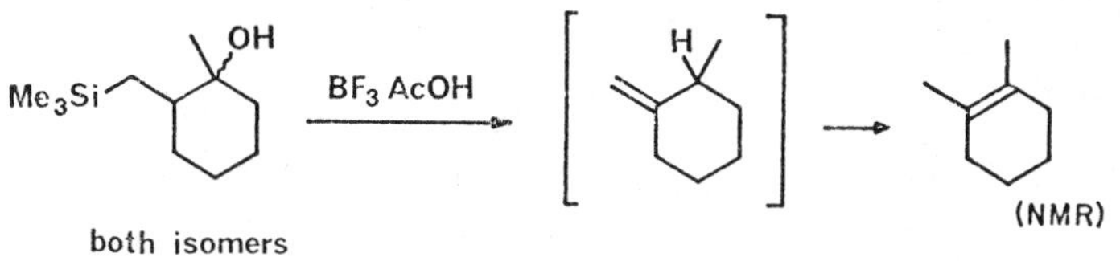

both isomers
(29)

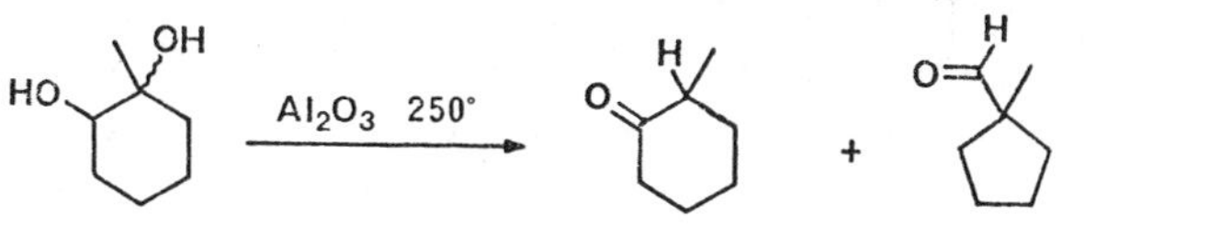

(30)   90%   'a little' from *cis*
                   70%   30% from *trans*

related epoxide (31) catalysed by lithium bromide takes a different course, giving largely alkyl migration (31 → 32 → 33)(39). This too has its parallel in silicon chemistry, where Magnus (40) has found that the paranitrobenzoate (34) of the corresponding secondary alcohol

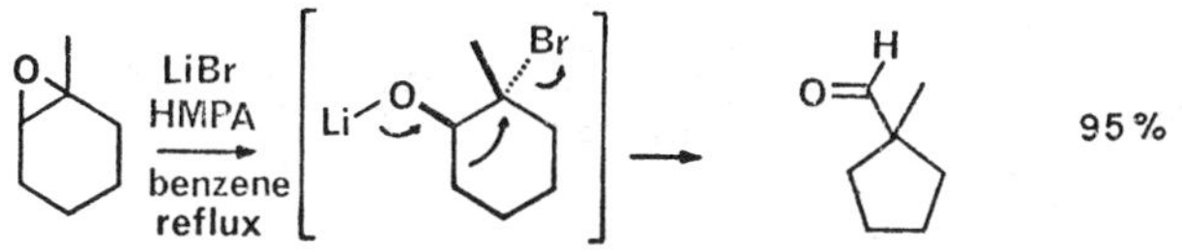

(31)   (32)   (33)

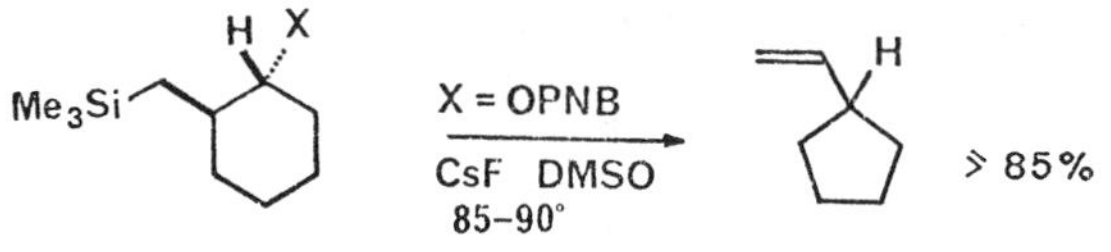

(34)

rearranges by ring contraction on treatment with fluoride ion. It seems likely that a different course is taken in this case, and in the lithium bromide promoted ring contraction (32, arrows) because the substrate is effectively 'pushed' into rearrangement, whereas the earlier acid-catalysed reactions were effectively 'pulled' into rearrangement. This must lead to a different distribution of electrons in the transition state: the more 'push' there is, the less important is it to have alkyl groups remaining behind to stabilise any electron deficiency developing at the migration origin.

The silicon controlled rearrangements described above are remarkably clean and high yielding, more so, in general, than the pinacol reactions which they otherwise so closely resemble. They should therefore recommend themselves as synthetically useful, the more so in that the starting materials are easily made by a variety of routes. The route to the alcohol (29a)

Scheme 22

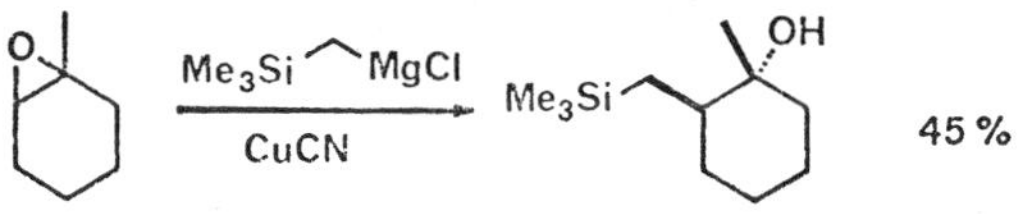

(29a)

(Scheme 22) and the routes to the alcohols (29b) and (37)(Scheme 23) correspond to the two possible senses of the disconnection (35). On the other hand, the alcohol (38) was prepared

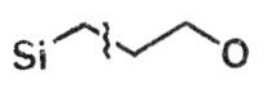

(35)    (36)

by a route (Scheme 24) which corresponds to the disconnection (36). The other γ-silyl alcohols described in this work were prepared by similar routes.

Scheme 23

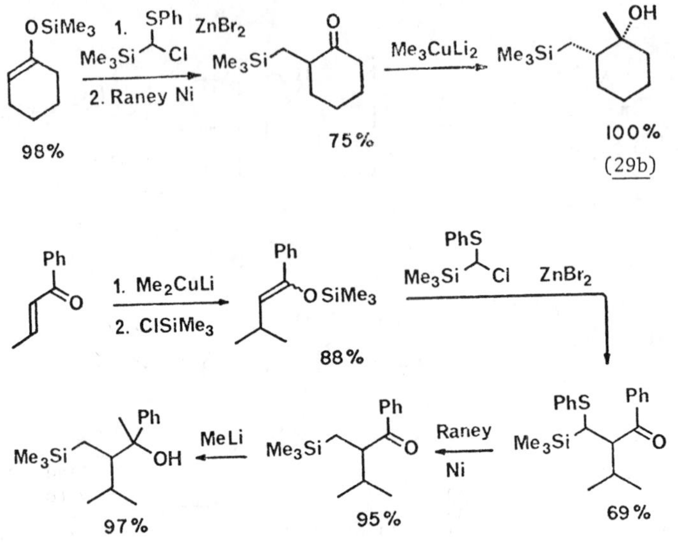

98%   75%   100%
                                                                                  (29b)

97%   95%   69%

(37)

Scheme 24

                                            92%

89%   78%   70%

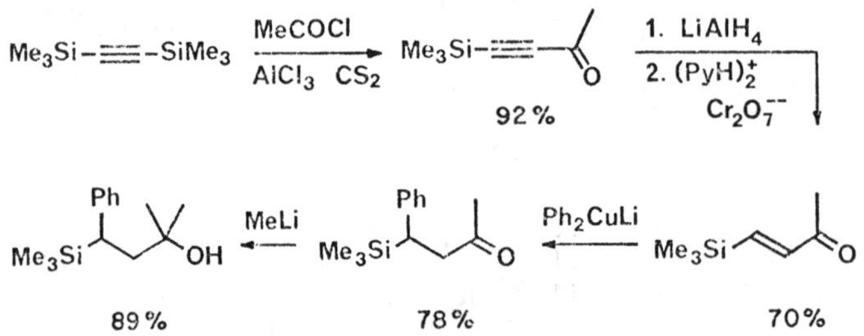

(38)

We are now in a position to return to the problem posed by our wish to develop an oxindole
group from the left-hand ketone group of the intermediate (2). As a model reaction, we wan-
ted to carry out the conversion of cyclohexanone to the corresponding oxindole (39). One way

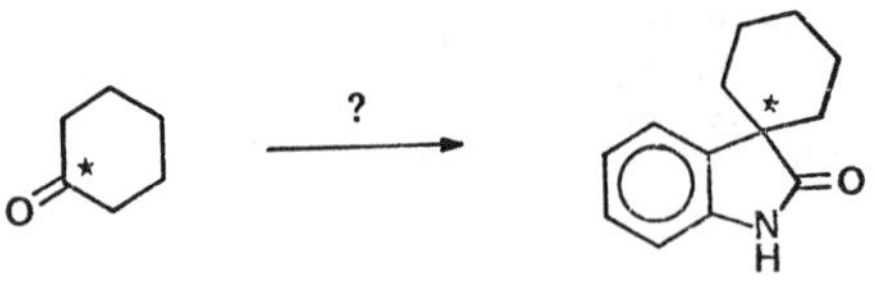

(39)

to do this was to use a pinacol type of rearrangement. The problem was that the known reaction (Scheme 25)(41) gave a poor yield of the aldehyde (40), because, as is so often the case
in the examples in the literature, it had not been done under the best conditions for getting
what ought to be the kinetically favoured product. We chose to look at this problem using an

Scheme 25

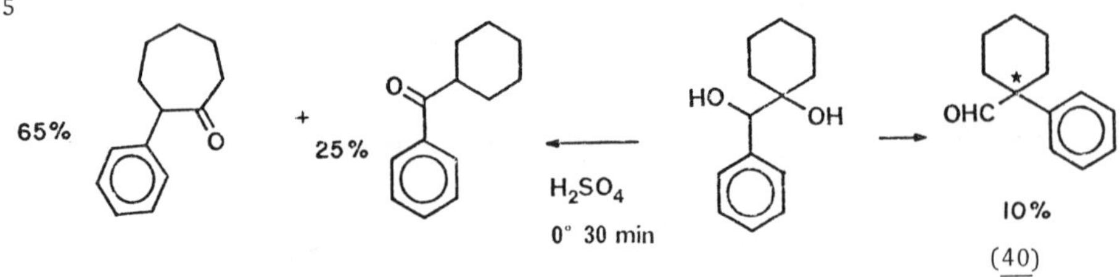

65%   +   10%

              25%   H₂SO₄   (40)

                                0° 30 min

ortho-functionalised benzene ring, and specifically the epoxide (41), which we were easily able to make (42) from cyclohexanone. Our early results (Scheme 26) were understandable but unproductive, and it took quite some time before we were able to find conditions which gave a

Scheme 26

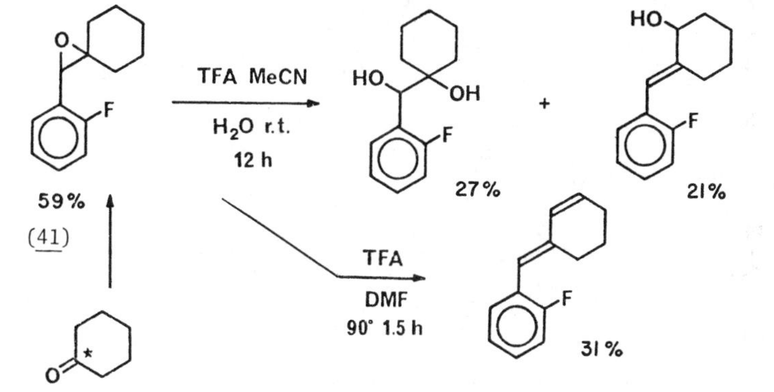

reasonable yield of the aldehyde (42)(Scheme 27). These difficulties, and the success we had enjoyed in our work on the silicon controlled rearrangements led us now to try to improve this route. In contrast to the long time we spent finding out how to do the pinacol rearran-

Scheme 27

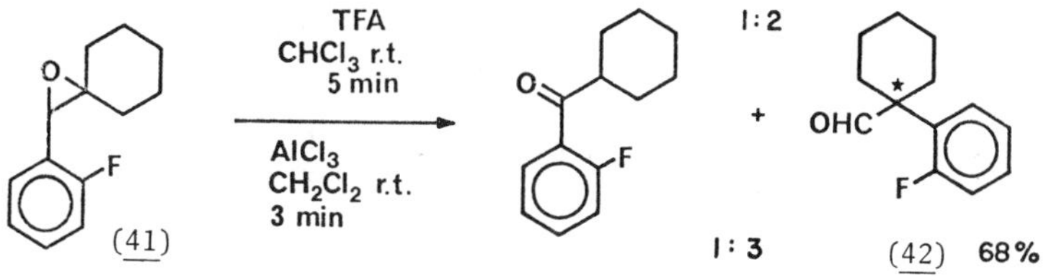

gements, we were almost immediately successful in preparing the alkene (43)(Scheme 28). This alkene can easily be converted to the aldehyde (42), and we have converted the aldehyde to the oxindole (39) by way of the amide (44), although not yet in an entirely satisfactory way.

Scheme 28

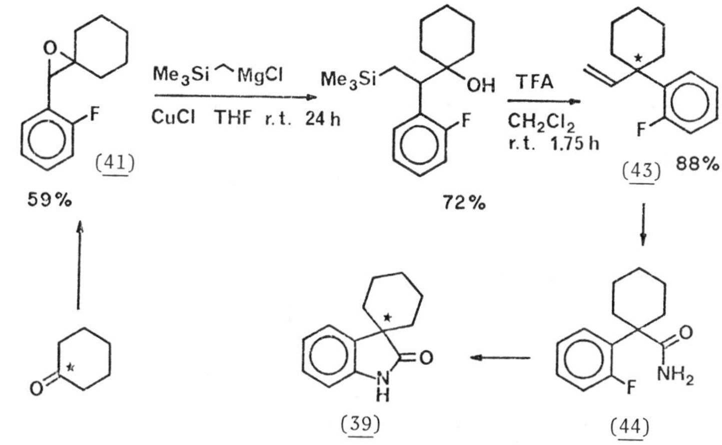

In conclusion, we have found again that a silicon-containing system brings considerable order into an area which has been in the past, because of lack of control, of limited usefulness to synthetic chemists. The unpublished work described in this lecture was carried out by my co-workers Dr. Tom Lee, Dr. Peter Gallagher, Dr. Decio Marchi, Mr. Shailesh Patel, and Mr. Ian Wallace, to each of whom I am most grateful.

REFERENCES

1. I. Paterson and I. Fleming, *Tetrahedron Letters*, 993 (1979).
2. I. Paterson and I. Fleming, *Tetrahedron Letters*, 995 (1979).
3. I. Paterson and I. Fleming, *Tetrahedron Letters*, 2179 (1979).
4. I. Fleming, J. Goldhill, and I. Paterson, *Tetrahedron Letters*, 3205 and 3209 (1979).
5. T. Mukaiyama and A. Ishida, *Chemistry Letters*, 319, 1167, and 1201 (1975), and *Bull. Chem. Soc. Japan*, **51**, 2077 (1978).
6. D. Seebach, *Angew. Chem. Internat. Edn.*, **18**, 239 (1979).
7. I. Fleming, *Frontier Orbitals and Organic Chemical Reactions*, Wiley, New York, 1976.
8. S. Danishefsky, M. Prisbylla, and B. Lipisko, *Tetrahedron Letters*, 805 (1980).
9. I. Fleming and A. Pearce, *J. C. S. Chem. Comm.*, 633 (1975).
10. I. Fleming and A. Pearce, *J. C. S. Perkin I*, in press, and I. Fleming, A. Pearce, and R. L. Snowden, *J. C. S. Chem. Comm.*, 182 (1976).
11. M. J. Carter and I. Fleming, *J. C. S. Chem. Comm.*, 679 (1976).
12. I. Fleming and A. Percival, *J. C. S. Chem. Comm.*, 681 (1976) and 178 (1978).
13. B.-W. Au-Yeung and I. Fleming, *Tetrahedron*, in press, and *J. C. S. Chem. Comm.*, 79 and 81 (1977).
14. R. L. Snowden and J. P. Michael, unpublished results.
15. Peter Gallagher, unpublished results.
16. S. F. Martin, *Tetrahedron*, **36**, 419 (1980).
17. I. Fleming and I. Paterson, *Synthesis*, 445 (1979).
18. D. Seyferth, K. R. Wursthorn, T. S. O. Lim, and D. J. Sepelack, *J. Organometallic Chem.*, **181**, 293 (1979).
19. I. Paterson, unpublished result.
20. D. J. Ager and I. Fleming, *J. C. S. Chem. Comm.*, 177 (1978).
21. I. Fleming and F. Roessler, *J. C. S. Chem. Comm.*, 276 (1980).
22. Decio Marchi Jr., unpublished result.
23. L. H. Sommer, R. E. Van Strien, and F. C. Whitmore, *J. Amer. Chem. Soc.*, **71**, 3056 (1949).
24. C. A. Grob and A. Waldner, *Helv. Chim. Acta*, **62**, 1736 and 1854 (1979).
25. A. H. Davidson, I. Fleming, J. I. Grayson, A. Pearce, R. L. Snowden, and S. Warren, *J. C. S. Perkin I*, 550 (1977).
26. I. Fleming, I. Paterson, and A. Pearce, *J. C. S. Perkin I*, in press, and P. Brownbridge, I. Fleming, A. Pearce, and S. Warren, *J. C. S. Chem. Comm.*, 751 (1976).
27. I. Fleming and J. P. Michael, *J. C. S. Chem. Comm.*, 245 (1978).
28. G. D. Hartmann and T. G. Traylor, *J. Amer. Chem. Soc.*, **97**, 6147 (1975).
29. H. Sakurai, T. Imai, and A. Hosomi, *Tetrahedron Letters*, 4045 (1977).
30. Shailesh Patel, unpublished results.
31. A. Behal and M. Sommelet, *Bull. Soc. chim. France*, 300 (1904).
32. J. W. Huffman and L. E. Browder, *J. Org. Chem.*, **27**, 3208 (1962).
33. H. Meerwein, *Annalen*, **419**, 121 (1919).
34. R. Lagrave, *Ann. Chim.*, [10]8, 363 (1927).
35. I. Elphimoff-Felkin, *Bull. Soc. chim. France*, 494 (1950).
36. I. Elphimoff-Felkin, *Bull. Soc. chim. France*, 497 (1950).
37. H. P. A. Groll and G. Hearne, U.S. Patent 2042224 (1936). W. Kirmse, J. Knist, and H.-J. Ratajczak, *Chem. Ber.*, **109**, 2296 (1976).
38. M. Tiffeneau and B. Tchoubar, *Compt. Rend.*, **199**, 1624 (1934).
39. B. Rickborn and R. M. Gerkin, *J. Amer. Chem. Soc.*, **93**, 1693 (1971).
40. P. Magnus, personal communication.
41. D. G. Botteron and G. Wood, *J. Org. Chem.*, **30**, 3871 (1965); J. Rouzaud, G. Cauquil, L. Giral, and J. Crouzet, *Bull. Soc. chim. France*, 4837 (1968).
42. Ian Wallace, unpublished results.

NEW STRATEGIC METHODS FOR SYNTHESIS

P. Magnus

Ohio State University, 140 West 18th Avenue, Columbus, Ohio 43210, USA

Most new methods of organic synthesis that have been developed during the past several years are usually tactical devices that are created to implement a particular strategy. Seldom are methods devised that lead to new general strategic solutions for the synthesis of complicated molecules. Even more rarely do new methods provide entirely new types of molecules.

It is the intention of this lecture to describe part of a research program that makes use of organosilicon chemistry for the development of new strategies and tactics in organic synthesis. An outcome of this program has been to realize the first synthesis of a primary helical molecule. This talk is divided into three sections as follows -
a) New tactical methods for the synthesis of ring A aromatic steroids and sesquiterpenes.
b) A general strategic solution for a sequence involving a ring expansion and multiple functionalization.
c) A reiterative reaction sequence for the first synthesis of a primary helical molecule.

The isolation of estrone in 1929 paved the way for many ingenious total syntheses, each in its own way reflecting, to some extent, the state of the art of synthesis at that time. The early routes culminated in the Torgov method[1] for connecting ring D to the AB-ring. Recently the use of o-xylylenes (o-quinodimethanes) has further simplified the synthesis of estrone. The

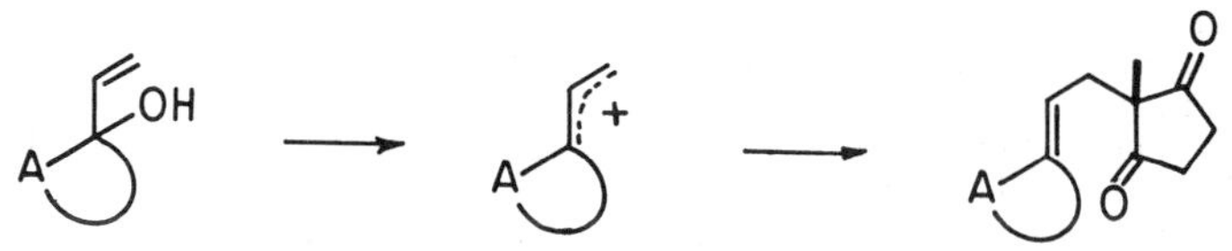

TORGOV METHOD

methods for generating the o-xylylene intermediates are usually based upon thermolysis of benzocyclobutenes or cheletropic extrusion of sulfur dioxide from a benzo[c]thiophene-2,2-dioxide.[2]

97

XYLYLENE METHOD

Combining these two strategies, and making use of certain aspects of organosilicon chemistry, we have developed a short, high yield synthesis of 11-oxo-
estrone. p-Methoxyphenyloxazoline 1 was treated with n-butyllithium in ether
at 0°C to give 1a which was quenched with methyliodide to give 1b (96%).[3] When

1b was further treated with n-butyllithium in ether at 0°C, 1c was formed, which on quenching with chlorotrimethylsilane gave 1d (92%). Unmasking 1d was accomplished by treatment of 1d with methyliodide followed by sodium borohydride, and work-up with 90% aqueous acetic acid to give the aldehyde 2 (97%). The aldehyde 2 was converted into the vinylcarbinol 3 (95%) on treatment with vinylmagnesium bromide. The synthesis of 3 proceeds in three steps, since the isolation of 1b is unnecessary, and the conversion of 1d into 2 can be carried out in one-pot; the overall yield of 3 from 1, with purification of 1b, 1d and 2 by distillation, is 81%.

2-Methylcyclopentenone was converted into the trimethylsilylenol ether 4 using described procedure.[2]

4

When a 1:1 mixture of 3 and 4 in dichloromethane at -78°C was treated with zinc bromide (0.1 equiv.) and warmed to -20°C, a clean transformation takes place to give 5 (88%), as a mixture of epimers at C-13, with the desired trans-D ring isomer predominating (ca. 5:1). Note that the benzylsilane 3 is stable to the electrophilic conditions used to make 5; indeed treatment of 5 with acids leads to polymerization without loss of the trimethylsilyl group.

Treatment of 5 with m-chloroperbenzoic acid, buffered with sodium bicarbonate, or benzonitrile hydrogen peroxide, gave the epoxide 6. When the epoxide 6 was treated with cesium fluoride in diglyme it cleanly produced 11-hydroxy-estrone methyl ether 7 as a mixture of epimers at C-11. Oxidation (PCC) gave 11-oxoestrone methyl ether 8, thus completing a short direct synthesis (seven steps).

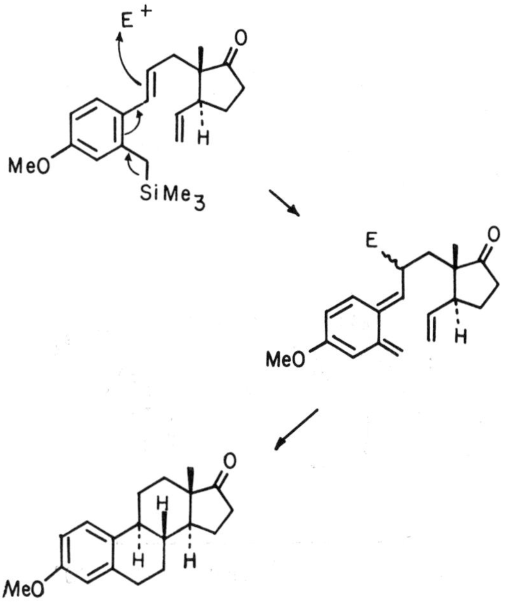
E+
O
MeO
SiMe3
H
E
O
MeO
H
O
H
MeO
H
H

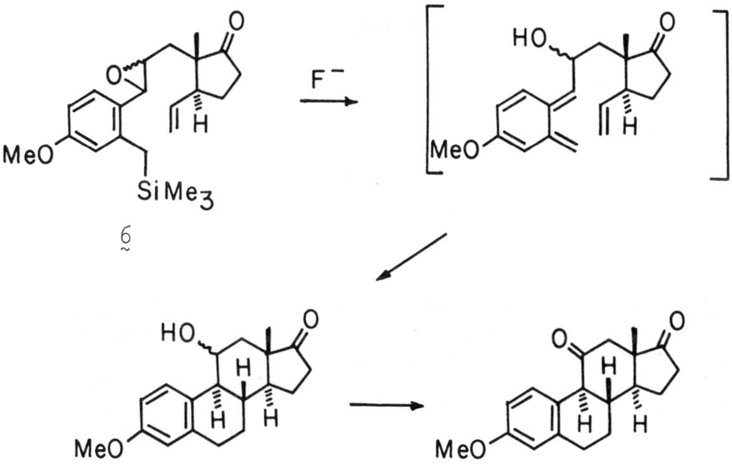
O
O
MeO
H
SiMe3
F−
HO
O
MeO
H
HO
O
H
MeO
H
H
O
O
H
MeO
H
H

6

7

8

We have devised a new cyclopentenone annulation reaction that leads directly
to a functionalized cyclopentenone. The following scheme outlines our
progress, to date, in using this new reaction for a short synthesis of the
sesquiterpene hirsutene.

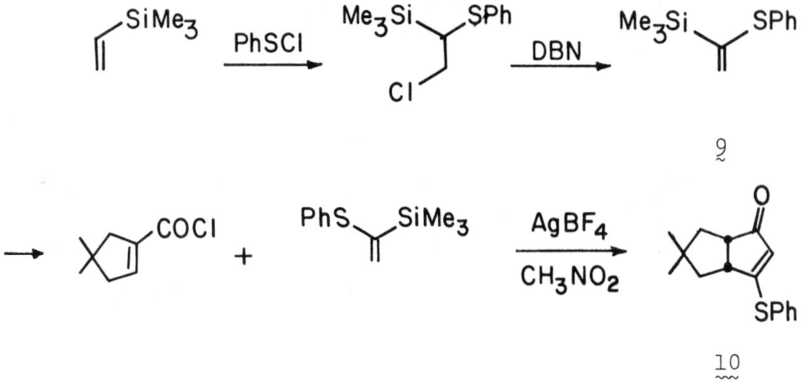

The reagent 9 reacts with α,β-unsaturated acid chlorides in the presense of
silver tetrafluoroborate to give, for example, 10. Mechanistically this
reaction may be rationalized as follows.

It should be emphasized that using vinylphenylsulfide in place of 9 did not produce 10. The trimethylsilyl group is essential.

HIRSUTENE

This synthesis uses organosilicon in three discrete steps to provide a short, direct synthesis of hirsutene.

The general formulation shown below depicts a complicated overall transformation involving one carbon ring expansion of a cyclic enone, combined with the introduction of an electrophile and nucleophile in a 1,4- or 1,2-relationship to one another.

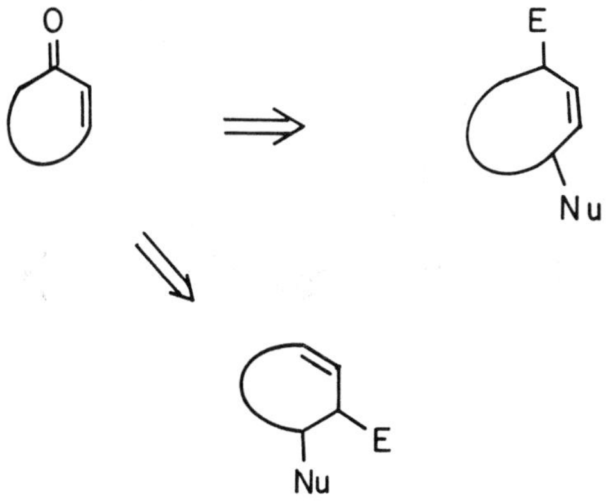

Here we describe a short and flexible way of carrying out these transformations that utilizes the combined chemistry of silylcyclopropanes, cyclopropyl-carbinyl rearrangements, and allylsilane electrophilic substitution.

The first part of the sequence involves the synthesis of silylcyclopropanes, and is outlined below. [4]

$$Me_3SiCH_2SMe_2I \xrightarrow{\quad n-BuLi \quad} Me_3Si-Bu^n + {}^{\ominus}CH_2-\overset{\oplus}{S}Me_2$$

with s-BuLi giving the silylcyclopropane product (top), and t-BuLi giving $Me_3SiCH_2CH_2SMe$ (bottom), using $Me_3Si\overset{\ominus}{C}H\overset{\oplus}{S}Me_2$ and cyclohexenone.

$$Me_3SiCH_2CH_2SMe$$

Treatment of the silylcyclopropane 11, derived from cyclohexenone, with trimethylsilyliodide at −20°C cleaved the cyclopropane, without ring expansion, to give 12. The structure of 12 was demonstrated by removal of the iodine atom and comparison of the product with an authentic sample.

Reduction of 11 with sodium borohydride gave a single crystalline alcohol 13, which underwent the desired ring expansion when treated with acetic acid/dichloromethane containing a catalytic amount of perchloric acid, to give the allylsilane 14. Further exposure of 14 to the above conditions results in protodesilylation to give cycloheptenyl acetate in good yields.

The allylsilane may be intercepted by other electrophiles under extremely mild conditions. For example treatment of 13 with AcOH/Ac₂O/H⁺ at 0°C results in acylation of 14 to give 15 which eliminates acetic acid to give the dienone 16.

To prevent elimination to a dienone the methylcarbinol 17 was exposed to the above conditions, to give 18 as a mixture of cis- and trans-isomers.

When the silylcyclopropyl carbinol 13 was treated with peracetic acid in
acetic acid containing a catalytic amount of perchloric acid, the allylsilane
14 was initially formed, and was cleanly converted into the allylic alcohol
19.   Oxidation of 19 gave the γ-acetoxyenone 20, where the original conjugated
enone functionality of cyclohexenone has been restored, but now in a seven
membered ring with a γ-acetoxy functional group.

The ring expansion sequence can be combined with a 1,3-enone trans-position merely by exposure of the allylsilane 14 to diphenylseleninic anhydride.

Since the methodology described here started with an enone, and finished with
an enone, it is of course possible to carry out the same sequence of reactions
again, and by this <u>reiterative</u> process construct unusual molecules.

This simple notion of reiterative reactions may be used to construct types of molecules that have no analogies. Starting with a given functional group, one carries out one or more transformations, and modifies the molecule in such a way that the same functional group is still present, ready to enter into the sequence again. This is a more elaborate version of homologation, since instead of just adding a methylene group, a considerably more complex series of changes has taken place.

In general terms for a spirocyclic annulation sequence we can represent this reiterative process as follows.

GENERAL REITERATIVE REACTION

We have examined in detail one such sequence which has resulted in the synthesis of the first primary helical molecules.

The synthesis is outlined below[5]-

The suggestion in the early fifties that the secondary structure of polypeptides have an α-helix conformation led the way for extensive studies of helical topology, especially in biopolymers. In particular the far reaching double stranded helical conformation for DNA, and the helical nature of many polymers highlight the fact that helical molecules have the pre-eminent place in macromolecular chemistry. Molecules that have helical topology but do not fall into the above classes are helicenes, skewed paracyclophanes, and helical triphenylmethanes systems.

All of the above helical molecules owe their helical topology to their secondary and tertiary structure. To date there is no molecule that is helical because of the primary bonding structure.

Here we report a rational synthesis of the first primary helical molecules, based upon the shape (bond angles and bond lengths) of the tetrahydrofuran ring system.

The adduct 21 between cyclopentanone (a starting block) and α-lithio-α-methoxyallene, on treatment with potassium t-butoxide (0.2 equiv.) in t-butanol containing 18-crown-6 (0.05 equiv.) heated at reflux for 15 h. gave 22. Acid hydrolysis (6N H_2SO_4) of 22 gave 23. The choice of conditions for the release of the dihydrofuranone carbonyl group is crucial in this step and in the subsequent hydrolyses of the described enol ethers, since 1N H_2SO_4 or two phase systems (oxalic acid/dichloromethane) gave 23 contaminated with decomposition products, whereas increasing the strength of the acid to 6N H_2SO_4 gave no degradation, and only a clean high yield conversion of 22 into 23. (82% on a 8.0 g. scale, overall yield.)

Since we started with a carbonyl compound, cyclopentanone, and spiroannulated
a dihydrofuranone on to it, we have ended with a new carbonyl compound,
namely 1-oxaspiro[4.4]nonan-4-one 23, and in principle it is capable of being
subjected to another spiroannulation sequence. The carbonyl group of 23 is
relatively hindered, but it reacts cleanly with α-lithio-α-methoxyallene to
give the adduct 24, after aqueous ammonium chloride work-up, in 95-98% yield.
The adduct 24 on treatment with KOBut (0.1 equiv.)/HOBut/18-crown-6 (0.05
equiv.) heated at reflux for 15 h. gave 25, which on acid hydrolysis (6N
H₂SO₄) gave 26 (70%).

At this point we had reached a crucial stage. The first and only stereochemical issue arises. The carbonyl group in 26 has two faces available for
the addition of a nucleophilic species. Dreiding models and space filling
models indicate that one face of 26 is severely hindered by the two five-
membered rings already present. In particular the cyclopentane ring sits
under one face of the carbonyl group in 26. In the event 26 adds α-lithio-
α-methoxyallene to give a single crystalline adduct 27 m.p. 133°-137° in ≥
90% yield. Treatment of this adduct with KOBut (0.2 equiv.)/HOBut/18-crown-6
(0.05 equiv.) heated at reflux gave 28, which on acid hydrolysis (6N H₂SO₄)
gave the beautifully crystalline cyclopentyl[3]helixane 29, m.p. 88.5°-90°C
(67% overall from 27.)

We hoped that as we extended the spiro-rings the allenyl adducts would undergo ring closure under increasingly mild conditions. This optimistic view is
based upon the idea that the steric compression of the alkoxide of 30 would
favor cyclization since covalent association of a metal counterion (Li or K
in the cases described) is sterically hindered, and furthermore 30 may act as
its own crown ether-ionophore and effectively remove the counterion (Li or K)
from covalent association with the alkoxide.

Treatment of 29 with α-lithio-α-methoxyallene gave 30, m.p. 92°-99°C (45%)
which cyclized to 31 at room temperature (28°C) when treated with KOBut (0.2
equiv.)/HOBut/18-crown-6 (0.05 equiv.) to give cyclopentyl[4]helixane 32,
m.p. 123°-124°C (67%) after acid (6N H_2SO_4) hydrolysis. Models (CPK) (see
photograph) show that cyclopentyl[4]helixane is a helical shaped molecule.
The oxygen atoms and methylene groups spiral around the central core of the

molecule. To unequivocally designate the structure of 32 and confirm
speculations concerning stereochemistry a single crystal X-ray analysis was
conducted. The results are as follows-

Cyclopentyl[4]helixane crystallized in the triclinic space group P1‾ with four molecules per unit cell. Accurate lattice parameters are a = 13.881 (4), b = 8.853 (2) and c = 13.668 (3) Å and a = 78.712 (2) °. The structure was solved by direct methods and has currently refined to a standard crystallographic residual of 0.08 for the 3085 diffractometer measured intensities. Both molecules in the asymmetric unit have the same geometry and Figure 1 is a computer generated drawing of them.

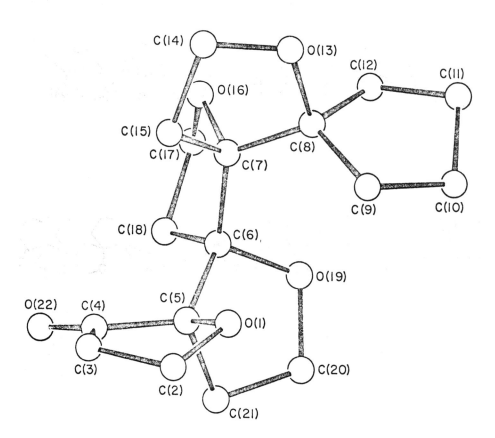

Figure 1

Cyclopentyl[4]helixane 32 was treated with α-lithio-α-methoxyallene to give the adduct 33, m.p. (90%) which cyclized to 34 when exposed to KOBut (0.1 equiv.)/HOBut/18-crown-6 (0.05 equiv.), acid hydrolysis (6N H₂SO₄) of 34 gave cyclopentyl 5 hilixane 35, m.p. 195°-197°C. Again 32 has only one face of the carbonyl group available for nucleophilic addition, and as a result 35 is a single, stereochemically pure compound. It should be noted that 35 has five quaternary carbons adjacent to one another in the carbon backbone, and each quaternary carbon is part of a tertiary ether function.

One can,in principle,continue this reiterative reaction and make extended versions of 35, such as the cyclopentyl[12]helixane 36 (see CPK model photograph)where the helicity is more readily seen. We are pursuing this objective and the more general concept of synthesizing primary helical molecules,along with studies of their physical and chemical properties.

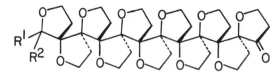

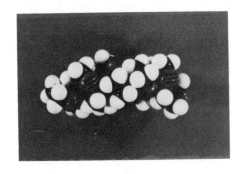

36

I wish to thank my coworkers, Dr. Ed Ehlinger, Dr. Tarun Sarkar, Dr. Stevan Djuric, Dave Gange, Glenn Roy, James Schwindeman, Dominic Quaglioto and Jack Venit for their skill and patience. Professor Jon Clardy is warmly thanked for his interest and expertise in carrying out the X-ray determination. Finally, the National Science Foundation, National Institute of Health and UpJohn Company are thanked for their financial support.

References

1. Ananchenko, S. N.; Torgov, I. V.; Tetrahedron Letters, 1963, 1553.

2. For a review of o-quinodimethanes in synthesis see - Oppolzer, W.;
    Synthesis, 1978, 793; For steroid syntheses see - Oppolzer, W.; Battig,
    K.; Petrzilka, M.; Helv. Chim. Acta., 1978, 61, 1945 and 1977, 60, 2964;
    Oppolzer, W.; Roberts, D. A.; Bird, T. G. C.; Ibid., 1979, 62, 2017;
    Kametani, T.; Nemoto, H.; Ishikawa, H.; Shiroyama, K.; Fukumoto, K.;
    J. Am. Chem. Soc., 1976, 98, 3378; Kametani, T.; Nemoto, H.; Ishikawa,
    H.; Shiroyama, K.; Matsumoto, H.; Fukumoto, K.; Ibid., 1977, 99, 3461.
    Nicolaou, K. C.; Barnette, W. E.; Ma, P.; J. Org. Chem., 1980, 45, 1463.
    Funk, R. L.; Vollhardt, K. P. C.; J. Am. Chem. Soc., 1979, 101, 215; and
    1977, 99, 5483.

3. Meyers, A. I.; Mihelich, E. D.; Angew. Chem. Int. Ed., 1976, 270; Meyers
    A. I.; Lutomski, K.; J. Org. Chem., 1979, 44, 4463. Gschwend, H. W.;
    Rodriquez, H. R.; Organic Reactions, Vol. 26, Wiley-Interscience, New
    York, 1980, Gschwend, H. W.; Hamden, A.; J. Org. Chem., 1975, 40, 2008.

4. Bundy, G. L.; Cooke, F.; Magnus, P.; J.C.S. Chem. Comm., 1978, 714.

5. Gange, D.; Magnus, P.; Bass, L.; Arnold, E. V.; Clardy, J.; J. Am. Chem.
    Soc., 102, 2134 (1980).

THE RICH CHEMISTRY OF VINYLIC ORGANOBORANES*

H. C. Brown

Department of Chemistry, Purdue University, West Lafayette, IN 47907, USA

Abstract - Vinylic organoboranes, many of which are readily available via
the hydroboration of acetylenes with appropriate hydroborating agents,
exhibit an exceptionally rich chemistry with great potential for synthetic
organic chemistry.

INTRODUCTION

Many vinylic organoboranes are now readily available by the hydroboration of acetylenes with
suitable borane derivatives (Strs. 1 & 2).

1   2

The corresponding cis derivative (Str. 3) is now also available by a modified procedure.

3

These vinylic derivatives exhibit an exceptionally rich chemistry of great promise in synthetic organic chemistry. Accordingly, it appears appropriate at this time to review their
synthesis and application.

The hydroboration of olefins with diborane in ether solvents provides a convenient synthesis
of aliphatic and alicyclic organoboranes (Ref. 1). The resulting organoboranes have proven
to be exceptionally useful synthetic reagents (Ref. 2).

Early work which attempted to extend hydroboration with borane to alkynes as a route to the
vinylic organoboranes met with limited success. While the reaction of internal alkynes with
diborane provides the corresponding trialkenylboranes in moderate yields (Eqn. 1), terminal

$$3 \ CH_3CH_2C{\equiv}CCH_2CH_3 + H_3B{\cdot}THF \longrightarrow \underset{H}{\overset{CH_3CH_2}{\diagdown}}C{=}C\underset{)_3B}{\overset{CH_2CH_3}{\diagup}} \tag{1}$$

alkynes yield little or none of the desired alkenylborane (Ref. 3). (In this discussion it
is convenient to utilize the more general term alkenyl as a synonym for the substituted vinyl
moiety.) Apparently dihydroboration competes with monohydroboration in the case of internal
acetylenes, becoming the predominant reaction in the case of terminal acetylenes (Eqn. 2).

---

* Based on a lecture presented at the Third IUPAC Conference on Synthesis, Madison, Wisconsin,
  June 15-20, 1980.

$$3 \ CH_3(CH_2)_3C{\equiv}CH + H_3B{\cdot}THF \longrightarrow CH_3(CH_2)_3CH_2\overset{\displaystyle B}{\underset{\displaystyle B}{C}}H + 1.5 \ CH_3(CH_2)_3C{\equiv}CH \qquad (2)$$
(unreacted)

The large steric requirements of certain olefins apparently hinder reaction with $H_3B{\cdot}THF$ beyond the formation of the corresponding mono- or dialkylboranes (Ref. 2a,d). This feature makes possible the convenient synthesis of several mono- and dialkylboranes and their utilization as desirable hydroborating reagents. Many of these reagents exhibit highly selective behavior in the hydroboration of olefins. Consequently, they appeared attractive as precursors for the controlled monohydroboration of acetylenes for the synthesis of the desired vinylic boranes. In fact, all of the mono- and dialkylborane reagents indicated in Fig. 1 have demonstrated utility for such monohydroboration of alkynes.

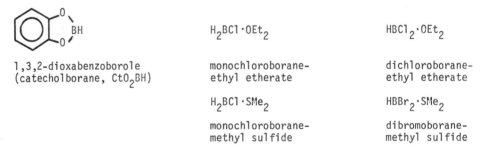

$$[(CH_3)_2CH\overset{\displaystyle CH_3}{\underset{}{CH}}]_2BH$$

bis-(3-methyl-2-butyl)borane
(disiamylborane, Sia_2BH)

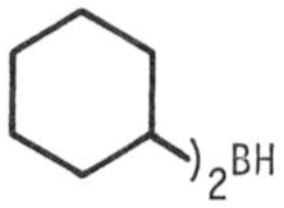

dicyclohexylborane
(Chx_2BH)

$$(CH_3)_2CHC(CH_3)_2BH_2 \equiv \quad \text{—}BH_2$$

2,3-dimethyl-2-butylborane
(thexylborane, $ThxBH_2$)

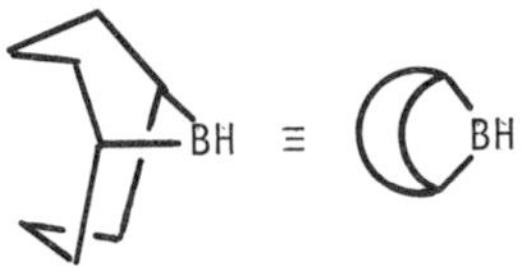

9-borabicyclo[3.3.1]nonane
(9-BBN)

Fig. 1. Mono- and dialkylborane hydroborating agents.

The later introduction of several other heterosubstituted boranes as hydroborating reagents further expanded the availability of reagents for preparing vinylic boranes from acetylenes (Fig. 2).

1,3,2-dioxabenzoborole
(catecholborane, CtO_2BH)

$H_2BCl{\cdot}OEt_2$

monochloroborane-
ethyl etherate

$H_2BCl{\cdot}SMe_2$

monochloroborane-
methyl sulfide

$HBCl_2{\cdot}OEt_2$

dichloroborane-
ethyl etherate

$HBBr_2{\cdot}SMe_2$

dibromoborane-
methyl sulfide

Fig. 2. Heterosubstituted boranes as hydroborating agents.

All of the reagents illustrated permit the controlled monohydroboration of both terminal and internal alkynes to provide the corresponding vinylic boranes. The mildness of the reagents tolerates the presence of a wide variety of functional groups, such as ester, ether, halogen, and nitrile. The stereospecific cis nature of hydroboration gives exclusively the trans-vinylic boranes, often also in high regioisomeric purity. Subsequent reactions of the vinylic boranes usually proceed by stereodefined pathways, thus allowing highly stereo- and regio-specific syntheses.

Many of the mono- and difunctional hydroborating reagents exhibit diverse reactivity characteristics toward different unsaturated substrates. Thus the availability of an array of hydroborating reagents to convert alkynes to the vinylic boron compounds expands the synthetic capability immensely. A wide spectrum of possible selective transformations of an alkyne in the presence of either functional groups, alkenes, or even structurally different alkynes, may then be conducted via hydroboration.

With the evolution of each new reagent came an advance in the capability of performing selective transformations of alkynes by hydroboration. Thus each new reagent will be discussed in a more or less chronological perspective. A separate section will examine directive effects in the hydroboration of several unsymmetrically substituted alkynes. Finally, representative transformations of the vinylic boranes or the diboraalkanes, readily available now via the hydroboration of alkynes, will be presented.

HYDROBORATION OF ALKYNES WITH BORANE DERIVATIVES

A. Disiamylborane

Reaction of either terminal or internal alkynes with disiamylborane at 0° proceeds rapidly to form the disiamylalkenylboranes (Ref. 3) (Eqn. 3).

$$CH_3(CH_2)_2C\equiv CH + Sia_2BH \longrightarrow \underset{H}{\overset{CH_3(CH_2)_2}{>}}C=C\underset{BSia_2}{\overset{H}{<}} \qquad (3)$$

Competing dihydroboration is insignificant even with an excess of the borane present, thus overcoming the difficulties associated with $H_3B\cdot THF$ as the hydroborating agent. Apparently the high steric requirements of disiamylborane minimize further reaction with the initially formed alkenylborane.

While diborane in THF is fairly non-discriminating among unsaturated substrates, disiamylborane reveals a far more selective reactivity. In fact, an internal or terminal alkyne can be selectively hydroborated with disiamylborane in the presence of all but unhindered terminal olefins (Refs. 4 & 5a) (Eqn. 4).

$$\underset{H}{\overset{H_3C(CH_2)_2}{>}}C=C\underset{CH_2C\equiv CH}{\overset{H}{<}} + Sia_2BH \longrightarrow \underset{H}{\overset{H_3C(CH_2)_2}{>}}C=C\underset{CH_2}{\overset{H}{<}}\overset{H}{>}C=C\underset{H}{\overset{BSia_2}{<}} \qquad (4)$$

B. Dicyclohexylborane

In many cases, dicyclohexylborane may be substituted for disiamylborane. However, the slightly lower steric requirements of dicyclohexylborane permit more potential dihydroboration of the alkyne to occur (Ref. 6). However, careful control of the reaction conditions affords the corresponding dicyclohexylalkenylboranes in excellent yields (Ref. 7) (Eqn. 5).

$$CH_3C\equiv CCOEt + (C_6H_{11})_2BH \longrightarrow \underset{H}{\overset{CH_3}{>}}C=C\underset{B(-\bigcirc)_2}{\overset{COEt}{<}} \qquad (5)$$

Although the relative reactivity of olefinic and acetylenic substrates toward dicyclohexylborane has not yet been quantitatively established, selective hydroborations are also achievable (Ref. 8) (Eqn. 6).

$$\underset{}{\overset{CH_3}{H_2C=C-C\equiv CH}} + (C_6H_{11})_2BH \longrightarrow \underset{H}{\overset{H_2C=C}{>}}\overset{CH_3}{\underset{}{}}C=C\underset{B(-\bigcirc)_2}{\overset{H}{<}} \qquad (6)$$

Hydroboration of 1-alkynes with two equivalents of dicyclohexylborane results in exclusive formation of the 1,1-diboraalkanes (Ref. 6) (Eqn. 7).

$$CH_3(CH_2)_2C\equiv CH + 2(C_6H_{11})_2BH \longrightarrow CH_3(CH_2)_2CH_2CH\underset{B(-\bigcirc)_2}{\overset{B(-\bigcirc)_2}{|}} \qquad (7)$$

Thus, direct access to such geminal organometallics can be accomplished virtually quantitatively from any 1-alkyne and dicyclohexylborane.

C. Thexylborane and Thexylmonoalkylboranes

Thexylborane is unique among the available alkylborane hydroborating reagents because of its difunctional nature (Ref. 5a). Reaction of two equivalents of a 1-alkyne with thexylborane cleanly produces the thexyldialkenylborane (Ref. 9) (Eqn. 8).

$$
\begin{array}{l}
\text{Thexyl–}BH_2 + 2\ CH_3(CH_2)_6C\equiv CH \longrightarrow \text{thexyldialkenylborane}
\end{array}
\tag{8}
$$

With the exception of simple terminal olefins, the controlled low temperature hydroboration of all other olefins with thexylborane provides the corresponding thexylmonoalkylboranes in nearly quantitative yield. Subsequent addition of an alkyne then gives the mixed thexyl-alkylalkenylboranes (Ref. 10) (Eqn. 9).

$$
\text{Thexyl–}BH_2 + CH_2=C(CH_3)CH_2CH_2CH_3 \xrightarrow{-20^\circ C} \text{product} \xrightarrow{CH_3(CH_2)_3C\equiv CH} \text{product}
\tag{9}
$$

Although reaction of thexylborane with one equivalent of a 1-alkyne does not cleanly afford the thexylmonoalkenylborane, reaction with either a 1-chloro- or 1-bromoalkyne does provide the thexyl-1-halo-1-alkenylborane (Ref. 11) (Eqn. 10).

$$
\text{Thexyl–}BH_2 + ClC\equiv CCH_2CH_3 \longrightarrow \text{product} \xrightarrow[HC\equiv CCH_2CH_3]{H_2C=CHCH_3} \text{products}
\tag{10}
$$

This monofunctional alkenylborane may then be used to prepare either mixed thexyldialkenyl- or thexylalkylalkenylboranes.

D. 9-BBN

Of the alkyl-substituted hydroboration reagents, 9-BBN possesses by far the greatest thermal and oxidative stability (Ref. 12). The reagent, a crystalline solid, may be stored for long periods of time under nitrogen and is in fact commercially available (Ref. 5b). Reaction of 9-BBN with internal alkynes affords the B-alkenyl-9-BBN derivatives in good yields (Ref. 13) (Eqn. 11).

$$CH_3CH_2C{\equiv}CCH_2CH_3 + \quad \overset{}{\bigcirc}BH \longrightarrow \quad \underset{H}{\overset{CH_3CH_2}{\diagdown}}C{=}C\underset{B}{\overset{CH_2CH_3}{\diagup}} \tag{11}$$

However, addition of 9-BBN to a 1-alkyne in stoichiometric quantities leads to the formation of substantial amounts of the dihydroboration product along with the desired vinylic borane. Presumably the openness of the boron atom in the 9-BBN moiety permits further reaction with the intermediate alkenylborane to give the 1,1-diboraalkane. However, use of a considerable excess of 1-alkyne, usually 100%, suppresses dihydroboration, yielding the desired vinylic borane in excellent yield (Ref. 13) (Eqn. 12). The excess alkyne is generally easily recovered.

$$2 \; Cl(CH_2)_3C{\equiv}CH + HB\bigcirc \longrightarrow Cl(CH_2)_3C{\equiv}CH + \underset{H}{\overset{Cl(CH_2)_3}{\diagdown}}C{=}C\underset{B}{\overset{H}{\diagup}} \tag{12}$$

The resulting B-alkenyl-9-BBN derivative is stable and can be isolated by vacuum distillation, if desired, to obtain the pure alkenylborane.

Addition of two equivalents of 9-BBN to the 1-alkyne readily produces the 1,1-diboraalkane in nearly quantitative yield (Eqn. 13) providing an alternative route to such derivatives (Ref. 14).

$$CH_3C{\equiv}CH + 2 \; \overset{}{\bigcirc}BH \longrightarrow CH_3CH_2\underset{B}{\overset{B}{\underset{|}{\overset{|}{C}H}}} \tag{13}$$

Unlike many of the other dialkylborane hydroborating reagents, 9-BBN demonstrates a significantly different reactivity toward unsaturated substrates. In general, unhindered terminal olefins react more readily than terminal and internal alkynes, while terminal alkynes react more readily than internal alkynes (Ref. 15) (Eqn. 14).

$$H_3C(CH_2)_2C{\equiv}CCH_2C{\equiv}CH + \overset{}{\bigcirc}BH \longrightarrow \underset{H}{\overset{H_3C(CH_2)_2C{\equiv}CCH_2}{\diagdown}}C{=}C\underset{B}{\overset{H}{\diagup}} \tag{14}$$

major product

E. Catecholborane

Addition of one equivalent of catechol to an equivalent of borane in THF gives the monofunctional reagent, catecholborane (Refs. 16 & 5b). The reaction of alkynes with catecholborane proceeds quite sluggishly at room temperature. However, at elevated temperatures, in refluxing tetrahydrofuran, the reaction proceeds smoothly to give the catecholalkenylboranes in excellent yields (Ref. 17) (Eqn. 15).

$$\bigcirc{-}C{\equiv}CH + \overset{O}{\underset{O}{\bigcirc}}BH \longrightarrow \underset{H}{\overset{\bigcirc}{\diagdown}}C{=}C\underset{B}{\overset{H}{\diagup}}\overset{O}{\underset{O}{\bigcirc}} \tag{15}$$

These vinylic boranes are quite stable to air and can be isolated by simple distillation or recrystallization. Such access to the catecholalkenylboranes also allows direct entry into the class of stereo defined vinylic boronic acids and esters via hydroboration (Eqn. 16).

$$\underset{H}{\overset{BrCH_2}{>}}C=C\underset{H}{\overset{H}{<}}\overset{O}{\underset{O}{>}}B\overset{O}{\underset{O}{<}} \xrightarrow{H_2O} \underset{H}{\overset{BrCH_2}{>}}C=C\underset{B(OH)_2}{\overset{H}{<}} \tag{16}$$

F. Mono- and dichloroborane-ethyl etherates

The preparation of the mono- and dichloroborane-ethyl etherates and their use as hydroborating reagents for olefins subsequently led to their development as precursors to the alkenylchloroboranes. Reaction of monochloroborane etherate (Ref. 18) with two equivalents of an internal alkyne cleanly affords the dialkenylchloroboranes in excellent yields (Ref. 19) (Eqn. 17).

$$CH_3CH_2C\equiv CCH_2CH_3 + H_2BCl\cdot OEt_2 \longrightarrow \underset{H}{\overset{CH_3CH_2}{>}}C=C\underset{)_2BCl}{\overset{CH_2CH_3}{<}} \tag{17}$$

Attempts to extend the reaction to 1-alkynes cause complications due to competing dihydroboration. However, use of a modest excess ($\sim$ 40%) of the 1-alkyne gives nearly quantitative yields of the desired bis-(1-alkenyl)chloroboranes (Ref. 18) (Eqn. 18).

$$2 \text{<cyclohexyl>}-C\equiv CH + H_2BCl\cdot OEt_2 \longrightarrow \underset{H}{\overset{\text{<cyclohexyl>}}{>}}C=C\underset{)_2BCl}{\overset{H}{<}} \tag{18}$$

Hydrolysis or alcoholysis of these boranes then provides a simple, direct synthesis of dialkylborinic acids or esters of known stereochemistry (Eqn. 19).

$$\underset{H}{\overset{CH_3(CH_2)_4}{>}}C=C\underset{)_2BCl}{\overset{H}{<}} \xrightarrow{CH_3OH} \underset{H}{\overset{CH_3(CH_2)_4}{>}}C=C\underset{)_2BOCH_3}{\overset{H}{<}} \tag{19}$$

The reaction of dichloroborane-ethyl etherate with alkynes proceeds slowly. Apparently the strong bond between the ether and the Lewis acid, dichloroborane, impedes the reaction. However, addition of one equivalent of boron trichloride to a mixture of dichloroborane-etherate and alkyne in pentane results in rapid hydroboration with deposition of the $Cl_3B\cdot OEt_2$ adduct (Ref. 20) (Eqn. 20).

$$(CH_3)_3CC\equiv CH + HBCl_2\cdot OEt_2 + BCl_3 \longrightarrow \underset{H}{\overset{(CH_3)_3C}{>}}C=C\underset{BCl_2}{\overset{H}{<}} + Cl_3B\cdot OEt_2\downarrow \tag{20}$$

Presumably, the stronger Lewis acid (Ref. 21), BCl_3, effectively removes the complexing ethyl ether, permitting the liberated dichloroborane to undergo immediate reaction. Hydrolysis or alcoholysis of the resultant alkenyldichloroborane then provides a simple preparation of the desired alkenylboronic acid or esters via hydroboration.

G. Dibromoborane-methyl sulfide

Quite recently, dibromoborane-methyl sulfide was prepared (Ref. 22) and demonstrated to react directly with olefins without addition of a stronger Lewis acid (Ref. 23). This is puzzling since one would expect dibromoborane to form an especially strong adduct which should exhibit a diminished reactivity. Nevertheless, reaction of dibromoborane-methyl sulfide with alkynes cleanly affords the corresponding alkenyldibromoboranes in excellent yields (Ref. 24) (Eqn. 21).

$$(CH_3)_2CHC\equiv CCH_3 + HBBr_2\cdot SMe_2 \longrightarrow \underset{H}{\overset{(CH_3)_2CH}{>}}C=C\underset{BBr_2\cdot SMe_2}{\overset{CH_3}{<}} \tag{21}$$

Fortunately, with 1-alkynes, the reaction proceeds readily to the vinylic borane stage, with no significant complications arising from competitive dihydroboration.

Moreover, dibromoborane-methyl sulfide exhibits an unusually rapid reaction with internal alkynes, far faster than the reaction of the reagent with terminal double or triple bonds.

This offers considerable promise. The relative reactivity data suggests that selective hydroboration of internal alkynes in the presence of 1-alkynes or any olefin should be feasible. The markedly different selectivity of 9-BBN (Ref. 15) and $HBBr_2 \cdot SMe_2$ (Ref. 24) should be noted (Eqn. 22).

$$
RC \equiv C(CH_2)_x CH = CH_2 \quad
\begin{cases}
\xrightarrow{\text{9-BBN}} & RC \equiv C(CH_2)_x CH_2 CH_2 \!-\! B \\
\\
\xrightarrow{HBBr_2 \cdot SMe_2} & \underset{H}{\overset{R}{\diagdown}} C = C \underset{BBr_2 \cdot SMe_2}{\overset{(CH_2)_x CH=CH_2}{\diagup}}
\end{cases}
\tag{22}
$$

DIRECTIVE EFFECTS

Availability of the considerable number of different hydroborating reagents for the preparation of alkenyl- and dialkenylboranes permits a number of various valuable selective hydroborations. For example, 9-BBN permits the selective hydroboration of a terminal alkene in the presence of an internal alkyne (Ref. 15) (Eqn. 22). On the other hand, dibromoborane-methyl sulfide selectively hydroborates an internal alkyne in the presence of a terminal alkyne (Ref. 24) (Eqn. 22).

The various hydroborating reagents also provide a regioselectivity spectrum in the hydroboration of unsymmetrically substituted alkynes. Nearly all of the reagents place boron exclusively at the terminal position in 1-alkynes. However, many internal alkynes involve a balance between steric and electronic effects in the placement of the boron.

In Table 1, the directive effects encountered in the hydroboration of 2-hexyne and 4-methyl-2-pentyne are summarized for the hydroborating reagents discussed.

TABLE 1. Directive effects in the hydroboration of 1-substituted propynes[a]

Hydroborating reagent	$CH_3(CH_2)_2 C \equiv CCH_3$	$\underset{CH_3}{\overset{CH_3}{\diagdown}} CH-C \equiv CCH_3$
Diborane	40 60	25 75
Thexylborane[b]	39 61	19 81
Disiamylborane[b]	39 61	7 93
Dicyclohexylborane[b]	33 67	8 92
$HBBr_2 \cdot SMe_2$[c]	25 75	4 96
9-BBN[d]	22 78	4 96

[a]Determined by oxidation of the alkenylboranes to the carbonyls. [b]Data from Ref. 8. [c]Data from Ref. 24. [d]Data from Ref. 13.

Clearly, all of the reagents show some sensitivity to steric effects, placing boron predominantly at the least hindered position. Both 9-BBN and $HBBr_2 \cdot SMe_2$ appear to be even more selective than the highly hindered reagents, disiamylborane and dicyclohexylborane.

The effect of a phenyl group, which alters the electronic requirements of the triple bond, is significant, as is evidenced by a comparison of the data for 1-phenyl-1-propyne with those for 1-cyclohexyl-1-propyne (Table 2).

TABLE 2. Directive effects in 1-cyclohexyl- and 1-phenyl-1-propyne[a]

Hydroborating agent	⬡—C≡CCH$_3$		⬡—C≡CCH$_3$	
	↑	↑	↑	↑
Disiamylborane[b]	9	91	19	81
Dicyclohexylborane[b]	8	92	29	71
HBBr$_2$·SMe$_2$[c]	9	91	64	36
9-BBN[d]	4	96	65	35

[a],[b],[c],[d]Same as in Table 1.

In 1-cyclohexyl-1-propyne, all of the reagents show a marked tendency to place boron at the least hindered position. However, the presence of the phenyl group directs both 9-BBN and HBBr$_2$·SMe$_2$ primarily to the position adjacent to the ring. Contrariwise, both dicyclohexyl-borane and disiamylborane still show a large preference to locate next to the smaller methyl group. Both 9-BBN and HBBr$_2$·SMe$_2$ appear to be sensitive to both electronic and steric effects and can be strongly influenced by electronic effects. On the other hand, disiamylborane and dicyclohexylborane are apparently controlled mainly by steric effects, with considerably smaller sensitivity to electronic factors.

The diverse regioselectivity exhibited by the several hydroborating reagents has important implications for the regioselective transformations of alkynes. Based upon the steric and electronic environment proximate to an alkyne, the appropriate choice of hydroborating reagent could provide a range of regiospecific hydroborations.

SYNTHESIS OF OTHER VINYLIC ORGANOBORANES

As was pointed out earlier, hydroboration of simple terminal acetylenes by a variety of borane derivatives provides a convenient route to the corresponding trans-vinylic boranes (Str. 2). Fortunately, the related cis derivatives (Str. 3) are now also readily available.

Originally, Negishi and his coworkers introduced a simple synthesis involving the hydrobora-tion of 1-halo-1-alkynes with dialkylboranes, followed by treatment with alkali metal trialkyl-borohydrides (Ref. 25a). An improved procedure was recently reported in which tert-butyl-lithium was used to transfer hydride to the intermediate which subsequently undergoes a trans-fer process, forming the desired cis-vinyldialkylborane (Ref. 25b) (Eqn. 23).

(23)

In order to realize the full potential of the chemistry of vinylic organoboranes, methods are needed to synthesize the remaining types (Strs. 4, 5 & 6).

Fortunately, chemistry exists which promises convenient routes to these structural types.

REPRESENTATIVE APPLICATIONS OF VINYLIC BORANES

All of the resulting classes of vinylic boranes prepared by the hydroboration of alkynes have demonstrated considerable utility in organic synthesis (Ref. 2). Many of the reactions are, in fact, extensions of known alkylborane chemistry. However, the vinylic boranes also exhibit their own unique characteristics in many cases. Moreover, since the hydroboration of an alkyne produces the trans-alkenylborane solely, the stereodefined nature of many of the subsequent reactions often permits the precise prediction of the stereochemistry of the product. Representative transformations of vinylic organoboranes are reviewed below.

A. Protonolysis

Addition of acetic acid to a vinylic borane results in a mild protonolysis of the boron-carbon bond to yield the corresponding alkenes (Ref. 3). The reaction proceeds stereospecifically with retention of configuration (Ref. 26) (Eqn. 24).

$$
\begin{array}{c}
\text{C}_2\text{H}_5\text{C}{\equiv}\text{C}\diagdown \quad \diagup\text{H} \\
\text{C=C} \\
\text{H}\diagup \quad \diagdown(\text{CH}_2)_6\text{OAc}
\end{array}
\quad\xrightarrow[\text{2. HOAc,0}^\circ]{\text{1. Sia}_2\text{BH}}\quad
\begin{array}{c}
\text{H}\diagdown \quad \diagup\text{H} \\
\text{C=C} \\
\text{C}_2\text{H}_5\diagup \quad \diagdown \text{C=C} \\
\qquad \text{H}\diagup \quad \diagdown(\text{CH}_2)_6\text{OAc}
\end{array}
\tag{24}
$$

sex pheromone of
Lobesia botrana

Thus, hydroboration-protonolysis of an alkyne provides a non-catalytic hydrogenation of triple to double bonds. In the case of internal alkynes, this procedure provides a valuable route to the pure cis-alkenes. The mildness and selectivity allows the presence of many functional groups and ready adaptability to the synthesis of many natural products. Deuterioacetic acid provides a simple, stereospecific preparation of deuterated olefins (Ref. 8) (Eqn. 25).

$$
\text{(cyclohexenyl)}-\text{C}{\equiv}\text{CH}\ \xrightarrow{\text{Chx}_2\text{BH}}\
\begin{array}{c}\text{C=C, B(Chx)}_2\end{array}\ \xrightarrow{\text{DOAc}}\
\begin{array}{c}\text{C=C, D}\end{array}
\tag{25}
$$

87%

B. Oxidation

Oxidation of vinylic boranes can be easily achieved with alkaline hydrogen peroxide to produce the corresponding carbonyl compounds (Refs. 3 & 13) (Eqn. 26).

$$
\begin{array}{c}
\text{CH}_3\text{CH}_2\diagdown \quad \diagup\text{CH}_2\text{CH}_3 \\
\text{C=C} \\
\text{H}\diagup \quad \diagdown\text{B}
\end{array}
\quad\xrightarrow[\text{H}_2\text{O}_2]{\text{NaOH}}\quad
\text{CH}_3\text{CH}_2\text{CH}_2\overset{\displaystyle O}{\overset{\|}{\text{C}}}\text{CH}_2\text{CH}_3
\tag{26}
$$

97%

For oxidation of 1-alkenylboranes, the addition of a pH 7 buffer is desirable to minimize base-promoted condensations of the resulting aldehydes (Refs. 3 & 25) (Eqn. 27).

$$
\begin{array}{c}
\text{(CH}_3)_3\text{C}\diagdown \quad \diagup\text{H} \\
\text{C=C} \\
\text{H}\diagup \quad \diagdown\text{BBr}_2{\cdot}\text{SMe}_2
\end{array}
\quad\xrightarrow[\substack{\text{2. pH 7}\\ \text{3. H}_2\text{O}_2}]{\text{1. 2 NaOH}}\quad
\text{(CH}_3)_3\text{CCH}_2\overset{\displaystyle O}{\overset{\|}{\text{C}}}\text{H}
\tag{27}
$$

77%

Hydroboration of 1-trimethylsilylacetylenes, followed by oxidation, provides a method of converting 1-alkynes to the corresponding carboxylic acids (Ref. 27) (Eqn. 28).

$$
\text{C}_6\text{H}_5-\text{C}{\equiv}\text{CH}\ \xrightarrow[\text{2. (CH}_3)_3\text{SiCl}]{\text{1. n-BuLi}}\ \xrightarrow{\text{Chx}_2\text{BH}}\
\begin{array}{c}
\text{C}_6\text{H}_5\diagdown \quad \diagup\text{Si(CH}_3)_3 \\
\text{C=C} \\
\text{H}\diagup \quad \diagdown\text{BChx}_2
\end{array}
\tag{28}
$$

$$
\xrightarrow[\text{2. H}^+]{\text{1. [O]}}\ \text{C}_6\text{H}_5\text{CH}_2\text{CO}_2\text{H}\quad 91\%
$$

The 1-trimethylsilylacetylenes may be prepared and used directly *in situ* to prepare the carboxylic acids.

C. Halogenation

A variety of stereochemically pure vinylic halides may be prepared from vinylic borane precursors. Iodination of a vinylic boronic acid in the presence of base gives excellent yields of the trans-1-alkenyl iodide (Ref. 28a). Use of $HBBr_2 \cdot SMe_2$ permits a simple one-pot conversion of 1-alkynes to stereodefined alkenyl iodides without isolation of any intermediates (Ref. 25) (Eqn. 29).

$$CH_3(CH_2)_3C \equiv CH + HBBr_2 \cdot SMe_2 \longrightarrow$$

(29)

67% isolated

The corresponding cis-alkenyl iodide may be prepared by first treating the alkenylboronic acid with excess iodine, allowing sufficient time to form the diiodo derivative, followed by the addition of base (Ref. 28b).

The faster addition of bromine to the double bond makes the preparation of cis-1-bromoalkenes easier by a process involving the reaction of bromine with the alkenylcatecholborane, followed by addition of sodium methoxide (Ref. 29) (Eqn. 30).

(30)

95%

The reaction of such vinylic boronic acids with bromine in methanol at -78°, in the presence of sodium methoxide, provides a convenient synthesis of α-bromo acetals (Ref. 30) (Eqn. 31).

(31)

D. Conjugate addition

B-Alkenyl-9-BBN derivatives undergo 1,4-addition of the alkenyl group to acyclic enones, yielding γ,δ-unsaturated ketones (Ref. 31) (Eqn. 32).

(32)

The addition occurs with strict retention of configuration in the initial vinylic borane. The reaction evidently proceeds through a cyclic transition state, so that transoid enones, such as 2-cyclohexenone, cannot be utilized. On the other hand, cisoid enones, including very sensitive and easily polymerized derivatives, such as methyl vinyl ketone, react without difficulty.

E. 1,2-Addition to aldehydes

Unlike alkylboranes, B-alkenyl-9-BBN derivatives add across the carbonyl group of aldehydes to produce stereochemically pure allylic alcohols (Ref. 32) (Eqn. 33).

$$
\begin{array}{l}
\underset{H}{\overset{Cl(CH_2)_3}{>}}C=C\underset{B}{\overset{H}{<}} + CH_3CH_2\overset{O}{\overset{\|}{C}}H \longrightarrow \underset{H}{\overset{Cl(CH_2)_3}{>}}C=C\underset{\underset{OH}{\overset{|}{CHCH_2CH_3}}}{\overset{H}{<}} \tag{33}
\end{array}
$$

Since many functional groups, such as ester, halogen, and nitrile, are tolerated by hydroboration, a "Grignard-like" synthesis of such allylic alcohols with reactive substituents present in the organometallics is now possible.

F. Mercuration

The mercuration of vinylic boranes provides easy access to stereochemically defined alkenylmercurials (Ref. 33). Addition of mercuric acetate to a catecholalkenylborane results in clean formation of the alkenylmercuric compound in excellent yields (Ref. 33b) (Eqn. 34).

$$
\begin{array}{c}
\overset{\overset{\overset{CH_3}{|}}{H_2C=C}}{\underset{H}{>}}C=C\underset{B}{\overset{H}{<}}\overset{O}{\underset{O}{\bigcirc}}
\xrightarrow[\text{2. NaCl}]{\text{1. Hg(OAc)}_2}
\overset{\overset{\overset{CH_3}{|}}{H_2C=C}}{\underset{H}{>}}C=C\underset{HgCl}{\overset{H}{<}} \tag{34}
\end{array}
$$

The resulting organomercurials have since been shown to be exceptionally useful synthetic reagents, undergoing a variety of carbon-carbon bond forming reactions (Ref. 34).

G. Transmetallation to copper

Recently, dialkenylchloroboranes were reported to undergo methyl copper induced coupling to give trans,trans-1,4-dienes in excellent yields and high stereochemical purity (Ref. 35) (Eqn. 35).

$$
2 \quad \underset{H}{\overset{CH_3CH_2}{>}}C=C\underset{)_2BCl}{\overset{CH_2CH_3}{<}}
\xrightarrow[0°]{3\ CH_3Cu}
\underset{H}{\overset{CH_3CH_2}{>}}C=C\underset{\underset{H_3CCH_2}{\overset{C=C}{<}}{\overset{/}{\underset{CH_2CH_3}{\backslash}}}}{\overset{CH_2CH_3}{<}}\ H \tag{35}
$$

The reaction presumably involves initial formation of an alkenylcopper reagent which undergoes thermal dimerization with retention of configuration to give the 1,3-diene. An alternative procedure employing milder conditions proceeds via the sodium methoxide addition compound of an alkenyldialkylborane (Ref. 36a) (Eqn. 36).

$$
\underset{H}{\overset{(CH_3)_3C}{>}}C=C\underset{B}{\overset{H}{<}}
\xrightarrow[\text{2. CuBr·SMe}_2]{\text{1. NaOMe,0°}}
\underset{H}{\overset{(CH_3)_3C}{>}}C=C\underset{\underset{H}{\overset{C=C}{<}}{\overset{/H}{\underset{C(CH_3)_3}{\backslash}}}}{\overset{H}{<}} \tag{36}
$$

98%

The stability of the vinylic copper intermediate should be greater at lower temperatures. Indeed, by working at -15°, the decomposition of the copper intermediate is retarded and it can be trapped by allylic halides to afford a stereochemically defined synthesis of 1,4-dienes (Ref. 36b) (Eqn. 37).

$$
\underset{H}{\overset{AcO-(CH_2)_2}{>}}C=C\underset{B(-\bigcirc)_2}{\overset{H}{<}}
\xrightarrow[\substack{\text{2. CuBr·SMe}_2 \\ -15°}]{\text{1. NaOMe}}
\left[\quad\right]
$$

$$
\xrightarrow{\quad\overset{Br}{\diagup\!\!\diagdown\!\!\diagup}\quad}
\underset{H}{\overset{AcO-(CH_2)_2}{>}}C=C\underset{CH_2CH=CH_2}{\overset{H}{<}} \tag{37}
$$

73% isolated

Again the mildness of the reaction allows a wide variety of functional groups to be tolerated. Finally, conjugate addition of the copper reagent to cyclic enones can be effected (Ref. 36c), supplementing the 1,4-addition reaction of B-alkenyl-9-BBN derivatives (Sect. D) (Eqn. 38).

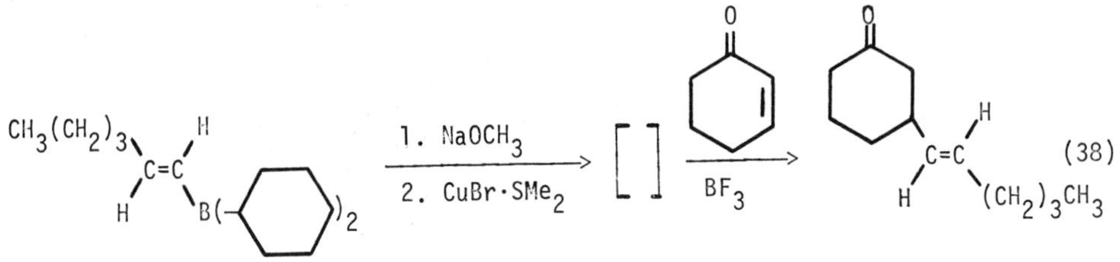

$$\text{(38)}$$

H. cis-Olefin synthesis

Addition of iodine to an alkenyldialkylborane in the presence of base results in the exclusive formation of a cis-olefin derived from transfer of an alkyl group from boron to the adjacent carbon (Ref. 37) (Eqn. 39).

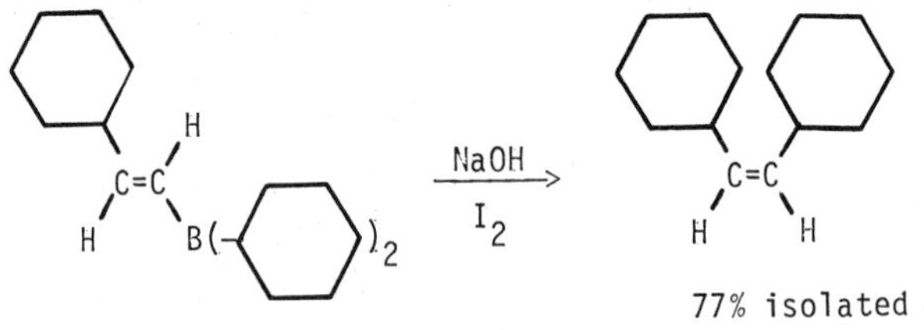

77% isolated

$$\text{(39)}$$

Migration of the alkyl group from boron occurs with strict retention of configuration, defined in the hydroboration step. Because of the known stereochemical outcome, the reaction has proven to be of value in the preparation of prostaglandin analogs (Ref. 38) (Eqn. 40).

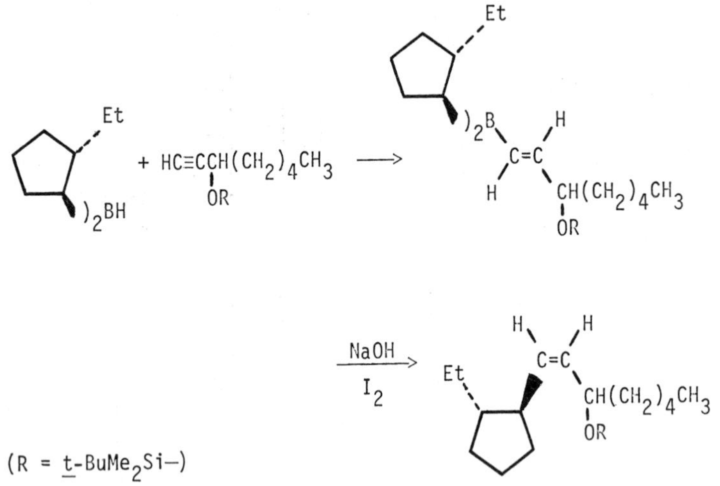

$$\text{(40)}$$

(R = t-BuMe$_2$Si—)

I. trans-Olefin synthesis

Reaction of an alkenyldialkylborane with cyanogen bromide produces the trans-olefin, again derived from alkyl group transfer from boron (Ref. 39) (Eqn. 41).

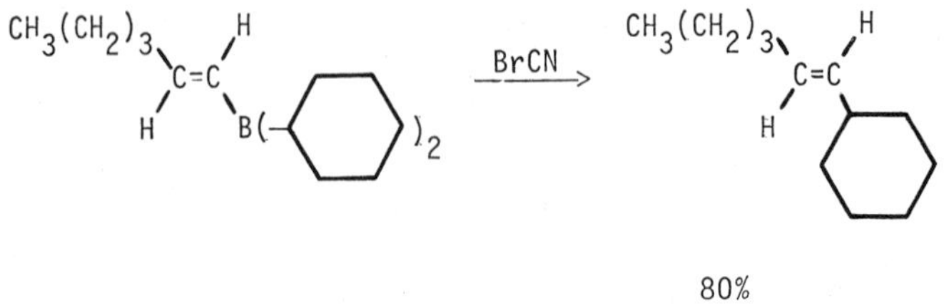

$$\text{(41)}$$

80%

Alternatively, hydroboration of 1-halo-1-alkynes with a dialkylborane followed by treatment with sodium methoxide provides trans-olefins (Ref. 40) (Eqn. 42).

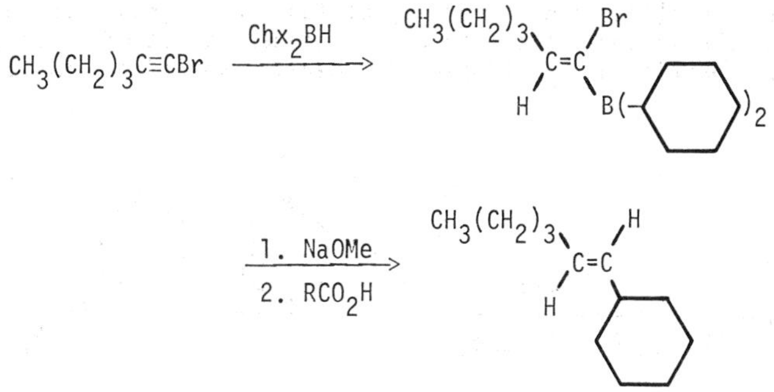

$$CH_3(CH_2)_3C\equiv CBr \xrightarrow{Chx_2BH}$$

(42)

$$\xrightarrow[\text{2. } RCO_2H]{\text{1. NaOMe}}$$

Unfortunately, these reactions are limited by the availability of dialkylborane hydroborating reagents. Also, only one of the two available groups on boron transfers.

Use of the thexylmonoalkylboranes, readily synthesized from thexylborane, overcomes these difficulties, allowing the introduction of many alkyl groups. An alkyl-1-haloalkenylthexyl-borane (Ref. 10) is perfectly set up for an alkyl group migration, and, indeed, treatment with sodium methoxide, followed by protonolysis, releases the desired trans-olefin (Eqn. 43).

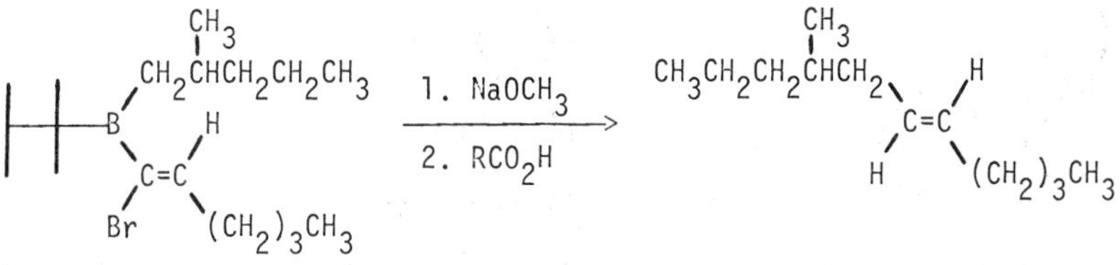

$$\xrightarrow[\text{2. } RCO_2H]{\text{1. NaOCH}_3}$$

(43)

The applications of such a stereodefined reaction to the synthesis of natural products is again illustrated by the preparation of a prostaglandin model (Ref. 41) (Eqn. 44).

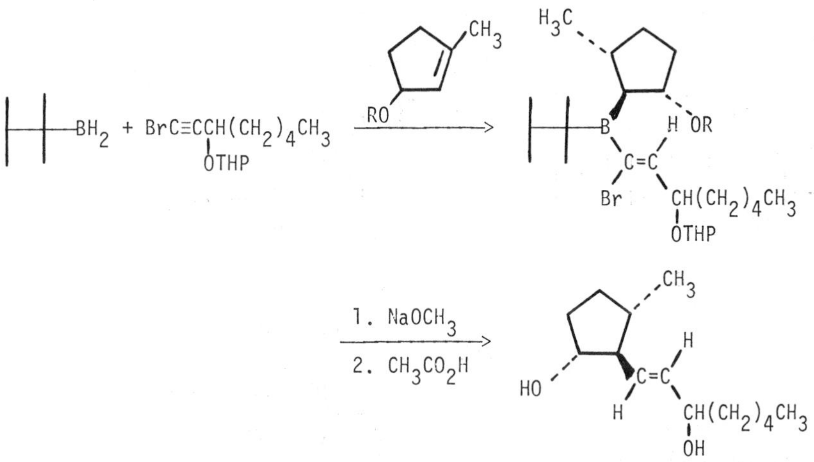

(44)

$$\xrightarrow[\text{2. } CH_3CO_2H]{\text{1. NaOCH}_3}$$

(R = t-BuMe$_2$Si—)

J. Synthesis of conjugated trans,trans-dienes

Reaction of thexylborane, first with a 1-haloalkyne followed by addition of a second alkyne, gives the mixed thexyldialkenylborane (Ref. 11). Treatment first with sodium methoxide followed by protonolysis provides the conjugated trans,trans-diene in good yields (Ref. 11) (Eqn. 45).

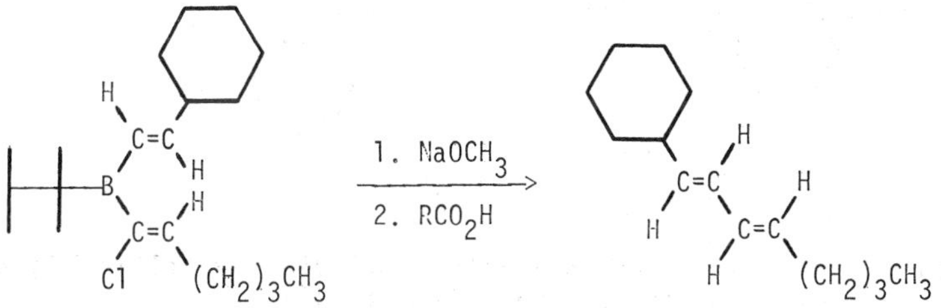

$$\xrightarrow[\text{2. } RCO_2H]{\text{1. NaOCH}_3}$$

(45)

This procedure also permits the synthesis of unsymmetrical trans,trans-dienes. An alternative preparation of symmetrical trans,trans-dienes has been discussed earlier (Sect. G).

K. Synthesis of conjugated cis,trans-dienes

Preparation of symmetrical cis,trans-dienes may be performed by iodination of dialkenylboronic acids in the presence of base, a procedure analogous to that described for the synthesis of cis-olefins (Ref. 42) (Eqn. 39). The requisite dialkenylborinic acids are most conveniently formed by basic hydrolysis of the dialkenylchloroboranes (Eqn. 46), readily produced by hydroboration of alkynes with chloroborane-etherate (Ref. 43).

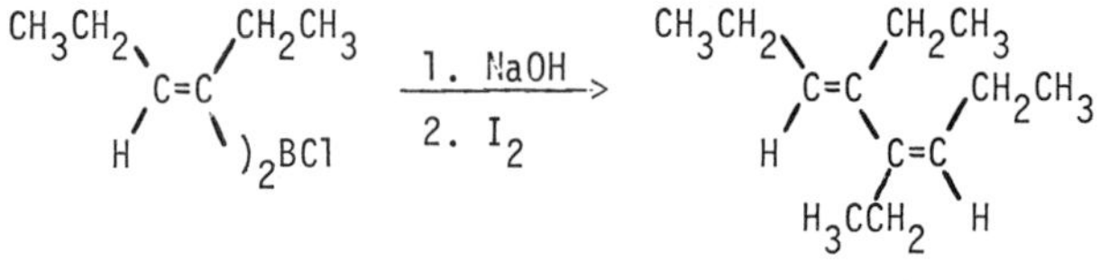

$$\tag{46}$$

84%

Thus, hydroboration of alkynes with $H_2BCl\cdot OEt_2$, followed by sequential treatment with base and iodine, provides a direct route to such dienes.

Another more general procedure for preparing cis,trans-dienes, which can be utilized for the synthesis of unsymmetrical derivatives, involves stepwise treatment of an alkenyldialkylborane with a lithium alkynylide, followed by boron trifluoride etherate (Ref. 44) (Eqn. 47).

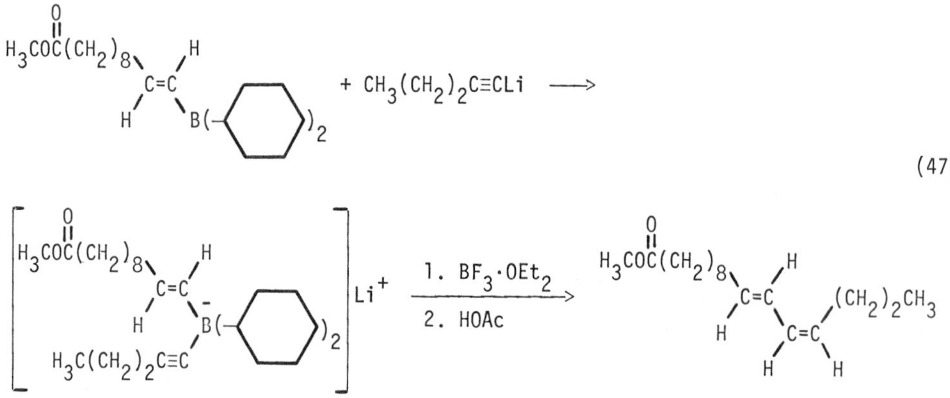

$$\tag{47}$$

65% isolated
sex pheromone Bombyx mori

Protonolysis of the intermediate releases the isomerically pure unsymmetrical cis,trans-diene in good yield.

L. Synthesis of conjugated cis,cis-dienes

The oxidative coupling of 1-alkynes in the presence of copper salts provides a convenient route to the symmetrical conjugated diynes (Ref. 45) (Eqn. 48).

$$2\ RC{\equiv}CH \xrightarrow[\substack{pyridine\\acetone}]{O_2,CuCl} RC{\equiv}C{-}C{\equiv}CR \tag{48}$$

Hydroboration-protonolysis of such diynes makes readily available the symmetrical conjugated cis,cis-dienes (Ref. 46) (Eqn. 49).

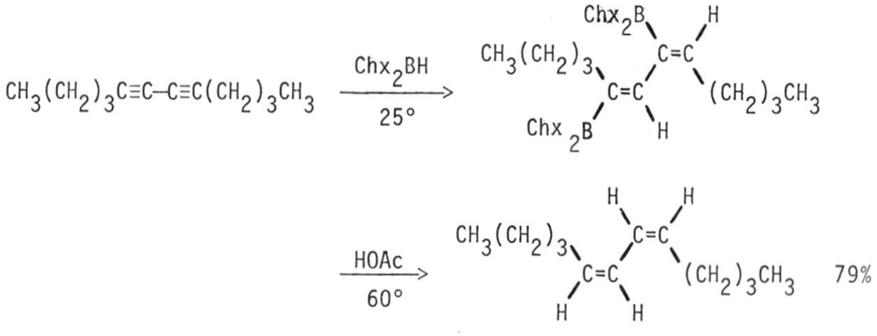

$$\tag{49}$$

79%

Borane chemistry now provides two synthetic routes to the synthesis of unsymmetric conjugated diynes, one proceeding through dicyclohexylmethylthioborane (Ref. 47) and the other proceeding through disiamylmethoxyborane (Ref. 48). The latter will be illustrated here (Eqn. 50), although both appear equally satisfactory.

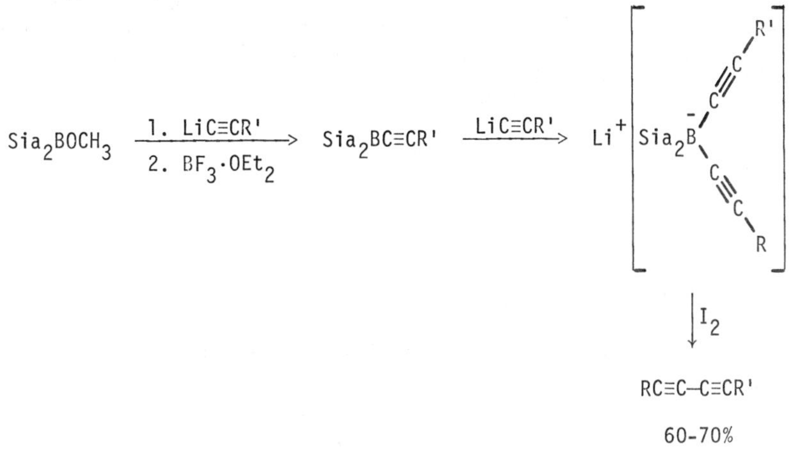

$$Sia_2BOCH_3 \xrightarrow[\text{2. } BF_3 \cdot OEt_2]{\text{1. } LiC \equiv CR'} Sia_2BC \equiv CR' \xrightarrow{LiC \equiv CR'} Li^+ \left[Sia_2B \underset{\underset{R}{\overset{C}{\underset{\|}{C}}}}{\overset{\overset{R'}{\overset{C}{\underset{\|}{C}}}}{<}} \right]$$

$$\downarrow I_2 \qquad\qquad (50)$$

$$RC \equiv C-C \equiv CR'$$

60-70%

Hydroboration-protonolysis of this unsymmetrical conjugated diyne by the Zweifel-Polston procedure (Ref. 46) should provide the corresponding conjugated cis,cis-diene (Str. 7).

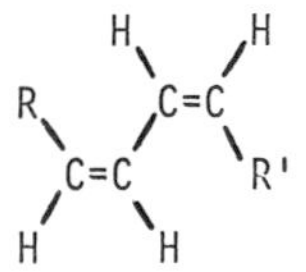

7

M. 1,1-Dibora compounds
The often troublesome 1,1-dibora compounds accompanying the hydroboration of alkynes have themselves, in fact, revealed some interesting synthetic potential. Treatment of a 1,1-diboraalkane with one equivalent of CH₃Li (Ref. 49), followed by excess of an alkyl halide, gives a substituted secondary alcohol upon oxidation (Ref. 14) (Eqn. 51).

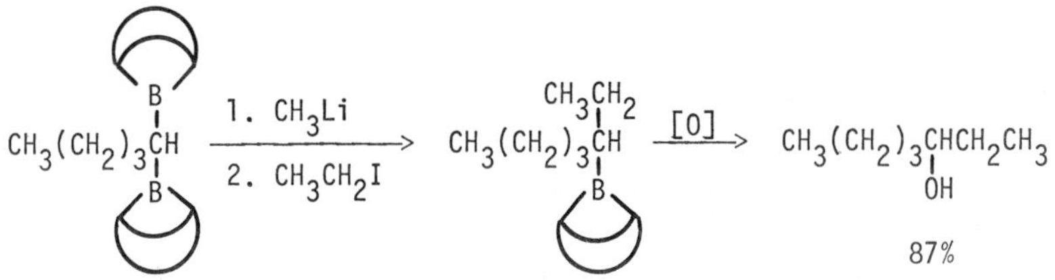

$$(51)$$

87%

Additions of the organometallic reagent derived from 1,1-diboraalkanes and methyllithium to carbonyl compounds give a "Wittig-like" olefination (Ref. 50) (Eqn. 52).

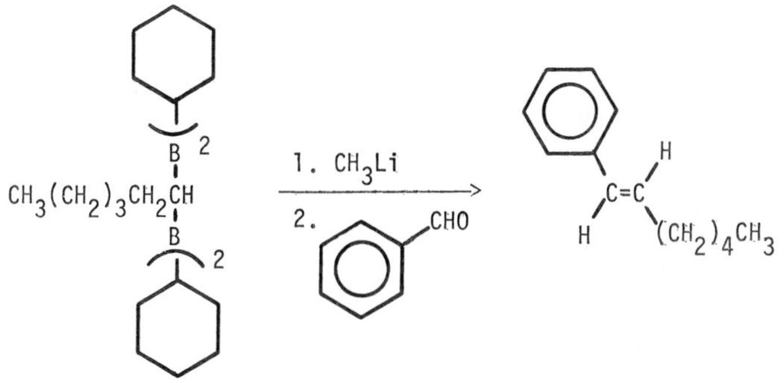

$$(52)$$

CONCLUSION

The controlled monohydroboration of both internal and terminal alkynes provides a stereo-specific synthesis of vinylic organoboranes. In some cases the reaction proceeds to the formation of 1,1-diboraalkanes. However, highly selective and regiochemically distinct transformations of alkynes are possible depending upon the proper choice of hydroborating reagent. The alkenylboranes thus produced can undergo a variety of stereoselective reactions with defined stereochemical results.

Undoubtedly, the chemistries of vinylic boranes and 1,1-diboraalkanes are still in their infancy, awaiting further exploitation of the remarkably versatile derivatives made available by the hydroboration of acetylenes. Further development of novel hydroborating reagents and of convenient synthetic routes to other vinylic organoboranes (Strs. 4-6) will expand the horizon further, allowing many more selective transformations of acetylenes.

Acknowledgement - The support of the National Science Foundation (Grants CHE 76-20846 and CHE 79-18881) and the assistance of Dr. James B. Campbell, Jr., and of Dr. Surendra U. Kulkarni in the preparation of this manuscript are gratefully acknowledged.

REFERENCES

1. H. C. Brown, Hydroboration, W. A. Benjamin, New York (1962).
2. (a) H. C. Brown, Boranes in Organic Chemistry, Cornell University Press, Ithaca, New York (1972); (b) G. M. L. Cragg, Organoboranes in Organic Synthesis, Marcel Dekker, New York (1973); (c) T. Onak, Organoborane Chemistry, Academic Press, New York (1975); (d) H. C. Brown, G. W. Kramer, A. B. Levy and M. M. Midland, Organic Syntheses via Boranes, Wiley-Interscience, New York (1975).
3. H. C. Brown and G. Zweifel, J. Am. Chem. Soc. 83, 3834-3840 (1961).
4. H. C. Brown and A. W. Moerikofer, J. Am. Chem. Soc. 85, 2063-2065 (1963).
5. (a) Kits are available from the Aldrich Chemical Company; (b) The reagent is available from the Aldrich Chemical Company.
6. G. Zweifel and H. Arzoumanian, J. Am. Chem. Soc. 89, 291-295 (1967).
7. J. Plamondon, J. T. Snow and G. Zweifel, Organomet. Chem. Syn. 1, 249-252 (1971).
8. G. Zweifel, G. M. Clark and N. L. Polston, J. Am. Chem. Soc. 93, 3395-3399 (1971).
9. G. Zweifel and H. C. Brown, J. Am. Chem. Soc. 85, 2066-2072 (1963).
10. E. Negishi, J.-J. Katz and H. C. Brown Synthesis 555-556 (1972).
11. E. Negishi and T. Yoshida, Chem. Commun. 606-607 (1973).
12. (a) E. F. Knights and H. C. Brown, J. Am. Chem. Soc. 90, 5280-5281 (1968); (b) C. G. Scouten and H. C. Brown, J. Org. Chem. 38, 4092-4094 (1973).
13. H. C. Brown, C. G. Scouten and R. Liotta, J. Am. Chem. Soc. 101, 96-99 (1979).
14. L. Brener, Final Report, Purdue University (1976).
15. C. A. Brown and R. A. Coleman, J. Org. Chem. 44, 2328-2329 (1979).
16. H. C. Brown and S. K. Gupta, J. Am. Chem. Soc. 93, 1816-1818 (1971).
17. (a) H. C. Brown and S. K. Gupta, J. Am. Chem. Soc. 94, 4370-4371 (1972); (b) H. C. Brown and S. K. Gupta, Ibid. 97, 5249-5255 (1975).
18. H. C. Brown and P. A. Tierney, J. Inorg. Nucl. Chem. 9, 51-55 (1959).
19. H. C. Brown and N. Ravindran, J. Am. Chem. Soc. 98, 1785-1798 (1976).
20. H. C. Brown and N. Ravindran, J. Am. Chem. Soc. 98, 1798-1806 (1976).
21. H. C. Brown and R. R. Holmes, J. Am. Chem. Soc. 78, 2173-2176 (1956).
22. (a) K. Kinberger and W. Siebert, Z. Naturforsch. B 30, 55-59 (1975); (b) H. C. Brown and N. Ravindran, Inorg. Chem. 16, 2938-2940 (1977).
23. N. Ravindran and H. C. Brown, J. Am. Chem. Soc. 99, 7097-7098 (1977).
24. H. C. Brown and J. B. Campbell, Jr., J. Org. Chem. 45, 389-395 (1980).
25. (a) E. Negishi, R. M. Williams, G. Lew and T. Yoshida, J. Organometal. Chem. 92, C4-C6 (1975); (b) J. B. Campbell, Jr. and G. A. Molander, J. Organometal. Chem. 156, 71-79 (1978).
26. E. Negishi and A. Abramovitch, Tetrahedron Lett. 411-414 (1977).
27. G. Zweifel and S. J. Backlund, J. Am. Chem. Soc. 99, 3184-3185 (1977).
28. (a) H. C. Brown, T. Hamaoka and N. Ravindran, J. Am. Chem. Soc. 95, 5786-5788 (1973); (b) Unpublished results with T. Hamaoka.
29. H. C. Brown, T. Hamaoka and N. Ravindran, J. Am. Chem. Soc. 95, 6456-6457 (1973).
30. T. Hamaoka and H. C. Brown, J. Org. Chem. 40, 1189-1190 (1975).
31. P. Jacob, III and H. C. Brown, J. Am. Chem. Soc. 98, 7832-7833 (1976).
32. P. Jacob, III and H. C. Brown, J. Org. Chem. 42, 579-580 (1977).
33. (a) R. C. Larock and H. C. Brown, J. Organometal. Chem. 36, 1-12 (1972); (b) R. C. Larock, S. K. Gupta and H. C. Brown, J. Am. Chem. Soc. 94, 4371-4373 (1972).
34. (a) D. Seyferth, ed., J. Organometal. Chem., Lib. 1 257-303 (1976); (b) R. C. Larock, Angew. Chem., Int. Ed. Engl. 17, 27-37 (1978).
35. H. C. Brown and N. Ravindran, J. Am. Chem. Soc. 98, 1785-1798 (1976).
36. (a) J. B. Campbell, Jr. and H. C. Brown, J. Org. Chem. 45, 549-550 (1980); (b) J. B. Campbell, Jr. and H. C. Brown, Ibid. 45, 550-552 (1980); (c) Unpublished results with J. B. Campbell, Jr.

37. G. Zweifel, H. Arzoumanian and C. C. Whitney, J. Am. Chem. Soc. 89, 3652-3653 (1967).
38. D. A. Evans, T. C. Crawford, R. C. Thomas and J. A. Walker, J. Org. Chem. 41, 3947-3953 (1976).
39. G. Zweifel, R. P. Fisher, J. T. Snow and C. C. Whitney, J. Am. Chem. Soc. 94, 6560-6561 (1972).
40. G. Zweifel and H. Arzoumanian, J. Am. Chem. Soc. 89, 5086-5088 (1967).
41. E. J. Corey and T. Ravindranathan, J. Am. Chem. Soc. 94, 4013-4014 (1972).
42. G. Zweifel, N. L. Polston and C. C. Whitney, J. Am. Chem. Soc. 90, 6243-6245 (1968).
43. H. C. Brown and N. Ravindran, J. Org. Chem. 38, 1617-1618 (1973).
44. G. Zweifel and S. J. Backlund, J. Organometal. Chem. 156, 159-170 (1978).
45. I. D. Campbell and G. Eglinton, Org. Syn. 45, 39-42 (1965).
46. G. Zweifel and N. L. Polston, J. Am. Chem. Soc. 92, 4068-4071 (1970).
47. A. Pelter, R. Hughes, K. Smith and M. Tabata, Tetrahedron Lett. 4385-4388 (1976).
48. J. A. Sinclair and H. C. Brown, J. Org. Chem. 41, 1078-1079 (1976).
49. G. Zweifel and H. Arzoumanian, Tetrahedron Lett. 2535-2538 (1966).
50. G. Cainelli, G. Dal Bello and G. Zubiani, Tetrahedron Lett. 4315-4318 (1966).

SYNTHESIS OF POLYETHER ANTIBIOTICS. ADJACENT AND REMOTE ASYMMETRIC INDUCTION VIA CYCLIC HYDROBORATION

W. Clark Still

Department of Chemistry, Columbia University, New York, NY 10027, USA

Abstract: It is shown that boranes may be used to hydroborate a number of acyclic 1,3-, 1,4- and 1,5-dienes to give diols in which high degrees of 1,2-, 1,3-, 1,4- and 1,5-asymmetric induction are produced. The results are consistent with an intramolecular 4-membered cyclic transition state for the second hydroboration. Product distributions were measured by ^{13}CMR and stereochemical assignments were made by alternative syntheses of diastereomerically pure materials. The utility of the method is illustrated by stereoselective syntheses of the Prelog-Djerassi lactonic acid, a naturally-occurring dihydromyoporone and the vitamin E sidechain.

INTRODUCTION

In considering the synthesis of stereochemically complex acyclic molecules such as the poly-ether antibiotic lysocellin (below), two synthetically significant types of diastereomeric relationships may be distinguished. Sets of asymmetric centers may be either directly connected (adjacent, e.g. C3 and C4) or they may be separated by one or more atoms (remote, e.g. C4 and C6). The efficient addition of new asymmetric centers relative to preexisting ones

(relative asymmetric induction) is one of the most important problems facing would-be architects of complex acyclic structures, and although several useful methods for building up adjacent stereocenters with high 1,2-asymmetric induction have been devised, general approaches to the construction of remote stereorelationships by efficient 1,>2-asymmetric induction are rare. Previous solutions to the remote stereochemistry problem have largely avoided remote asymmetric induction and have relied on the coupling of optically active fragments or on increasing the separation of proximate asymmetric centers by some form of chirality transfer. Recently however P. Bartlett's phosphate chain-extended epoxidation (1,3-asymmetric induction) and Y. Kishi's bishomoallylic alcohol epoxidation (1,4-asymmetric induction) have demonstrated that general methods for acyclic remote asymmetric induction were within grasp and could be synthetically useful processes. During the past several months, we have had occasion to examine another reaction, the cyclic hydroboration of dienes, as a potential route to stereochemically well-defined acyclic diols. Our initial studies in this area involved the synthesis of the C3-C9 fragment of lysocellin via a cyclic hydrobor-ation involving 1,2-asymmetric induction. Although the result of this work was a surprisingly efficient synthesis of the desired molecule, the most interesting rewards came when we began to look at the application of cyclic hydroboration to remote asymmetric induction. In the pages which follow I will describe some of the results we have gleaned from a most interesting reaction and hopefully convince you that cyclic hydroboration is not only a synthetically useful pathway to remote asymmetric induction but also suggests a more general strategy by which acyclic stereochemistry may be controlled.

OS - F*

DISCUSSION

The effect upon which we wish to focus is the asymmetric induction which occurs during an intramolecular hydroboration where the cycle being closed contains one or more asymmetric centers (*). As illustrated below, this operation can create new asymmetry (*') and may proceed via either fused or bridged intermediates (A or B respectively). Since the object

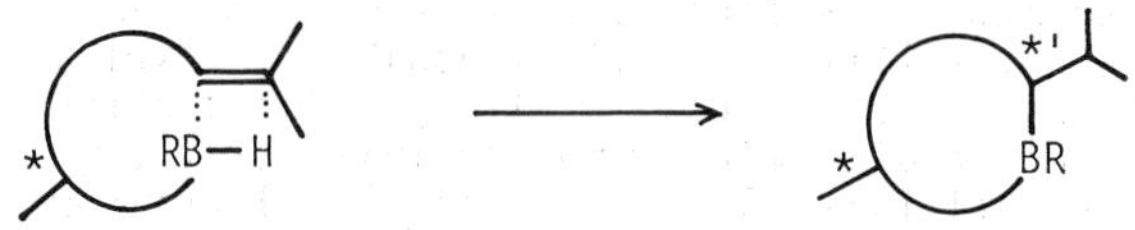

of this study was exploration of methodology for the remote control of stereochemistry, we hoped that the diastereomeric relationship of the asymmetric centers thus produced could be made to rely less on a direct and necessarily weak interaction between widely separated substituents and more on the overall conformation(s) of the transition state(s) leading to the boracycles. Factors which determine the detailed geometry of such transition states are not well understood; however, important first approximation considerations may include the energetically accessible conformations of the ring being closed and any stereoelectronic constraints inherent to hydroboration itself.

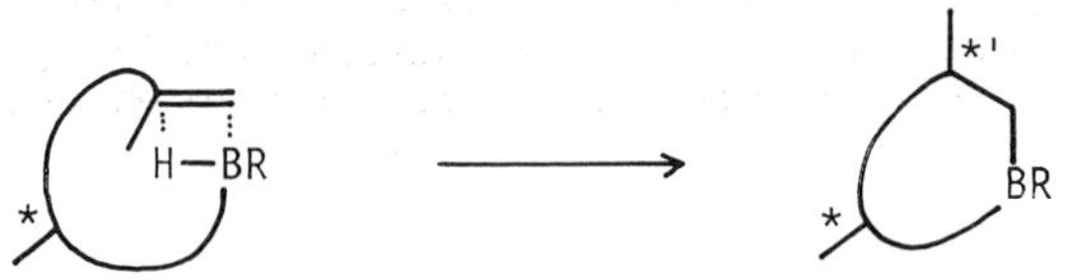

Before moving to our results on remote asymmetric induction, let me first describe the study which kindled our interest in cyclic hydroboration as a method for acyclic stereoselection. That work involves the preparation of the segment of lysocellin which stretches from C3 to C9. When this piece is excised in a suitable oxidation state, it is seen to be the now-famous Prelog-Djerassi lactone:

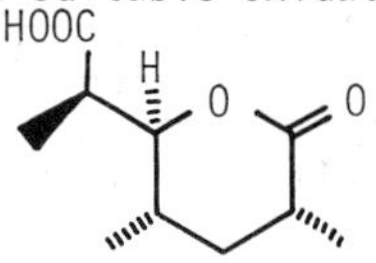

Our original plan was to simply produce the compound by one of the previously described syntheses - the recent and relatively concise approaches by S. Masamune, G. Stork and P. Bartlett for example would seem to offer access to synthetically useful quantities of material. It soon became apparent however that a more interesting and efficient solution to the Prelog-Djerassi lactone problem could be devised and based on an intramolecular hydroboration. The scheme we used is shown below:

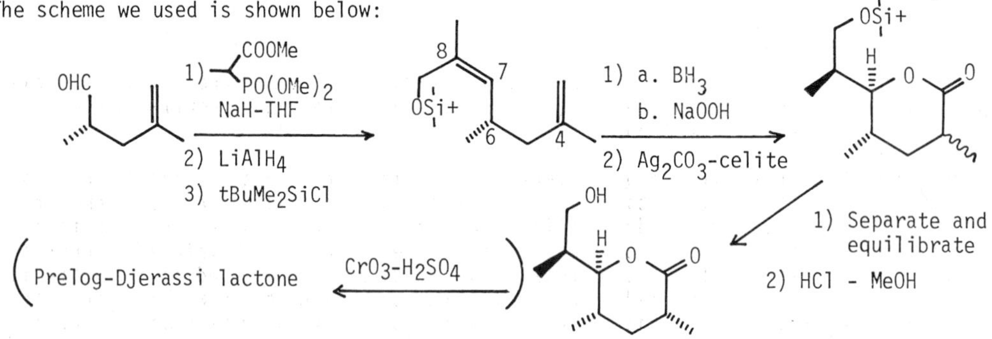

This reaction sequence leads to the desired C3-C9 fragment of lysocellin and contains two points of interest. First, the overall yield is in excess of 50%. Second, the crucial cyclic hydroboration gives virtually complete selection for the desired asymmetry at C7 and C8 with control by the preexisting chiral center at C6. This high 1,2-asymmetric induction is of course to be expected for the intramolecular reaction since the chain bearing the intermediate monoalkylborane is sterically constrained to the β-face of the C7,C8 double bond. Although the scheme incorporates an undesirable separation/equilibration sequence in lieu of stereo-control at C4 (cf. P. Grieco's Prelog-Djerassi synthesis), it provides a ready supply of the required lactonic alcohol and more importantly suggests a new way to use hydroboration for the efficient construction of acyclic stereochemistry.

Though the very high 1,2-asymmetric induction which was obtained in the Prelog-Djerassi lactone synthesis described on the previous page was most appreciated, it was essentially a trivial result - to the extent that the hydroboration was intramolecular, the conformational situation at C6 and C7 forced production of the desired stereorelationship. The application of cyclic hydroboration to remote asymmetric induction would seem, however, somewhat less predictable since the asymmetric center which controls the olefin face-selection must exert its effect in an indirect way. In particular, we hoped that face-selection for the intra-molecular hydroboration step could be controlled by a conformational effect similar to an equatorial-axial preference by a preexisting substituent in the cycle being closed. Shortly after we began our laboratory investigation of remote asymmetric induction via cyclic hydro-boration, a very interesting result was obtained. This result involved hydroboration of 2,6-dimethyl-1,5-heptadiene. Although the diene reacted rapidly in tetrahydrofuran with borane almost without stereochemical control, a similar reaction with thexylborane produced substantial 1,4-asymmetric induction to give a 6:1 mixture of diastereomers (81% yield) in which i was the major product. This ratio as well as most of the others given here was

determined by high resolution ^{13}CMR with calibration by an authentic mixture of diastereomers and a pure diastereomer prepared without stereochemical ambiguity by a different route. In the example above, authentic samples of both i and ii were prepared from isomenthone and menthone respectively (a. MCPBA; b. LDA, TMSCl; c. O$_3$; d. LiAlH$_4$). The identification of i as the major product provides evidence for the operation of a substantial stereoelectronic effect in the hydroboration reaction. Of the two most likely transition states, it will be seen that the major product i results from the boat-like geometry if it is assumed that non-hydrogen substituents will predominantly select equatorial environments. The thexyl substituent turns out to be lost as tetramethylethylene (cf. H.C. Brown, 1975) prior to cyclization and the preference for the boat over the chair may well reflect the situation that the boron-hydrogen bond eclipses the olefin π-system in the former conformation but not

in the latter. The transition state for the second, intramolecular hydroboration would thus be a semiplanar, 4-center one as suggested by Brown some years ago and recently by some molecular orbital calculations (but not by others).

Related diene hydroborations proceed by formation of a 5-membered boracycle and do not appear to incorporate the unfavorable aspects of a boat-like transition state. In fact, some of H.C. Brown's early work on cyclic hydroboration showed that the preference for transition states of this type is so great that the normal preference for anti-Markovnikov hydroboration may be largely overridden. Although the scope of the reaction as presented here is somewhat limited by the requirement of a well-defined initial hydroboration, the examples which follow demonstrate the crucial point that the second (intramolecular) hydroboration again occurs with substantial remote asymmetric induction:

c) 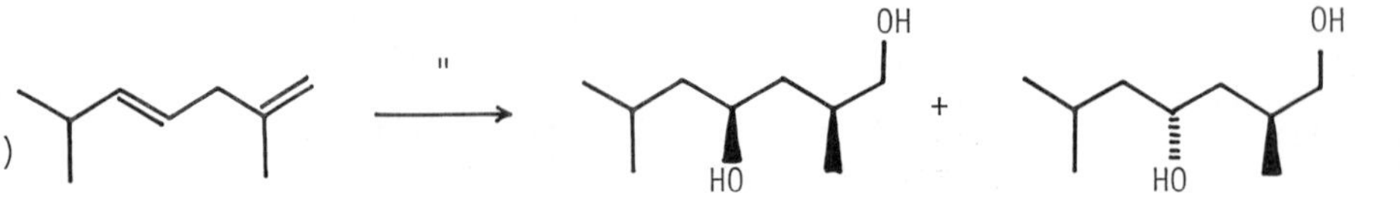   15:1
                                                          (77%)

d) 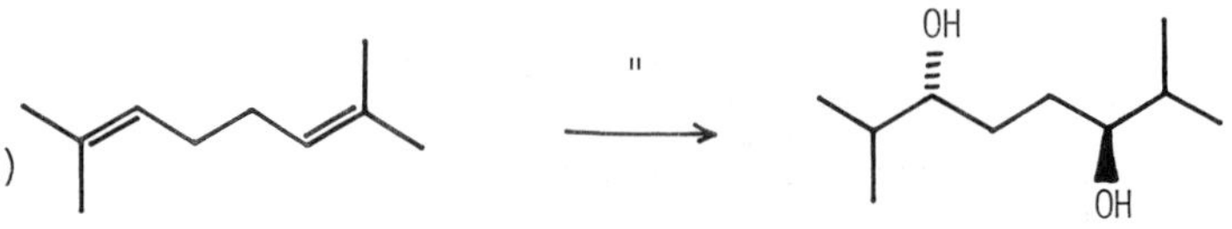   >20:1
                                                          (30%)

In hydroboration b), the relatively poor stereocontrol is due in part to competitive hydroboration of the cis-disubstituted olefin as the initial step. This type of selectivity is
of course well-documented and is supported here by isolation of the isomeric 1,5-diol (5%).
It may also be noted that the low yield associated with hydroboration d) is also precedented
in the literature by the previous observation that thexylborane reacts only sluggishly with
trisubstituted olefins. It is, however, interesting to note that in each instance, the major
product may be rationalized in terms of a transition state having eclipsed B-H and C=C bonds
and what would appear to be the least strained conformation of the connecting chain. As in
the related cyclopentanoid systems, this preferred conformation would be expected to consist
of trans adjacent substituents and pseudoequatorial nonadjacent ones. Only in the case of
hydroboration c) is some ambiguity involved and this is presumably due to a steric interaction between the isopropyl and the thexyl substituents.

A simple application of this methodology involves the synthesis of a potato stress metabolite
known as dihydromyoporone (iii). The product of hydroboration c) was monotosylated (pTsCl,
C_5H_5N), silylated (t-BuMe$_2$SiCl), and iodide exchanged (NaI, acetone) to give the silyloxy
iodide shown below. β-Ketosulfoxide alkylation (NaH, DMF) followed by reductive desulfur-
ization (Al-Hg, THF-H$_2$O) as described by E.J. Corey and finally deprotection (HOAc-H$_2$O)
gave (±)-dihydromyoporone (iii). An analogous sequence starting from the product of hydro-

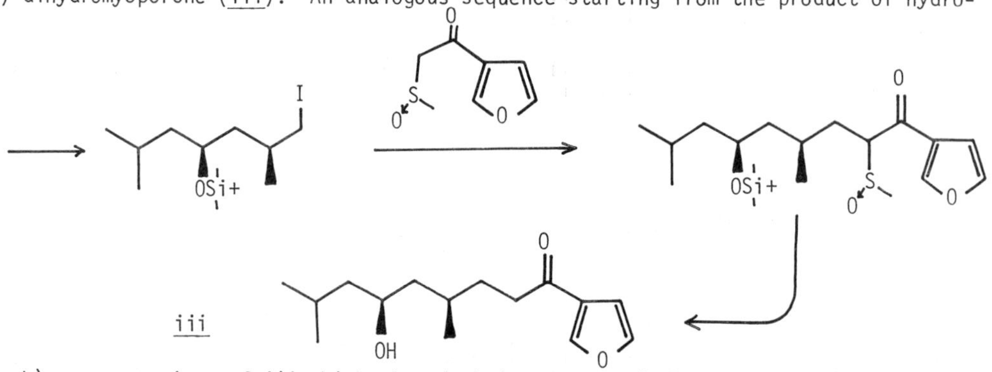

boration b) gave an epimer of iii which showed obvious spectral differences with a sample of
natural dihydromyoporone provided by L.T. Burka only by [13]CMR. Since the relative stereochemistry of the two asymmetric centers in dihydromyoporone has not been previously reported,
this synthesis establishes the stereochemistry of the natural product as threo.

Although the extension of this method to the formation of more widely separated asymmetric
centers seemed possible, we encountered substantial difficulties in accomplishing this goal.
In particular, examination of the hydroboration of several homologs of the dienes above
at various dilutions revealed no obvious stereochemical control. Although there could be
several reasons for this lack of control, a reasonable explanation would be that the macrocyclic rings being formed are simply too flexible to allow good face-selection. It is perhaps
useful to note here that the hydroborations thus far described are ones which proceed via a
fused transition state (A, above). It seemed possible that some of the related bridged
transition states (B, D) would be less flexible and might improve stereocontrol. Further
consideration of this proposition revealed another potentially interesting feature of these
bridged transition states: the geometry of the hydroboration reaction would seem to resemble
a low energy conformational substructure found in a number of rings having more than six
atoms (C). This geometrical similarity would suggest that transition state D should not
experience substantially more strain than that inherent to the ring being formed, and, to the
extent that the hydroboration is intramolecular, the product should be meso:

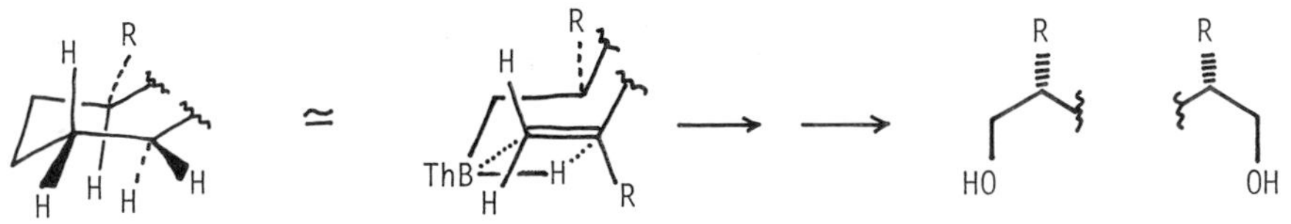

Results with thexylborane (0.1M in tetrahydrofuran) seem to bear out this prediction. Thus hydroboration of the dienes shown below proceed by formation of 7- and 8-membered boracycles and give the expected meso diols with high 1,4- and 1,5-asymmetric induction. To the best of our knowledge, these reactions give the most extended remote asymmetric induction yet observed in acyclic systems. Product identities here were established by the preparation

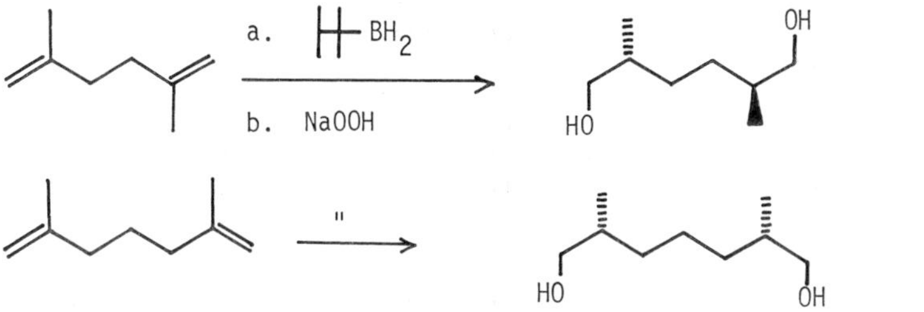

meso:dl (Yield)

>20:1   (86%)

15:1   (73%)

of authentic mixtures with the dienes shown and 9-BBN. Samples of the pure diols were obtained by coupling enantiomerically pure fragments derived from (+)-β-hydroxyisobutyric acid. Product compositions could only be determined by high resolution ^{13}CMR.

A direct application of the last reaction is the preparation of the vitamin E sidechain (iv) which has been used previously by researchers at Hoffmann-La Roche. The synthesis is carried out in a straightforward way by statistical monotosylation (pTsCl, C$_5$H$_5$N; 0° C) of the 1,7-diol above followed by bromide exchange (LiBr, DMF) and coupling with isoamylmagnesium bromide (THF, Li$_2$CuCl$_4$; 0° C). The racemic alcohol was identical by ^{13}CMR with authentic (+)-iv kindly provided by N. Cohen.

iv

Although numerous cyclic hydroborations are shown above to proceed with synthetically useful remote asymmetric induction, diene hydroboration as presented here provides only a partial solution to the general problem of remote stereocontrol. The most serious limitation is that relative asymmetric induction operates here only in a mechanistic sense since the first, controlling asymmetric center is produced in the same overall reaction which subsequently creates the second. The key point, however, is that the stereochemistry associated with the intramolecular step can be efficiently controlled by a remote chiral center. We believe that the generality of the cyclization approach to remote asymmetric induction is now firmly established and that examination of other intramolecular reactions in this context will reveal similar stereochemical controls.

It is with a deep sense of appreciation that I acknowledge the contributions of my coworkers upon whose skill and persistence the success of this project rests. Mr. Kenneth Shaw solved the Prelog-Djerassi lactone problem as an undergraduate research associate last summer and is now engaged in a single-handed assault on lysocellin itself. Dr. Kevin Darst, whose graduate training was in transition metal coordination chemistry, attacked the remote asymmetric induction problem not only with the skill of a seasoned organic chemist but also with a tenacity which propelled us rapidly through some very difficult waters. It is both a pleasure and an honor to work with such men and to be able to describe their work to you.

NEW METHODS FOR FORMATION OF CARBON – CARBON BONDS

L. Ghosez

Laboratoire de Chimie Organique de Synthèse, Université Catholique de Louvain,
Place L. Pasteur, 1, B - 1348 Louvain-La-Neuve, Belgium

Abstract - *New methods will be presented which employ activated forms of tertiary amides in reaction forming carbon-carbon bonds. Emphasis will be put on the rules governing the reactivity of the new synthons.*

INTRODUCTION

Carboxylic acids play an important rôle in organic chemistry as ubiquitous functional groups of numerous natural compounds but also as commonly used synthons for the construction of carbon skeletons. Acyl halides or anhydrides are frequently used as sources of acyl cations while esters are often employed as sources of enolate anions. In contrast, the synthetic potential of carboxamides has been rather neglected. It is the purpose of this lecture to show how they can be transformed by simple structural manipulations into a variety of activated forms capable of forming carbon-carbon bonds. Time will not allow for a detailed discussion of each system. I rather intend to give you the rules of this chemical game with a few illustrations taken from our recent studies.

Scheme 1

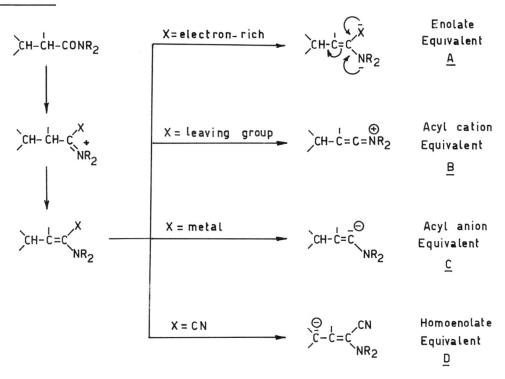

Our approach results from the consideration that tertiary amides can be readily transformed into α-heterosubstituted iminium salts which, in turn, are
potential sources of α-heterosubstituted enamines. These molecules provide us
with two handles which should permit control of their chemical behaviour.
One is the nitrogen substituent which can be made more or less electrondona-
ting by a variation of substituent R. The other is the X group which is expected to play a crucial rôle on the reactivity of the enamine. Thus, provided that X is not an electronwithdrawing group , the compound should exhibit
the typical behaviour of an enamine and is thus the synthetic equivalent of
an *enolate anion*. If X is a suitable leaving group, ionization to a keteniminium salt becomes feasible, leading to an *acyl cation* equivalent. A third
synthetic intermediate C, equivalent to an *acyl anion*, would result from the
replacement of X = halogen by an electropositive element such as Li or Mg.
Finally, with X being an anion-stabilizing group, removal of the allylic proton should become more facile leading to an interesting *homoenolate* equivalent.
Thus according to this concept, carbon-carbon bond-forming reactions should be
possible on the first three carbon atoms of an amide.

SYNTHESIS OF α-HALOENAMINES

A common precursor for synthons A-D could be an α-chloroenamine or the corresponding α-chloroiminium chloride (amide chloride) as we could anticipate that,
in these molecules, the chlorine substituent at C-1 should be exchangeable for
other atoms or groups. Amide chlorides are readily obtained from tertiary
amides and phosgene. On treatment with triethylamine, these were readily converted into α-chloroenamines (Ref. 1,2). The method works very well and is
highly practical for large scale preparation, provided that R^1 and R^2 are
different from hydrogen. These β-disubstituted α-chloroenamines are stable and
readily isolable compounds. They can be easily transformed into the corresponding α-fluoro, bromo and iodoenamines (Scheme 2)(Ref.3).

Scheme 2

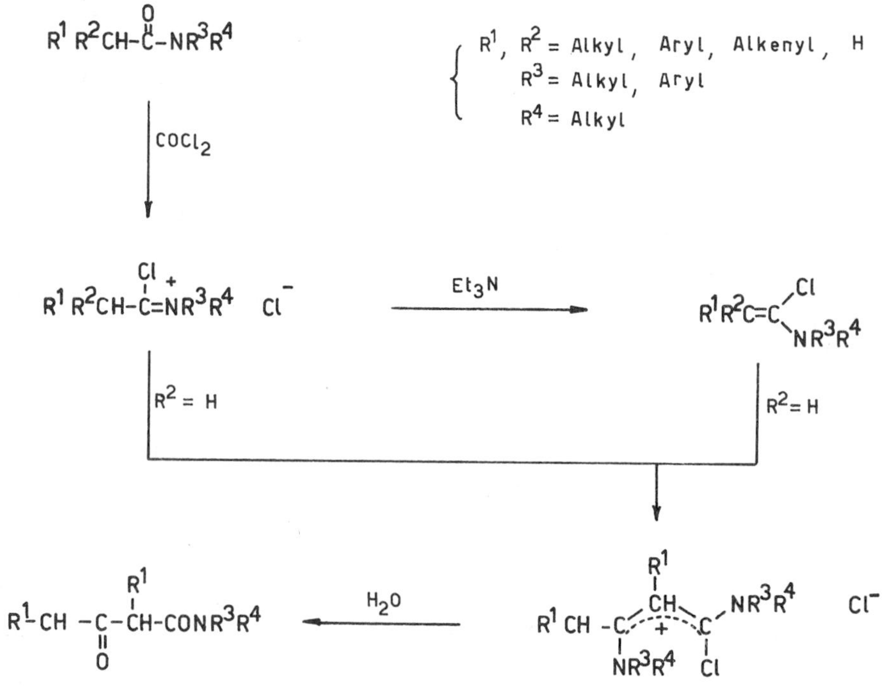

The experimental problem associated with the synthesis of β-monosubstituted
α-chloroenamines (R^2=H) arises from their greater reactivity as enolate equivalents.   In solvents like dichloromethane and chloroform, the elimination
step is accompanied by a condensation reaction linking the nucleophilic carbon
atom C-2 of the α-chloroenamine and the electrophilic carbon atom C-1.   This
equivalent of a Claisen condensation occurs in very mild conditions and can
be effected efficiently in one-pot using the appropriate quantity of triethylamine.   The reaction can be stopped at the stage of the intermediate bis-elec-
trophile which can be used in heterocyclic synthesis. The problem associated
with an unwanted condensation can be circumvented by using solvents like ether,
CCl$_4$ or petroleum ether : stable solutions (up to 1 molar) of monosubstituted
α-chloroenamines were obtained under these conditions (Ref.4).   These α-chloroenamines were often too unstable to be distilled and, therefore, were used
in situ.

α-HALOENAMINES AS ENOLATE EQUIVALENTS

The reactivity of α-haloenamines as masked enolates is summarized in the representative transformations illustrated in Scheme 3.(Ref.5).   The obvious
dependence of nucleophilic character at C-2 on the basicity of the amine substituent deserves no special comment.

Scheme 3

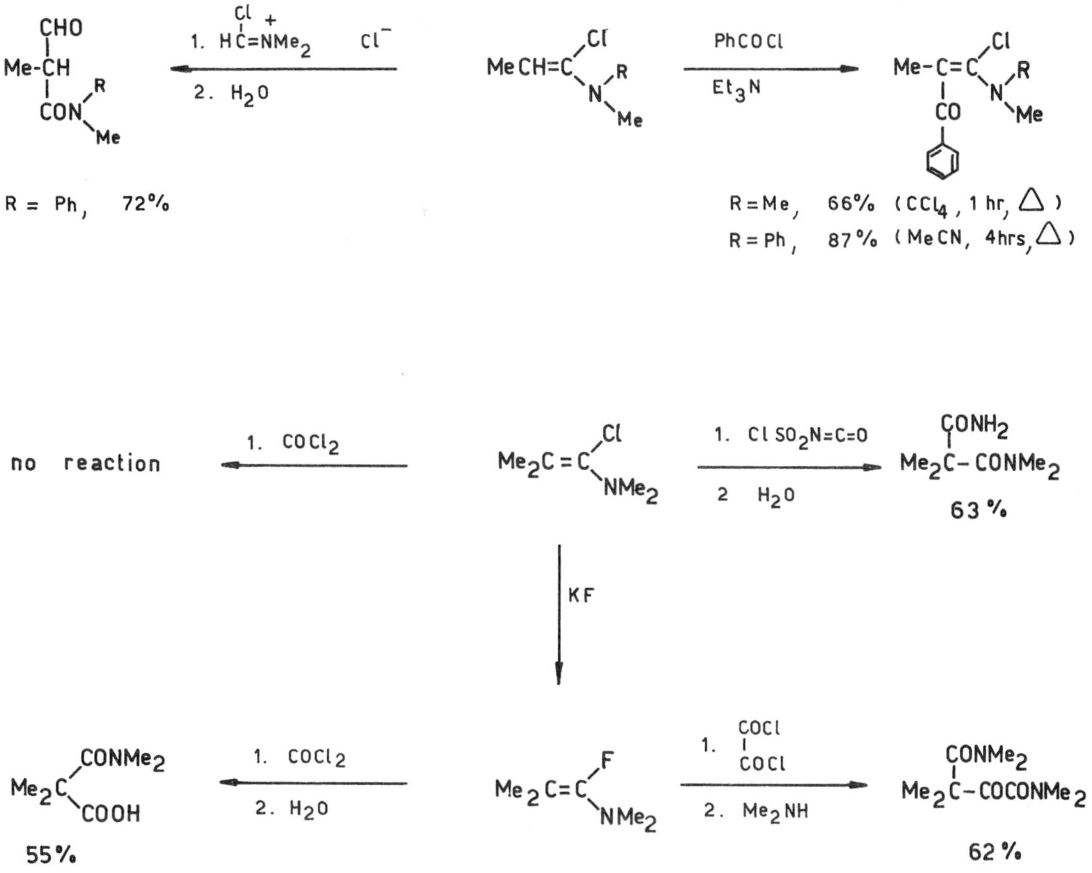

β-Disubstituted-α-chloroenamines are much less nucleophilic than the monosubstituted derivatives or the related enamines. An X-ray diffraction analysis has shown that, in these β-disubstituted molecules, the lone pair on the nitrogen is practically orthogonal to the π bond. The molecule (Scheme 4) is a cryptoiminium salt as shown by the shortening of the C-N bond and the lengthening of the C-Cl bond (Ref. 6).

Scheme 4

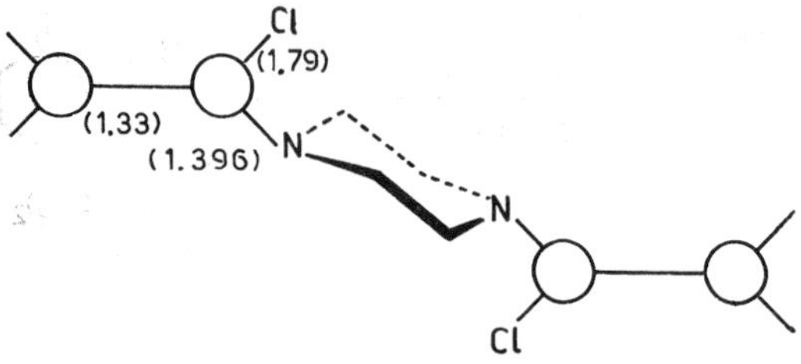

Substitution of chlorine by fluorine, a more electronegative atom but also a better π donor, could be easily done. The new enamine showed increased nucleophilicity at C-2 and could be used successfully as an enolate equivalent (Ref.7). Endocyclic α-chloroenamines showed a much higher nucleophilicity than their acylic counterparts as a result of the conformations of these rings which force the lone pair to interact with the π bond. It is significant that the nucleophilicity at C-2 decreases with increasing ring size and conformational mobility (Scheme 5)(Ref.8).

Scheme 5

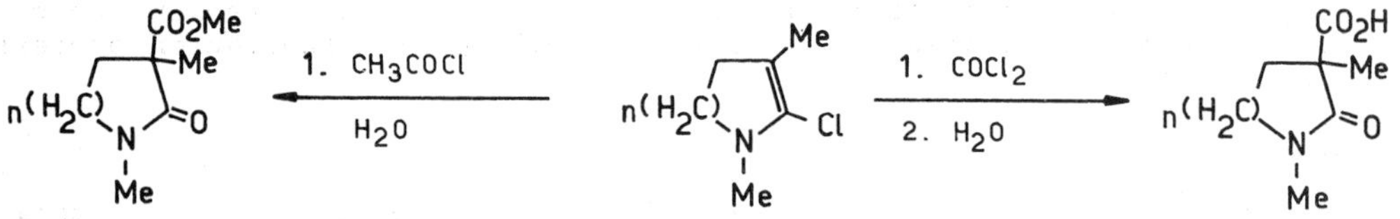

n = 1 ,   87% (0°C)

n = 2 ,   59% (20°C 7 hrs)

n = 3 ,   0% (△, 24 hrs)

n = 1 ,   87% (0°C)

n = 3 ,   76% (0°C)

n = 4 ,   72% (20°C, 24 hrs)

These simple synthetic transformations set up simple rules which should allow to select the appropriate enolate equivalents for a desired modification of the carbon skeleton of an amide. To illustrate the utility of these ideas, I would like to turn your attention to a spectacular example of such control of reactivity. 1-Chloro-1-dimethylaminoisoprene was prepared in 80% yields from tiglic or angelic amide. This activated isoprene behaved has a vinylketeneiminium chloride and readily underwent nucleophilic substitution reactions. This provides a simple and practical route toward a series of activated isoprenes (Scheme 6).

Scheme 6   **1,4   Pull-Push   Isoprenes**

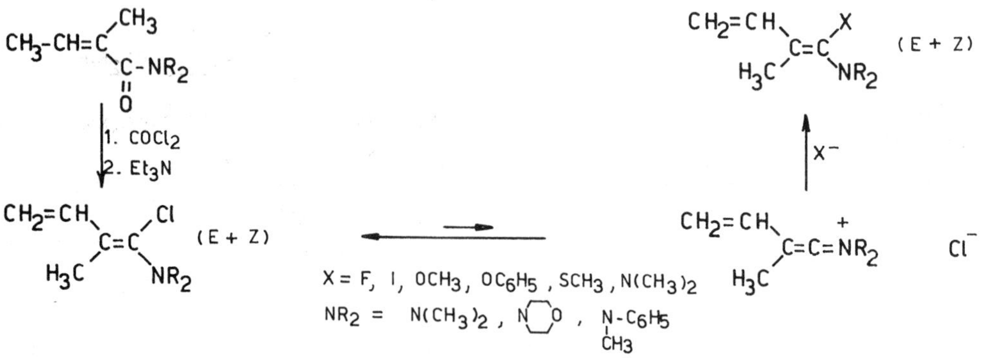

X = F, I, OCH$_3$, OC$_6$H$_5$, SCH$_3$, N(CH$_3$)$_2$

NR$_2$ =   N(CH$_3$)$_2$, N⬡O ,   N-C$_6$H$_5$
                                          |
                                        CH$_3$

Annulation of these new dienes with electron-poor olefins might open a facile route toward substituted cyclohexenones. The first experiments were quite frustrating since our 1-chloro-1-aminodiene did not react with a dienophile like acrylonitrile. However, when X=F, OMe or NMe$_2$, the annulation took place very smoothly to yield a cyclohexenone after hydrolysis (Scheme 7). The substitution of chlorine by iodine in the diene by addition of KI into the

Scheme 7

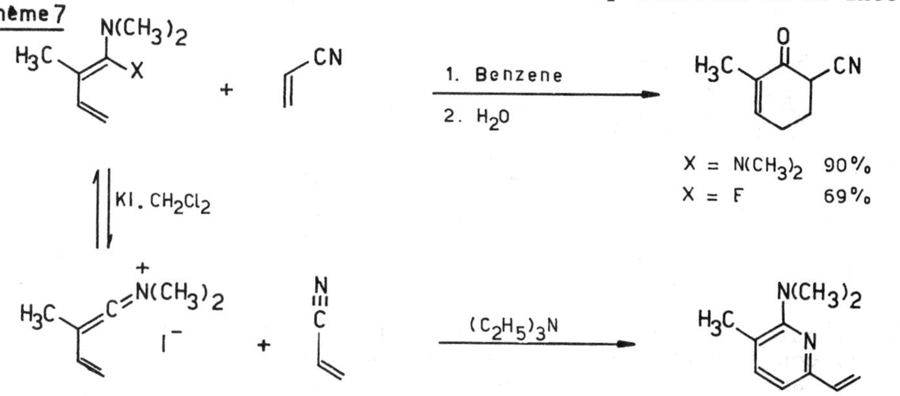

X = N(CH$_3$)$_2$ 90%
X = F   69%

mixture lead to a dramatic reversal of the reactivity : the reaction with acrylonitrile occured accross the C≡N bond to yield, after aromatization, a substituted pyridine. This unusual preference of a diene for the C≡N bond of acrylonitrile is readily explained by the propensity of the 1-iodo-1-amino-diene to ionize to a vinylketeneiminium iodide. Such an electrophilic diene is expected to be very reactive in Diels-Alder reaction with inverse demand. The full synthetic potential of this class of versatile dienes remains to be explored. A few more examples of adducts obtained by this annulation process are shown in Scheme 8.

Scheme 8

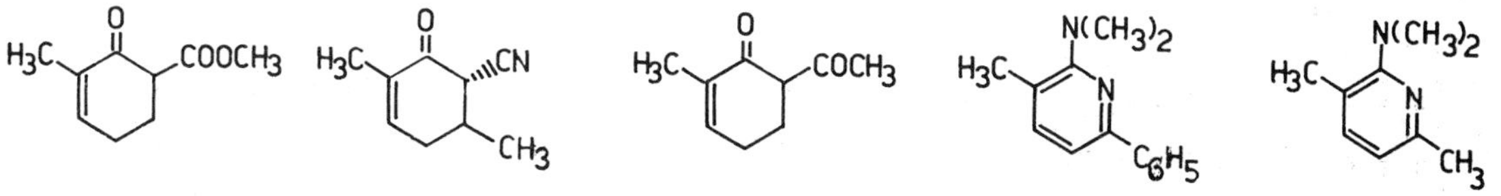

The mechanistic intricacies of the cycloadditions with electrophilic olefins are yet to be revealed. However the unexpected result of the cycloaddition of 1-dimethylamino-1-methoxy-isoprene with dimethyl acetylenedicarboxylate shed some light on this apparently trivial problem (Scheme 9). The substitution

Scheme 9

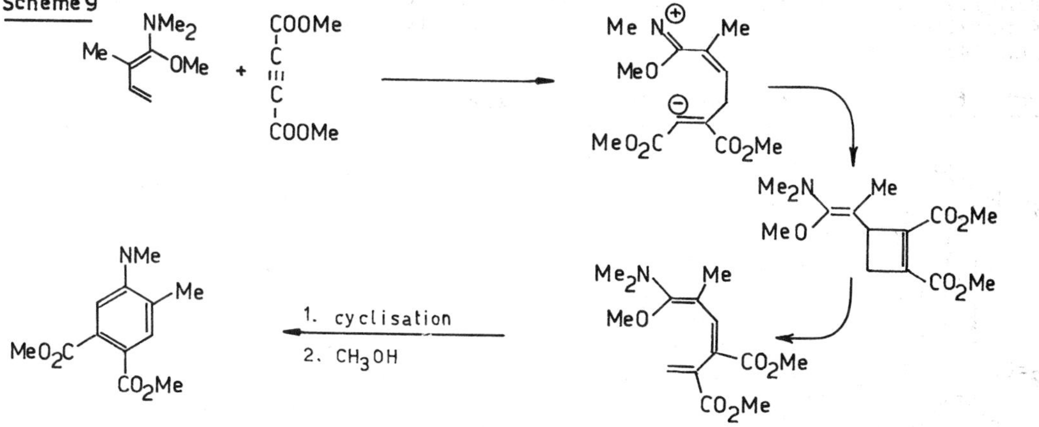

1. cyclisation
2. CH$_3$OH

pattern of the aromatized adduct was incompatible with a normal Diels-Alder
cycloaddition. However it can be explained by a mechanism involving (a) the
formation of a stabilized 1,6 dipole, (b) a preferential cyclization to a cy-
clobutene, (c) an electrocyclic opening of the four-membered ring, (d) an
electrocyclization to a 6-membered ring followed by aromatization (Ref.9).

KETENEIMINIUM SALTS AS ACYL CATION EQUIVALENTS

We now turn to a consideration of the use of α-haloenamines as sources of keteniminium salts. Some of the transformations discussed above have already demonstrated the capacity of α-chloroenamines to undergo nucleophilic substitution at C-1. The reaction of α-chloroenamines with carbon nucleophiles offers
potentially a *connective route to enamines*. This approach named "Electrophilic Aminoalkenylation" contrasts with the unusual methods of formation of enamines based on a functional group transformation (Ref.10). Numerous applications of the aminoalkenylation process may be readily perceived. A few examples are outlined in Scheme 10. Here also the success of the methodology resulted from the possibility of regulating the electrophilic reactivity of the
α-haloenamine. This could be done by varying the basicity of the amine substituent and by adding KI or $ZnCl_2$ to the α-chloroenamine.

Scheme 10   **Electrophilic Aminoalkenylation**

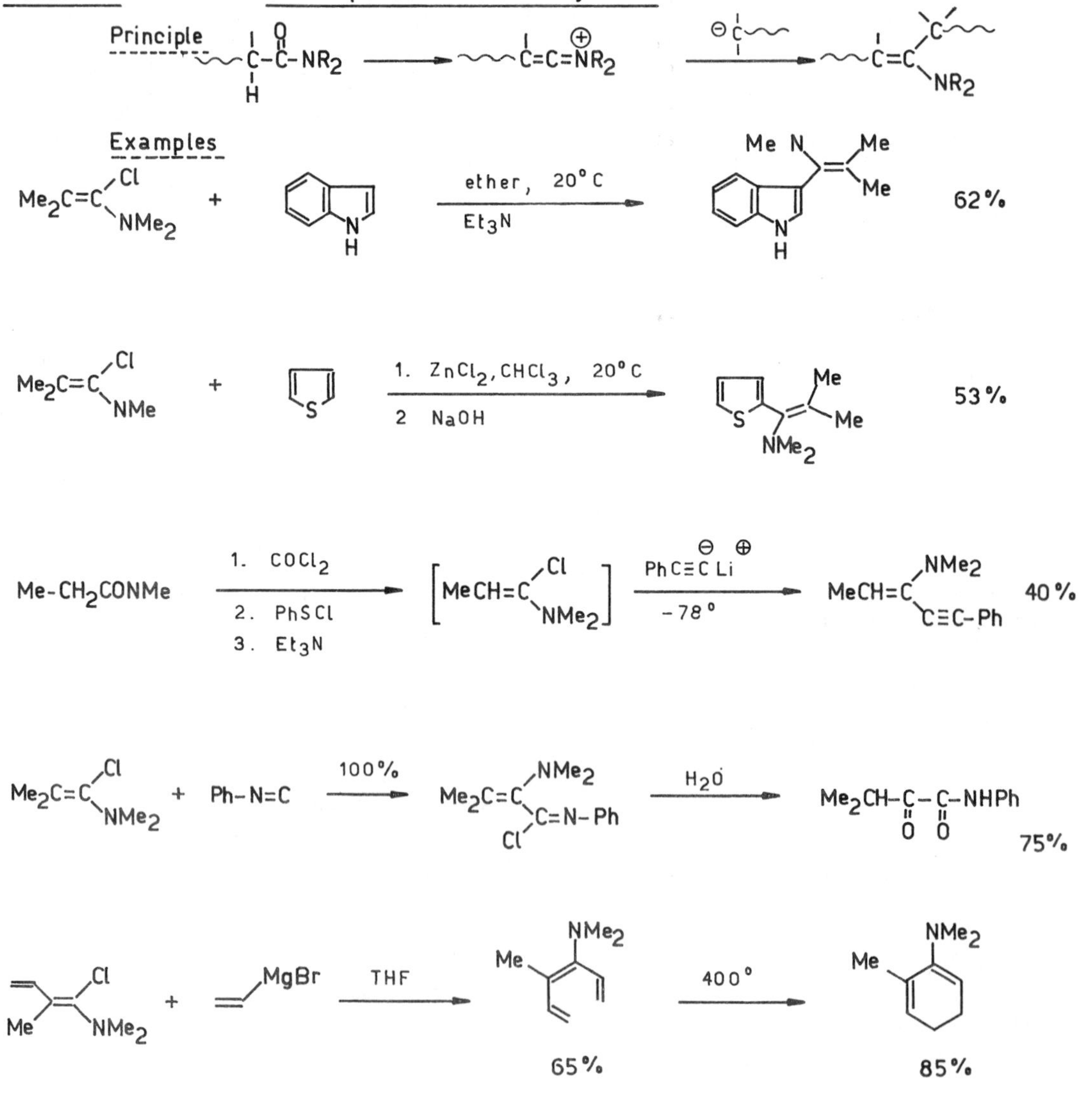

We have already reported on the remarkable ability of tetramethylketeneiminium salts to form [2+2] cycloadducts with olefins and acetylenes. Let me just remind you of their capacity to react with ethylene and acetylene at room temperature and atmospheric pressure (Ref.11).

Scheme 11

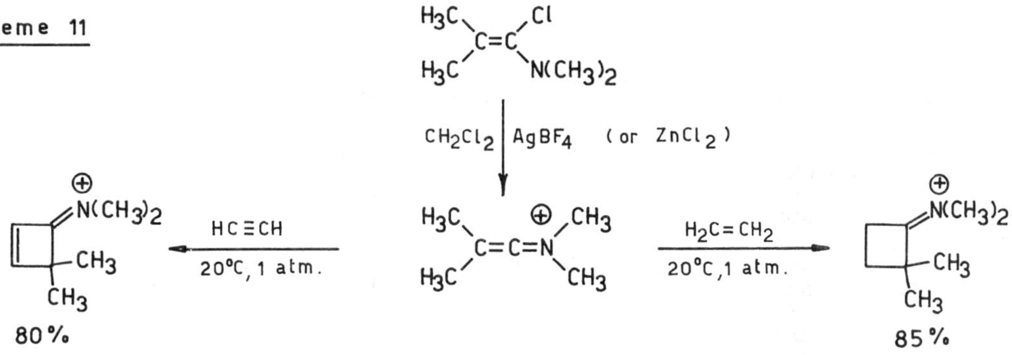

These results had generated real enthusiasm for their application in synthesis. Indeed we did not find too difficult to extend the reaction to other "keto" keteneiminium salts. However the results with the "aldo" keteniminium salts were quite disappointing. Apparently the generated aldoketeniminium salt reacted faster with its precursor α-chloroenamine than with the olefinic partner. Clearly the success of these cycloadditions lay upon the availability of a less nucleophilic precursor for the aldo-keteneiminium salts. We anticipated that this could be achieved by placing a triflate group (X=OSO₂CF₃) α to the amine. I must say that it was quite gratifying to obtain 60% yields (not optimized) of a pure [2+2] adduct by reacting N-dimethyl-propionamide with trifluoromethanesulfonyl anhydride and collidine in the presence of styrene (Scheme 12) (Ref. 12).

Scheme 12

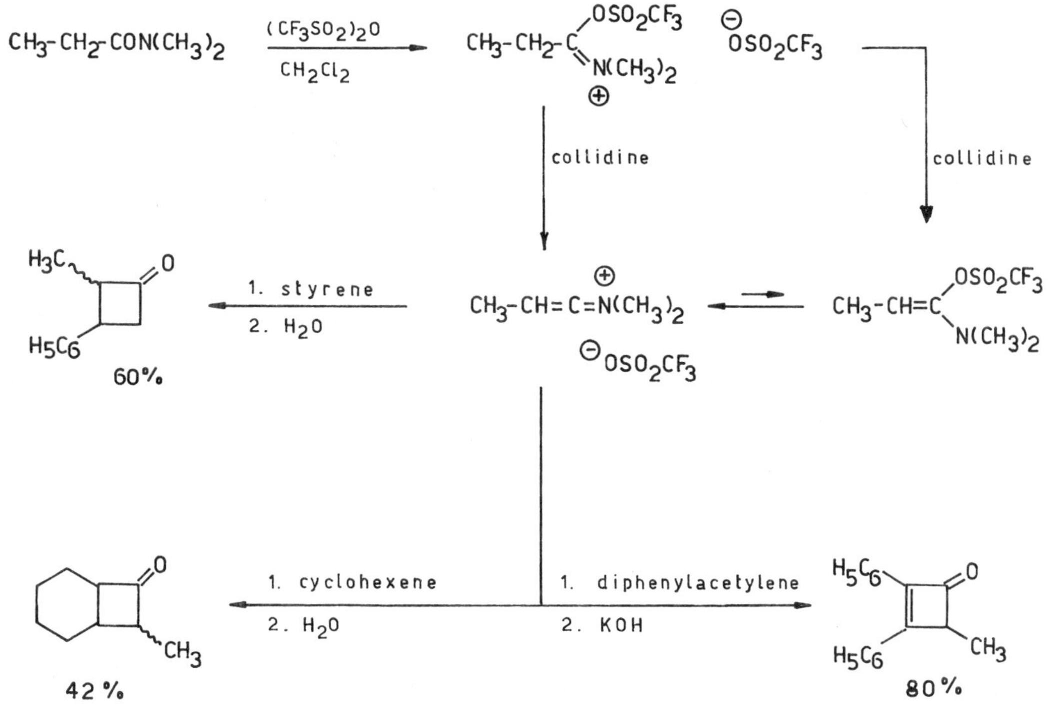

L. Ghosez

A few more cycloadditions involving rather poor ketenophiles have been
successfully performed using this experimental method. We feel thus rather
secure in predicting a broad scope for this method of synthesis of four-membered rings.

At this point I should perhaps justify our interest for these small carboxylic
rings. In recent years we have tried to develop a methodology for the vicinal
alkylation of olefins and acetylenes (Ref.13). The method involves the formation of a cyclobutanone followed by ring cleavage. Both steps should be
regio-and stereoselective. Representative examples (Scheme 13) illustrate the
facile and highly convergent construction of substituted cyclopentenes using
ketenes bearing excellent anion-stabilizing groups.

Scheme 13

Vicinal Alkylation of Olefins

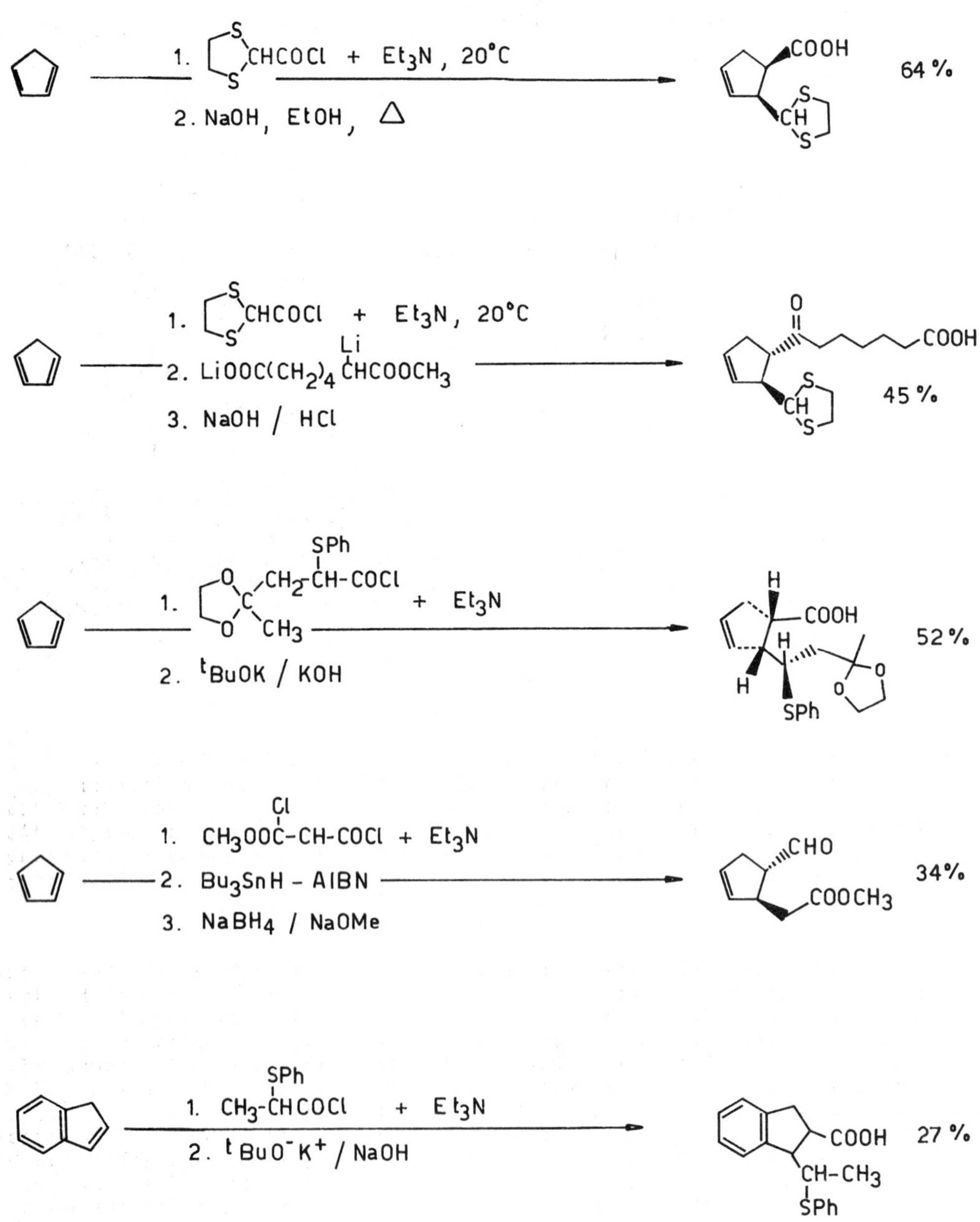

With reactive dienes, the sequence proceeded in good yields and with total
regio-and stereoselectivity. However, in spite of these encouraging results
leading to potentially useful prostanoid synthons, we did not feel entirely
satisfied because simple unactivated olefins reacted only sluggishly with
our ketene reagents. A partial solution to this problem originated from the
availability of α-acyloxyenamine from α-chloroenamines (Scheme 14) and carboxylic acids (Ref. 14). The rate of fragmentation of α-acyloxyenamines to
ketenes and amides was found to be strongly dependent on the amine substituent.

Scheme 14

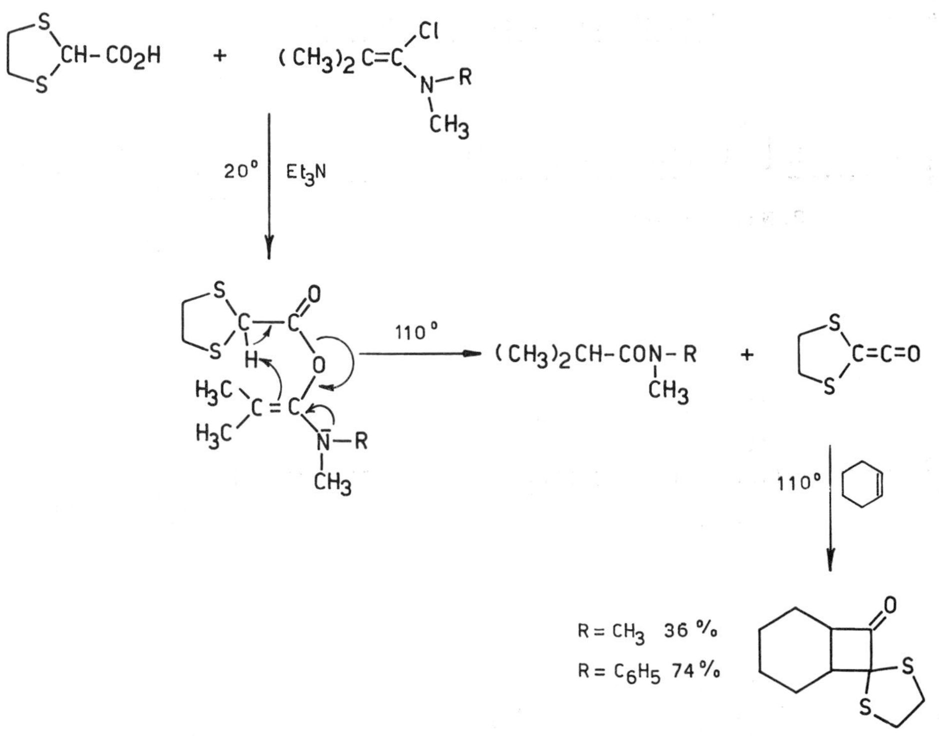

$R = CH_3$ 36 %

$R = C_6H_5$ 74 %

This offers a possibility of controlling the rate at which the ketene is released at a given temperature and thus, of favouring the cycloadditions by
minimizing the polymerization of the ketene. In practice, it was found that,
in going from R = Me to R = Ph, yields in adducts increased from 36% to 74%.
This method is applicable when the α-acyloxyenamines bear anion-stabilizing
substituents α to the carbonyl group. Otherwise they undergo a thermal re-
organization leading to a β-ketoamide (Ref. 1,13).

Keteneiminium salts were obvious candidates for the elaboration of these four-
membered rings. Let me illustrate further their synthetic potential by describing an example of vicinal alkylation of an acetylene (Scheme 15). The
desired α-chloroenamine was prepared from propionamide by successive reactions
with phosgene, isopropylsulfenyl chloride and triethylamine. Treatment of
this α-chloroenamine with silver tetrafluoroborate or zinc chloride gave
disapointing results (mostly tars). This probably results from competing
attacks of Lewis acids on sulfur and chlorine atoms. The use of the corresponding α-fluoroenamine and a harder acid such as BF_3 lead to a selective
ionization reaction to a keteneiminium salt and the formation of adducts with
substituted acetylenes (Ref. 15). With all these possible variations on this
theme of vicinal alkylation, we believe the method should be of wide applicability.

Scheme 15

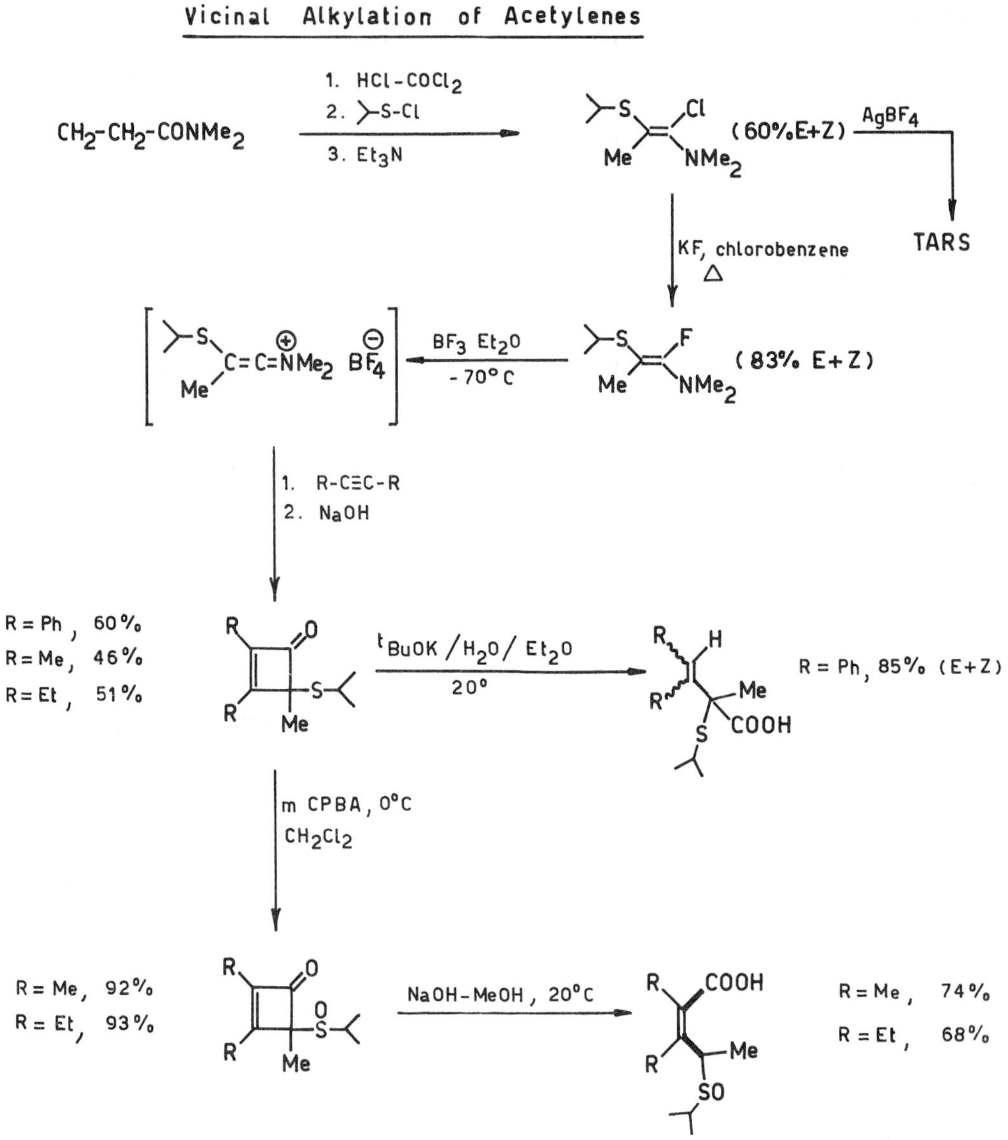

Vicinal Alkylation of Acetylenes

$$CH_2-CH_2-CONMe_2 \quad \xrightarrow[\text{3. Et}_3N]{\substack{\text{1. HCl-COCl}_2 \\ \text{2. }\rangle\text{-S-Cl}}}$$

(60% E+Z)

KF, chlorobenzene
Δ

(83% E+Z)

BF_3 Et$_2$O
-70°C

TARS

AgBF$_4$

1. R-C≡C-R
2. NaOH

R = Ph , 60%
R = Me , 46%
R = Et , 51%

^{t}BuOK /H$_2$O/ Et$_2$O
20°

R = Ph, 85% (E+Z)

m CPBA, 0°C
CH$_2$Cl$_2$

R = Me, 92%
R = Et, 93%

NaOH-MeOH, 20°C

R = Me,   74%
R = Et ,   68%

α-METALLED ENAMINES AS ACYL ANION EQUIVALENTS

We now turn to a discussion of an attractive possibility that consists of inverting the polarity of the C-1 atom of α-chloroenamines by exchanging the chlorine for an electropositive substituent. The new synthon, an enamine anion, is equivalent to an acyl anion. Its reactions with electrophilic carbons can be viewed as another potential connective synthesis of enamines that we may call "the Nucleophilic Aminoalkenylation Reaction". (Scheme 16).

Scheme 16

Nucleophilic Aminoalkenylation

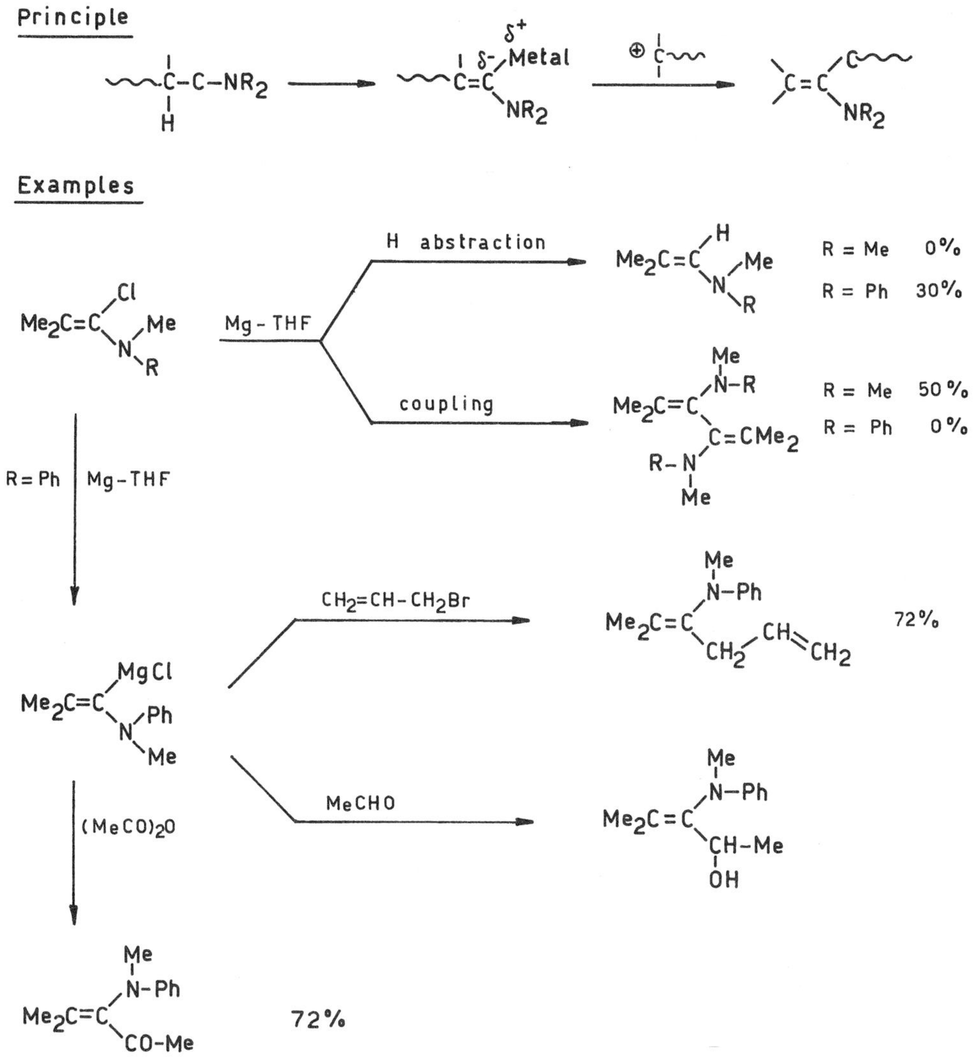

Principle

Examples

Our expectations were only partially fulfilled. With the highly electrophilic
tetramethyl-α-chloroenamine, a reaction with magnesium took place in boiling
THF but a coupling product was obtained. The replacement of the dimethylamino
substituent by a less basic group was known to decrease the electrophilic character to the α-chloroenamine. Correspondingly the reaction of less basic
α-chloroenamines with magnesium produced less coupling and yielded more of the
Grignard reagents. The sequence *N-methylanilino > morpholino > N-dimethylamino*
is just in the opposite order of the keteniminium character of the α-haloenamine. The reaction of the aniline derivative with magnesium gave the Grignard
reagent (78%, titration) and an enamine resulting from hydrogen abstraction
from the solvent. This new Grignard reagent could be coupled to a variety of
carbon electrophiles (Ref. 16).

The scope of the reaction appears to be somewhat limited since we were unable
to generate Grignard reagents in practical yields from less substituted enamines. However, recent results rose again our interest for this problem (Ref.17).
We found that Grignard reagents could be readily prepared from α-chloroenamines
derived from monoesters of malonamides (Scheme 17).

Scheme 17

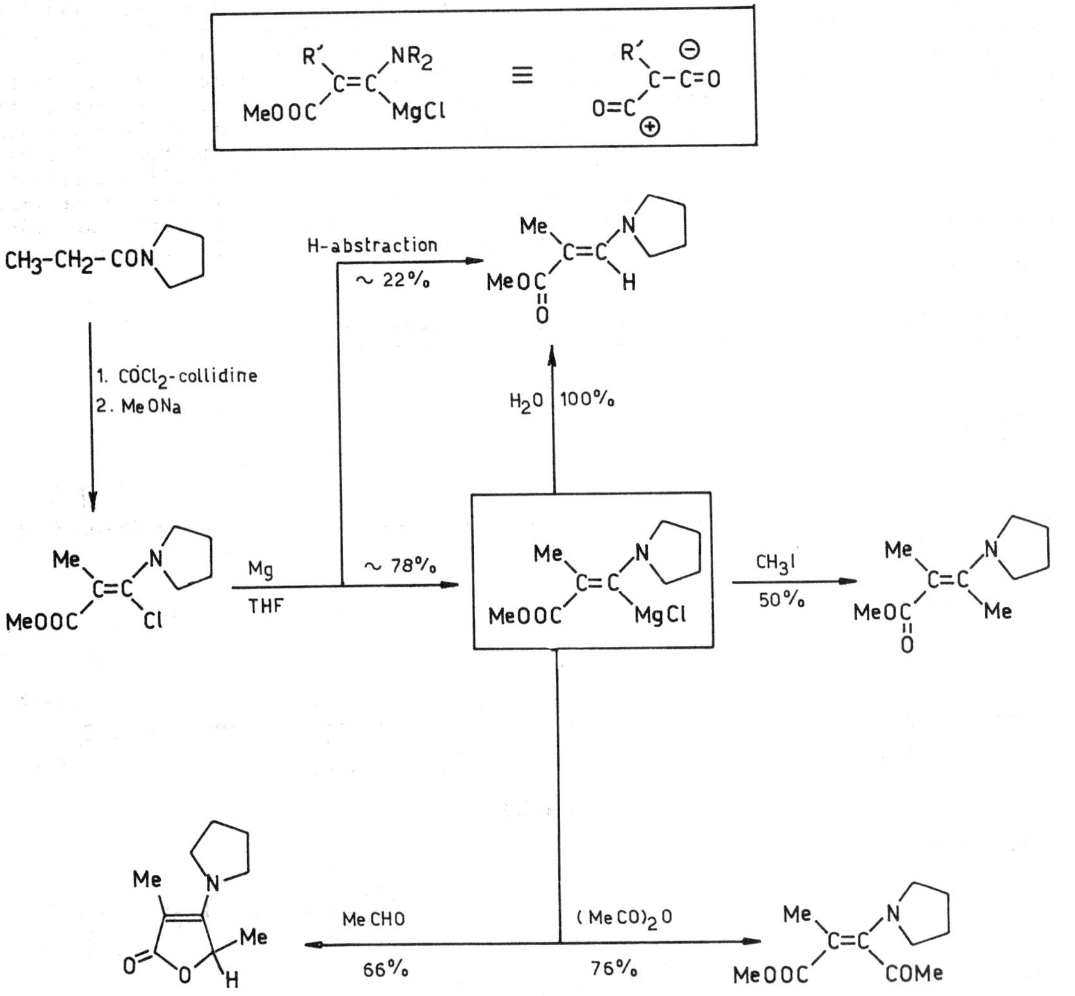

However in this series, the highest yields were obtained with the most basic
α-chloroenamines, the sequence being *pyrrolidino (72%) > dimethylamino (66%)
> morpholino (33%) > N-methylanilino (< 10%)*. The enamines resulting from
hydrogen abstraction always accounted for the remaining material. Thus,
whereas more ionic character (n→σ* interaction) in the α-chloroenamine disfa-
vours formation of the Grignard reagent, stronger delocalization of the nitrogen lone pair into the π bond (n→π* interaction) favours this formation. It
is worth noting that the ester group remained unchanged in these reactions.
The Grignard reagent reacted readily with various carbon electrophiles and
could thus be used as a malonyl anion equivalent. The reaction with aldehydes
or ketones takes benefit from the presence of the electrophilic ester group
and leads to a substituted butenolide.

α-CYANOENAMINES AS HOMOENOLATE EQUIVALENTS

Through the various reagents discussed above, carbon-carbon bond could be
formed at the C-1 and C-2 atoms of a tertiary amide. An important advantage
of these synthons results from the presence of a double bond between C-1 and
C-2 : the C-3 atom which is unreactive in the original amide has become an
allylic carbon atom. It is clear that removal of a proton from C-3 should be
greatly facilitated when X is an anion-stabilizing group. Therefore we set
ourselves to prepare α-cyanoenamines which fulfill this condition. A simple
method of synthesis of α-cyanoenamines involves the treatment of tertiary amides with phosgene, zinc cyanide and triethylamine. The method applies to the
synthesis of a wide variety of structures with the exception of the derivatives of acetamides which are nevertheless available by other methods (Scheme
18). It was found that organo lithium or Grignard reagents readily added to
the nitrile group of α-cyanoenamines. Hydrolysis of these adducts gave α-diketones. The overall sequence provides a simple general and versatile process for the stepwise construction of the carbon skeleton of α-diketones
(Ref. 18).

Scheme 18   A   Synthesis of α-Diketones

Principle

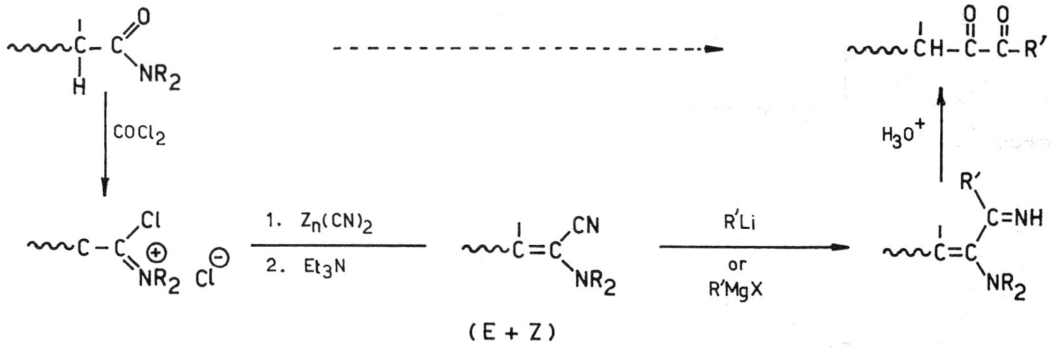

Examples

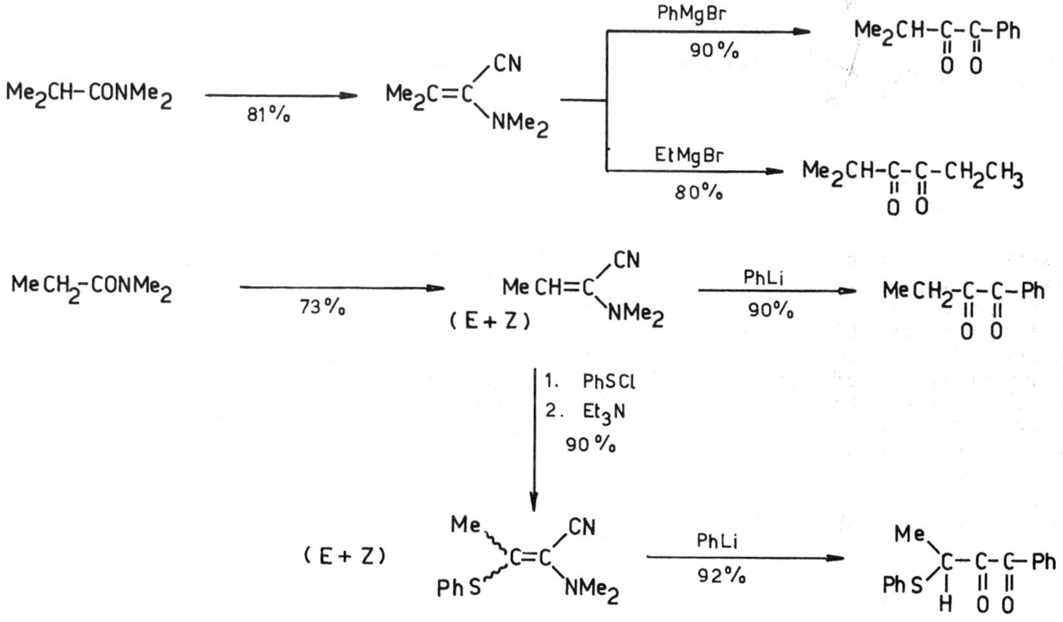

The metallation of α-cyanoenamines could be easily effected with non nucleophilic strong bases. Lithium diisopropylamide in THF at -30° gave a lithiocyanoenamine formed of two interconvertible E and Z isomers. These new organolithium reagents were treated with various carbon electrophiles (Scheme 19) (Ref.19).

Scheme 19

Three Carbons Chain Extensions with Homoenolate Equivalents

Principle

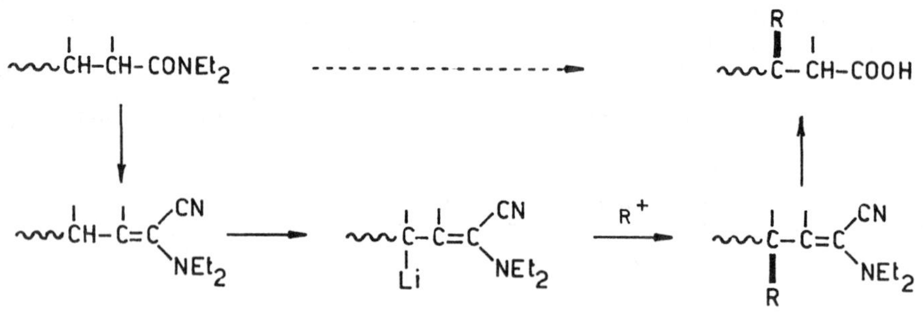

Examples

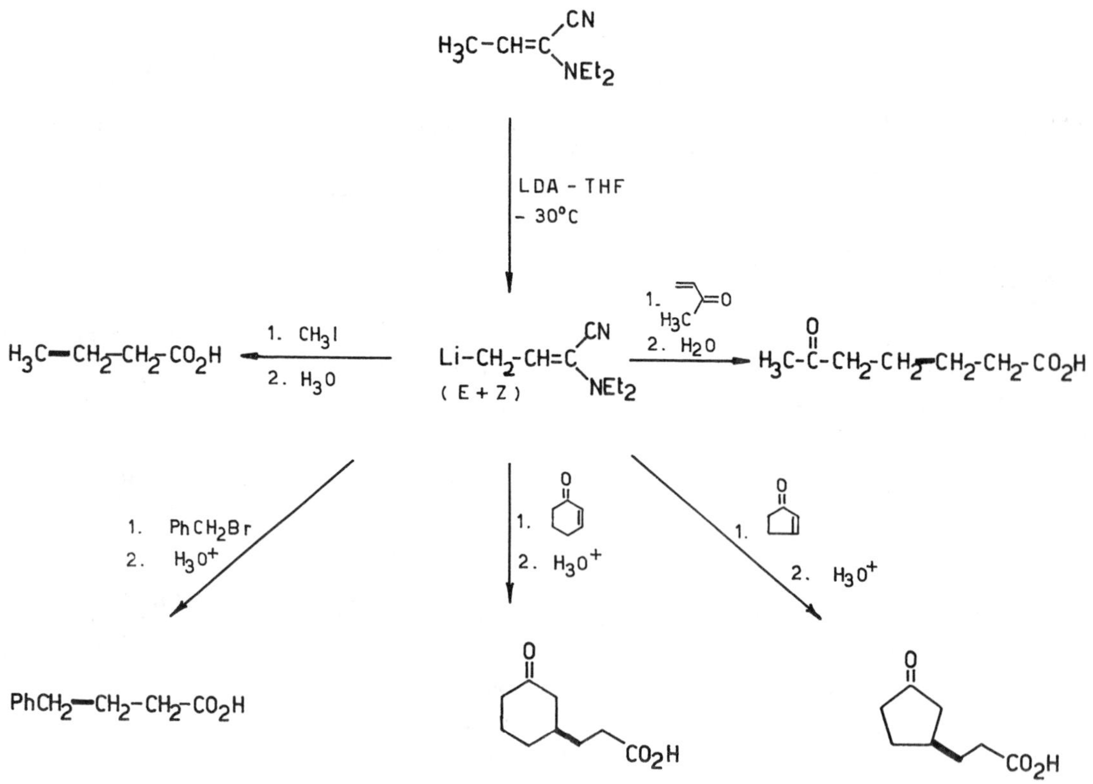

The results can be summarized as follows :

1. The site of alkylation was exclusively the γ-carbon atom when the amino substituents were ethyl groups. A mixture of γ-(predominant) and α-alkylated products were obtained when the substituents on nitrogen were methyl groups. This behaviour contrasts with the exclusive α-alkylation of the related α-cyanoethers (Ref.20).

2. α-enones such as cyclohexenone, cyclopentenone or methylvinylketone yielded mainly 1,4 adducts. With the more substituted enones, more 1,2 addition was observed. However it could be minimized by allowing the kinetic 1,2 adducts to equilibrate to the more stable 1,4 adducts.

3. Hydrolysis of the alkylated α-cyanoenamines in 6N HCl smoothly yielded the corresponding acids.

Thus, we have at our disposal a sequence which allows for a three carbon chain extension at the β-carbon atom of tertiary amides. We believe that these lithio cyanoenamines are versatile homoenolate equivalents which should find wide use in organic synthesis.

To further extend the scope of these homoenolate equivalents, we set ourselves to prepare α-cyanoenamines bearing a phenylthio group at the γ-position (Scheme 20). These are potentially equivalent to a vinyl anion at the

Scheme 20

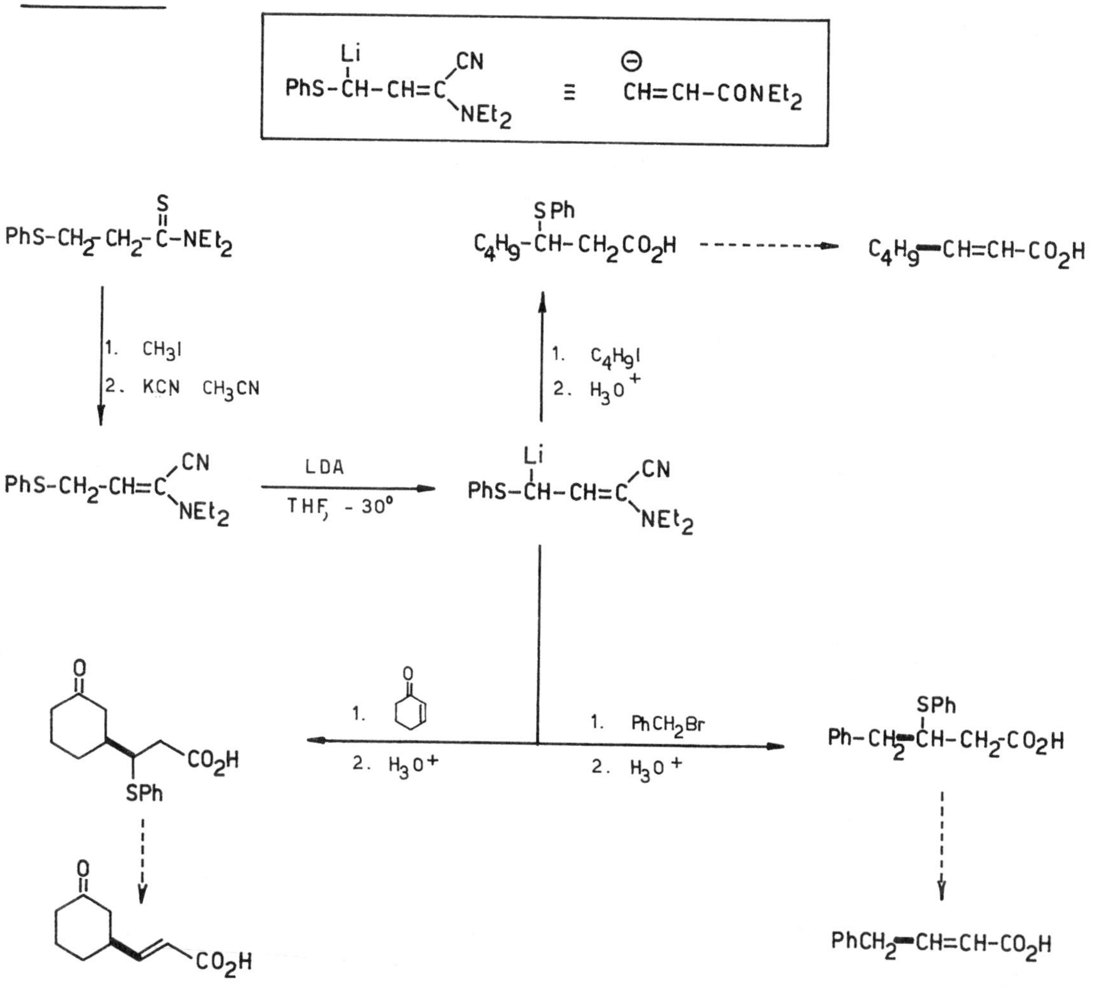

β-position of an unsaturated amide. A representative of this class of synthors
was easily prepared from the appropriate thioamide. Metallation occured
smoothly to yield the corresponding lithio cyanoenamine which reacted exclusively at the γ-position (Ref.21). Hydrolysis gave an acid bearing a phenylthio
group at the β-carbon atom, allowing thus the introduction of the conjugated
double bond by a well-established procedure (Ref.22).

α,β DEHYDROGENATION OF CARBOXAMIDES

The final study to be described in this lecture relates to the dehydrogenation
of saturated fatty acids to the synthetically more useful α,β unsaturated derivatives. They are only few viable methods to effect such a transformation.

As stated before and illustrated by the formation and use of lithio cyanoenamines, the synthons described above allow the activation of the C-3 carbon
atom of the original amide. This consideration established the ground for
a new method of dehydrogenation of carboxamides. The desired synthon should
be an α-heterosubstituted enamine where X is an oxygen atom linked to a suitable leaving group, e.g. an ammonium or a sulfonium ion (Scheme 21).

Scheme 21

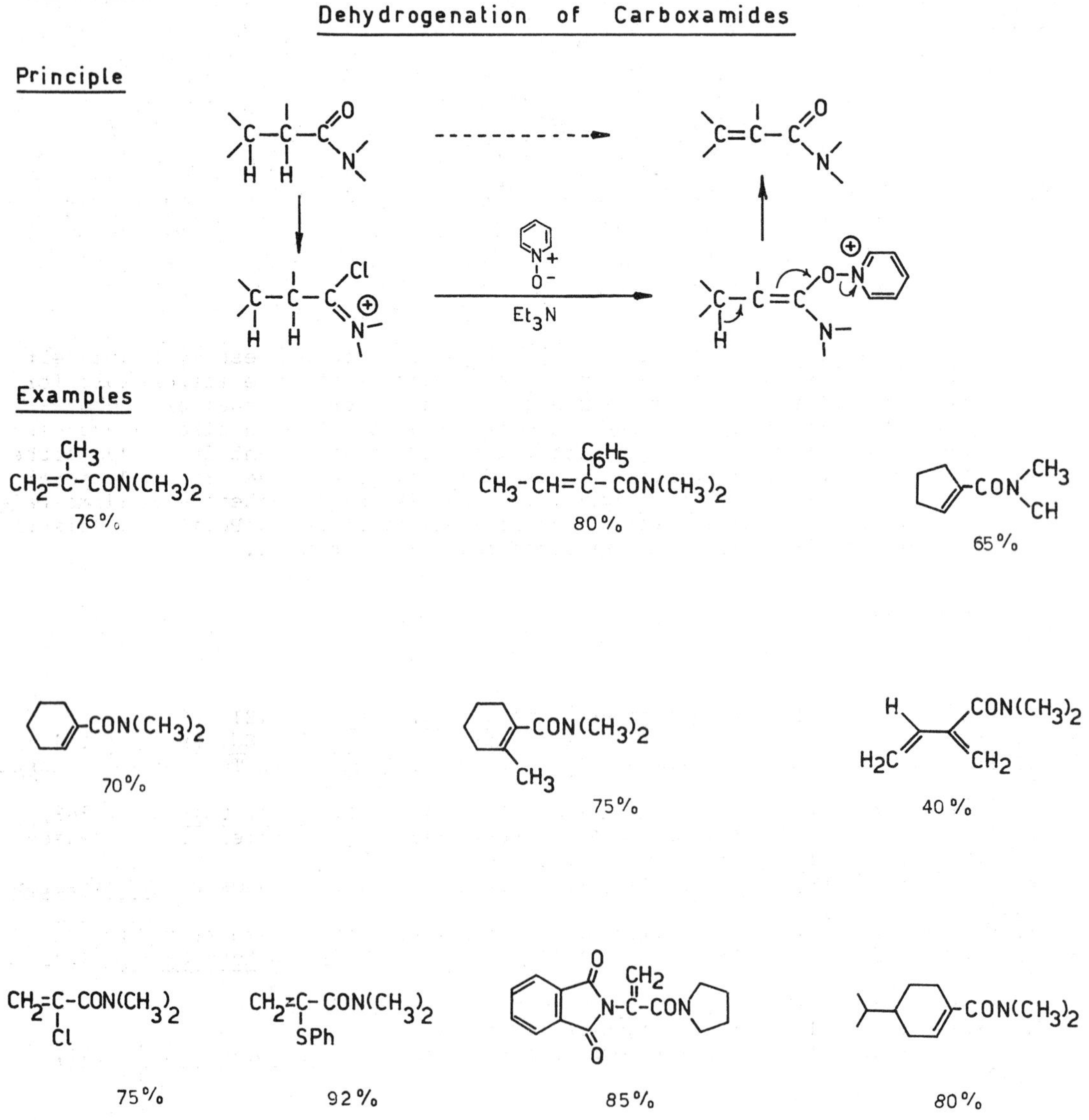

Dehydrogenation of Carboxamides

Principle

Examples

The reaction of pyridine-N-oxide with α-chloroenamine was found to produce the required synthon which, as expected, gave the unsaturated amide on treatment with triethylamine. This result lead us to develop a one-flask procedure for α,β dehydrogenation of various amides (Ref. 23). The amides were converted into the amide chlorides with phosgene. Treatment of the crude amide chlorides with pyridine-N-oxide and triethylamine gave α,β unsaturated amides in good yields. These can be easily transformed into the α,β unsaturated acids by the mild procedure described by Professor Gassman (Ref. 24). The method could be applied to the preparation of useful functionalized acrylamides. However it does not presently apply to the dehydrogenation of monosubstituted acetamides. In these cases α-addition of a chloride ion competes with the elimination reaction.

FINAL REMARKS

Here ends this survey of new methods for the formation of carbon-carbon bonds based on simple transformations of amides. I intended to show you how to manipulate the structure of α-heterosubstituted enamines in order to effect a given transformation with high chemo- and regioselectivity. By applying the rules we have discussed today,one should be able to effect specifically various modifications of the skeleton of carboxamides as shown in Scheme 22.

Scheme 22

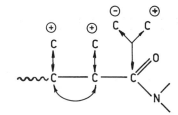

Acknowledgements - In closing, I would like to express my heart-felt thanks to my many co-workers and co-thinkers who are responsible for the evolution of this research programme. Most of them are acknowledged individually in the cited references. I wish also to express my appreciation to the "Institut pour l'Encouragement de la Recherche Scientifique dans l'Industrie et l'Agriculture", the "Fonds National de la Recherche Scientifique", the "Fonds de la Recherche Fondamentale et collective", the "Service de Programmation de la Politique Scientifique" and "U.C.B." who supported most of this work.

REFERENCES

1. L. Ghosez and J. Marchand-Brynaert, Adv. Org. Chem. 9, 421 (1976).
2. L. Ghosez, B. Haveaux and H.G. Viehe, Angew. Chem. Int. Ed. Engl. 8,454 (1969); B. Haveaux, A. Dekoker, M. Rens, A.R. Sidani, J. Toye and L. Ghosez, Org. Syn. in press.
3. A. Colens, M. Demuylder, B. Téchy and L. Ghosez, Nouv. J. Chim., 1, 369, (1977); A. Devos, J. Remion, A.M. Frisque-Hesbain, A. Colens and L.Ghosez, J. Chem. Soc., Chem. Comm., 1180 (1979).
4. M. Henriet, M. Houtekie, B. Téchy, R. Touillaux and L. Ghosez, Tetrahedron Lett., 223 (1980)
5. M. Houtekie, Li-Te-Fu, B. Téchy and L. Ghosez, unpublished results.
6. M. Van Meerssche, G. Germain, J.P. Declercq, A. Colens, Acta Cryst. B.35 983 (1979).
7. A. Colens and L. Ghosez, Nouv. J. Chim. 1, 37I (1976).
8. M. Staelens and L. Ghosez, unpublished results.
9. M. Gillard C. t'Kint, E. Sonveaux and L. Ghosez, J. Am. Chem. Soc. 101, 5837 (1979).

10. J. Marchand-Brynaert and L. Ghosez, J. Am. Chem. Soc. 94, 2869 (1972);
    Bui-Xuan-Hoa and L. Ghosez, unpublished results.
11. J. Marchand-Brynaert and L. Ghosez, J. Am. Chem. Soc. 94, 2870 (1972);
    A. Sidani, J. Marchand-Brynaert, and L. Ghosez, Angew. Chem. Int. Ed. 13,
    267 (1974); C. Hoornaert, A.M. Hesbain-Frisque and L. Ghosez, Angew. Chem.
    Int. Ed. 14, 569 (1975).
12. J. B. Falmagne and L. Ghosez, unpublished results.
13. L. Ghosez, "Stereoselective Synthesis of Natural Products", Proceedings of
    the Seventh Workshop Conference Hoechst, Ed. W. Bartmann and E. Winter-
    feldt, Excerpta Medica, Amsterdam-Oxford (1979); E. Cossement, R. Binamé,
    and L. Ghosez, Tetrahedron Lett. 997 (1974); Ph. Michel, M. O'Donnell,
    R. Binamé, A.M. Hesbain-Frisque, L. Ghosez, J.P. Declercq, G. Germain and
    M. Van Meerssche, Tetrahedron Lett. in press; P. Vannes, S. Goldstein, C.
    Houge, C. Zamar-Wiaux and L. Ghosez, unpublished results.
14. J. Rémion, J. P. Dejonghe and L. Ghosez, unpublished results.
15. S. Sahraoui-Taleb and L. Ghosez, unpublished results.
16. C. Zamar-Wiaux, J. P. Dejonghe, L. Ghosez, J. F. Normant and J. Villieras,
    Angew. Chem. Int. Ed. 15,371 (1976).
17. E. Arguedas-Chaverri and L. Ghosez, unpublished results.
18. J. Toye and L. Ghosez, J. Am. Chem. Soc. 97, 2276 (1975); M. Baudhuin and
    L. Ghosez, unpublished results.
19. B. Lesur, J. Toye, M. Chantrenne, and L. Ghosez, Tetrahedron Lett. 2835
    (1979).
20. G. Stork, L. Maldonado, J. Amer. Chem. Soc. 93, 5286 (1971); V. Herten-
    stein, S. Hunig, M. Oller, Synthesis, 416 (1976).
21. B. Lesur, S. De Lombaert, M. Chantrenne and L. Ghosez, unpublished results.
22. B.M. Trost, T.N. Salzmann and (in part) J. K. Hiroi, J. Am. Chem. Soc. 98,
    4887 (1976).
23. R. Da Costa, M. Gillard, J. B. Falmagne and L. Ghosez, J. Am. Chem. Soc.
    101, 4381 (1979).
24. P.G. Gassman, P.K.G. Hodgson, R.J. Balchunis, J. Am. Chem. Soc. 98, 1275
    (1976).

NEW REAGENTS BASED ON HETEROATOM-FACILITATED LITHIATIONS

H. W. Gschwend

Research Department, Pharmaceuticals Division, CIBA-GEIGY Corp., Summit,
NJ 07901, USA

Heteroatom facilitated lithiations have proven to be exceptionally useful in the regio-selective and very often regiospecific functionalization of carbocyclic and heterocyclic aromatic as well as olefinic substrates. Our attention has focused predominantly on the utility and improvement of methodologies for the efficient ortho lithiation of benzenoid systems as well as the exploitation and development of new and synthetically useful direct-ing groups. A schematic illustration of these reactions is shown in Fig. 1 (Ref. 1).

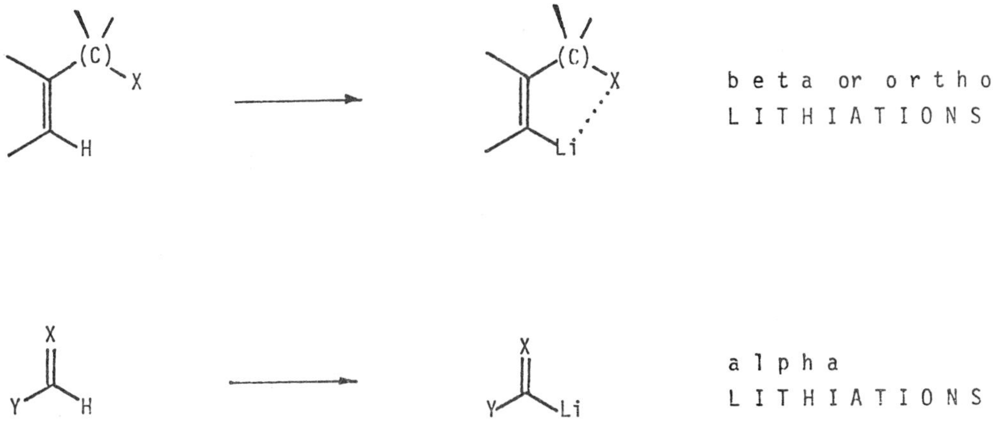

Fig. 1

Beta or ortho lithiations are reactions in which a heteroatom or group of heteroatoms attached directly or indirectly to an olefinic or aromatic π-system directs the metalating agent to deprotonate the beta or ortho position. Conversely, in alpha lithiations depro-tonation occurs on the same sp^2-carbon to which the heteroatom X is attached.

A rather general and in many instances synthetically useful side reaction in ortho lithia-tions is the deprotonation of an ortho alkyl group as illustrated in Fig. 2. (Ref. 2)

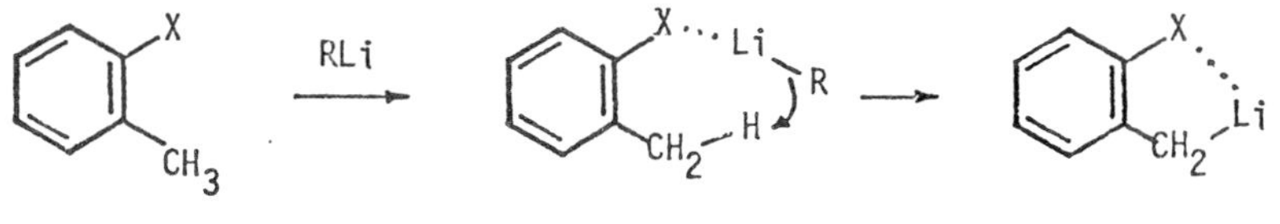

Fig. 2

It is noteworthy that most of these ortho'deprotonations occur quite readily and compete successfully with regular ortho lithiations. Whereas in the derivatives of o-toluic acid, the benzylic hydrogen atoms exhibit an increased acidity relative to toluene, this is most certainly not the case in the substrates where the heteroatom X is directly attached to the aromatic nucleus. The metalation must be envisaged to proceed <u>via</u> precoordination of the metalating agent with the heteroatom followed by internal protophilic attack, resulting in the formation of an internally chelated organolithio species.

In an effort to further investigate the scope of these ortho' or beta' lithiations, our recent interest has focused on the potential ability of heteroatoms or groups of heteroatoms to direct and/or facilitate deprotonation of alkyl groups in allylic systems. Specifically we had an active interest in the generation of allylic anions derived from methacrylic acid and reduced versions thereof. It certainly appeared feasible to utilize a heteroatom X (Fig.3)

163

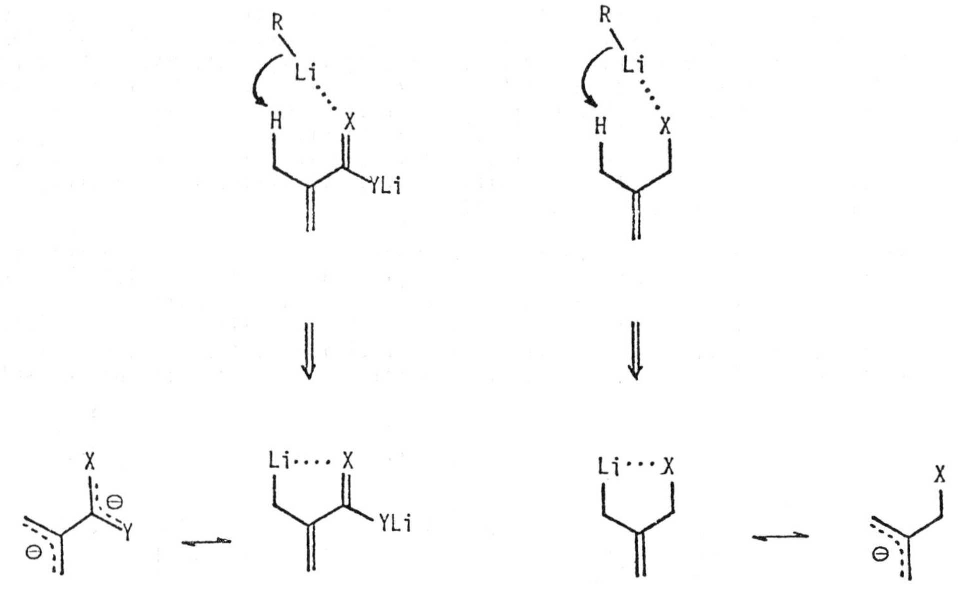

Fig. 3

in both of these substrates to facilitate deprotonation of the allylic position to form the synthetically useful allylic anions. Previous efforts towards the generation of a dianion of methacrylic acid or a derivative thereof have not been encouraging from a preparative point of view. Double deprotonation of methacrylic acid itself resulted in the rapid formation of dimers and polymers. Such observations were made by Carlson (Ref. 3) as well as ourselves. Derivatives of the carboxylic acid function, which have served rather well as directing groups in ortho, ortho' and beta metalations of aromatic systems, are the oxazoline and carboxamides (Ref. 1). Oxazolines have been particularly useful largely because of their inertness towards the metalating agents. The oxazoline of methacrylic acid (Fig. 4)

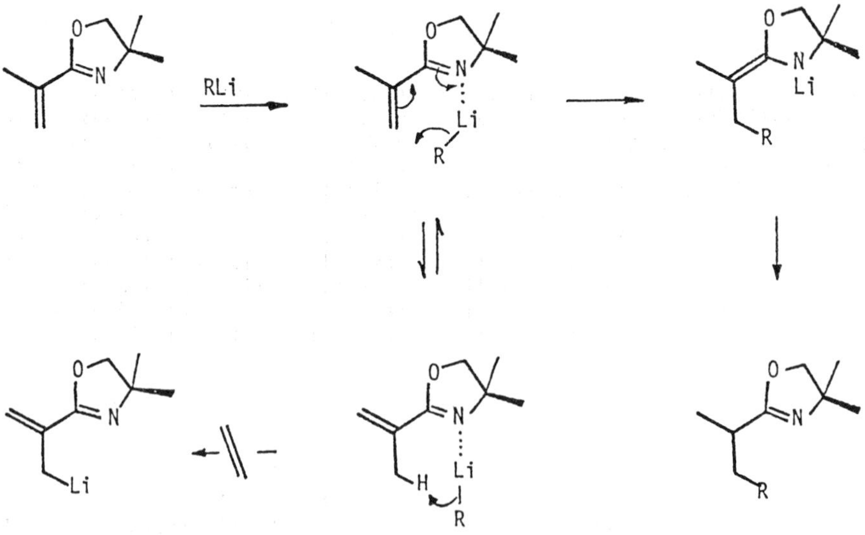

Fig. 4

can be obtained in pure form as a volatile liquid with considerable polymerization tendencies. However, it is clearly not a useful substrate for the generation of the desired anion, since 1,4-addition of an organolithium reagent, including the sterically demanding lithium tetramethylpiperidide, is the predominant pathway. This mode of reaction is the basis of Meyers' exceedingly useful and efficient asymmetric induction via 1,4-addition to chiral oxazolines (Ref. 4).

The only remaining readily accessible derivatives of carboxylic acids which provided some hope of achieving the practical generation of the desired anion are the carboxamides. In aromatic systems, certain tertiary carboxamides have proven extremely effective directing groups in facilitating ortho metalation (Ref. 5). Hauser's ortho lithiation of secondary arylcarboxamides in fact constitutes the first and original approach towards the direct and regiospecific ortho functionalization of benzoic acids (Ref. 6). Utilization of amides of methacrylic acid for the generation of the desired allylic anion is indeed feasible. Kauffmann and coworkers (Ref. 7) were able to effect deprotonation of the methyl group of the diisopropylamide of methacrylic acid with lithium diisopropylamide to generate the corresponding allylic anion. Its synthetic utility, though, is thwarted by an exceedingly rapid polymerization via intermolecular Michael addition.

With this rather extensive background of negative findings, it was concluded that we needed a carboxylic acid derivative which would not only preclude 1,2- but also 1,4-addition. A sterically hindered, secondary carboxamide seemed to have the most promise. Indeed, upon treatment of the t-butylamide of methacrylic acid 1 with n-butyllithium/TMEDA in tetrahydrofuran, first at -70° then at -20° for one hour, an optimal amount of the desired dianion 2 is formed (Fig. 5). This is illustrated by its addition to benzophenone and the isolation

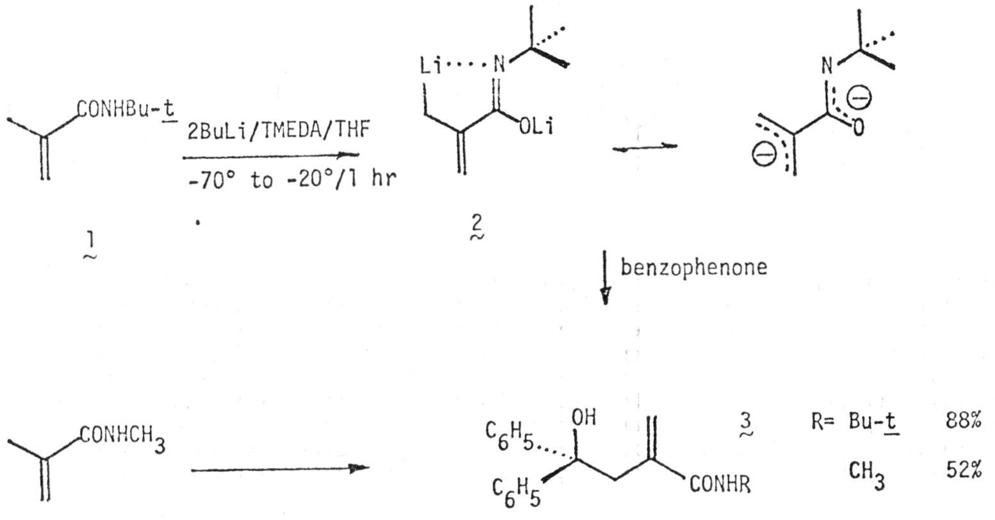

Fig. 5

of the corresponding adduct 3 in 88% yield. In the absence of TMEDA no anion is formed at -70°, whereas at -20° the yield is considerably lower. The choice of the t-butyl group confers not only stability to the starting material itself but also to the dianion. This is reflected in a lower yield of product in the case of the N-methyl amide, which is an oily substance with appreciable tendency to polymerize. The t-butylamide on the other hand is a crystalline compound which can be stored at ambient temperature for several weeks. As expected, the dianion 2 is an allylic species in the classical sense which is corroborated by a labeling experiment. Dilithiation of the corresponding amide with a deuterium content of better than 90% at the CH₃ group followed by reaction with benzophenone produces the known adduct 3 with the deuterium label in all three possible positions. The scope of the reaction of the methacrylic amide dianion with various electrophiles is illustrated in Fig.6.

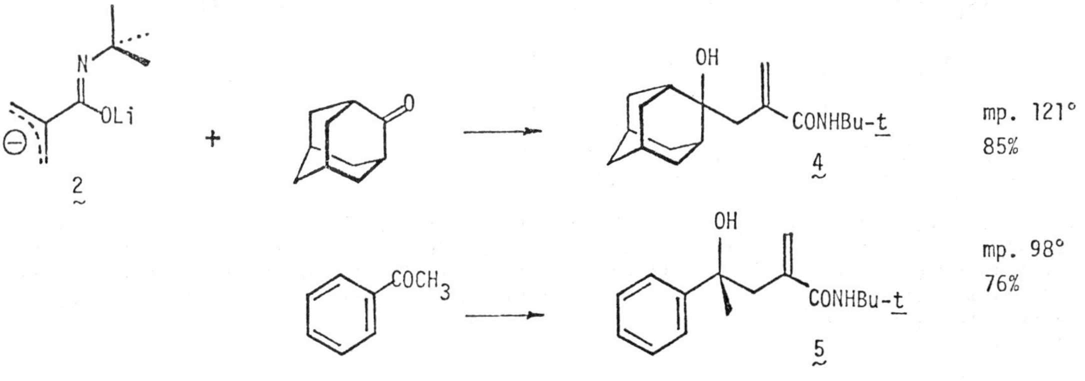

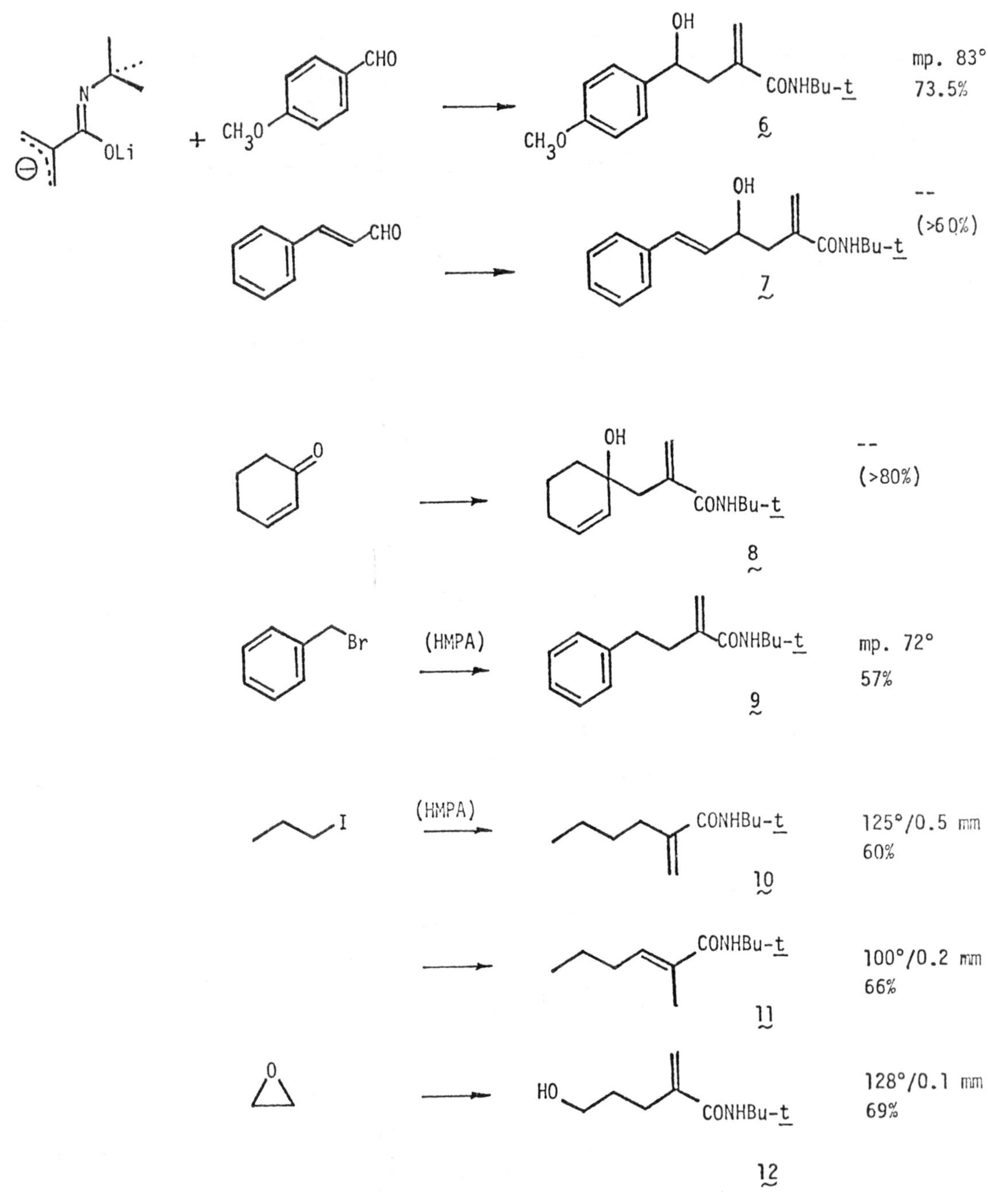

Fig. 6

It is evident that the reaction of the dianion 2 with ketones, aldehydes, halides, and epoxides proceeds fairly well and is optimal in the case of non-enolizable ketones. An interesting facet in the alkylation of the dianion 2 with alkyl halides is the location of the double bond in the product. In the absence of hexamethylphosphoramide (HMPA), the alkylated product 11 is exclusively the one with the trisubstituted double bond, whereas additon of one equivalent of HMPA prior to the alkylating agent produces the 'normal' products 9 and 10. The presence of the other isomer could not be detected under these experimental conditions. Thus, through a judicious selection of the reaction conditions, either of the two possible isomers can be obtained.

Although selection of the t-butyl amide 1 brought about the desired advantages in the formation of the dianion 2, further elaboration of the carbinols 3 - 8 was rendered somewhat more difficult because of the relative inertness of the bulky amide group. Hydrolysis under a variety of acidic and basic conditions proved to be unsuccessful (Fig. 7). Acid catalyzed dehydration leads to a diene 13 which, under the conditions, undergoes a (2 + 4)-cycloaddi-

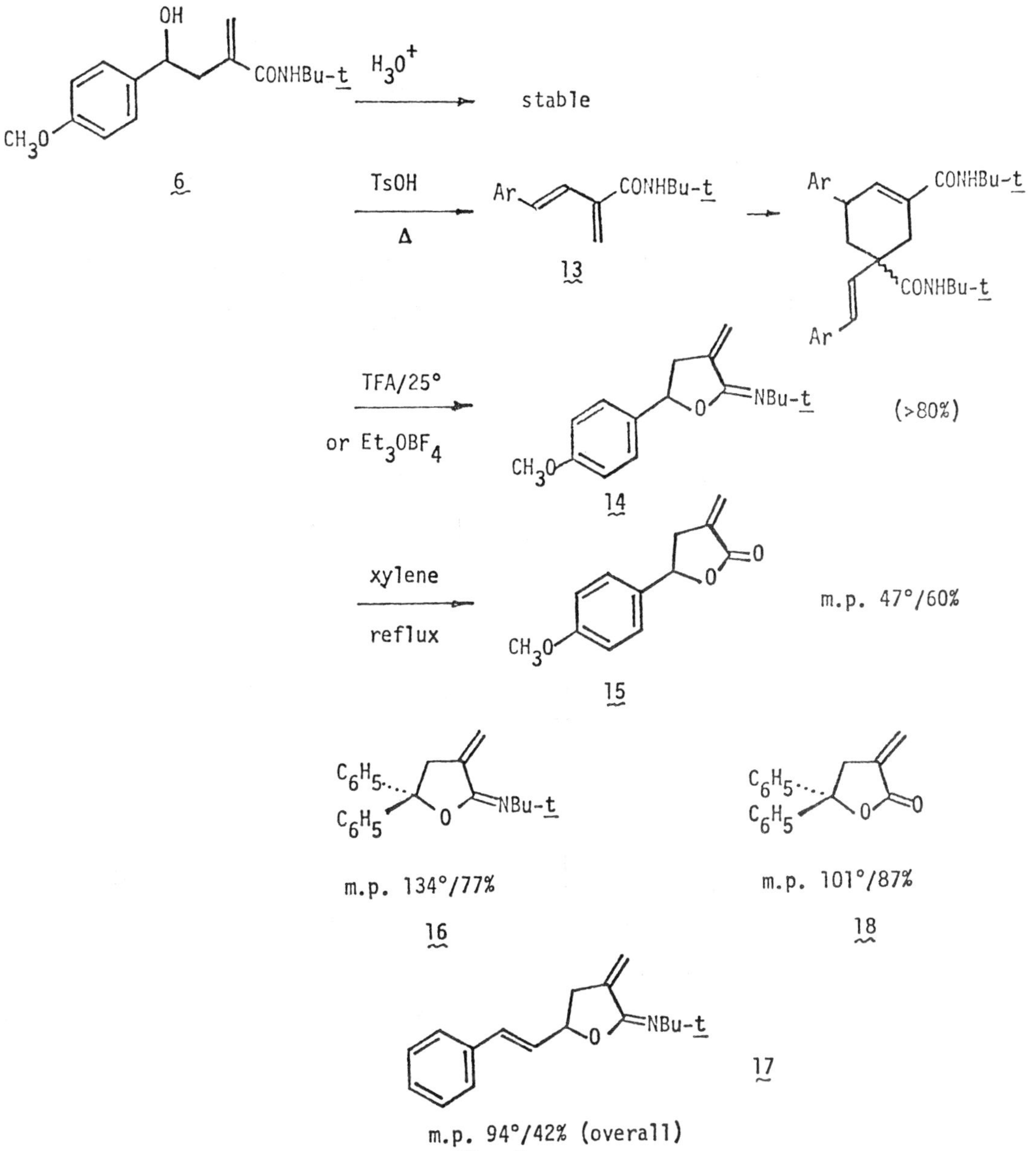

Fig. 7

tion with itself. The mildest method of amide activation, namely O-alkylation with tri-
ethyloxonium tetrafluoroborate in methylene chloride, leads to instant internal nucleophilic
attack by the carbinol oxygen and formation of the corresponding crystalline α-methylene
iminolactone 14. While these species serve as effective Michael acceptors towards good
nucleophiles (e.g. thiophenol), their stability towards aqueous acids is quite remarkable
(Ref. 8). In no instance was it possible to obtain the corresponding lactones. These,
however, are accessible by prolonged heating of the carbinol amides in xylene 15. Similarly
obtained were the respective iminolactones 16 and 17 from 3 and 7, and the lactone 18 from 3.

Alternate methods to prepare α-methylene lactones from carbonyl compounds through the use of
less readily accessible reagents have been reported (Ref. 9, 10). It should be added that
the same type of dimetalation of α,β-unsaturated carboxamides has independently been developed by Beak and coworkers (Ref. 11).

In the second part we will discuss the metalation of a reduced methacrylamide, in particular
the lithiation of N,N-dimethylmethallylamine (Fig. 8). A pertinent precedent for such a
reaction is the dimetalation of methallylalcohol reported by Carlson (Ref. 3). The reported
yields in the reaction of the allylic dianion with carbonyl compounds, though, are rather
low. Other relevant cases are the deprotonation of N,N-dimethyl o-toluidine (Ref. 2e) and
o-methyl-N,N-dimethylbenzylamine respectively (Ref. 2f). Although not exactly analogous to
the methallylamine situation, these examples illustrate the ability of a basic nitrogen atom

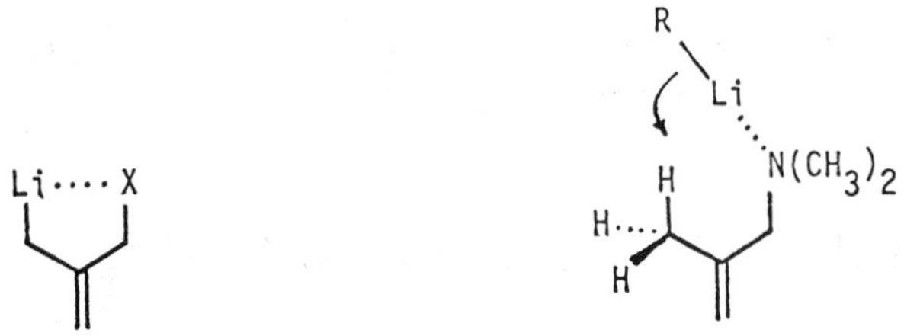

Fig. 8

to facilitate deprotonation of an aromatic, non-acidic methyl group. It has been shown that quite generally the addition of tertiary amines or diamines such as TMEDA vastly accelerate metalation reactions with alkyl lithiums. This phenomenon is the result of a coordination of the amine with the metalating agent followed by its attendant depolymerization and a resulting increase in the kinetic basicity (Ref. 1).

Although allylamines are known to undergo a base catalyzed isomerization to enamines, such a reaction does apparently not interfere in the case of N,N-dimethylmethallylamine (Fig. 9).

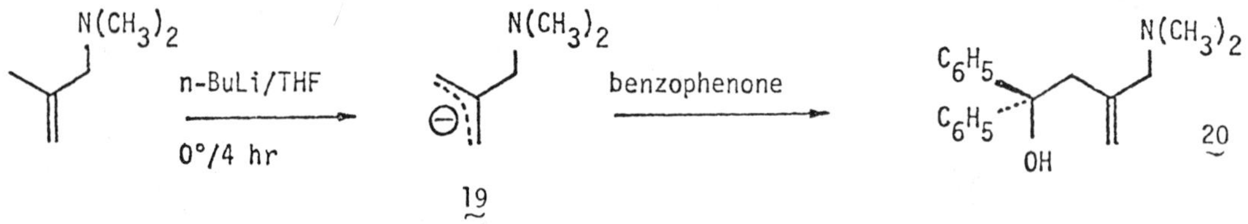

mp. 199° as HCl salt

Fig. 9

Treatment with n-butyllithium in tetrahydrofuran at 0° for four hours leads to the formation of the desired allylic anion 19. Its reaction with benzophenone produces the expected adduct 20 which can be isolated as a crystalline hydrochloride salt in 86% yield under optimal conditions. In the course of the reaction a considerable amount of butyllithium is consumed, most likely by the known reaction of the metalating agent with the solvent which is accelerated by the depolymerizing amine. The reaction of the anion 19 with various electrophiles (Fig. 10) is again quite general and proceeds in yields which compare quite favorably with those observed by Carlson in the case of methallylalcohol (Ref. 3). With the exception of the product derived from N,N-dimethylbenzamid 24 in which the double bond under our experimental conditions moves into conjugation with the ketone, the terminal location of the olefin in maintained.

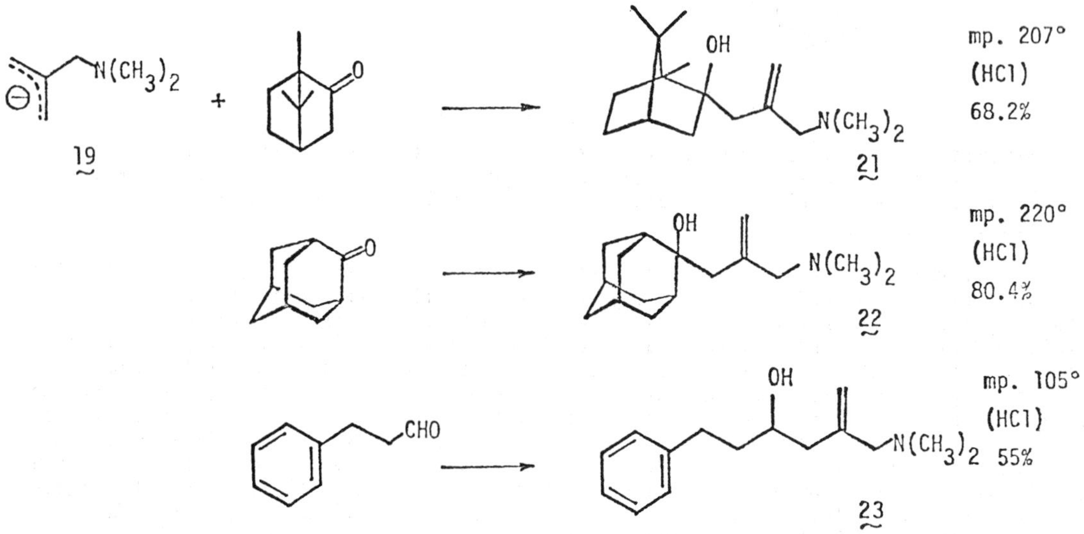

mp. 207°
(HCl)
68.2%

21

mp. 220°
(HCl)
80.4%

22

mp. 105°
(HCl)
55%

23

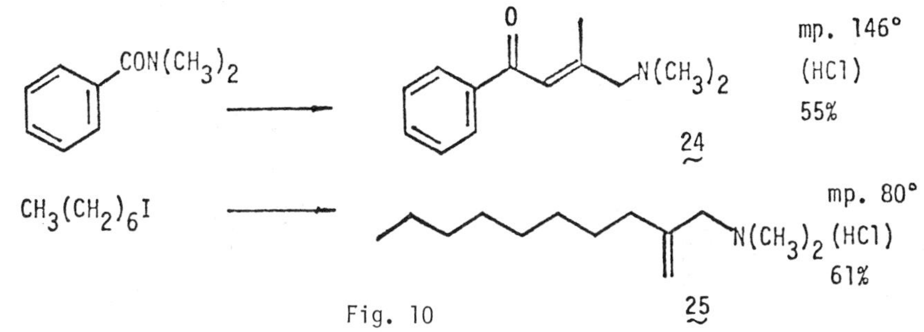

Fig. 10

The versatility of the allylic dimethylamino group as a functional handle for further synthetic elaborations of the primary products is illustrated in Fig. 11. Cleavage of the allylic bond with cyanogen bromide proceeds readily to give access to a reactive halide 26.

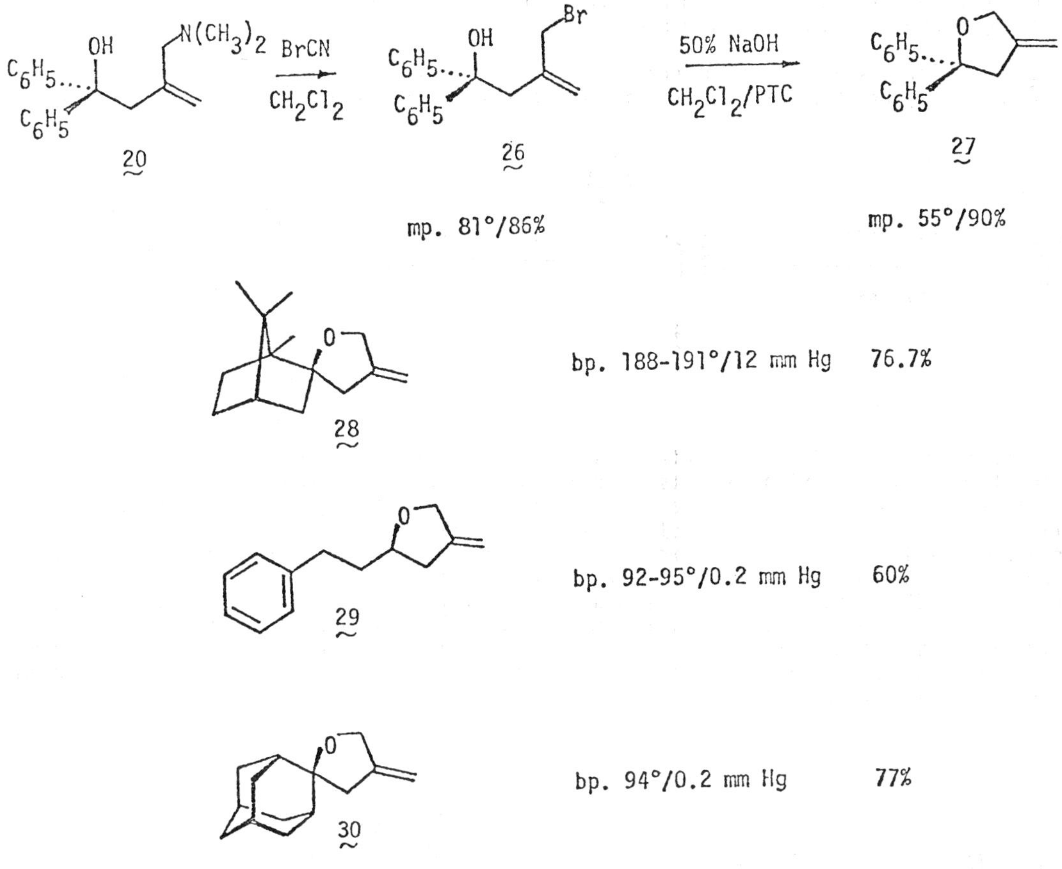

Fig. 11

Its treatment with base, for example, leads to internal displacement and formation of a β-methylene tetrahydrofuran 27. This reaction is quite general as illustrated by the three corresponding products derived from camphor (28), 3-phenylpropionaldehyde (29) and adamantanone (30).

In the third vignette I would like to present a somewhat more practical solution to the alpha lithiation of vinyl ethers. These, of course, are now well established sources of acyl anions. All acyclic alkyl vinyl ethers require the use of the highly pyrophoric t-butyl-lithium as well as low temperature to achieve alpha lithiation. As we have experienced in the lithiation of other unrelated substrates, the presence of an additional heteroatom in the directing group quite often markedly facilitates metalation. Accordingly, we set out to prepare a vinyl ether with an additional oxygen ligand. It was our expectation that such a substrate could be metalated under more practical conditions. We chose vinyloxytetrahydro-pyran 31 as a reagent which was readily accessible by base catalyzed elimination of hydrogen bromide from the tetrahydropyranyl derivative of 2-bromoethanol. Lithiation indeed proceeds readily with n-butyllithium at 0° to give the rather stable acetyl anion equivalent 32 (Fig. 12). It is likely that the oxygen of the protecting group provides additional

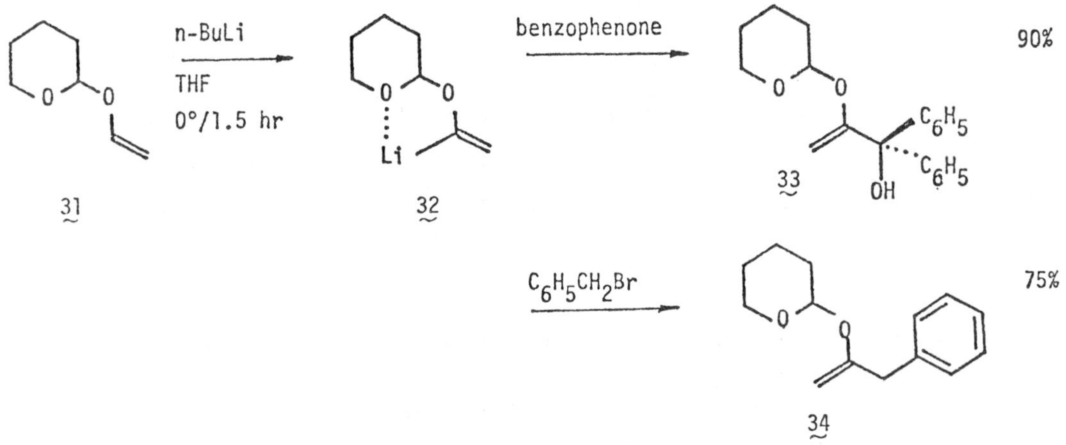

Fig. 12

chelation for the metalating agent as well as stability to the anion. Reaction with benzophenone and benzyl bromide, for example, gives the respective products 33 and 34 in yields
which are quite comparable to those obtained with similar substrates from α-methoxyvinyl-
lithium (Ref. 12). Reference should be made to the work of Schlosser (Ref. 13), who has
metalated under more complex conditions a homolog of 31 derived from the tetrahydropyranyl
ether of allyl alcohol via base catalyzed isomerization of the double bond.

And, in a fourth and last part I would like to present our efforts to apply heteroatom
facilitated lithiation in a new approach to the synthesis of an old natural product,
Nuciferin 35 (Fig. 13). This isoquinoline alkaloid has the typical aporphine skeleton.

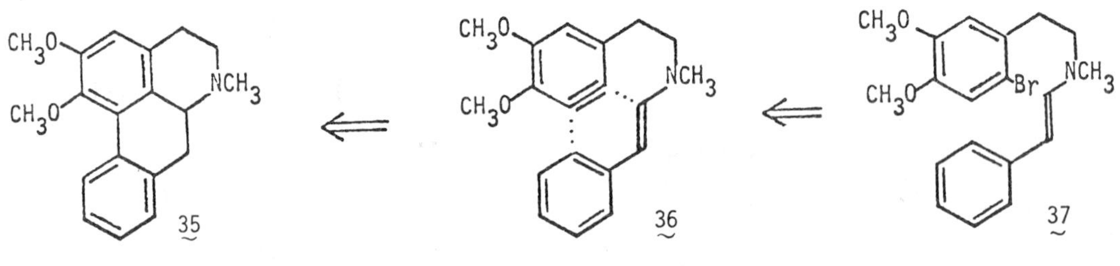

Fig. 13

Faced with the not particularly challenging task of producing a 10 g quantity of this natural
product, we set out to scrutinize the syntheses reported in the literature. Clearly, the
poorest step in any of the classical sequences is the formation of the aryl-aryl bond
achieved either via Pschorr reaction or the more elegant photochemical cyclization. In view
of the rather low and impractical yields in both the photochemical approaches and in the
Pschorr reaction, we decided to look into entirely different synthetic strategies. One
rather attractive possibility presented itself after the realization that the tetracyclic
molecule could be assembled via an intramolecular Diels-Alder reaction (36). Methodolog-
ically, such a reaction appears quite intriguing as it involves an aryne as the dienophile
in an intramolecular situation. To our knowledge there is only one rather special precedent
in the literature (Ref. 14). The intramolecular Diels-Alder reactions of an acetylenic
dienophile to a styrene on the other hand is quite well documented (Ref. 15). It was felt
that the requisite precursor for this crucial intramolecular (2 + 4)-cycloaddition should
be accessible from the enamine 37. Again, based on literature precedent (Ref. 16), it was
assumed that a heteroatom facilitated lithiation should occur regioselectively between the
bromine and the methoxy group as the most likely of the four possibilities. The synthesis
of the bromo-enamine 37 is outlined in Fig. 14. The only critical step was a selective
reduction of the amide carbonyl in 38 without concomitantly affecting the halogen or the
double bond. This transformation was readily achieved through the use of aluminum hydride.
Although the final product 37 is not crystalline, it has been well characterized by spectroscopic means leaving no doubt about its constitution. Upon treatment of this precursor 37
with an excess of lithium diisopropylamide, a major product was isolated with a melting
point corresponding to that of racemic Nuciferin. Gratifyingly, the mass spectrum indicated
the expected molecular ion. Upon closer inspection of the spectroscopic data, in particular
the HMR and CMR, it became quite apparent that we had in fact obtained a compound which was
isomeric with the natural product. The conspicuous presence of a methyl group attached to
an unsaturated carbon together with a UV spectrum characteristic of isoindoles led us to

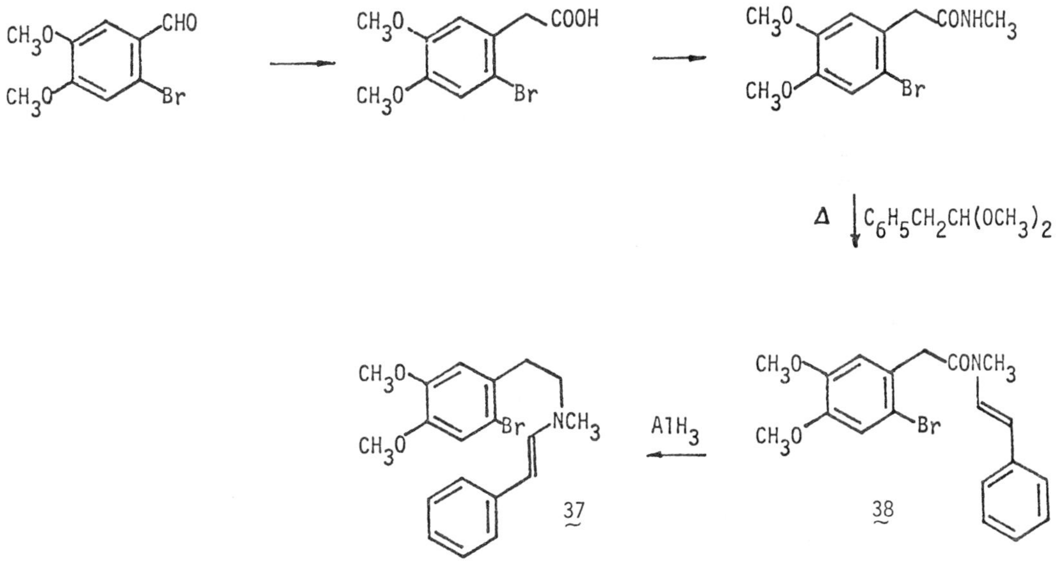

Fig. 14

propose structure 39 for this product. At this time we can only speculate as to the possible

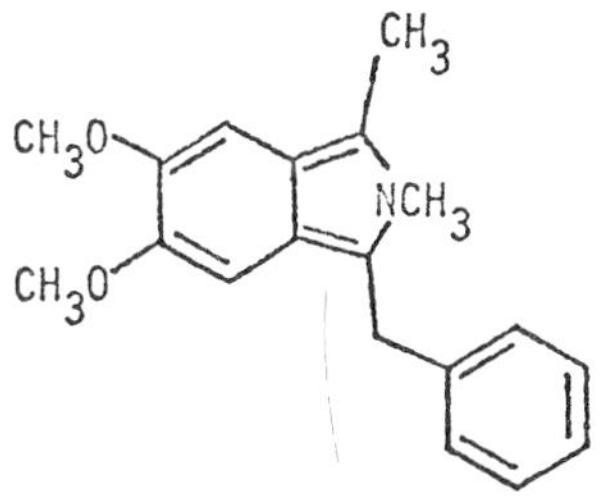

mechanism of its formation. Clearly, the presumed sequence of events must have been pre-
cipitated by the fact that metalation did not take place in the desired position but most
likely at the sp^2-carbon alpha to the enamine nitrogen. This was followed by formation of
the isoquinoline skeleton ether via direct nucleophilic displacement of the bromide or via
elimination/addition. Subsequent deprotonation of the benzylic carbon followed by fragmentation may then lead to a styrene system which, upon work-up, can be protonated at the terminal
position and captured by the enamine nitrogen leading to the isoindole skeleton.

With the failure of this synthetic sequence, a more practical synthesis of Nuciferin was
still to be achieved. Careful reinvestigation of the Pschorr reaction for this particular
case led to the conclusion that the presence of external nucleophiles and of reagents pro-
moting radical formation were rather detrimental. Ideal conditions had to be such as to
permit effective generation of a carbonium ion which would predominantly lead to carbon-
carbon bond formation. Our optimal experimental conditions are diazotization with i-amyl
nitrite in trifluoroacetic acid followed by thermal decomposition of the diazonium salt. In
this manner a 59% yield of ($\pm$)-Nuciferine hydrochloride 35 can be reproducibly achieved.

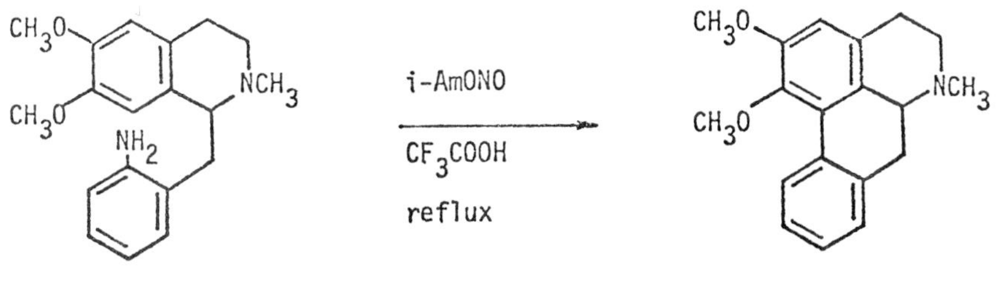

i-AmONO

CF$_3$COOH

reflux

59% as HCl salt

REFERENCES

1. H. W. Gschwend and H. R. Rodriguez, *Organic Reactions* 26, 1-475 (1979).

2. a) P. L. Creger, *J. Am. Chem. Soc.* 92, 1396-1397 (1970).

   b) R. L. Vaulx, W. H. Puterbaugh and C. R. Hauser, *J. Org. Chem.* 29, 3514-3517 (1964).

   c) H. W. Gschwend and A. Hamdan, *J. Org. Chem.* 40, 2008-2009 (1975).

   d) R. L. Letsinger and A. W. Schnizer, *J. Org. Chem.* 16, 869-873 (1951).

   e) R. E. Ludt, G. P. Crowther, and C. R. Hauser, *J. Org. Chem.* 35, 1288-1296 (1970).

   f) R. L. Vaulx, F. W. Jones, and C. R. Hauser, *J. Org. Chem.* 29, 1387-1391 (1964).

3. R. M. Carlson, *Tetrahedron Lett.* 111-114 (1978).

4. a) A. I. Meyers and C. E. Whitten, *Tetrahedron Lett.* 1947-1950 (1976).

   b) A. I. Meyers, R. K. Smith, and C. E. Whitten, *J. Org. Chem.* 44, 2250-2256 (1979).

5. P. Beak, G. R. Brubaker and R. F. Farney, *J. Am. Chem. Soc.* 98, 3621-3627 (1976).

6. R. L. Vaulx, W. H. Puterbaugh and C. R. Hauser, *J. Org. Chem.* 29, 3514-3517 (1964).

7. W. Bannwarth, R. Eidenschink and Th. Kauffmann, *Angew. Chem. Int. Ed.* 13, 468-469 (1974).

8. M. Riediker and W. Graf, *Helv. Chim. Acta* 62, 1586-1602 (1979).

9. a) A. Löffler, R. D. Pratt, J. Pucknat, G. Geibard and A. S. Dreiding, *Chimia* 23, 413-416 (1969).

   b) E. Oehler, K. Reininger, and U. Schmidt, *Angew. Chem. Int. Ed.* 9, 457-458 (1970).

10. A. Hosomi, H. Hashimoto and H. Sakurai, *Tetrahedron Lett.* 951-954 (1980).

11. P. Beak and D. J. Kempf, presentation at ACS Natl. Meeting, March 23-28, 1980, Houston, submitted for publication. We thank Dr. Beak for informing us of his observations.

12. T. E. Baldwin, G. A. Höfle and O. W. Lever, Jr., *J. Am. Chem. Soc.* 96, 7125-7127 (1974).

13. J. Hartmann, M. Stähle and M. Schlosser, *Synthesis* 888-889 (1974).

14. U. S. Patent 4,036,978 (July 19, 1977).

15. W. Oppolzer, *Angew. Chem.* 89, 10-24 (1977).

16. H. Lida, Y. Yuasa and C. Kibayashi, *J. Am. Chem. Soc.* 100, 3598-3599 (1978).

SYNTHESIS AND REACTIONS OF NOVEL 3-OXO-1,2-DIAZETIDINIUM YLIDES

E. C. Taylor*, R. B. Greenwald, N. F. Haley, H. Yanigasawa, and R. J. CLemens

Department of Chemistry, Princeton University, Princeton, NJ 08544, USA

Abstract: Ketone α-chloroacylhydrazones cyclize with non-nucleo-
philic bases to give 1-(disubstituted methylene)-3-oxo-1,2-diazeti-
dinium ylides. These ylides undergo 1,3-dipolar cycloaddition reactions with dimethyl acetylenedicarboxylate and with ketene, are reduced with hydride reagents to 1-substituted diazetidinones, and can
be hydrolyzed under carefully controlled conditions to give stable
salts of 1,2-diazetidinone itself. These latter salts are useful for
the synthesis of pyrazoles, pyrazolines, 3-oxo-1,2-diazetidinium
ylides not available by cyclization, and for the construction of
mono- and bi-cyclic aza-β-lactam analogs of the β-lactam antibiotics.

Several years ago, as a result of another research project, we had on hand a
generous supply of the chloroacetyl hydrazone of benzophenone. It occurred to
us that this compound might serve as an ideal precursor to aziridinone, the
simplest α-lactam, since the initial product of intramolecular dehydrohalogenation should be capable of facile N-N hydrogenolysis under conditions which
might permit survival of the fragile α-lactam functionality. We therefore
treated the chloroacetyl hydrazone of benzophenone with sodium hydride in
anhydrous THF and were gratified to find that an intramolecular dehydrohalogenation product was indeed formed in quantitative yield. Its spectral properties, however, immediately eliminated the anticipated α-lactam structure. The
strong infrared absorption carbonyl band at 1780 cm^{-1} seemed low for an α-
lactam, but more strikingly, its ultraviolet spectrum displayed an absorption
maximum at 325 nm (ε = 26,200), demonstrating the introduction of a potent
chromophore. Its nmr spectrum revealed that the resonances of two phenyl
protons were shifted downfield from the remaining eight aromatic hydrogens,
but that the methine hydrogens were clearly intact (see Scheme 1).

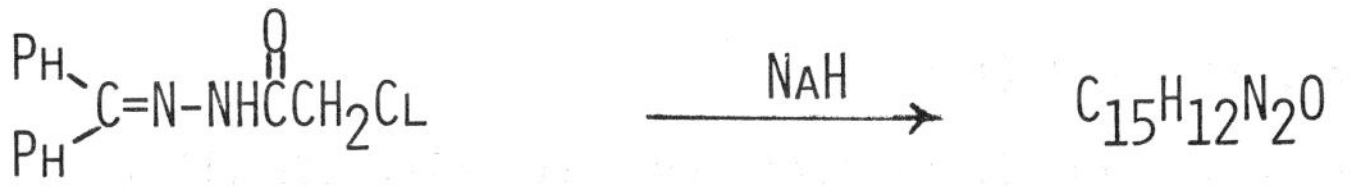

$$\text{Ph}\diagdown\,\underset{\text{Ph}\diagup}{\text{C}}=\text{N}-\text{NH}\overset{\overset{\text{O}}{\|}}{\text{C}}\text{CH}_2\text{CL} \xrightarrow{\text{NaH}} C_{15}H_{12}N_2O$$

IR: 1780, 1760, 1750 cm^{-1}

UV: λ_{MAX} (EtOH) 245 (ε 17,000) nm

325 (ε 26,200) nm

NMR: 8.02 (m, 2H)

7.55 (m, 8H)

5.40 (s, 2H)

SCHEME 1

The chloroacetyl derivatives of both the E and the Z isomers of p-bromobenzophenone hydrazone were individually treated with sodium hydride in anhydrous
THF to give isomeric and non-interconvertible intramolecular dehydrohalogenation products (see Scheme 2). Clearly, the geometrical integrity of the

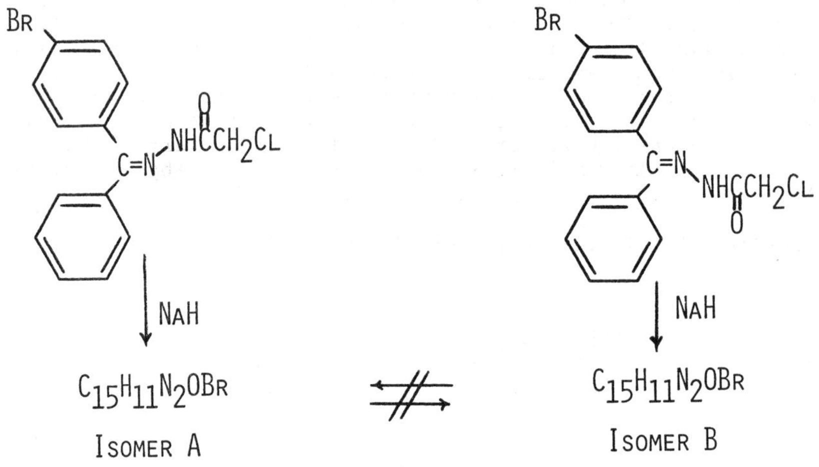

SCHEME 2

precursors had been retained in the products.

There were four reasonable structures for these intramolecular dehydrohalogenation products (Scheme 3). The originally anticipated α-lactam structure

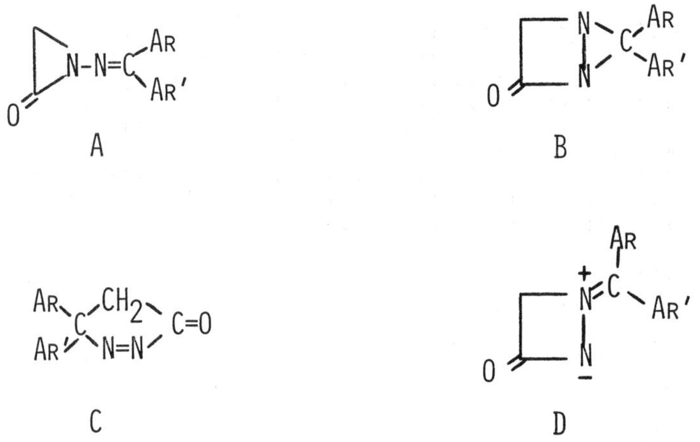

SCHEME 3

A is eliminated, since both the observed infrared and ultraviolet absorption
spectra are incompatible with this formulation. Although structure B might be
expected to have an infrared absorption band in the neighborhood of 1780 cm^{-1},
it does not contain a chromophore which could be responsible for the ultraviolet absorption maximum at 325 nm. Structure C does contain a chromophore,
but it is readily eliminated both by the observed infrared absorption band at
1780 cm^{-1} and by the observation that isomeric, non-interconvertible dehydrohalogenation products were obtained from the E and Z chloroacetyl hydrazones
of p-bromobenzophenone. Structure D, unusual as is may appear at first glance,
is compatible with all of the above experimental data, and it has been confirmed by an X-ray study on (Z)-1-(p-bromophenyl-phenylmethylene)-3-oxo-1,2-
diazetidinium inner salt. The chemistry of these unusual, highly reactive and
synthetically versatile ylides is the subject of this lecture.

A number of experiments were carried out in an attempt to delineate the scope and limitations of this intramolecular dehydrohalogenation reaction leading to 3-oxodiazetidinium ylides. It was found, for example, that steric bulk at the imine carbon was necessary for a successful cyclization. Thus, the chloro-acetyl hydrazones of diaryl ketones, bulky dialkyl ketones, and aralkyl ketones all underwent intramolecular dehydrohalogenation to give 3-oxodiaze-tidinium ylides, although the latter conversions required unusual reaction conditions. On the other hand, we were unsuccessful in converting the chloro-acetyl hydrazones of less hindered dialkyl ketones or of aldehydes (either alkyl or aryl) to 3-oxodiazetidinium ylides (Scheme 4).

STERIC BULK REQUIRED AT THE IMINE CARBON

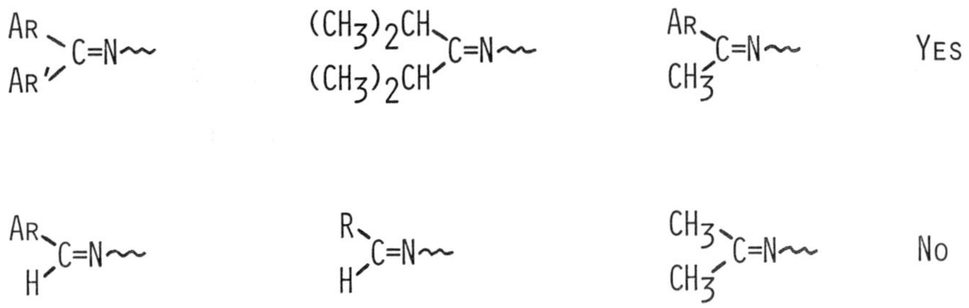

SCHEME 4

Any mechanism advanced for this transformation must accomodate the observation that halide loss occurs by S_N2 displacement with complete inversion of con-figuration. Pure <u>erythro</u> and <u>threo</u> α-bromo-β-methoxybutyrylbenzophenone hydrazones were prepared from <u>trans</u> and <u>cis</u> crotonic acid respectively by reaction with methyl hypobromite, followed by conversion of the resulting acids to their respective acid chlorides with $SOCl_2$ and subsequent reaction with benzophenone hydrazone. The reaction of each pure diastereomeric α-bromoacyl hydrazone with sodium hydride in anhydrous THF, as shown in Scheme 5, gave only one diastereomeric ylide which was > 98% pure as determined by

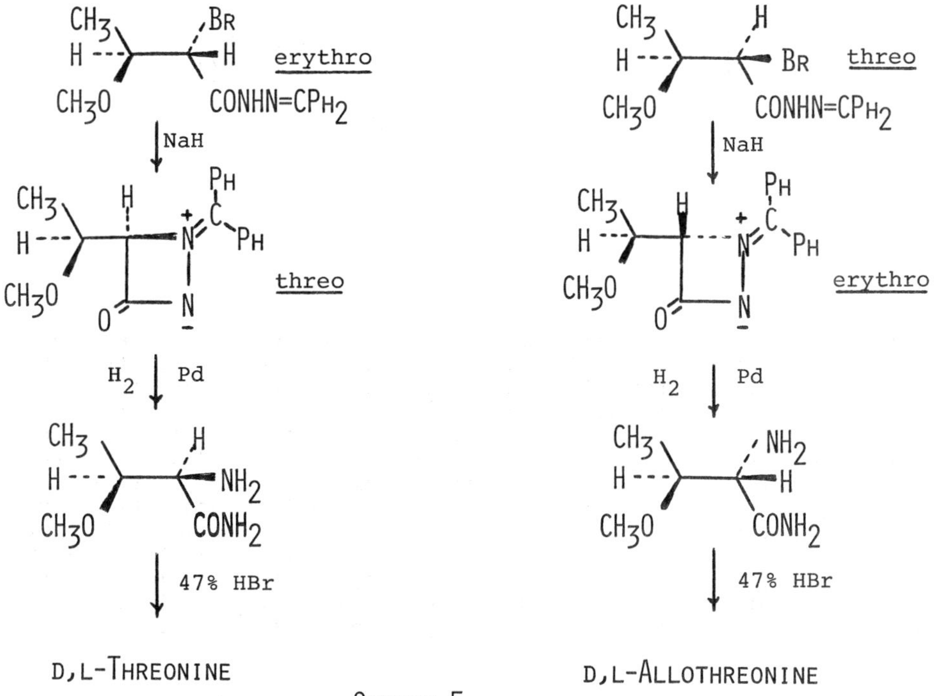

SCHEME 5

nmr. The configurations of these ylides were determined by degradation to
d,l-threonine and d,l-allothreonine by hydrogenolysis of the iminium C=N and
N-N bonds followed by hydrolysis of the resulting amides with 47% HBr. (It is
known that hydrolysis of methoxy amino amides with HBr proceeds without racemization).   These degradation results show conclusively that the <u>erythro</u>
hydrazone gives only the <u>threo</u> ylide, and that the <u>threo</u> hydrazone gives only
the <u>erythro</u> ylide; displacement of halide from the acylated hydrazones therefore occurs exclusively with inversion of configuration.

These observations serve to eliminate a number of conceivable reaction pathways for ylide formation, and we are currently left with the two mechanisms
depicted in Scheme 6. Certainly the first step is proton abstraction by sodium

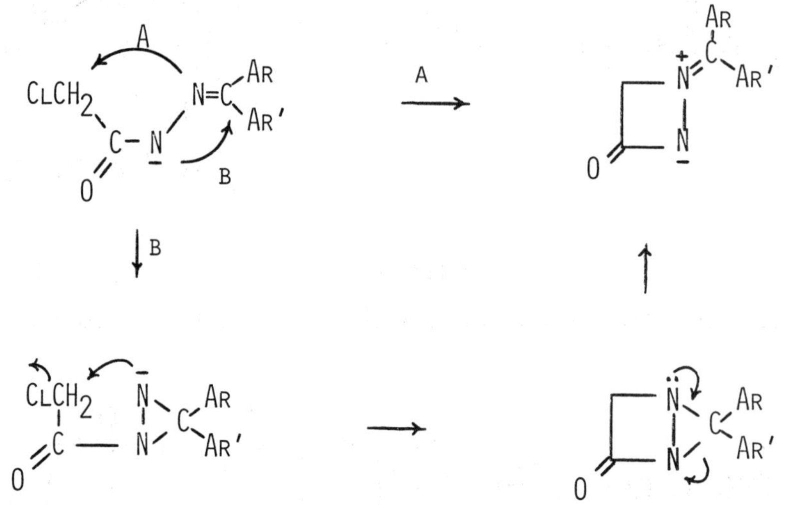

SCHEME 6

hydride to form the amide anion shown in the upper left.   Pathway A represents
direct intramolecular S_N2 displacement of the α-chloro group by the imine
nitrogen, which is rendered more nucleophilic by the adjacent amide anion.
This mechanism readily accomodates all of the substituent effects noted previously, and we prefer it to mechanism B, which requires an initial intramolecular addition of the amide nitrogen to the imine carbon to give the
diaziridine anion shown, which then undergoes intramolecular S_N2 halide displacement. Electrocyclic ring opening of the resulting bicyclic diaziridine,
shown in the lower right, would then give the observed ylide. We have no
concrete evidence which eliminates pathway B, but we are currently exploring
an independent synthesis of the chloroacetyl derivatives of 3,3-disubstituted-
1,2-diaziridines in order to examine the possible intermediacy of their anions
in this transformation.

The observations summarized in Scheme 7 are also consistent with pathway A.

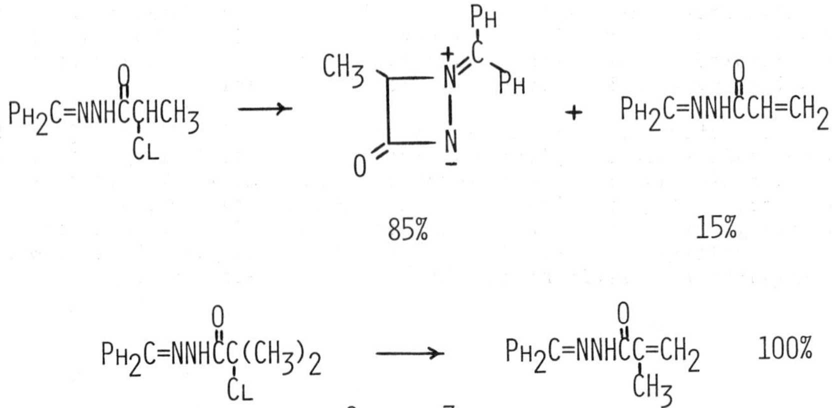

SCHEME 7

Increasing steric bulk at the α-carbon atom has a predictably deleterious effect upon intramolecular halide displacement. Thus, the α-chloropropionyl hydrazone of benzophenone underwent intramolecular dehydrohalogenation to give the ylide shown, but this was accompanied by a small amount of the product of side-chain dehydrohalogenation. The latter pathway was the only one observed when two methyl groups were present on the α-carbon atom.

The β-chloropropionyl hydrazone of benzophenone likewise undergoes intra-molecular halide displacement to give the homologous 5-membered ylide shown in Scheme 8, although it is again accompanied by a small amount of the isomeric

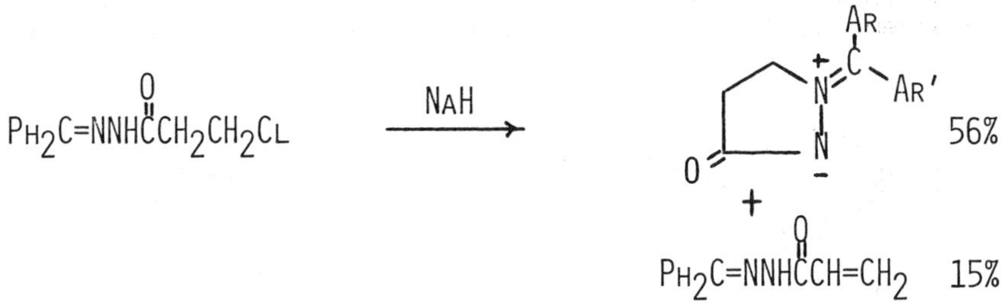

SCHEME 8

side-chain olefin. Steric bulk at the α-carbon, as illustrated in Scheme 9,

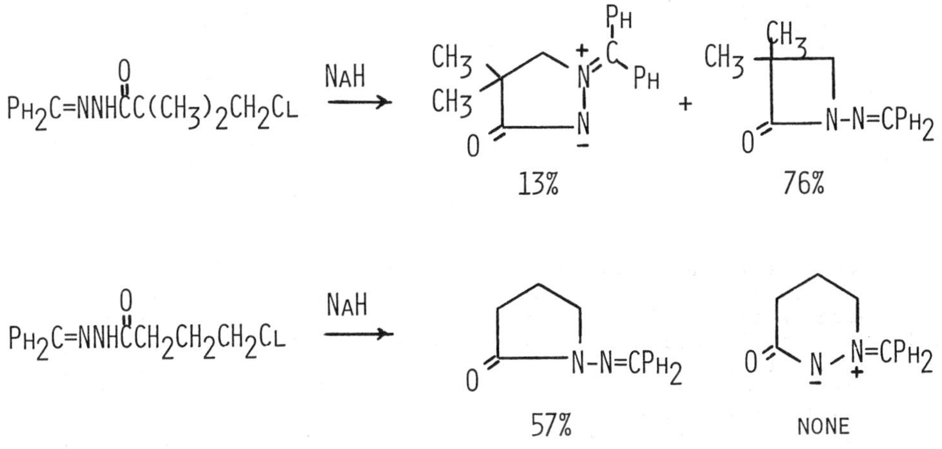

SCHEME 9

diminishes but does not eliminate ylide formation; the major product in this case arises from chloride displacement by the amide nitrogen to give the azetidinone derivative shown in the upper right. This pathway is the ex-clusive one followed by the γ-chlorobutyryl hydrazone of benzophenone; none of the 6-membered ylide could be detected from this reaction.

It is unfortunate that our initial explorations of 1,3-dipolar cycloaddition reactions were made with the ylide obtained from benzophenone, since we have now found that these reactions are strongly inhibited by steric hindrance at the iminium carbon. Nevertheless, our experiments using this less-than-ideal substrate clearly reveal the synthetic potential of 3-oxodiazetidinium ylides in 1,3-dipolar cycloaddition reactions. For example, the ylide shown in Scheme 10 (prepared by dehydrohalogenation of the α-chloropropionyl hydrazone

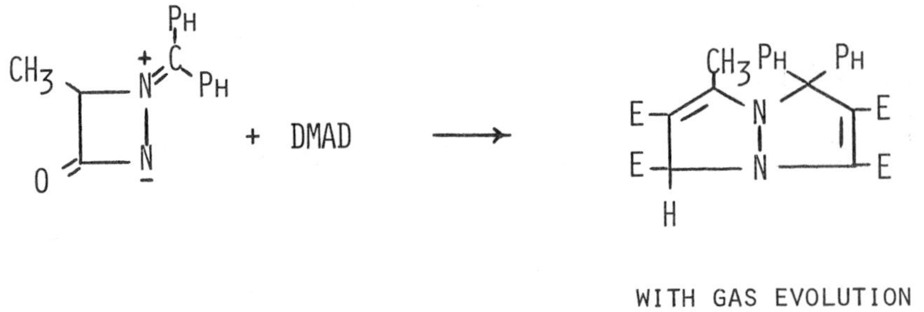

WITH GAS EVOLUTION

SCHEME 10

of benzophenone) reacted with dimethyl acetylenedicarboxylate on heating to give a bicyclic tetracarboxylate adduct possessing two bridgehead nitrogen atoms. This reaction, which was accompanied by gas evolution, appears to occur as depicted in Scheme 11. Thus, stirring the 3-oxodiazetidinium ylide

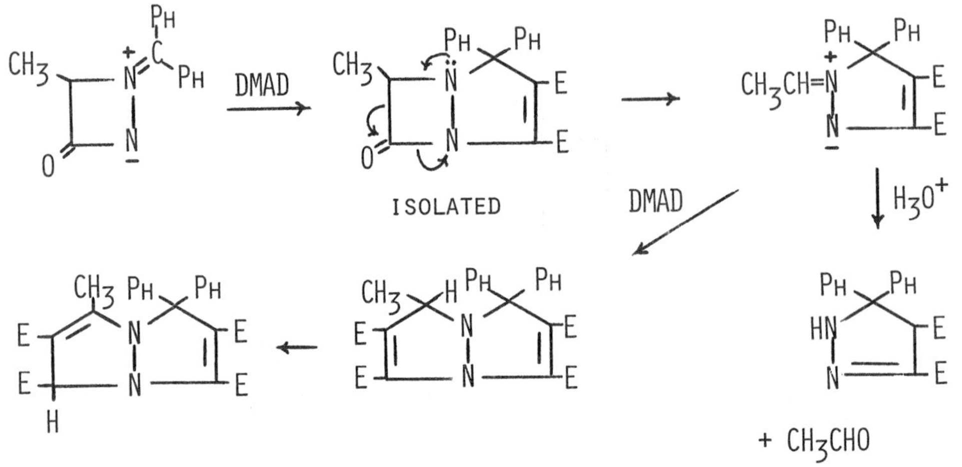

SCHEME 11

with dimethyl acetylenedicarboxylate in dry methylene chloride for 5 days at room temperature resulted in the formation of a 1:1 adduct. This fascinating bicyclic aza-β-lactam exhibits an infrared absorption band at 1850 cm^{-1}. It loses CO on heating, presumably to give a new 1,3-dipolar species which is shown at the upper right of Scheme 11. This is captured by dimethyl acetylenedicarboxylate to give a bicyclic adduct which must undergo a rapid prototropic rearrangement to give the observed tetracarboxylate product. Evidence for the intermediate dipolar species shown was obtained by heating the isolated initial bicyclic aza-β-lactam adduct until gas evolution commenced; addition of aqueous acid then gave acetaldehyde (identified as its DNP derivative) and 5,5-diphenyl-3,4-dicarbomethoxy-Δ^2-pyrazoline, identical with an authentic sample prepared from dimethyl maleate and diphenyl diazomethane).

Our diphenylmethylene 3-oxo-1,2-diazetidinium ylide reacted with ketene to give a bicyclic aza-β-lactam which is shown at the lower left of Scheme 12. We assume that initial acylation at N-2 gives a charge-separated ylide which must undergo proton transfer from C-4 to give the 1,3-diplar species shown at the lower right. This must then be captured by a normal cycloaddition reaction with a second equivalent of ketene to give the observed product. This transformation is particularly intriguing since it indicates the feasibility of base-catalyzed functionalization of 3-oxodiazetidinium ylides at C-4, and initial experiments (see Scheme 13) appear to confirm this expectation.

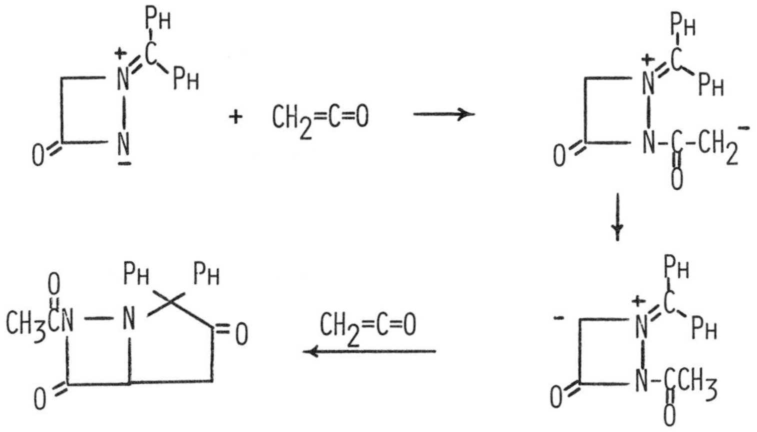

SCHEME 12

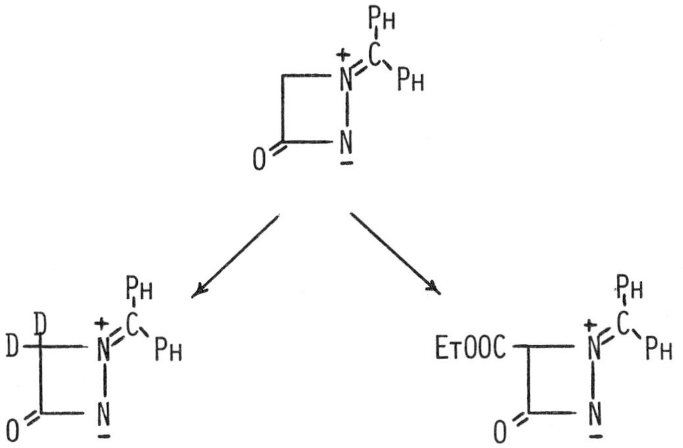

SCHEME 13

Deuterium exchange at C-4 is readily accomplished with NaOD, and experiments are currently underway to effect condensation reactions at C-4.

These 3-oxodiazetidinium ylides can be reduced to give a variety of different products depending upon the nature of the reducing agent. These reductions are illustrated in Scheme 14 with our diphenylmethylene ylide. Sodium boro-

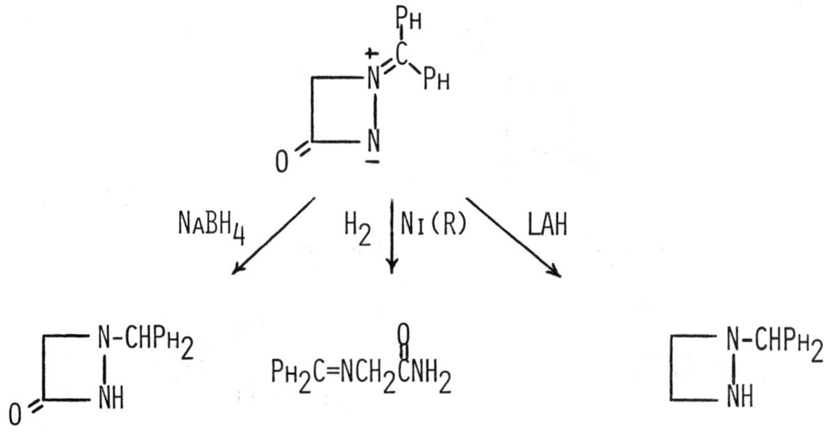

SCHEME 14

hydride selectively reduces the iminium bond to give 1-diphenylmethyl-3-
diazetidinone. Under mild conditions, reduction with hydrogen in the presence
of Raney nickel results in hydrogenolysis of the N-N bond to give the diphenylmethylene imine of aminoacetamide; further reduction to diphenylmethane
and aminoacetamide is observed under more stringent conditions. Lithium
aluminum hydride reduces both the iminium bond and the carbonyl group to give
1-diphenylmethyldiazetidine. It should be noted in passing that these 1-
alkyl-3-diazetidinones can be readily substituted at N-2 by reaction of their
thallium(I) salts with alkyl halides (see Scheme 15), and this facile trans-

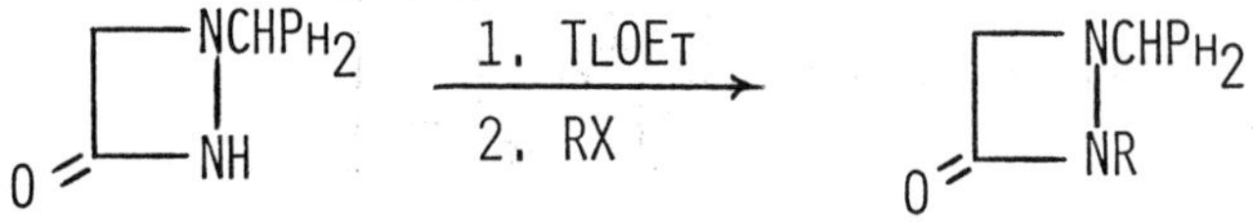

SCHEME 15

formation suggests a route to bicyclic aza-β-lactams by intramolecular alkylation which we are currently exploring.

Although these ylides are surprisingly stable to base, they are much more
sensitive to acid. Aqueous acid rapidly hydrolyses 1-diphenylmethylene-1,2-
diazetidinium inner salt to benzophenone and hydrazinoacetic acid, as shown in
Scheme 16. However, it is possible to effect selective cleavage of the iminium bond by acid-catalyzed hydrolysis with one molar equivalent of water,
conditions which are most conveniently satisfied by the use of one equivalent
of p-toluenesulfonic acid monohydrate in dry methylene chloride, and which
lead to the quantitative formation of the tosyl salt of diazetidinone itself.
This is a beautifully crystalline, colorless solid which melts sharply with
decomposition at 145° and exhibits a strong infrared carbonyl band at 1820
cm^{-1}.

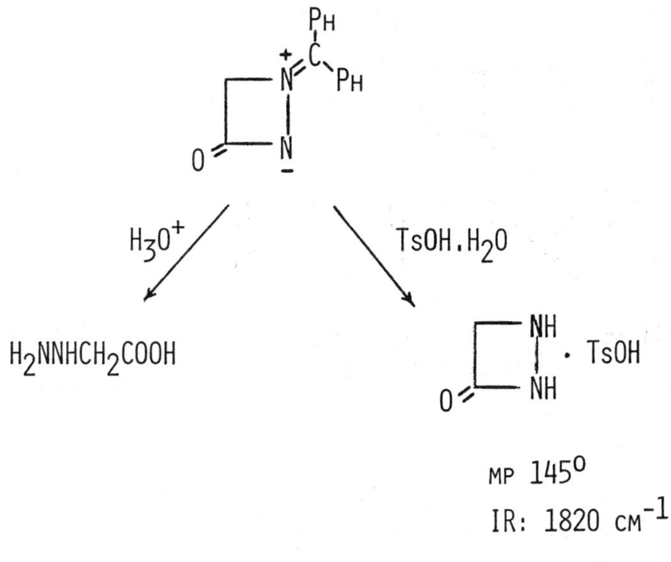

SCHEME 16

Diazetidinone tosylate is an extraordinarily interesting, highly reactive, and synthetically versatile compound. It reacts with carbonyl compounds in DMF solution in the presence of one equivalent of sodium bicarbonate to regenerate diazetidinium ylides (Scheme 17); this reaction is particularly useful because

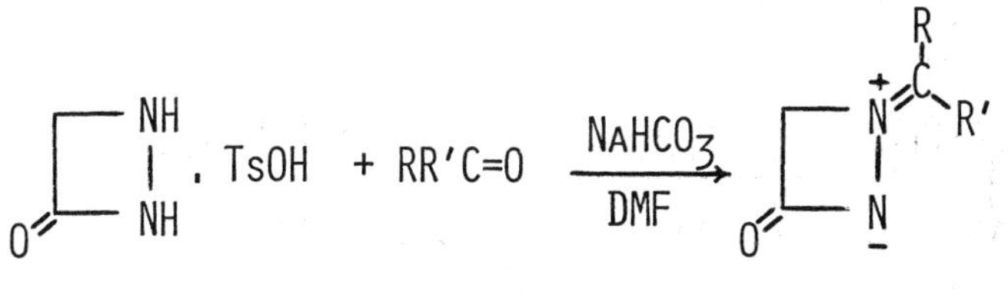

SCHEME 17

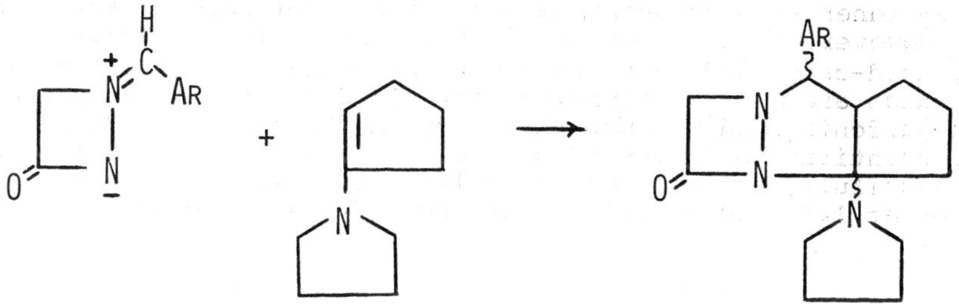

it provides a route to ylides from aliphatic ketones, and from both aliphatic and aromatic aldehydes, which are inaccessible by intramolecular dehydro-halogenation of the corresponding α-haloacyl hydrazones. The ylide prepared from diazetidinone tosylate and benzaldehyde, which possesses the Z-structure shown in Scheme 18, reacts readily with enamines to give 1,3-dipolar cycloadducts.   Reaction with 1-pyrrolidinocyclopentene, for example, is rapid and almost quantitative to produce a tricyclic aza-β-lactam.   The structural assignment given in Scheme 18 is still tentative, since we have not yet determined with certainty the position of the bridgehead pyrrolidine substituent. It is already clear, however, that 3-oxodiazetidinium ylides prepared from aldehydes rather than ketones react with exceptional facility in 1,3-dipolar cycloaddition reactions; their utility as intermediates for the construction of bicyclic aza-β-lactams, of obvious interest as potential antibiotics, is under active investigation.

Another potential route to fused 4-5 aza-β-lactam systems with two bridgehead nitrogen atoms is suggested by observations made on the ylide prepared from diazetidinone tosylate and cinnamaldehyde.   Sodium borohydride effects regio-selective reduction of the iminium bond to give the 1-alkenyldiazetidinone shown in Scheme 19, and it is attractive to speculate that functionalization

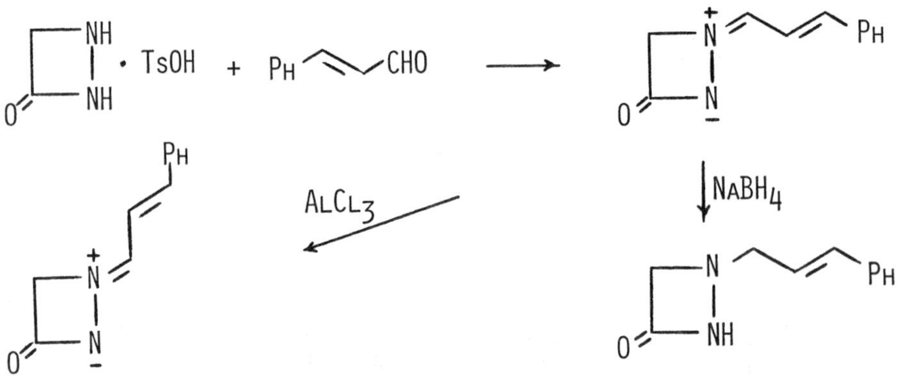

SCHEME 19

of the remaining olefinic bond, perhaps by epoxidation or halogenation, might be followed by intramolecular alkylation to close the second ring.

Surprisingly, only one of the two possible geometrical isomers of this ylide is formed in the DMF/NaHCO$_3$ condensation of diazetidinone tosylate and cinn-amaldehyde, and nmr shows conclusively that it is the one with the negatively charged ylide nitrogen and the iminium hydrogen _trans_ to each other.   This isomer rearranges in the presence of AlCl$_3$ to the isomer in which the nega-tively charged nitrogen and the iminium hydrogen are now _cis_ to each other, and this observation, which has not yet been extended to other ylides, sug-gests a possible general method for arranging stereochemistry at the iminium bond.

1,3-Dicarbonyl compounds, illustrated in Scheme 20 by acetylacetone, react similarly with diazetidinone tosylate in DMF solution in the presence of sodium bicarbonate to give ylides, but we have not yet effected condensation of the amide nitrogen with the side-chain carbonyl group.   By contrast, the reaction between diazetidinone tosylate and acetylacetone in methanol solution led directly to the formation in high yield of the methyl ester of 3,5-di-methylpyrazole-1-acetic acid.   As shown in Scheme 21, this unlikely-appearing transformation results from methanolysis of diazetidinone to give methyl hydrazinoacetate, which is then trapped by acetylacetone.   Diazetidinone tosylate is, in fact, a convenient precursor of hydrazinoacetic acid or its esters; methanolysis of diazetidinone tosylate, for example, represents a method of choice for the preparation of methyl hydrazinoacetate.

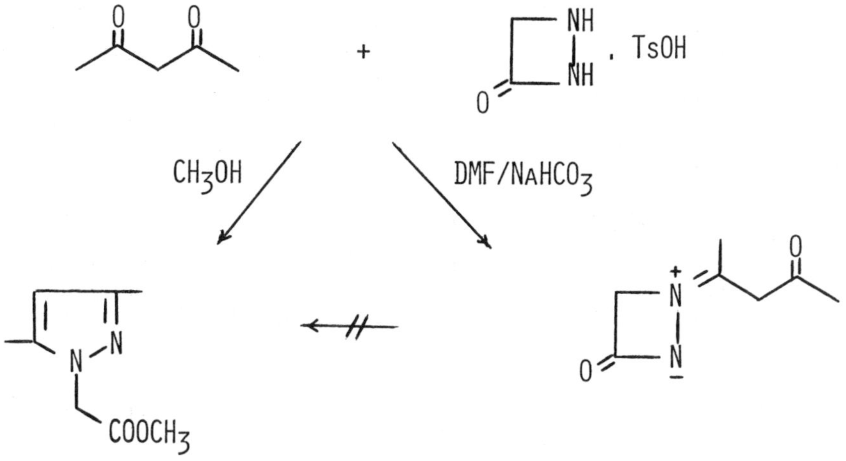

SCHEME 20

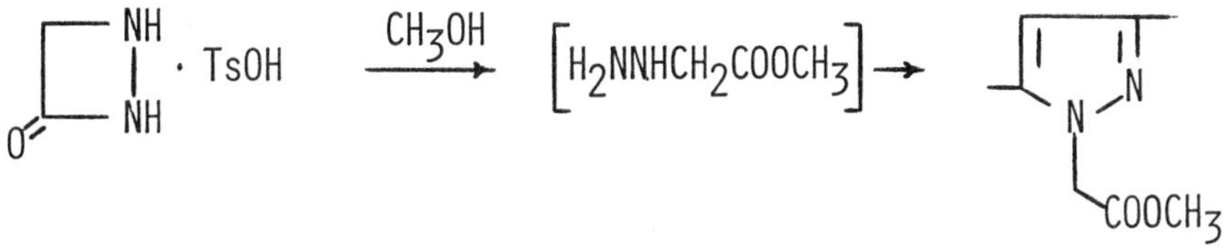

SCHEME 21

Similarly, the reaction of diazetidinone tosylate with methyl vinyl ketone in methanol solution led to the methyl ester of 3-methyl-Δ^2-pyrazoline acetic acid, again by ring opening to methyl hydrazinoacetate, which was trapped by Michael addition to methyl vinyl ketone (Scheme 22). In DMF solution in the

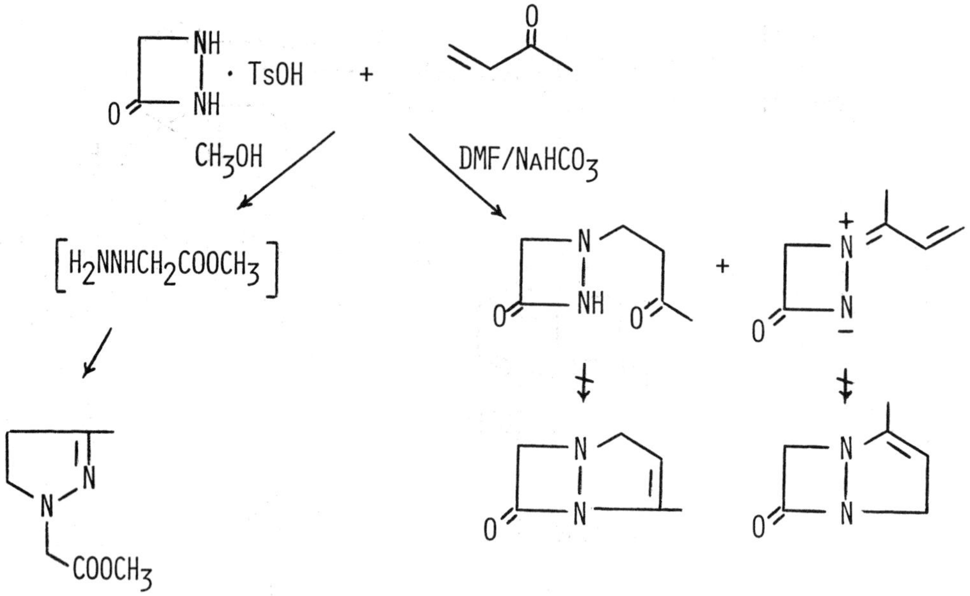

SCHEME 22

presence of one equivalent of sodium bicarbonate, however, diazetidinone
tosylate underwent both Michael addition at N-1 and ylide formation. We have
not yet observed cyclization of either of these adducts to bicyclic aza-β-
lactams.

Since ring strain in the above target bicyclic systems certainly militates
against ring closure, we constructed a homologous intermediate by alkylation
of diazetidinone tosylate with γ-bromobutyrophenone (Scheme 23). The desired

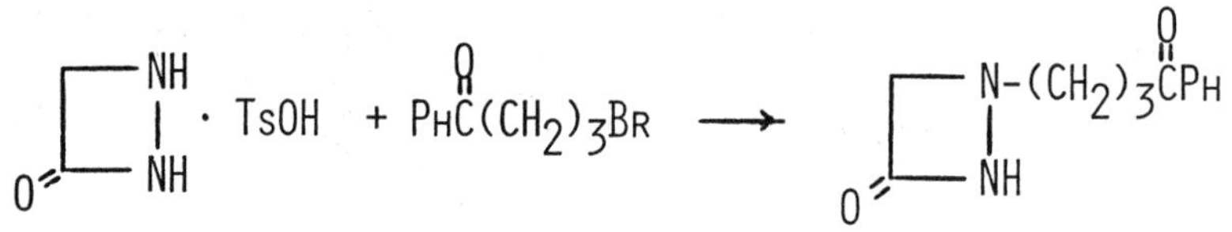

SCHEME 23

1-alkyldiazetidinone was formed in low yield, but again no cyclization could
be effected, perhaps because of a combination of the low nucleophilicity of
the amide nitrogen and the poor reactivity of the aralkyl ketone carbonyl
group. Success was finally achieved, however, by increasing the electrophilic
character of the side-chain carbonyl group. Thus, condensation of diazetidinone tosylate with ethyl 5-bromo-2-oxovalerate led directly to the fused
aza-β-lactam system shown in Scheme 24. Efforts are in progress to effect
dehydration of this compound.

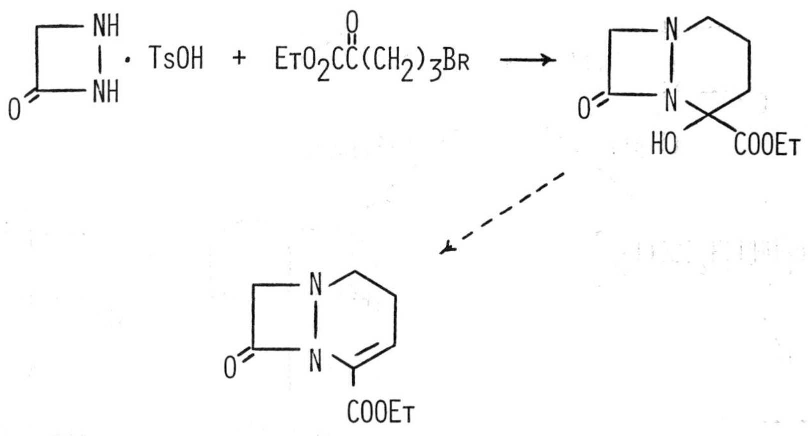

SCHEME 24

It thus appears that both diazetidinone itself, and its derived ylides, may serve as versatile intermediates for the construction of a variety of highly strained mono- and bi-cyclic aza analogs of the β-lactam antibiotics.

SYNTHETIC STUDIES TOWARDS CALCIMYCIN

P. A. Grieco, E. Williams and Ken-ichi Kanai

Department of Chemistry, Indiana University, Bloomington, Indiana 47405, USA

Abstract - Calcimycin, a divalent cation ionophore isolated from cultures of Streptomyces chartreusensis, possesses, in addition to seven chiral centers, a novel 1,7-dioxaspiro[5,5]undecane ring system. This paper describes our progress toward a total synthesis of this unique natural product.

The unique divalent cation ionophore calcimycin (1) (A-23187) was isolated from cultures of Streptomyces chartreusensis. The structure of this complex natural product was established via single crystal x-ray analysis by Chaney and co-workers at the Lilly Research Laboratories.[1] Since the structure was announced in 1974, considerable attention has focused on this antibiotic as a result of its high Ca^{+2} specificity and its ability to transport ions across cell membranes.[2]

CALCIMYCIN   1

The structure of 1 reveals, in addition to a host of chiral centers, three basic components: an α-ketopyrrole, a substituted benzoxazole unit, and a novel 1,7-dioxaspiro[5,5]undecane ring system. The fact that each ring oxygen atom bears an axial relationship to the six-membered ring to which it is attached implies that, of the other possible spiro ketal arrangements, the 1,7-dioxaspiro[5,5]undecane ring system found in calcimycin is the most stable conformation. This point is indeed reasonable in view of the well known anomeric effect. Thus, analysis of 1 suggests that the acyclic keto diol 2 or its equivalent should close under thermodynamic, dehydrative conditions (acid catalysis) to the desired dioxospirane. This point has recently been demonstrated (cf. 3 → 4) by Evans and co-workers whose efforts have culminated in the first total synthesis of calcimycin.[3] Herein we describe our investigations in this area.

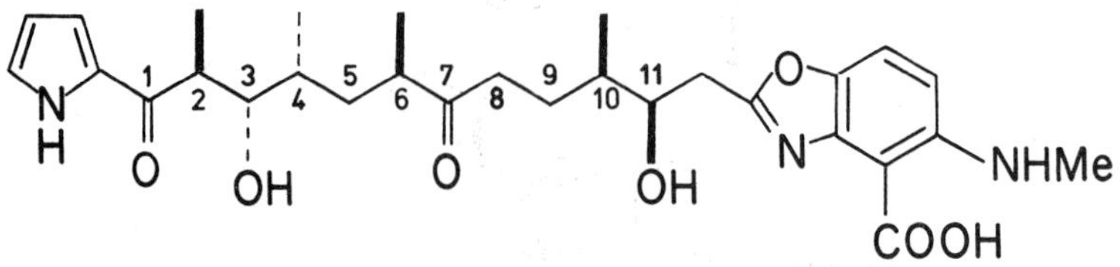

2

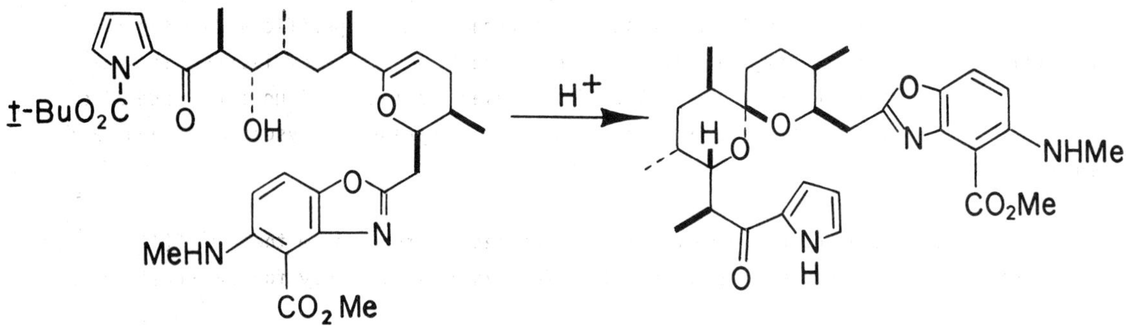

3   **4**

In our preliminary planning we set out to construct the aldol linkage in 2 from keto pyrrole 5 and a suitably protected form of aldehyde 6. Along these lines we have reported a

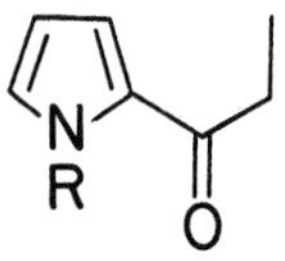

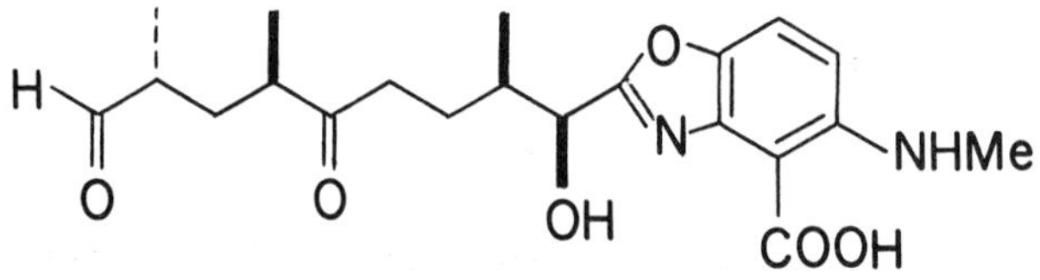

5   **6**

synthesis of 7 [4] which had previously been converted into aldehyde 8 and condensed with the zinc enolate derived from the t-BOC derivative of pyrrole 5 giving rise to a mixture of all

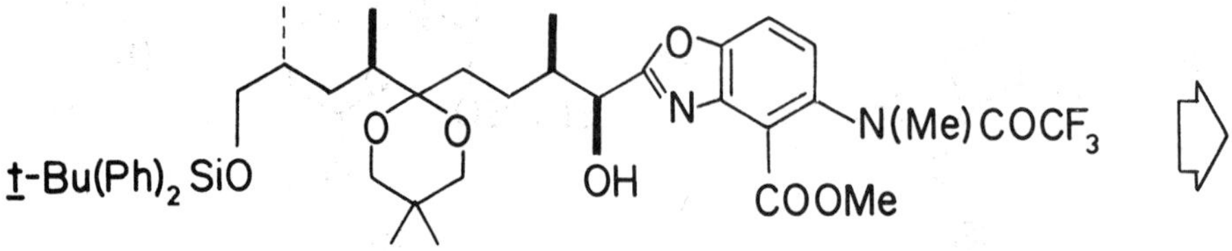

7

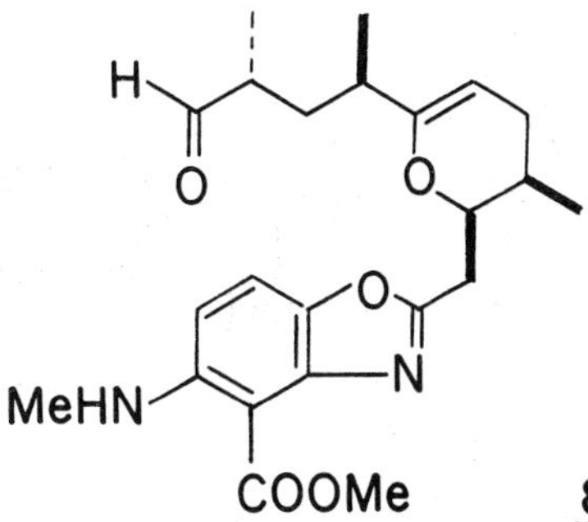

8

four possible diastereomers.[3] The lack of any significant acyclic stereoselection during
the critical aldol condensation led us to seek an alternate stereospecific method for the
construction of the three contiguous C(2)-C(4) chiral centers. It is of interest to note
that the three contiguous chiral centers C(2)-C(4) present in the aldol unit of 1 are also
found in other polyether antibiotics such as lasalocid A,[5] monensin,[6] lysocellin,[7] and sali-
nomycin.[8]

Intent on avoiding the aldol route to calcimycin, we set out to prepare the C(1)-C(11)
acyclic fragment 9 and the benzoxazole unit 10. The synthetic strategy for the construction

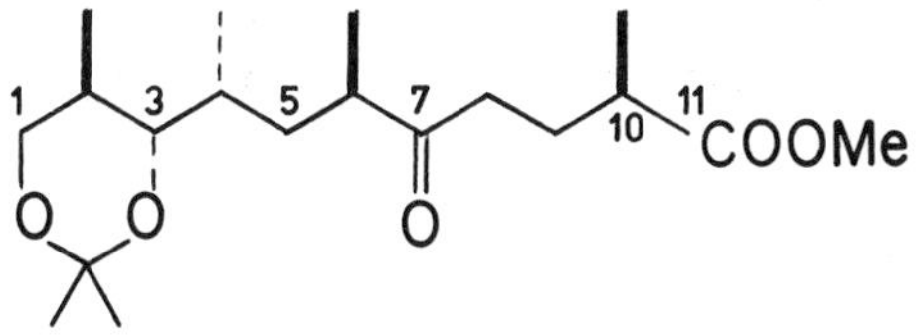

9

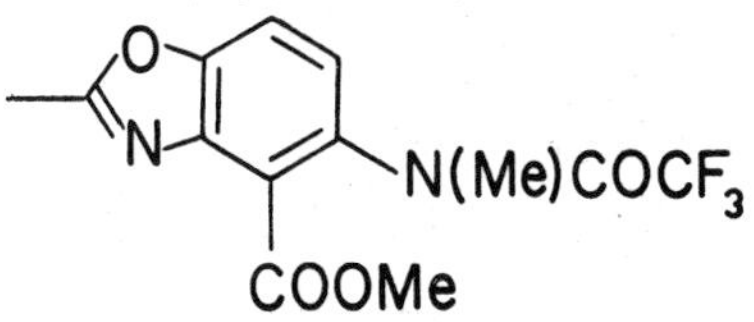

10

of fragment 9 was to be carried out in two stages: (1) preparation of the C(1)-C(7) segment
17 from the bicyclo[2.2.1]heptenone 11 (Scheme I) and (2) elaboration of the remaining four
carbon atoms C(8)-C(11) bearing the additional chiral center at C(10) onto the C(1)-C(7)
unit. We were particularly attracted to the bicyclo[2.2.1]heptane derivative 11
because of its conformational rigidity which would permit C(2) endo methylation[9] thereby
indirectly establishing the three contiguous chiral centers at C(2)-C(4) [equation 1].

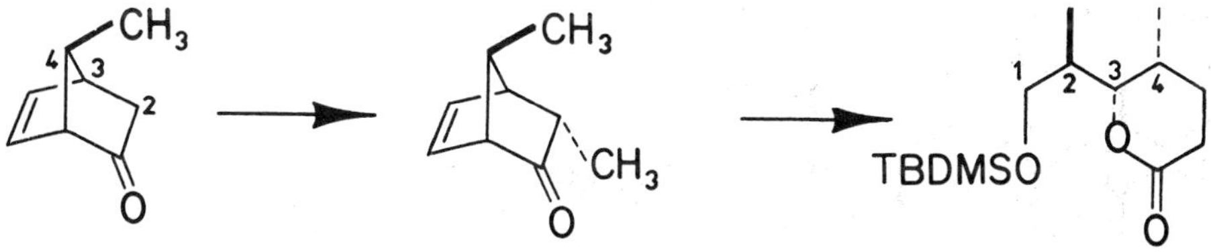

(1)

11   **18**   **15**

Treatment of lithium enolate of ketone 11 in tetrahydrofuran cooled to -78°C with methyl
iodide provided exclusively the expected product 18 in 75% isolated yield. Reaction of 18

with basic hydrogen peroxide and subsequent exposure to borontrifluoride etherate rearranged the intermediate hydroxy carboxylic acid to bicyclic lactone 13 in 85% overall yield. In an alternate sequence of reactions (Scheme I) ketone 11 was subjected to Baeyer-Villiger oxidation followed by acid-catalyzed rearrangement to lactone 12(71% overall). Alkylation of 12 from the more readily accessible β-face of the bicyclo[3.3.0]octane ring system generated, as the sole product, lactone 13 in very high yield. Needless to say, the bicyclic lactone 13 obtained by the latter route was identical in all respects with the sample of 13 prepared above by the former sequence.

SCHEME I

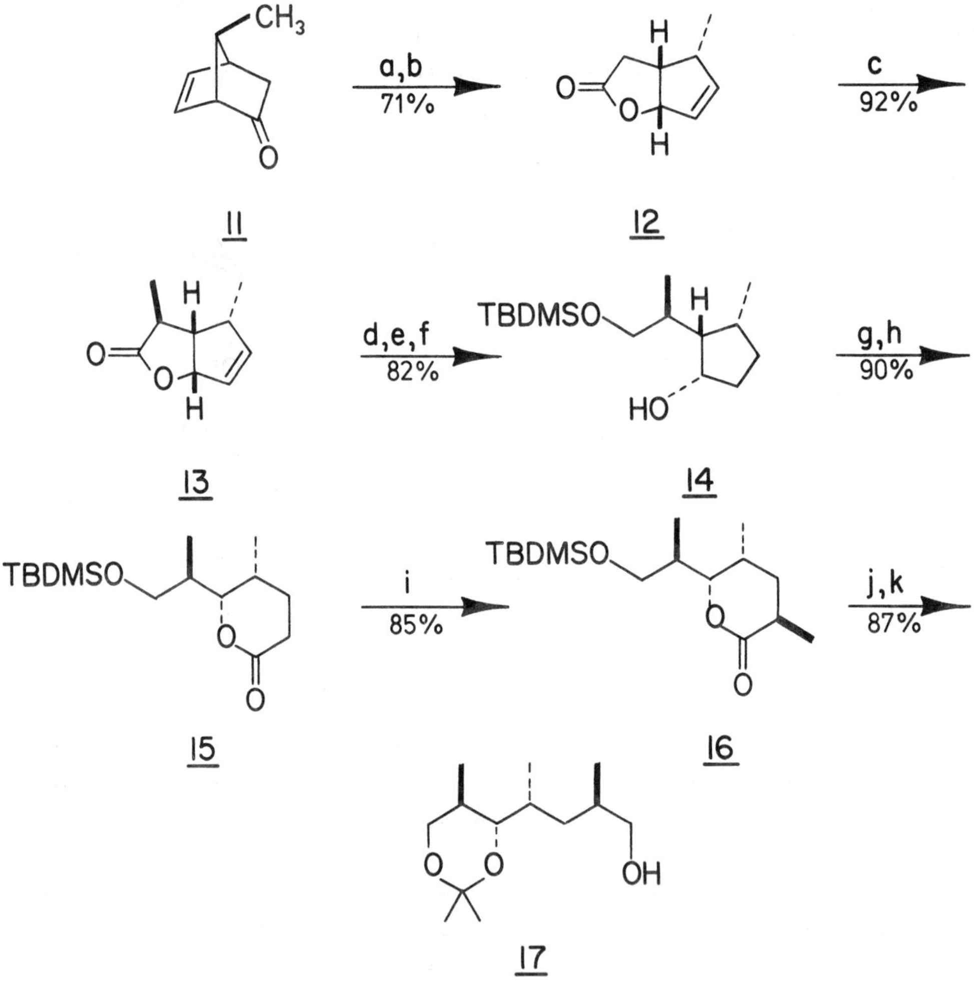

a, H_2O_2,OH⁻; b, $BF_3 \cdot Et_2O$; c, LDA, THF, MeI; d, LiAlH₄; e, H_2,
PtO₂, EtOAc; f, t-Bu(Me)₂SiCl, DMAP, Et₃N; g, CrO₃·2Py; h, MCPBA;
i, LDA, THF, MeI; j, LiAlH₄; k, acetone, CuSO₄, TsOH.

In a straightforward fashion, 13 was transformed into δ-lactone 15 (Scheme I) which set the stage for introduction of a fourth chiral center at C(16). Alkylation of lactone 15 proceeded cleanly, which is undoubtedly due to the presence of the axial methyl substituent at C(4) [calcimycin numbering] which directs the entry of the new methyl group from the β-face

of the molecule presumably via a pseudo boat conformation. This was indeed a pleasant surprise in view of our experience in the methymycin series where alkylation of lactone 19 led

to a 1:1 mixture of lactones 20.[10]

With four of the five chiral centers established, we proceeded to address the task of elabor-
ating the remaining four carbon atoms including the chiral center at C(10). Toward this end,
lactone 16 was reduced to the corresponding triol which was immediately converted into acetonide 17 leaving the hydroxyl bearing C(7) carbon exposed for incorporation of the remaining
carbon atoms. Note that during the lithium aluminum hydride reduction of lactone 16 the
tert-butyldimethylsilyl ether did not survive. Whereas silyl ethers are generally stable to
such reaction conditions, we have found that neighboring oxygen atoms will facilitate the
cleavage of silyl ethers during lithium aluminum hydride reductions.

Our plan for incorporating the remaining carbon atoms is illustrated in Scheme II. Addition
of vinyl magnesium bromide or vinyl lithium to aldehyde 21 derived from alcohol 17 was expected to provide vinyl carbinol 22 as the major product based on Cram's rule.[11] While it
would be unreasonable to expect complete (100%) selectivity, we were not particularly concerned since the isomeric vinyl carbinol 26 could also, in principle, be transformed into 24
(vide infra). Given the configuration at C(7) in vinyl carbinol 22, it was anticipated that
ester enolate Claisen rearrangement[12] of the E-O-silylketene acetal 25 derived from propionate ester 23 would provide access to the C(1)-C(11) fragment 24 bearing the additional
chiral center at C(10).

SCHEME II

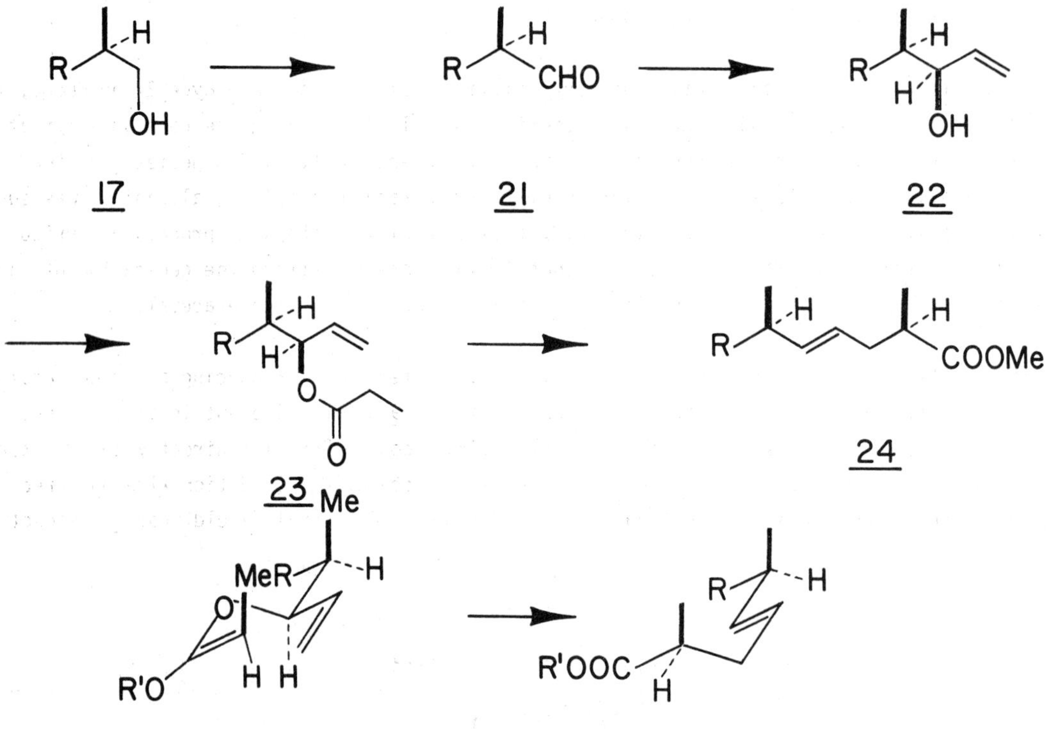

17

21

22

23

24

25

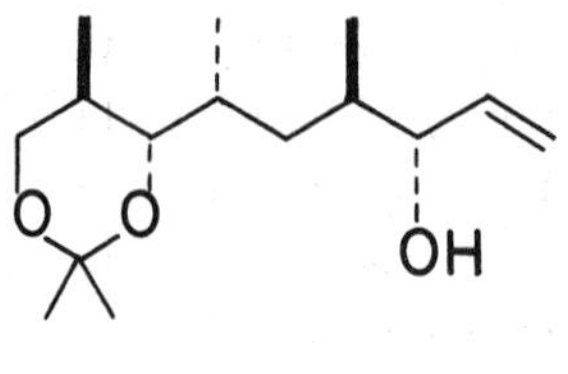

26

Collins oxidation of alcohol 17 and subsequent treatment of aldehyde 21 with vinyl magnesium
bromide afforded a 75% yield of vinyl carbinol 22 and the isomeric alcohol 26 in a 3:1 ratio.
Fortunately alcohol 22 could be separated from 26 by conventional chromatography on silica
gel. Reaction of pure vinyl carbinol 22 with propionyl chloride followed by application of

the Claisen rearrangement[12] utilizing the E-O-silylketene acetal 25 (R' = t-Bu(Me)$_2$Si), gen-erated from lithium diisopropylamide in tetrahydrofuran and subsequent addition of t-butyl-dimethylchlorosilane in HMPA, gave rise, via the chair-like transition state 25 to the silyl ester corresponding to 24. Exposure of the silyl ester to fluoride ion followed by treatment of the resultant acid with an etheral solution of diazomethane generated (90% overall from 22) the C(1)-C(11) fragment 24. Proton NMR analysis at 600 MHz of the crude methyl ester 24 derived from the ester enolate claisen rearrangement of 23 revealed a single substance as evidenced by only four distinct doublet methyls. In addition the carbon spectrum of 24 sug-gested the presence of a single compound.

As pointed out above, condensation of vinyl magnesium bromide with aldehyde 21 produced, in addition to the major product 22, the isomeric carbinol 26. Knowing the configuration at C(7) in compound 26 one can convert 26 into 24 by inverting the olefin geometry of the intermediate O-silylketene acetal prior to Claisen rearrangement. Thus, alcohol 26 was suc-cessfully transformed, in comparable yield, into 24 employing the same procedure carried out on vinyl carbinol 22 with the exception that HMPA was present during the generation of the ester enolate so as to ensure specific formation of the Z-O-silylketene acetal.[12]

With the C(1) to C(11) fragment 24 intact we concentrated on reintroducing an oxygen atom into the C(7) position. The transformation of 24 into 9 was carried out in an efficient manner. Glycolation (OsO$_4$) of the C(7)-C(8) olefinic double bond led directly to a mixture of two crystalline hydroxy γ-lactones which were not separated. Oxidation with buffered pyridinium chlorochromate[13] provided keto lactone 27 in 87% overall yield from 24. Reductive

27

cleavage of the C(8)-oxygen bond of 27 was carried out in calcium in liquid ammonia at -60°C. Esterification (CH$_2$N$_2$) of the resultant acid gave (87% for two steps) keto ester 9.

The primary objective of constructing the C(1)-C(11) acyclic fragment 9 was thus successfully completed in a stereospecific fashion. Prior to incorporating the benzoxazole unit 10 into the C(1)-C(11) fragment, it was necessary to protect the C(7) carbonyl function. Toward this end, 9 was treated with 0.01 M p-toluenesulfonic acid in methanol at -15°C. The anticipated ketal 28 was isolated in good yield. In contrast, when the reaction temperature was raised to 0°C or above, the bicyclo[3.3.1]nonane system 29 was isolated as the sole product. Attracted to this internally protected ketal, we set out to append the benzoxazole unit onto the C(11) carbon atom of 29.

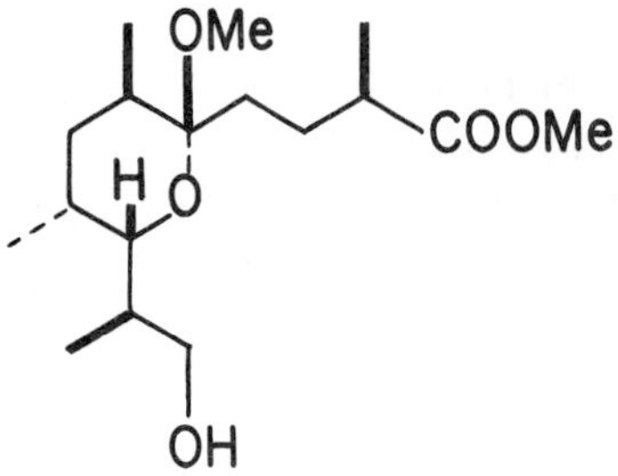

28 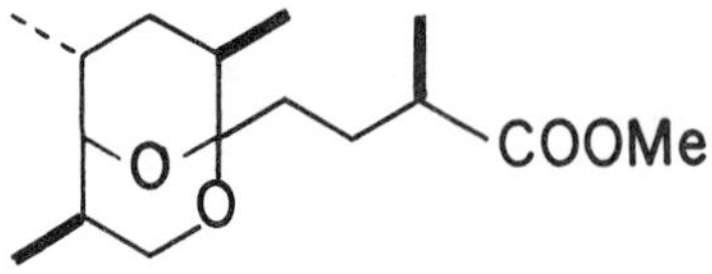   **29**

The requisite benzoxazole 10 was synthesized in a straightforward manner from the known methyl 5-hydroxyanthranilate (30) as previously described.[14] Trifluoroacetylation (TFAA, CH_2Cl_2, Py) of 30 provided the trifluoroacetamide 31, mp 140-141°C, in 95% yield. Mono-

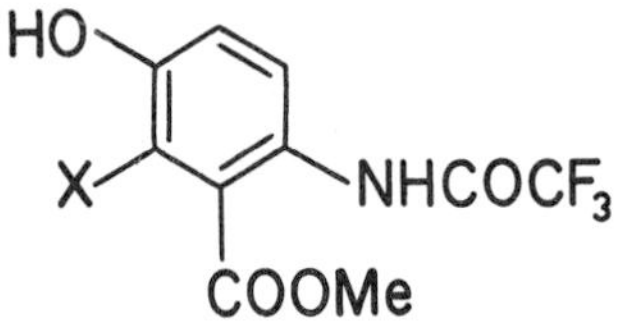

31   X = H

32   X = NO_2

33   X = NH_2

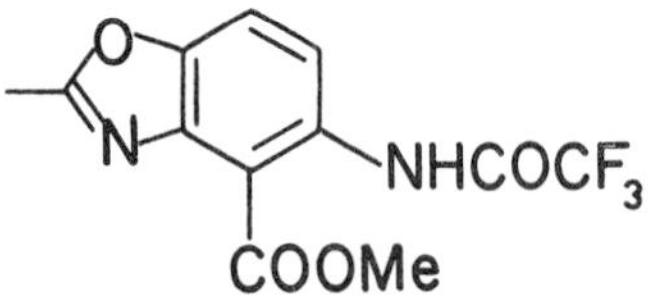

34

nitration of 31 (HNO_3, CH_3NO_2) gave 32 in 62% yield along with 31% of the corresponding 4-nitro derivative. Reduction (H_2, 10% Pd-C, EtoAc) of the nitro group afforded in 86% yield amine 33, mp 170-171°C, which upon treatment with acetyl chloride in refluxing xylene gave rise (94%) to benzoxazole 34, mp 156-157°C. Methylation (MeI, K_2CO_3, acetone) of 34 provided in near quantitative yield 10, mp 99-100°C.

With the ester 29 and benzoxazole 10 on hand, we addressed the question of combining the two fragments. Reduction of ester 29 with lithium aluminum hydride provided alcohol 35 in quantitative yield. Collins oxidation of 35 followed by immediate condensation with the lithiated derivative of benzoxazole 10 (prepared at -100°C in tetrahydrofuran using lithium diisopropylamide) gave rise to a 44% yield of alcohol 36 as the major (anticipated) product

35   **36**

in addition to the minor isomeric alcohol. Treatment of 36 with camphorsulfonic acid in
methylene chloride gave as the sole product spiro ketal 37 in 88% yield.

37

Completion of the synthesis of calcimycin requires elaboration of the α-ketopyrrole unit.
Model studies toward this end have been carried out. Treatment of alcohol 35 with camphorsulfonic acid in methylene chloride smoothly rearranged in very high yield to a single spiro
ketal which upon Jones oxidation provided (> 95%) carboxylic acid 38. Acid chloride formation using oxalyl chloride followed by addition of pyrrole magnesium bromide led to keto-

38   **39**

pyrrole 39 in modest yield. Attempts to apply this methodology to intermediate 37 have to
date been unsuccessful. Similarly, utilization of the corresponding thiol and selenol esters
have been without success. Studies are now underway to solve this temporary roadblock.

Acknowledgements: This research was supported by a Public Health Research Grant
from the National Institute of Allergy and Infectious Diseases, the National
Institutes of Health NMR Facility for Biomedical Studies (Mellon Institute),

and, in part, by a Grant from G.D. Searle & Co.

References:

1. M.D. Chaney, P.V. Demarco, N.D. Jones, and J.L. Occolowitz, J. Am. Chem. Soc., 96, 1932 (1974).

2. D.R. Pfeiffer, R.W. Taylor, and H.A. Lardy, Ann. N.Y. Acad. Sci., 402 (1978).

3. D.A. Evans, C.E. Sacks, W.A. Kleschick, and T.R. Taber, J. Am. Chem. Soc., 101, 6798 (1979).

4. P.A. Grieco, E. Williams, H. Tanaka, and S. Gilman, J. Org. Chem., 45, 0000 (1980).

5. J.W. Westley, J.F. Blount, R.H. Evans, Jr., A. Stempel, and J. Berger, J. Antibiot., 27, 597 (1974).

6. A. Agtarap, J.W. Chamberlin, M. Pinkerton, and L. Steinrauf, J. Am. Chem. Soc., 89, 5737 (1967); M. Pinkerton and L.K. Steinrauf, J. Mol. Biol., 49, 533 (1970).

7. N. Otake, M. Koenuma, H. Kinashi, S. Sato, and Y. Saito, J.C.S., Chem. Commun., 92 (1975).

8. H. Kinashi, N. Otake, H. Yonehara, S. Sato, and Y. Saito, Tetrahedron Lett., 4955 (1973).

9. Cf. P.A. Grieco, T. Takigawa, and D.R. Moore, J. Am. Chem. Soc., 101, 4380 (1979).

10. P.A. Grieco, Y. Ohfune, Y. Yokoyama, and W. Owens, J. Am. Chem. Soc., 101, 4749 (1979).

11. D.J. Cram and F.A. Abd Elhafez, J. Am. Chem. Soc., 74, 5828 (1952).

12. R.E. Ireland, R.H. Mueller, and A.K. Willard, J. Am. Chem. Soc., 98, 2868 (1976).

13. E.J. Corey, and J.W. Suggs, Tetrahedron Lett., 2647 (1975).

14. P.A. Grieco, K. Kanai, and E. Williams, Heterocycles, 12, 1623 (1979).

MACROLIDE SYNTHESIS:
6-DESOXYERYTHRONOLIDE B VIA ALDOL
CONDENSATIONS

S. Masamune

*Department of Chemistry, Massachusetts Institute of Technology, Cambridge,
Massachusetts 02139, USA*

Abstract - Macrocyclic natural products, represented by several well-known
antibiotics such as erythromycins, are large-sized lactones with an array of
substituents systematically attached to the monocyclic ring system. Despite
the structural and stereochemical complexity of these molecules, a general,
practical synthetic approach now appears feasible. A few newly developed
methods for the stereoselective aldol condensation have solved some of the
major synthetic problems posed by molecules of this type. This lecture
consists of a structural analysis, discussions of stereochemical control in
synthesis, and the application of the new methods to the construction of
6-desoxyerythronolide B.

BACKGROUND AND SYNTHETIC STRATEGIES

Macrolide antibiotics have attracted intense synthetic interest in recent years (Ref. 1).
These compounds include several polyoxo macrolides such as methymycin (1), pikromycin (2),
erythromycin A (3), leucomycin A_1 (4), and tylosin (5). As revealed by the unique structures

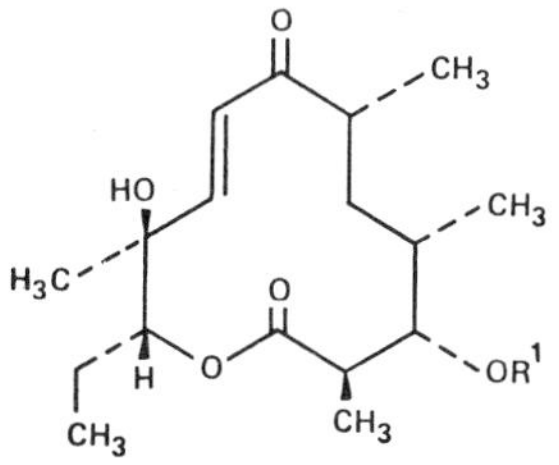

1 : R^1=desosaminyl

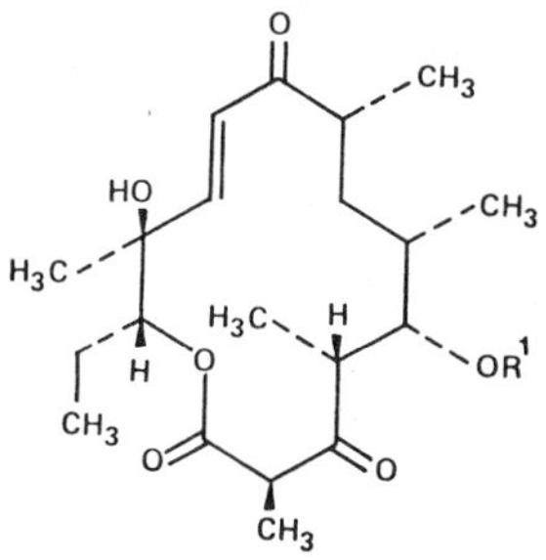

2 : R^1=desosaminyl

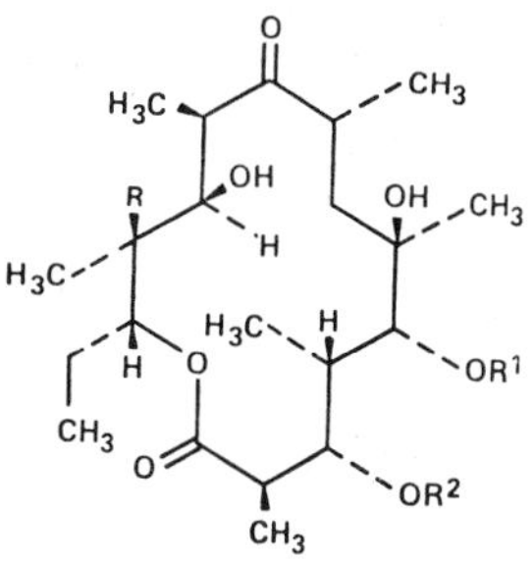

3 : R=OH
R^1=desosaminyl
R^2=cladinosyl

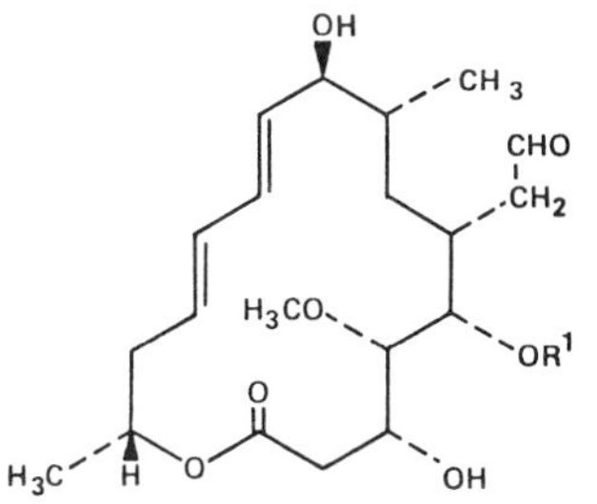

4 : R^1=mycarosyl-mycaminosyl

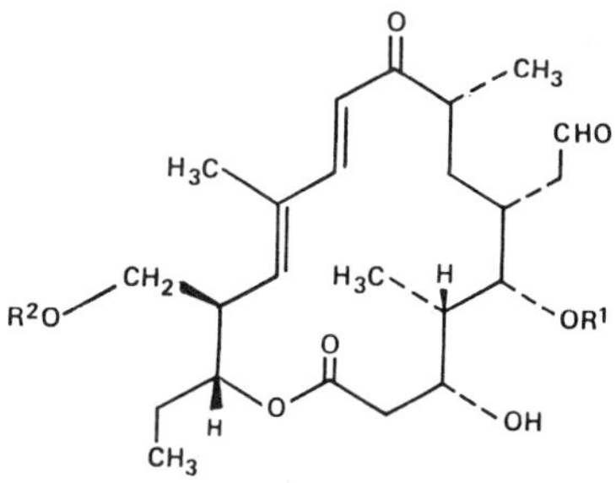

5 : R^1=mycarosyl-mycaminosyl
R^2=mycinosyl

shown above, the synthesis of these compounds poses two fundamental problems: (1) the construction of a medium- or large-sized lactone and (2) the stereochemical control of numerous
chiral centers embedded in the ring system. The attachment of a sugar (R^1) or sugars (R^1,R^2)
to an aglycone may present an additional challenge in many cases.

197

Many attempts have been made to solve the first problem (Ref. 1). Despite the size of macrolide rings, lactonization of the corresponding seco-acids has now proven to be a practical
solution. This is illustrated in the first synthesis of 1 in which the <u>tert</u>-butylthiol
ester (6) (Note a) of the methynolide seco-acid undergoes smooth ring closure with the aid
of a thiophilic reagent, $Hg(CF_3CO_2)_2$ (Ref. 2). This and several other methods to activate

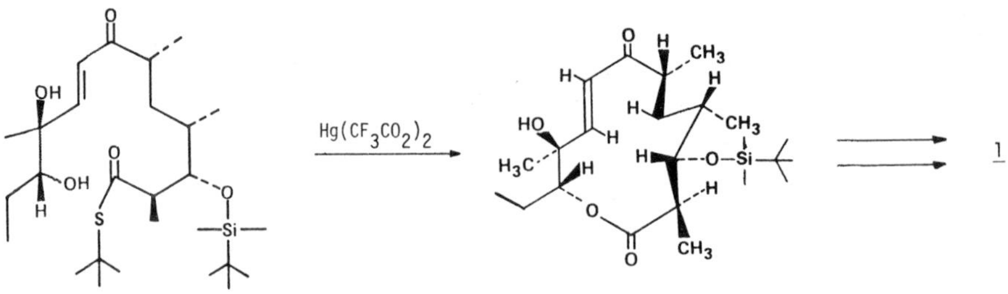

6

Scheme 1.

a carboxy group have been used successfully for the construction of many naturally occurring
macrocyclic lactones. Macrolides are conformationally rigid molecules owing to the heavy
substitution of the ring. Much of the rigidity is apparently retained in the corresponding
seco-acids, which thus may persist in a conformation favoring an intramolecular reaction
with proper activation. Synthetic schemes based on this seco-acid approach are now standard
in macrolide synthesis.

The synthetic methodology needed to overcome the second fundamental problem has been developed primarily along two major pathways. A conventional method makes use of the clearly
defined cis- and trans-relationships of substituents on a cyclic system. Ring cleavage
transfers this stereochemistry to the resulting acyclic system. In fact, this technique has
been utilized in the synthesis of methymycin (Ref. 2) and erythronolides A (R=OH, $R^1=R^2$=H in
3) (Ref. 3) and B (R=$R^1=R^2$=H in 3) (Ref. 4). In contrast, the construction of chiral
chain segments from acyclic precursors has not been widely explored, despite its obvious
advantage. Repetition of such a synthetic operation, if executed with high diastereoselection, would minimize the number of steps leading to a seco-acid.

We have selected as a synthetic target 6-desoxyerythronolide B (7), a metabolite leading to
all the erythromycins presently known (Ref. 5). Since 7 possesses a structural unit (C_3-C_9)
also found in two other macrolides 1 and 2, all three compounds can be synthesized from a

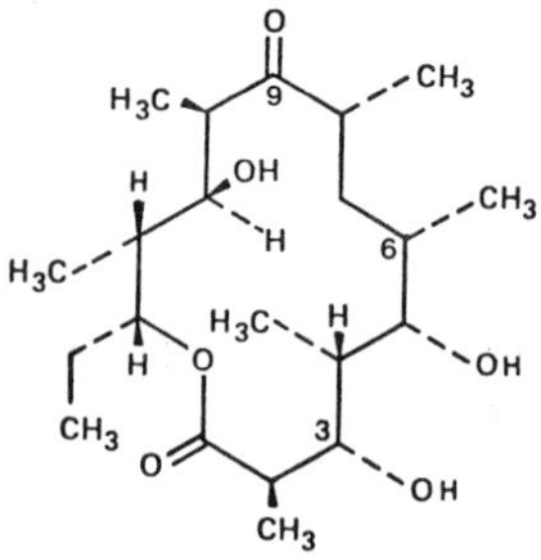

7

common intermediate. The presence of 1,3-diol and 1,3-hydroxycarbonyl moieties in 7 immediately suggests that the biosynthetic building blocks are exclusively propionates (one

Note a. This nomenclature is incorrect, but is used for simplicity and clarity throughout
        the lecture. The correct name is 2-methylpropane-2-thiol ester.

propionyl CoA and six methymalonyl CoA's). In chemical synthesis such units can be assembled in a simple and economical fashion through consecutive stereoselective aldol condensations, and this strategy may provide a general solution for the second problem mentioned earlier.

Once the seco-acid aldol approach is adopted for macrolide synthesis, a scheme usually can be designed in a straightforward manner. Thus, splitting the target seco-acid derivative (8), drawn in a zigzag fashion, into fragments A and B immediately suggests the order of aldol condensations to be utilized in the synthesis (Scheme 2). Aldol I produces fragment A, while aldol III and IV complete a synthesis of fragment B. Finally, A and B are combined via aldol II. Each aldol reaction creates two new chiral centers at the 2- and 3-positions of the product, relative to those pre-existing in the reacting aldehyde. The stereochemical problem, thus, is complex. The new chiral centers (at C_2 and C_3) must be formed with

correct relative stereochemistry and correct stereochemistry in relation to existing contiguous centers (e.g. C_4).

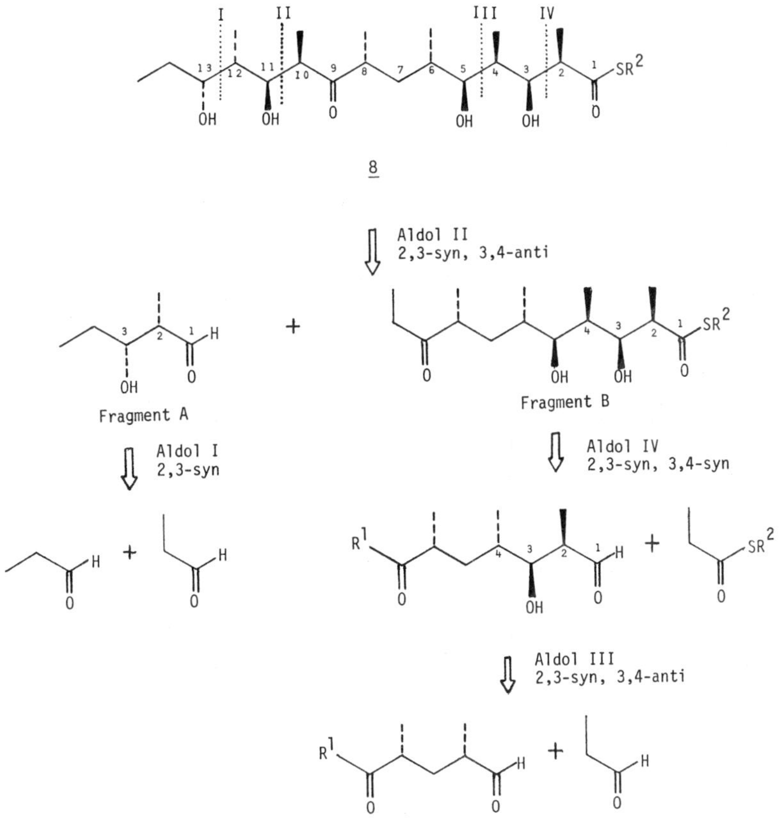

Scheme 2.

Since the conventional erythro/threo nomenclature is sometimes ambiguous due to the arbitrary choice of similar groups, we have resorted to a more convenient system of stereochemical designation. The main chain of a molecule is drawn in a zigzag fashion and two substituents on the same side (of the plane defined by the main chain) are designated syn and those on opposite sides anti. Thus the stereochemistry of an acyclic system is unambiguously defined.

All four aldol reactions mentioned above must proceed in the 2,3-syn fashion, while the 3,4-stereochemistry must be controlled in either the syn or anti fashion as indicated in Scheme 2. Moreover, the use of the optically active fragments A and B is necessary to avoid obvious complications. This lecture concerns recent developments in the stereochemical control of the aldol condensation reaction and its application to the synthesis of 6-desoxyerythronolide B (7).

STEREOCHEMISTRY OF THE ALDOL REACTION

The history of the aldol condensation probably dates back to the origins of organic chemistry. However, it was only in the mid-1960s that we witnessed a breakthrough in the control of this reaction. Thus, Wittig devised a procedure that allowed us to direct the reaction in such a manner as to preselect a specific carbonyl compound as the enolate (Ref. 6). Subsequently, the stereochemical aspects of this reaction have received the keen attention of organic chemists. As mentioned above, the aldol reaction creates two different types of stereochemistry, "internally" at the 2 and 3 positions and "relatively" at the 3 and 4 positions (Ref. 7). These two subjects are elaborated upon below.

2,3-Stereochemistry

The knowledge accumulated mainly by Dubois (Ref. 8), House (Ref. 9), and Heathcock (Ref. 10) and their coworkers, prior to the initiation of our present study, disclosed the following important aspects of the aldol reaction. When an aldehyde is reacted with a ketone-derived enolate under equilibrating conditions, the thermodynamically more stable 2,3-anti product predominates regardless of the geometry of the enolate. If, however, the reaction is kinetically controlled, the $\underline{Z}$- and $\underline{E}$-enolates preferentially provide the 2,3-syn and 2,3-anti aldol products, respectively.

Let us analyze these observations in terms of the commonly proposed 6-electron chair-type transition states, A-D (Scheme 3). Examination of A and B, assumed for the reaction with the $\underline{Z}$-enolate ($\underline{Z}$-9), reveals that B is less stable than A due to the steric interaction

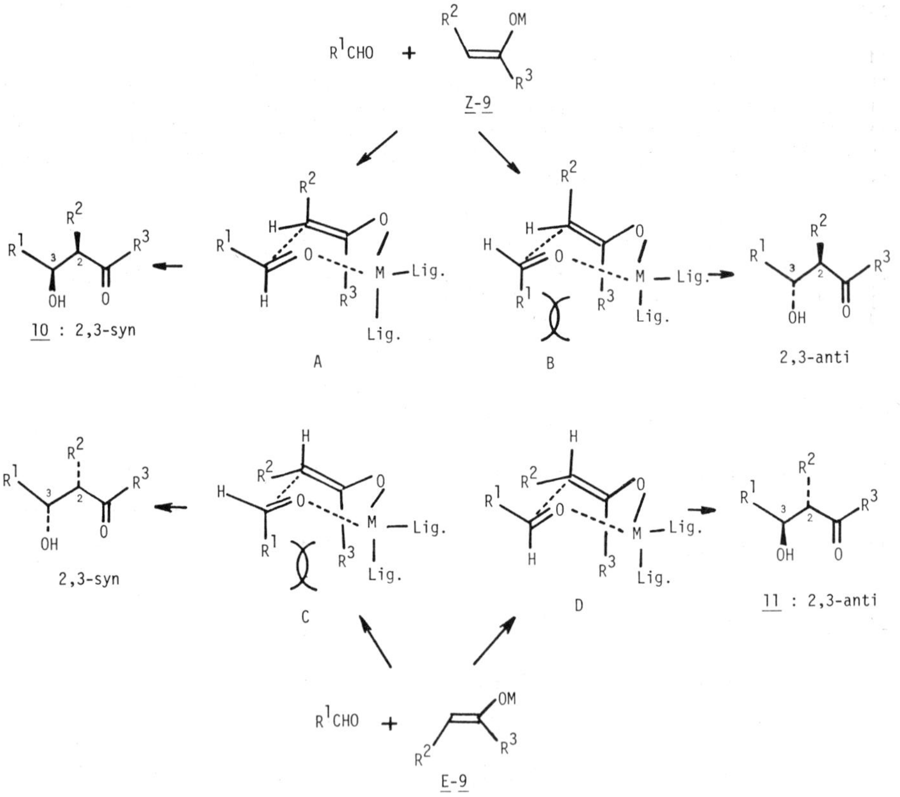

Scheme 3.

between R^1 and R^3; therefore, the kinetically controlled reaction proceeds through A, giving the 2,3-syn isomer ($\underline{10}$). In a similar manner the $\underline{E}$-enolate ($\underline{E}$-9) provides the anti-isomer ($\underline{11}$) through D. Although the stereoselection is excellent in certain cases, a search for improvement was warranted in order to achieve our objective in making the aldol condensation a generally applicable method in macrolide synthesis. If the transition state model used above is indeed correct, enhancement of the 2,3-stereoselectivity might be achieved by (1) compressing the transition state and (2) attaching bulky ligands (Lig.) to the metal.

Both approaches should result in increased interactions of R^1 with other groups.

A boron enolate appeared to be particularly suited for achieving these objectives: (1) both the B-C (1.5Å) and B-O (1.4Å) bonds are short (cf. Li-O, 1.9Å) (Ref. 11), (2) the affinity of boron toward an oxygen lone pair is high because of the empty orbital present on the metal, and (3) the attachment of a variety of ligands to the metal is possible. Our investigation was initiated with the examination of the boron-mediated aldol condensation.

Although several methods for generating boron enolates were known (Refs. 12-15), very little information was available concerning the stereoselective generation of $\underline{E}$- and $\underline{Z}$-boron enolates under mild conditions. Also, the stereochemical behavior of these enolates in aldol condensations was virtually unknown (Note b).

The reaction of the α-diazoketone ($\underline{12}$) with a trialkylborane (BR_3^1) achieves transfer of an alkyl group (R^1) from boron to the carbonyl α-position (Ref. 12). The final intermediate of this reaction is known to be a boron enolate ($\underline{13}$), so its stereochemistry has been

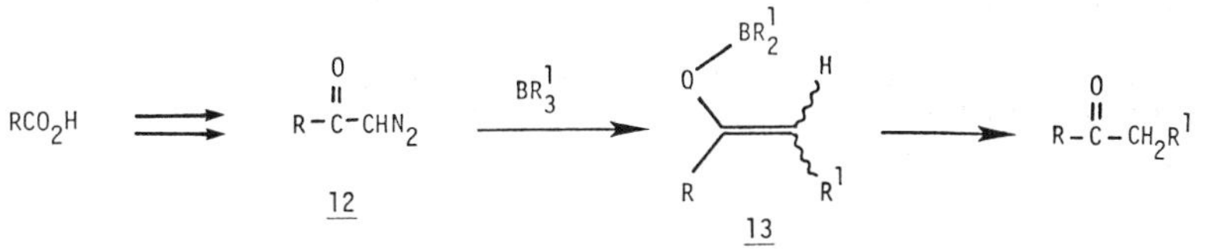

Scheme 4

scrutinized. Surprisingly and inexplicably, $\underline{13}$ is found to exist exclusively as the $\underline{E}$-isomer, which, in turn, can be isomerized to the corresponding $\underline{Z}$-isomer ($\underline{Z}$-$\underline{13}$) upon treatment with a trace amount of lithium phenoxide or pyridine. Moreover, $\underline{Z}$-$\underline{13}$ undergoes smooth aldol condensations with a variety of aldehydes to afford solely 2,3-syn products. The reaction with $\underline{E}$-$\underline{13}$ also proceeds to give predominantly the 2,3-anti aldol product, although the stereoselectivity is not as impressive as that for $\underline{Z}$-$\underline{13}$. The results are summarized in Table 1.

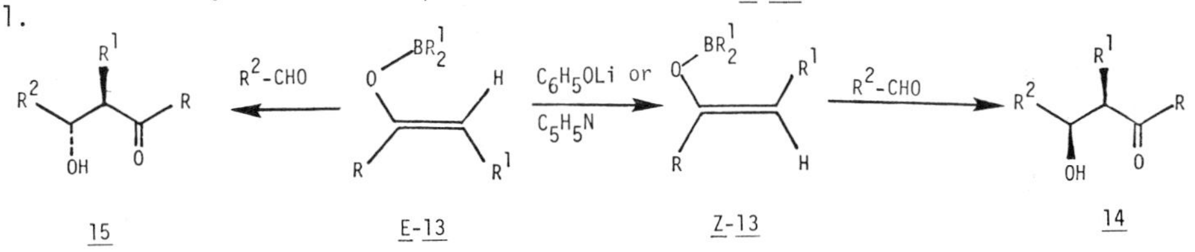

| 15 | E-13 | Z-13 | 14 |

Scheme 5. $R^1 = \underline{n}\text{-}C_4H_9$

TABLE 1. Aldol Reactions with $\underline{E}$-and $\underline{Z}$-Boron Enolates ($\underline{Z}$-$\underline{13}$ and $\underline{E}$-$\underline{13}$)

Enolate ($R^1 = \underline{n}\text{-}C_4H_9$)		Aldehyde	Ratio of 14/15	Combined Yield (%)
$\underline{Z}$-($\underline{13}$):	$R=CH_3$	$R^2=C_6H_5$	>20 : 1	90
	$R=CH_2C_6H_5$	$R^2=C_6H_5$	>20 : 1	84
	$R=C_6H_5$	$R^2=C_6H_5$	>20 : 1	85
	$R=C_6H_5$	$R^2=C_2H_5$	>20 : 1	87
	$R=C_6H_5$	$R^2=2\text{-}C_3H_7$	>20 : 1	86
$\underline{E}$-($\underline{13}$):	$R=CH_3$	$R^2=C_6H_5$	1 : 3	92
	$R=CH_2C_6H_5$	$R^2=C_6H_5$	1 : 4	88
	$R=C_6H_5$	$R^2=C_6H_5$	1 : 3	86

Note b. Köster, using stereodefined enolates, performed stereoselective aldol condensations in two isolated instances (Ref. 15).

The sequence of reactions outlined in Schemes 4 and 5 converts a carboxylic acid into a 2-alkyl-3-hydroxycarbonyl derivative, the type of system we have searched for. However, the usefulness of this sequence is limited in macrolide synthesis, as the reagents, trimethylborane or dialkylmethylborane, exhibit very little propensity to transfer methyl groups (R^1=CH$_3$ in 14 or 15) from the boron to the carbon atom, a prerequisite to a successful execution of the above reaction. Thus, exploration of another route to boron enolates was warranted at this stage.

Among several other methods, Mukaiyama's procedure is especially attractive (Ref. 14). Boron enolates are generated in high yield from ketones upon treatment with a dialkylboron trifluoromethanesulfonate and diisopropylethylamine, conditions mild enough to be applicable even to sensitive compounds. The aldol condensations planned in the synthesis of our target molecule 7 involve several enolates, one derived from an ethylketone, and the others from thiol esters. It has been found that the geometry of boron enolates, generated under Mukaiyama's conditions from these carbonyl compounds, is extremely sensitive to the kind of ligands attached to the metal. Thus, 9-borabicyclo[3.3.1]non-9-yl trifluoromethanesulfonate (9-BBN triflate 16) converts ethyl cyclohexyl ketone (17) and S-phenyl propanethioate (18) exclusively into enolates 17' and 18', respectively. Note that both enolates carry the methyl group and the oxygen functionality on the same side of the double bond and hence provide 2,3-syn adducts (19 and 20) as expected (Table 2).

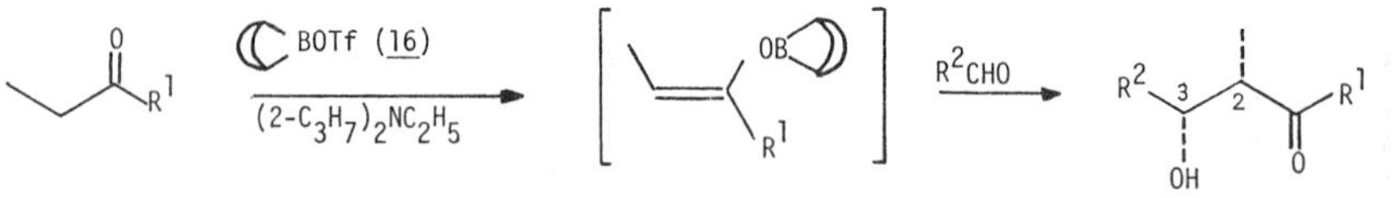

17 : R^1=c-C$_6$H$_{11}$

18 : R^1=SC$_6$H$_5$

17'

18'

19 : R^1=c-C$_6$H$_{11}$

20 : R^1=SC$_6$H$_5$

TABLE 2. Stereospecific Synthesis of syn-3-Hydroxy-2-methylcarbonyl Compounds

Carbonyl Compound	R^2CHO	Ratio of 19 (or 20) to its anti-isomer	Combined Yield (%)
17 : R^1=c-C$_6$H$_{11}$	R^2=C$_6$H$_5$	>30 : 1	79
	R^2=2-C$_3$H$_7$	>30 : 1	87
	R^2=CH$_2$CH$_2$C$_6$H$_5$	>30 : 1	82
18 : R^1=SC$_6$H$_5$	R^2=C$_6$H$_5$	~30 : 1	90
	R^2=2-C$_3$H$_7$	~30 : 1	79
	R^2=CH$_2$CH$_2$C$_6$H$_5$	>30 : 1	75
	R^2=c-C$_6$H$_{11}$	>30 : 1	77
	R^2=CH$_2$CH$_3$	>30 : 1	75

Surprisingly, replacement of 16 with dicyclopentylboron triflate (21) apparently reverses the geometry of the enolates, as treatment of 17 and of 22 with 21, both followed by aldol condensations, leads to the predominant formation of the 2,3-anti isomers, 23 and 24, respectively (see Table 3).

Incidentally, the Z-enolate 25 cannot be derived stereoselectively from 22, but the reaction of methylketene with tert-butyl di-n-butylthioborinate 26 (Ref. 14a,b) does provide 25

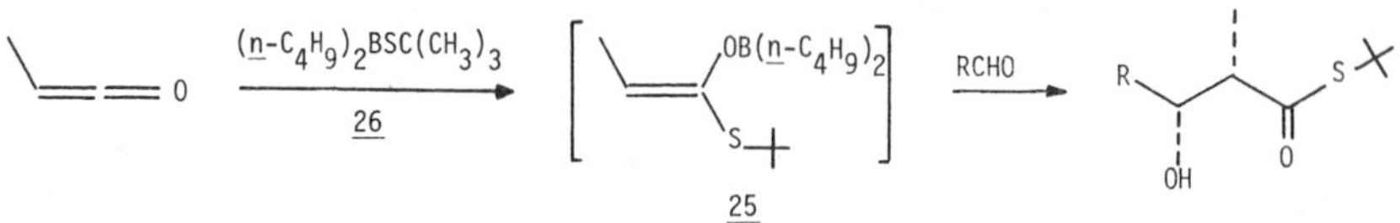

Scheme 6.

$17 : R^1=c-C_6H_{11}$

$22 : R^1=SC(CH_3)_3$

$17''$

$22''$

$23 : R^1=c-C_6H_{11}$

$24 : R^1=SC(CH_3)_3$

TABLE 3. Stereoselective Synthesis of anti-3-Hydroxy-2-methylcarbonyl Compounds

Carbonyl Compound	R^2CHO	Ratio of 23 (or 24) to its syn isomer	Combined Yield (%)
$17 : R^1=c-C_6H_{11}$	$R^2=C_6H_5$	>20 : 1	88
	$R^2=2-C_3H_7$	>20 : 1	86
	$R^2=CH_2CH_2C_6H_5$	>20 : 1	88
$24 : R^1=SC(CH_3)_3$	$R^2=C_6H_5$	6.1 : 1	83
	$R^2=2-C_3H_7$	7.3 : 1	79
	$R^2=CH_2CH_2C_6H_5$	5.7 : 1	68

almost exclusively (Note c).

Our findings described above disclose the uniqueness and utility of the boron enolates in the control of aldol 2,3-stereochemistry. The knowledge gained will be applied to the total synthesis of 7.

3,4-Stereochemistry
The planned total synthesis of 7 uses aldol condensations with several chiral aldehydes (26), all of which carry a methyl substituent at the 2 position. The 3,4-stereochemical problem that emerges in these instances is closely associated with the 1,2-asymmetric induction

26

observed in the reaction of a carbonyl compound with a nucleophile. Extensive studies on the latter subject have led to the proposal of several models (Cram, Conforth, Karabatsos, and Felkin) (Ref. 18) for the conformation of the carbonyl compound in the transition state. Theoretical evaluations of these models have also been made recently (Ref. 19). However, the problem is by no means fully resolved, and the aldol condensation represents a particularly complicated example.

In order to define several terms to be used in this lecture, the stereochemical consequences of the aldol reaction should be delineated at this point. Let us place the aldehydic group $-C{\displaystyle{O \atop H}}$ of 26 in the plane of the paper in order to define the α-(behind the plane) and β-side (in front of the plane) of the aldehyde. Further, let us maintain the assumption that the aldol proceeds through transition states of types A and D depicted in Scheme 3. It is now evident that Z-enolates attacking 26 from the α-side (A' in Scheme 7) will give (2,3-syn), 3,4-anti product 27, regardless of the relative conformation of R-CH(CH_3)- and the carbonyl

Note c. After our results summarized above were published (Ref. 16), Evans and coworkers disclosed their independent work related to ours (Ref. 17).

OS – H•

in 26. Similarly, approach of the same enolate from the opposite side (β-approach, A" in Scheme 7) will lead to the formation of (2,3-syn), 3,4-syn product 28. The stereochemical outcome of the α- or β-approach of an E-enolate on 26 will be the same with respect to the 3,4-stereochemistry (see D' and D").

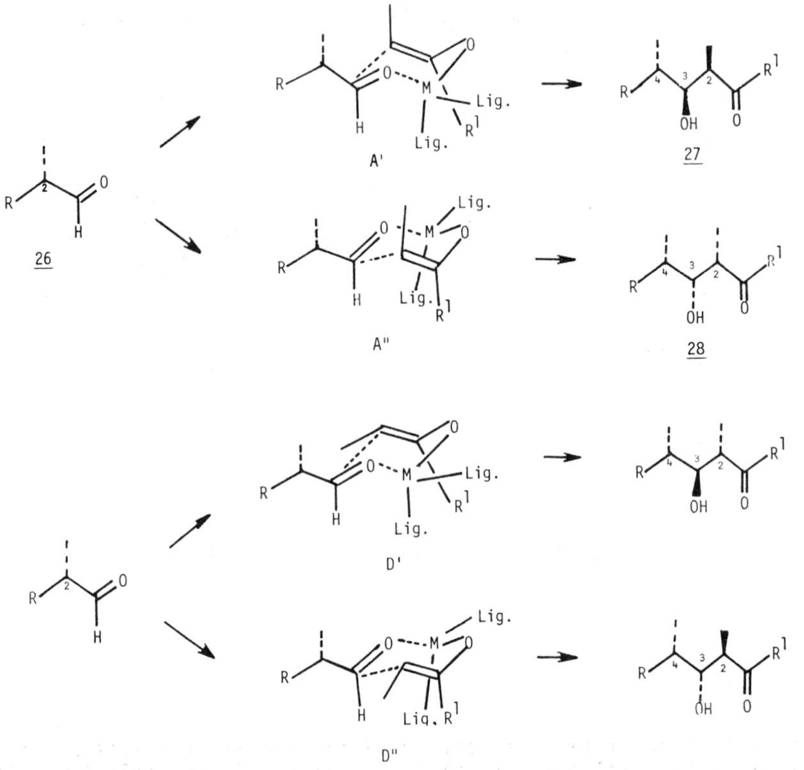

Scheme 7.

If a 3,4-syn product is formed, the reaction is said to follow Cram's rule (Note d) with the product being termed the Cram product, and the aldehyde being called Cram selective.

There are several factors which influence the relative stability of transition state A' as compared to A" (or D' to D"). A combination of these will make the α approach of the enolate more favorable than the β approach or vice versa. These factors are:

  (1) steric and/or stereoelectronic effects exerted by substituents R attached
      to the reacting aldehyde 26;

  (2) intramolecular chelation of a hetero-substituent present in 26 with the metal
      cation;

  (3) chirality of the enolate;

  (4) chiral ligands attached to the metal.

We have investigated these factors with the aim of enhancing the 3,4-stereoselectivity of the reactions to be utilized in the total synthesis. Some of our findings are summarized below.

Cram/Anti-Cram Selectivity

Aldol reactions of 2-phenylpropionaldehyde 29 with both lithium and boron enolates 30 and 18' provide predominant amounts of the Cram product (entry 1 of Table 4). Aldehyde 31 behaves similarly (entry 2). However, introduction of a methyl group at the rather remote 4 position of 31 (see 32) brings about a significant change resulting in the formation of the two isomers in almost equal quantities. In contrast with 29, 2-cyclohexylpropionaldehyde

---

Note d. Cram's acyclic model is intuitively conceived and does not appear to have any
         physical meaning. The presentation of A', A", D', and D" is not meant to
         confine the conformation of R-CH(CH$_3$)- in 26 to that presupposed in Cram's model.

TABLE 4. Cram/Anti-Cram Product Ratio (28 : 27) of Aldol Condensations:

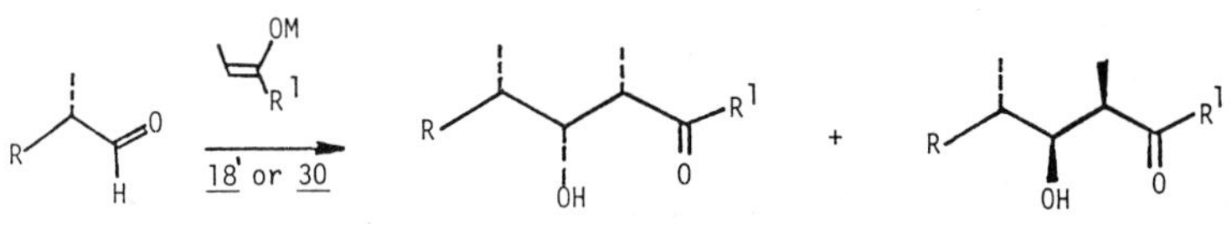

En-try	Aldehyde	Enolate	30 (OLi / OTMS)	$18'$ (OB / SC_6H_5)
			Ratio of 28 to 27 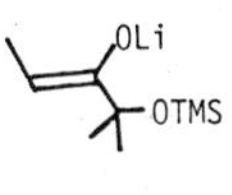	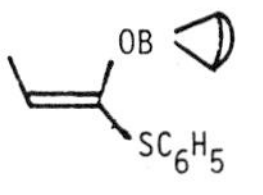
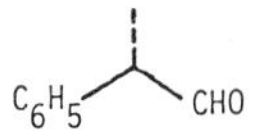 1	C_6H_5 CHO 29		4.3 : 1 (Ref. 10b)	2.3 : 1
2	H_3CO_2C CHO 31		2.8 : 1 (Ref. 10b)	
3	H_3CO_2C CHO 32		1 : 1.2	1 : 1.2
4	CHO 33		1 : 2.7	1 : 2.0

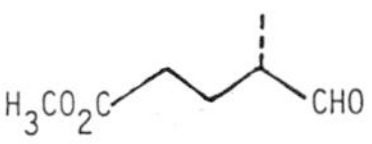

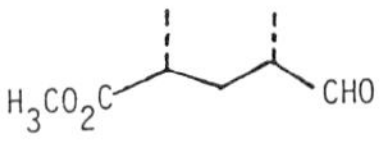

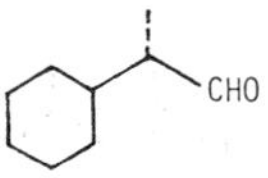

(33) clearly violates Cram's rule. The behavior of the last two aldehydes is, in fact, inexplicable by any model available at present. Although it is obviously difficult to predict the Cram/anti-Cram selectivity of a given aldehyde, the experimental results summarized in Table 4 show that variation of the (achiral) enolate has only a minor influence on the ratio of products derived from aldehydes 29, 32-33. Cram/anti-Cram selectivity is determined primarily by aldehyde rather than (achiral) enolate structure, a fact which becomes important later.

In the synthesis of a variety of macrolides, protected 3-hydroxyaldehydes 34 and 35 are

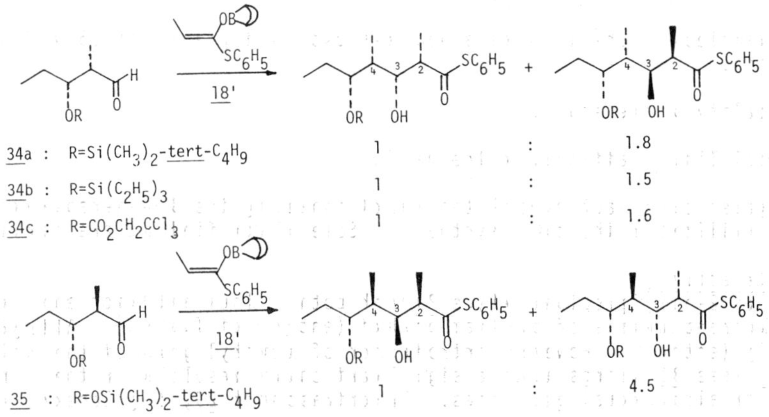

34a : R=Si(CH₃)₂-tert-C₄H₉ — 1 : 1.8
34b : R=Si(C₂H₅)₃ — 1 : 1.5
34c : R=CO₂CH₂CCl₃ — 1 : 1.6

35 : R=OSi(CH₃)₂-tert-C₄H₉ — 1 : 4.5

Scheme 8.

to be condensed with enolates; therefore, it is of prime importance to acquire information concerning the selectivity of these heteroatom-substituted aldehydes.

Interestingly, both syn- and anti-3-hydroxy-2-methylaldehydes favor the anti-Cram approach giving predominantly the 3,4-anti products as shown in Scheme 8, although not by a very substantial margin. These results emphasize that the 3-substituent of the aldehyde is as important as the 2-substituent in directing the enolate approach.

Attachment of a chelating methoxymethyl or benzyloxymethyl group to the 2,3-syn aldehyde, as in 36, enhances the anti-Cram selectivity, and now use of the lithium or magnesium enolate raises the ratio to more than 10:1 in favor of the 3,4-anti product (37).

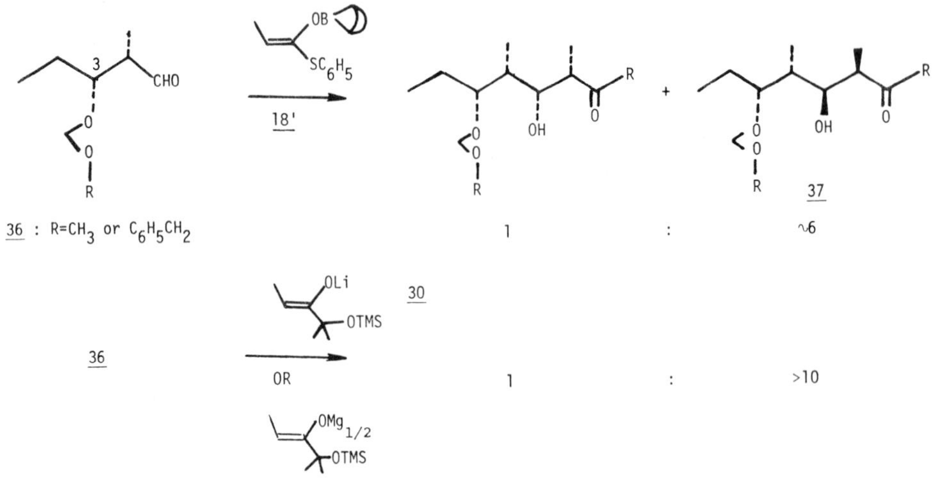

36 : R=CH$_3$ or C$_6$H$_5$CH$_2$

18'

36

30

OR

1   :   ~6

37

1   :   >10

Scheme 9.

It is very tempting to attribute this result to the type of intramolecular chelation used in Cram's cyclic model (Ref. 18b). In order to accommodate both the chelation with Li or Mg and the stereochemical outcome of the reaction, presumption of a (twist) boat-type 6-electron transition state (E) now appears to be required. The boron enolate presumably cannot chelate

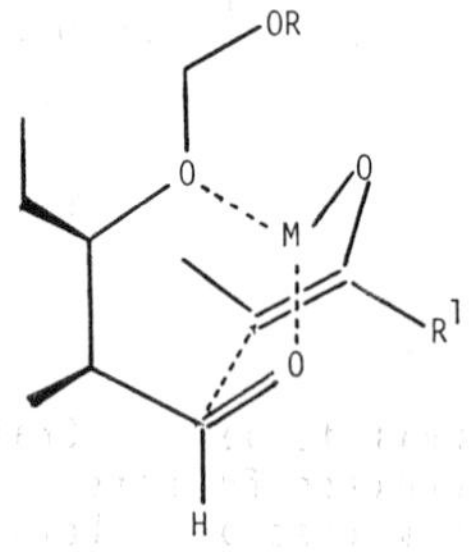

E

in a similar fashion as its coordination sites are filled by two alkyl and two oxygen substituents. The effect of the methoxymethyl or benzyloxymethyl groups in enhancing stereoselectivity with the boron enolates, thus, is not understood completely. The effect of substituents at the aldehyde 3 position can be used to advantage in some cases (Ref. 20).

Enantioselectivity
At an early stage of this project a small quantity of ethyl ketone 38 became available through the degradation of 6-desoxyerythronolide B (7). The corresponding boron enolate generated in the manner described earlier and reacted with benzaldehyde provides two products which appear to be isomeric as shown in 39 and 40. The fact that these isomers are produced

in unequal quantities (2:1 or 1:2) immediately drew to our attention the possibility of con-

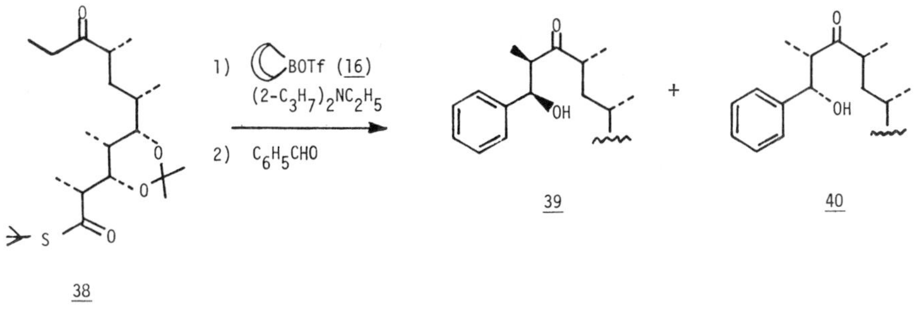

Scheme 10

trolling the 3,4-stereochemistry of aldol reactions with chiral enolate reagents. The chiral lithium enolate S-41, readily prepared from (+)-S-atrolactic acid, indeed provides a 3.5:1 ratio of 42 and 43, exhibiting what is tentatively termed enantioselectivity of the enolate (Note e). The β-side of the enolate (as defined in S-41) is apparently more accessible to the aldehyde owing to the preferred transition state conformation of the substituents at the indicated carbon(*) in S-41, resulting in the predominant formation of 42. 2-Cyclohexylpro-

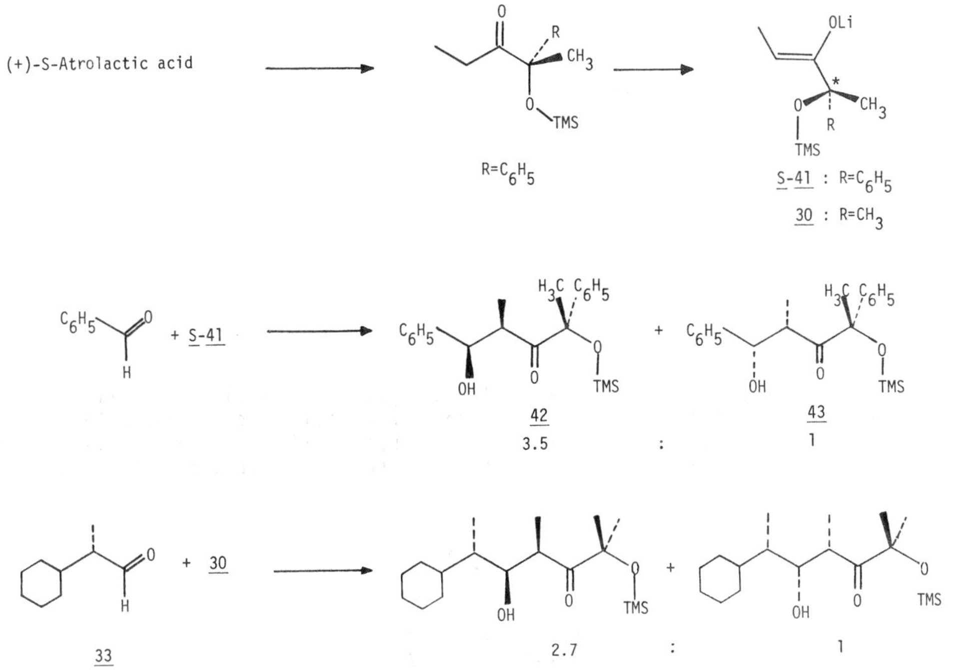

Scheme 11.

pionaldehyde (33) has already been shown to be anti-Cram selective, and therefore the S-enan-tiomer seems to prefer the enolate approach from the α-side of 33 (see A' in Scheme 7). Thus, upon combination of S-33 and S-41 (a matched pair) both the enantioselectivity of the enolate and the anti-Cram selectivity of the aldehyde should act in concert in such a manner as to enhance the 3,4-stereochemical selectivity of the reaction. On the other hand, the reaction of S-33 with the R-enantiomer of 41 (R-41) (a mismatched pair) should lead to inferior stereo-selection. These predictions are borne out by the results shown in Scheme 12 (8:1 in favor of the anti-Cram product (44) for the matched pair, and 1.5:1 in favor of the Cram product (45) for the mismatched counterparts). If it is naively assumed that these two selectivities operate multiplicatively, the ratios should be 9.5:1 and 1:1.3 in excellent (but obviously

Note e. Although the relative 2,3-stereochemistry in 42 and 43 is syn, the absolute config-
         uration of these centers has not been established beyond doubt. However, circum-
         stantial evidence from several related experiments indicates that the assignments
         shown in Scheme 11 are very likely to be correct.

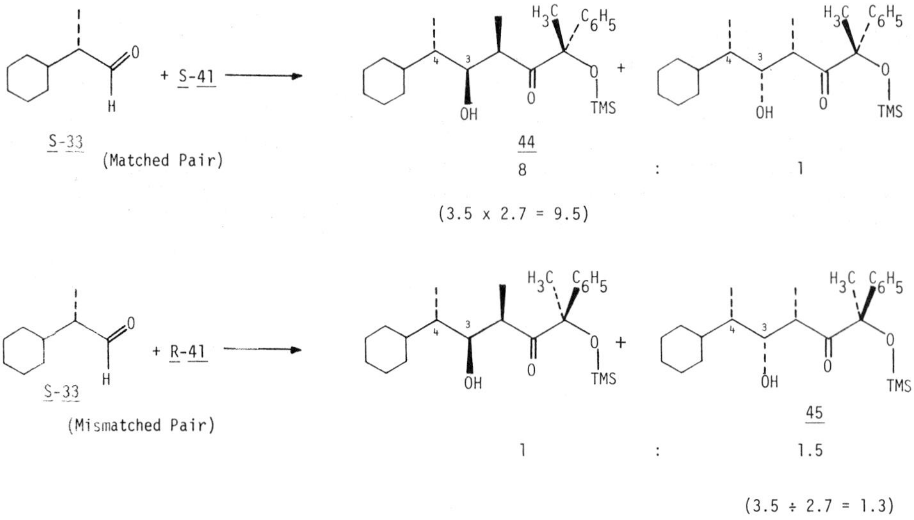

Scheme 12.

fortuitous) agreement with the experimental results (Note f). A corollary of the above
results is that if one can prepare a chiral enolate reagent which exhibits sufficiently high
enantioselectivity (i.e., 20:1), this should easily outweigh the normally small Cram/anti-
Cram selectivity to the extent that either the 3,4-syn or the 3,4-anti isomer could be
prepared as the predominant product. Indeed such reagents (e.g., 46 and 47 in Table 5) were
prepared and their uses will be presented below. The preparation of similar boron reagents
is currently in progress.

Table 5. Enantioselectivity of Chiral Enolates

Chiral Enolate	Enantio-selectivity	Aldehyde[a]
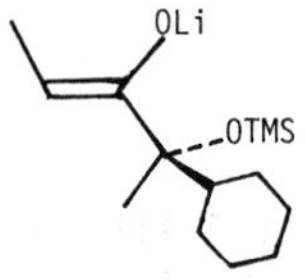 S-46	> 9 : 1	(C_6H_5CHO)
	> 15 : 1	$(C_6H_5CH_2CHO)$
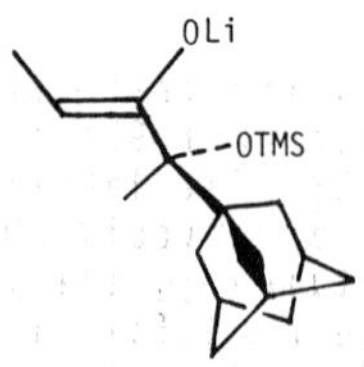 S-47	> 15 : 1	(C_6H_5CHO)
	> 25 : 1	$(C_6H_5CH_2CHO)$

[a]Aldehydes used to determine the enantioselectivity.

Note f. After these preliminary studies were completed (Ref. 21) and had been in fact
        presented in several public lectures, an elegant demonstration of the concept of
        double stereoselection in aldol condensation was disclosed by Heathcock et al.,
        (Ref. 22). More recently, closely related independent work by Hoffmann and Zeiss
        appeared in print (Ref. 23).

SYNTHESIS OF 6-DESOXYERYTHRONOLIDE B (<u>7</u>)

(1) Fragment B (see Scheme 2)

As described earlier, 6-desoxyerythronolide B (<u>7</u>), pikromycin (<u>2</u>), and methymycin (<u>1</u>) possess a common structural unit (C_3-C_9) which can be derived from Prelog-Djerassi lactonic acid (<u>48</u>) (Ref. 24). Four stereoselective syntheses of this latter compound reported in the recent literature (Refs. 2a, 25a-c) require more than a dozen steps. However, all chiral centers present in <u>48</u> can now be constructed stereoselectively through a single aldol reaction on the semi-aldehyde <u>32</u> derived from meso-dimethylglutaric acid, as shown in Scheme 13. This aldehyde shows a slight anti-Cram selectivity in the reaction with achiral enolates as shown

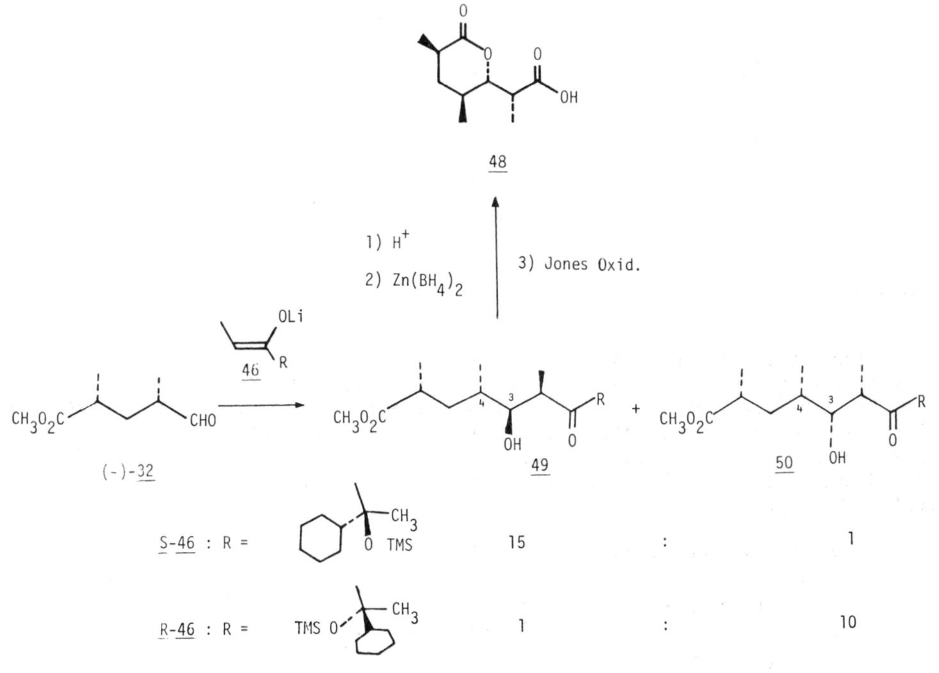

Scheme 13.

earlier (Table 4). However, reaction of (-)-<u>32</u> with the chiral lithium enolate of <u>S-46</u> discussed above proceeds with excellent diastereoselection to provide <u>49</u> and <u>50</u> in a ratio of 15:1. Furthermore, the combination of (-)-<u>32</u> with <u>R-46</u> results in the predominant formation of the 3,4-syn product (ratio 1:10), demonstrating that the small anti-Cram selectivity inherent to <u>32</u> can in fact be outweighed by a considerable margin. Conversion of the aldol product <u>49</u> into (+)-<u>48</u> is effected using the simple procedure shown above. Thus (+)-<u>48</u> is now readily available (Note g).

The next step in the sequence involves the addition of a propionate unit in the 3,4-syn fashion to either the lactonic aldehyde <u>51</u> (obtained from <u>48</u> in 95% yield via the Rosenmund reduction) or to the corresponding open-chain aldehyde <u>52</u>. The latter was not our choice because model studies indicated that the aldol reaction of <u>52</u> would almost certainly lead to the predominant formation of the 3,4-anti product. Fortunately, the Cram/anti-Cram selectivity of <u>51</u> with the boron reagent <u>18'</u> turned out to be 1.9-2.5:1 in favor of the desired product <u>53</u>, although the reason for this preference for the Cram product is not well understood at present. The use of a chiral reagent was anticipated to enhance the ratio to a great extent as in the previous case. However, the unusually high reactivity of the lactone group in <u>51</u> toward the basic lithium enolate <u>46</u> has created some complications which appear to demand the development of a new chiral boron, rather than lithium, reagent. This problem is currently being investigated.

Since the benzenethiol ester <u>53</u> has been found to be too labile to survive further synthetic transformations, <u>53</u> is converted with sodium tert-butylthiolate to the corresponding tert-butyl thiol ester <u>54</u> which is then subjected to a sequence of reactions: (1) lactone opening

Note g. The previous synthesis of methymycin (<u>1</u>) which utilized (±)-<u>48</u> as a key intermediate (Ref. 2) has thus been simplified considerably.

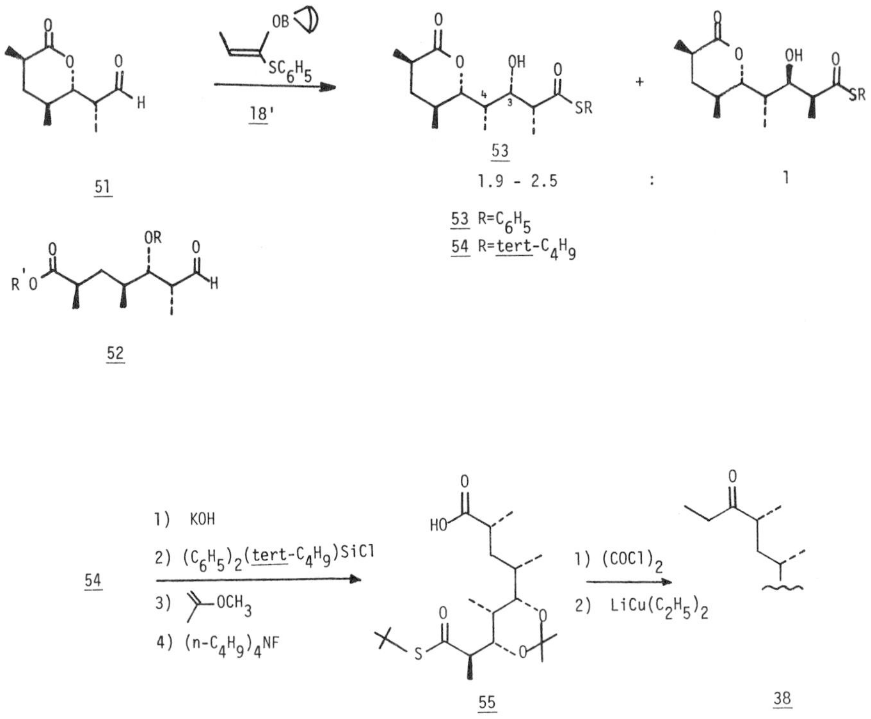

Scheme 14.

(KOH), (2) protection of the carboxylic acid, (3) preparation of the acetonide from the diol, and finally (4) liberation of the carboxylic acid. The resulting compound 55 has been found to be identical with the degradation product which represents the right-hand portion of 7 (see below). Further conversion of 55 into the corresponding ethyl ketone 38 completes this simple and efficient synthesis of the acetonide of fragment B in Scheme 2.

(2) Fragment A
The construction of fragment A is straightforward as it concerns only the 2,3-stereochemistry of a single aldol condensation. Thus, reaction of propanal with 18' provides the benzenethiol ester (±)-56, which is hydrolyzed to the racemic carboxylic acid (±)-57. Resolution of 57 with (+)-threo-p-nitrophenyl-2-amino-1,3-propane diol yields (+)-57, which has proven to possess the desired 2S,3R configuration by comparison with a degradation product of 7.

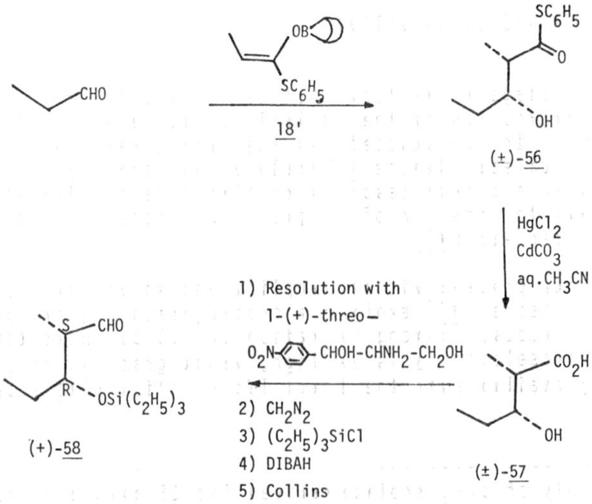

Scheme 15.

The proper choice of protecting group for the hydroxy group in 57 is critical at this stage. It must survive the conditions of the final aldol reaction combining fragments A and B, yet still be removed under conditions mild enough not to induce elimination of the hydroxy group β to the carbonyl function or hydrolysis of the acetonide. Unfortunately, the benzyloxy-methyl group cannot be used due to unexpected side reactions, although this group would probably enhance the anti-Cram selectivity of the β-hydroxyaldehyde as discussed earlier. The triethylsilyl group was finally chosen and the optically active fragment A, (+)-58, has been prepared in a conventional manner as shown in Scheme 15.

Degradation of 7 can be carried out as summarized in Scheme 16. After protection of the diol, dehydration, and reduction of the keto-group, the lactone ring is opened. The resulting carboxylic acid (59) is converted into the corresponding tert-butylthiol ester (60) using a standard method. Oxidation ($NaIO_4$-$KMnO_4$) provides two fragments, both of which are used to confirm the structures and stereochemistries of the corresponding synthetic materials (55 and 57).

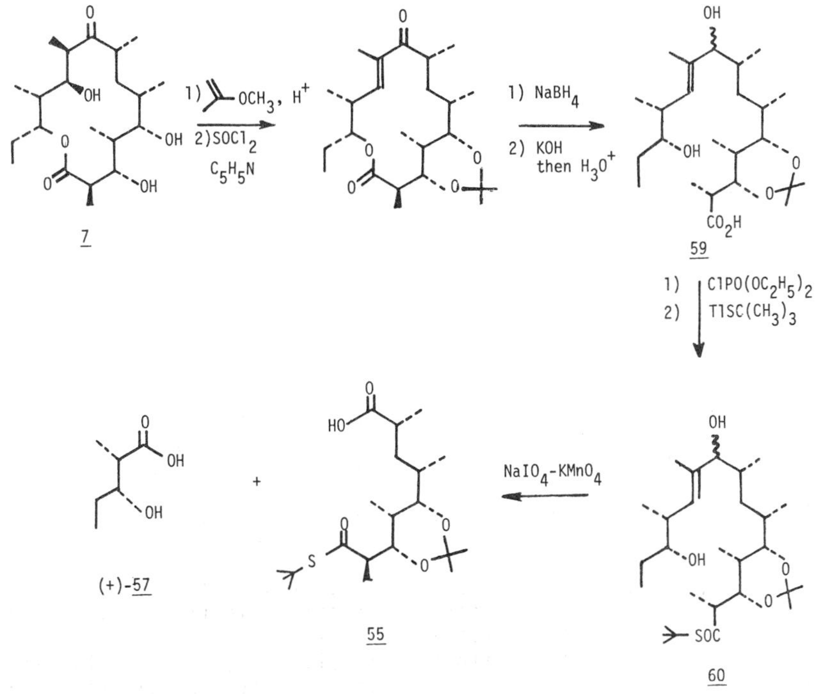

Scheme 16.

SYNTHESIS OF SECO-ACID AND LACTONIZATION

We have now reached the final stage of the total synthesis. An aldol condensation combining fragments A and B has been incorporated in the synthetic scheme based on information gained through earlier model studies. It is predicted that the chemo-, regio- and stereoselective generation of the Z-enolate from ethyl ketone 38 should proceed smoothly. Furthermore, since aldehyde (+)-58 is anti-Cram selective as described earlier (Note h), the expected course of the aldol reaction should lead to formation of the proper stereochemistry at the 10, 11, and 12 positions of the product, seco-acid 61.

In practice, the aldol reaction proceeds without complications to provide only two products (61 and 62). The predominant isomer (61) exhibits a proton-proton coupling pattern typical of 2,3-syn, 3,4-anti aldol products, a strong indication that 61 possesses the correct stereo-chemistry. Although the stereoselectivity is not high, we are gratified that the protected seco-acid (61) is now easily available and the direct lactonization of 62 appears to be imminent.

Note h.   The enantioselectivity of the Z-enolate derived from 38 appears to act in concert with the anti-Cram selectivity of (+)-58. However, a detailed discussion on the transition state conformation is abbreviated because of its complexity.

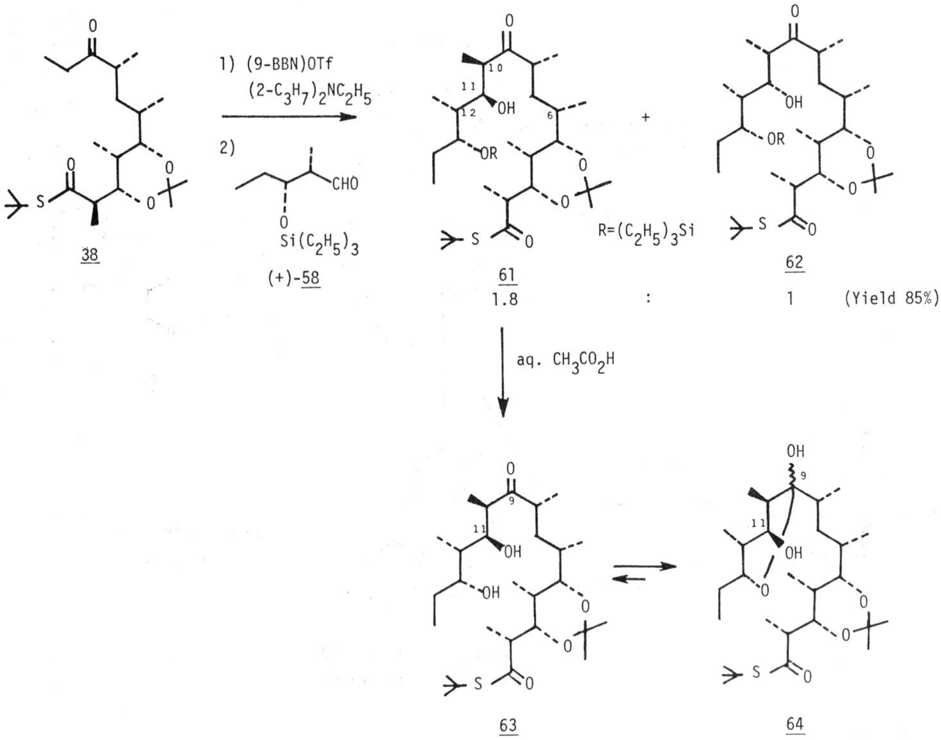

Scheme 17.

Removal of the triethylsilyl group from 61 with aqueous acetic acid yields the dihydroxy-ketone 63 which has been found to exist in equilibrium with the corresponding hemiketal 64 (Scheme 17). The equilibrium ratio of 63 and 64 in solution strongly favors the latter and remains virtually unchanged upon preparation of various C_{11}-hdyroxy derivatives. All

attempts to lactonize a mixture of 63 and 64 have been uniformly unsuccessful. This failure of direct lactonization has forced us to make synthetic detour via the C_9-hydroxy derivative, which, if successfully cyclized, could be oxidized to the C_9-keto compound preferentially in the presence of the free C_{11}-hydroxy group. It is already known that the C_{11}-hydrogen atom of erythronolide B (R = R^1 = R^2 = H in 3) is oriented (Ref. 26) toward the inside of the ring system and thus the C_{11}-hdyroxy group resists CrO_3 oxidation. Compound 61 has been reduced with $NaBH_4$ to give 1.4:1 mixture of the 9α- and 9β-hydroxy compounds (Note i) which, after separation, have been converted to the corresponding bis-dichloroacetates (65a,65b).

Lactonization of 65a to 66a and 65b to 66b is most efficiently achieved with excess CuOTf in benzene containing 2 equivalents of diisopropylethylamine to neutralize the strong acid liberated during the reaction (Scheme 18). Since the lactone formation proceeds with a delay relative to the disappearance of 65a and 65b, the mixed anhydride 67 shown below serves as a probable intermediate in the overall lactonization process. The noticeable differences in the cyclization yields between the two starting materials (41% from 65a and 23% from 65b) may well be attributed to the conformation of these compounds. Now that this critical step has been successfully executed, the ensuing transformations proceed rather uneventfully. The alkaline-labile dichloroacetate protecting group is removed, the C_9-

hydroxy group oxidized preferentially with pyridinium chlorochromate (see above), and finally the acetonide group hydrolyzed. This completes the total synthesis of 6-desoxy-erythronolide B.

Scheme 19 summarizes our achievement. Three aldol condensations effect the formation of the crucial C-C bonds (stout lines), and Cu(I) mediated thiol ester activation brings about

---

Note i.   As described in the following paragraph, both isomers were converted into 7 after
          lactonization and oxidation. Therefore, this low selectivity of the reduction is
          of no consequence.

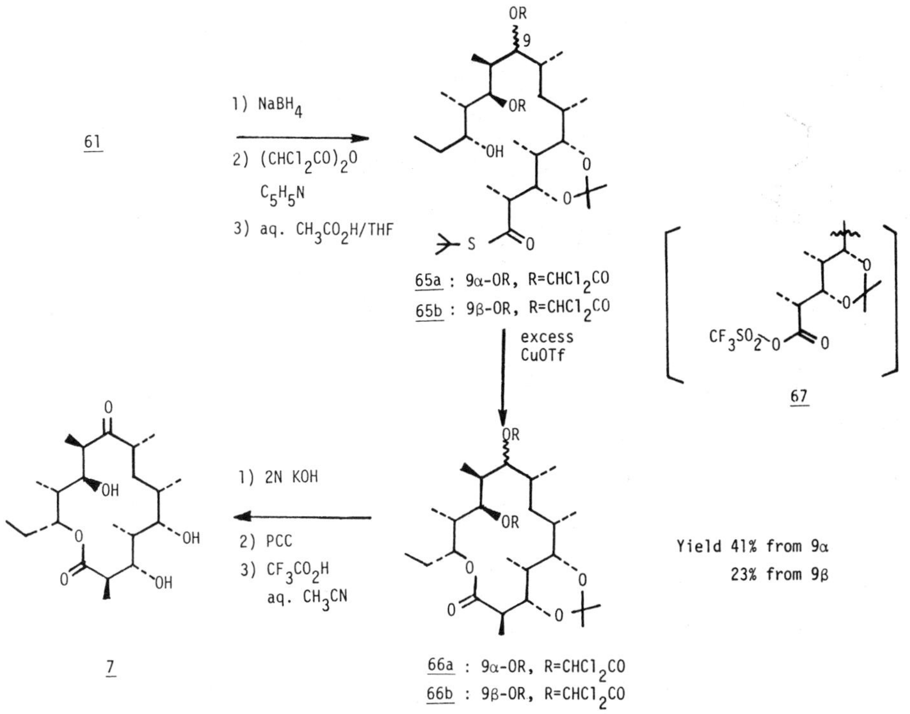

Scheme 18.

lactone ring formation, completing the synthesis of 6-desoxyerythronolide B. The total number of steps used to create 10 chiral centers is impressively small (approximately 20)

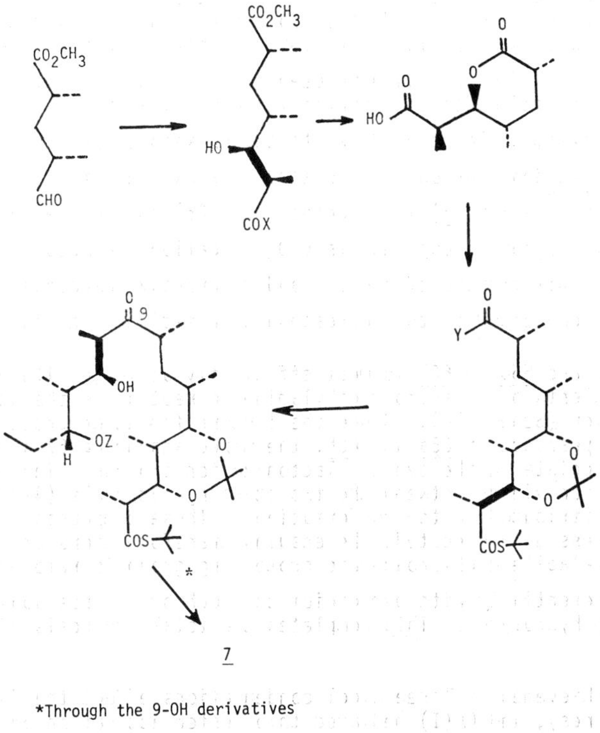

*Through the 9-OH derivatives

Scheme 19.

68

69

as compared with those required in recent syntheses of similarly complex macrolides. The number of steps would have been even less, had there been no complications in the lactonization stage.

The fundamentally important 3,4-stereochemical problem admittedly has not been fully explored, but is being actively investigated by us at present. Nonetheless, it is now evident that the aldol approach offers two distinct advantages: (a) the resulting schemes are concise and straightforward, involving no elaborate synthetic design, and (b) this approach is general and applicable to the synthesis of numerous compounds comprising β-hydroxycarbonyl moieties. In fact, more complex molecules such as amphotericin B (68) and rifamycin (69), which were considered almost impossible to approach several years ago, already have been chosen as the next targets. The routes leading to the construction of appropriate fragments are now well defined.

The recent surge of efforts directed toward the exploration of the aldol reaction is indeed noteworthy. There is reason to believe that a more thorough understanding of this funda- mental reaction will lead to more precise stereochemical control in the near future.

Acknowledgments - It is a pleasure to acknowledge gratefully the contribu- tions of my associates Drs. M. Hirama and S. Mori, who did the majority of the work presented above, and Drs. S.A. Ali, D.L. Snitman, D.E. Van Horn, and Mr. D.S. Garvey, who explored the aldol condensation. I would also like to thank Dr. A.O. Geiszler, Abbott Laboratories for a generous gift of 6-desoxyerythronolide B and finally the National Institutes of Health AI15403, NCI Training Grant 1T3 CA09112, and Hoffmann-La Roche for generous financial support.

REFERENCES

1. For recent reviews, see (a) S. Masamune, G.S. Bates, and J.W. Corcoran, Angew. Chem. Int. Ed. Engl. 16, 585 (1977); (b) K.C. Nicolaou, Tetrahedron 33, 683 (1977); (c) T.G. Back, ibid. 33, 3041 (1977).
2. (a) S. Masamune, C.U. Kim, K.E. Wilson, G.O. Spessard, P.E. Georghiou, and G.S. Bates, J. Am. Chem. Soc. 97, 3512 (1975); (b) S. Masamune, H. Yamamoto, S. Kamata, and A. Fukuzawa, ibid. 97, 3513 (1975); (c) S. Masamune, S. Kamata, and W. Schilling, ibid. 97, 3515 (1975).
3. E.J. Corey, P.B. Hopkins, S. Kim, S.-e Yoo, K.P. Nambiar, and J.R. Falck, J. Am. Chem. Soc. 101, 7131 (1979).

4. (a) E.J. Corey, E.J. Trybulski, L.S. Melvin, K.C. Nicolaou, J.A. Secrist, R. Lett, P.W. Sheldrake, J.R. Falck, D.J. Brunelle, M.F. Haslanger, S. Kim, and S. Yoo, J. Am. Chem. Soc. 100, 4618 (1978); (b) E.J. Corey, S. Kim, S. Yoo, K.C. Nicolaou, L.S. Melvin, D.J. Brunelle, J.R. Falck, E.J. Trybulski, R. Lett, P.W. Sheldrake, ibid. 100, 4620 (1978).
5. (a) J.R. Martin and W. Rosenbrook, Biochemistry 5, 2852 (1966); (b) J.R. Martin, T.J. Perun, R.L. Girolomi, ibid. 6, 435 (1967).
6. G. Wittig and H. Reiff, Angew. Chem. Int. Ed. Engl. 7, 7 (1968).
7. P.A. Bartlett, Tetrahedron 36, 3 (1980).
8. (a) J.E. Dubois and M. Dubois, Tetrahedron Lett. 4215 (1967); (b) J.E. Dubois and P. Fellman, ibid. 1225 (1975).
9. H.O. House, D.S. Crumrine, A.Y. Teranishi, and H.D. Olmstead, J. Am. Chem. Soc. 95, 3310 (1973).
10. (a) W.A. Kleschick, C.T. Buse, and C.H. Heathcock, J. Am. Chem. Soc. 99, 247 (1977); (b) C.T. Buse and C.H. Heathcock, ibid. 99, 8109 (1977).
11. (a) W.H. Zachariasen, Acta Crystallography 16, 385 (1963); (b) D. Groves, W. Rhine, and G.D. Stucky, J. Am. Chem. Soc. 93, 1553 (1971).
12. (a) J. Hooz and L. Linke, J. Am. Chem. Soc. 90, 5936, 6891 (1968); (b) J. Hooz and D.M. Gunn, Chem. Commun. 139 (1969); (c) J. Hooz and G.F. Morrison, Can. J. Chem. 48, 868 (1970); (d) J. Hooz and J.N. Bridson, J. Am. Chem. Soc. 95, 602 (1973), (e) D.J. Pasto and P.W. Wojtkowski, Tetrahedron Lett. 215 (1970).
13. (a) H.C. Brown, M.M. Rogic, M.W. Rathke, and G.W. Kabalka, J. Am. Chem. Soc. 90, 818, 1911 (1968); (b) A. Suzuki, A. Arase, H. Matsumoto, M. Itoh, H.C. Brown, M.M. Rogic, and M.W. Rathke, ibid. 89, 5708 (1967).
14. (a) T. Mukaiyama, K. Inomata, and M. Muraki, J. Am. Chem. Soc. 95, 967 (1973); (b) idem., Bull. Chem. Soc. Japan 46, 1807 (1973); (c) T. Mukaiyama and T. Inoue, Chemistry Lett. 559 (1976); (d) T. Inoue, T. Uchimaru, and T. Mukaiyama, ibid. 153 (1977).
15. (a) W. Fenzl and R. Köster, Liebigs Ann. Chem. 1322 (1975); (b) W. Fenzl, R. Köster, and H.-J. Zimmermann, ibid. 2201 (1975).
16. (a) S. Masamune, S. Mori, D.E. Van Horn, and D.W. Brooks, Tetrahedron Lett. 1665 (1979); (b) M. Hirama and S. Masamune, ibid. 2225 (1979); (c) D.E. Van Horn and S. Masamune, ibid. 2229 (1979); (d) M. Hirama, D.S. Garvey, L.D.-L. Lu, and S. Masamune, ibid. 3937 (1979).
17. D.A. Evans, E. Vogel, and J.V. Nelson, J. Am. Chem. Soc. 101, 6120 (1979).
18. For recent reviews see (a) H.B. Kagan and J.C. Fiaud in Topics in Stereochemistry 10, Ed. E.L. Eliel and N.L. Allinger, pp. 175-285, Wiley, New York (1978); (b) J.D. Morrison and H.S. Mosher, Asymmetric Organic Reactions, American Chemical Society, Washington, D.C. (1976).
19. (a) N. Trong-Anh and O. Eisenstein, Tetrahedron Lett. 155 (1976); (b) N. Trong-Anh and O. Eisenstein, Nouv. J. Chim. 1, 61 (1977).
20. D.B. Collum, J.H. McDonald, III, and W.C. Still, J. Am. Chem. Soc. 102, 2117 (1980).
21. S. Masamune, Sk. A. Ali, D.L. Snitman, and D.S. Garvey, Angew. Chem. 92, 573 (1980).
22. (a) C.H. Heathcock and C.T. White, J. Am. Chem. Soc. 101, 7076 (1979); (b) C.H. Heathcock, M.C. Pirrung, C.T. Buse, J.P. Hagen, S.D. Young, and J.E. Sohn, ibid. 7077 (1979).
23. R.W. Hoffmann and H.-J. Zeiss, Angew. Chem. Int. Ed. Engl. 19, 218 (1980).
24. (a) R. Anliker, D. Dvornik, K. Gubler, H. Heusser, and V. Prelog, Helv. Chim. Acta 39, 1785 (1956); (b) C. Djerassi and J.A. Zderic, J. Am. Chem. Soc. 78, 6390 (1956).
25. (a) J.D. White and Y. Fukuyama, J. Am. Chem. Soc. 101, 226 (1979); (b) G. Stork and T. Nair, ibid. 101, 1315 (1979); (c) P.A. Grieco, Y. Ohfune, Y. Yokoyama, and W. Owens, ibid. 101, 4749 (1979); (d) For a simple stereospecific synthesis of 48, see P.A. Bartlett and J.L. Adams, ibid. 102, 337 (1980).
26. E.J. Corey and L.S. Melvin, Jr., Tetrahedron Lett. 929 (1975).

THE TOTAL SYNTHESIS OF QUADRONE

S. Danishefsky*, K. S. Vaughan*, R. C. Gadwood*, K. Tsuzuki*
and J. P. Springer**

*Department of Chemistry, University of Pittsburgh, Pittsburgh, PA 15260, and
Department of Chemistry, Yale University, New Haven, CT 06511, USA
**Merck, Sharp, and Dohme, Rahway, NJ 07065, USA

Abstract - Experiments leading to the total synthesis of the anti-tumor agent, quadrone (1), are reported.

INTRODUCTION

In 1978, Calton et al (1,2) described the isolation of a sesquiterpene, quadrone, from *Aspergillus terreus*. Its structure was determined by strictly crystallographic means to be that shown in formula 1. In addition to the synthetic challenge implicit in its novel tetracyclic ring system, the attribution of promising anti-tumor properties to quadrone has undoubtedly added to the interest which has been lavished on this particular natural product.

Our laboratory was particularly attracted to this target since we were, at the time of the emergence of quadrone, struggling to complete the last phases of the total synthesis of pentalenolactone 2 (3,4), a structurally related sesquiterpene of reputed anti-tumor potential.

It was immediately recognized that the strategems which had brought us close to pentalenolactone would not be directly applicable to the quadrone problem. The secondary methyl group of pentalenolactone lies on the concave (endo) face of the bicyclo [3,3,0] octane system. In contrast, the corresponding ethano bridge in quadrone projects from the convex (exo) face of this ring system.

quadrone (1)   pentalenolactone (2)

In light of this major stereochemical difference, a new program was developed for the total synthesis of quadrone. The successful realization of this goal is described below.

SYNTHETIC STRATEGY

Although the efficacy and mode of pharmacologic action of quadrone as an anti-tumor agent remain to be defined, it seemed likely that its biologically active form might well be the unsaturated acid 3, related to quadrone in a formal manner by a retro-Michael elimination. Such α-methylene carbonyl systems have been implicated in the anti-tumor action of a wide variety of phytochemically derived sesquiterpenes (5).

Since essentially no chemistry had been carried out on quadrone, its relationship to the hypothetical 3 was not known. Nonetheless, it was assumed that the transformation of 3→1 could be achieved, if, indeed, it were not spontaneous. It will be recognized that in this transformation, a new chiral center is generated α to the cyclopentanone (see asterisk). Since this center is presumably subject to thermodynamic control, and since the cis fusion of the lactone and cyclopentanone rings of quadrone is undoubtedly preferred, we felt that this stereochemical issue might well resolve itself favorably. Hence, upon chemical and biological considerations, compound 3 and esters thereof emerged as attractive subgoals.

We had hoped to attain compound 3 via a system such as 4 wherein the "R" grouping would allow for creation of the α-methylene function in the proper regiochemical sense. A critical requirement of system 4 was the emergence of CO₂R' (R' = H or alkyl) in an axial position. As it will be seen, several of our approaches were designed to ensure this outcome. Unfortunately, all of these met with failure. Ironically, the scheme which was successful in practice did not, at the planning level, make convincing provision for the "axial" CO₂R' function (vide infra).

Of necessity, our notions as to how to construct the novel tricyclic system of 4 were not formulated in detail. It is clear that this is a rather strained ring system and that it would be wise at the planning stage to devise a program which offers maximum flexibility in achieving the required bond formations.

A generalized intermediate which we found attractive in this regard is system 5. Here, the enone system would provide for creation of the quaternary angular carbon. The β-ketoester arrangement seemed promising in allowing for orderly introduction of the needed α-methylene function. We leave unspecified the nature of X. Hence, the C₂ fragment which is required to go from 5→4 might be incorporated as part of the X function. Alternatively, it might initially be appended to the angular carbon (in a Michael sense) and then joined to the carbon bearing X.

System 5 is seen to be the aldolization product of 6. It was further conjectured that 6 would arise by alkylation of the site-specific metalloenolate 7 which would be derived from formal Michael addition of a nucleophile, "Nu", to enone 8 (6,7). A reasonable prospect as an alkylation agent in going from 7→6 would be bromide 9 (8). A large number of possibilities would result from the variety of nucleophiles which might be added in a Michael sense to 8. The only two restrictions on "Nu" were (i) that it be able to sustain the enolate 7 for site-specific alkylation of 9, and (ii) that it be readily converted to synthetically useful side chains of the type "CH₂CH₂-X".

The stereochemical outlook for this approach seemed particularly encouraging. It appeared that the required trans relationship of the hydrogens marked Hₐ and Hᵦ could be readily attained from either compound 6 or 5. Both systems should be subject to thermodynamic control. Indeed, it could be anticipated that even at the kinetic level, system 6 would emerge with Hₐ and Hᵦ trans to one another.

Seemingly, this plan also made ample provision for the stereochemistry of the quaternary

OMe

MeO$_2$C   Br

9

Nu

"Nu$^\ominus$" ?

Nu

O$^\ominus$ M$^\oplus$

8   7

H Nu H

MeO

Me O$_2$C

?

Ha Nu Hb

O

Me O$_2$C   6

? 5

carbon. Thus, if cyclization occurred via carbon functionality which was part of X, the 5:5 fusion would, per force, be cis, given the projected stereochemistry of intermediate 5. However, even if the two carbon fragment was introduced β to the enone via intermolecular reactions, it would be expected that a cis fusion would arise. In this case, the cis relationship required for closure of the bridge would also have been secured.

RESULTS

Some early setbacks

Enone 8 and bromide 9 were both known compounds (7,8) and were prepared accordingly. While bromide 9 has enjoyed some use as an alkylating agent (9), its potential for trapping site-specifically generated enolates had not been demonstrated.

In this exploratory phase of the inquiry, a variety of organometallic agents were added to the enone 8 in tetrahydrofuran under the influence of cuprous iodide catalysis. The resulting metalloenolates were subjected to the action of bromide 9, in the presence of hexamethyl-phosphorus triamide as an additive, giving rise to systems 10a-d in 40-60% yields. These were converted by relatively simple manipulations into bicyclooctenones 11a-d. Each of these compounds was prepared for potential utilization as a substrate of the type 5. Specifically, it was hoped that compounds 11a-d might serve as precursors to compounds 12a,b, or e, using standard reactions. Unfortunately, no product of the type 12 was detected or isolated in any case.

Two of the failures were particularly frustrating. Sakurai reactions (10) of type 11 compounds, including 11a, with allyltrimethylsilane afforded quite cleanly systems of the type 12c, yet the projected intramolecular counterpart 11a→12a failed dismally, leading only to protodisilylation. Similarly, a variety of compounds of the type 11, including 11b, underwent smooth photocycloaddition with allene in a typical Eaton reaction (11), giving adducts of the type 12d. The projected intramolecular counterpart 11b→12e, however, could not be reduced to practice even after attempts under a large variety of reaction conditions.

The "bogus" tricyclic compound

With our respect for the formidability of the synthetic challenge thus enhanced by these early skirmishes, we directed our attention to the possibilities inherent in compound 13, and the related possibility inherent in the reaction of 14 with dimethyl sodiomalonate. Although neither of these approaches afforded success, this phase of the inquiry will be described in some detail because it did serve to direct us toward a series of compounds from which success was eventually achieved. In addition, the results serve as poignant reminders of the capacity of organic reactions to provide major surprises for the unsuspecting.

Clearly, in the case of compound 13, we were hoping to reach the required ring system of 5 (cf. compound 15) by an intramolecular Michael reaction. With regard to the reaction of compound 14 with malonate anion, two mechanistic routes to 15 could be envisioned. Displacement of halide by the malonate nucleophile would, again, give rise to 13. Alternatively, Michael addition of malonate to 14 could give 16, which by cycloalkylation would give 15.

The reaction of vinylmagnesium bromide with enone 8 was carried out in the presence of Bu$_3$P.CuI in the usual manner. Alkylation of the resultant metalloenolate with bromide 9 afforded 17 in 40-55% yield. A significant yield improvement (55-65%) was realized using the corresponding iodide (12) as the alkylating agent. Compound 17 was converted into its

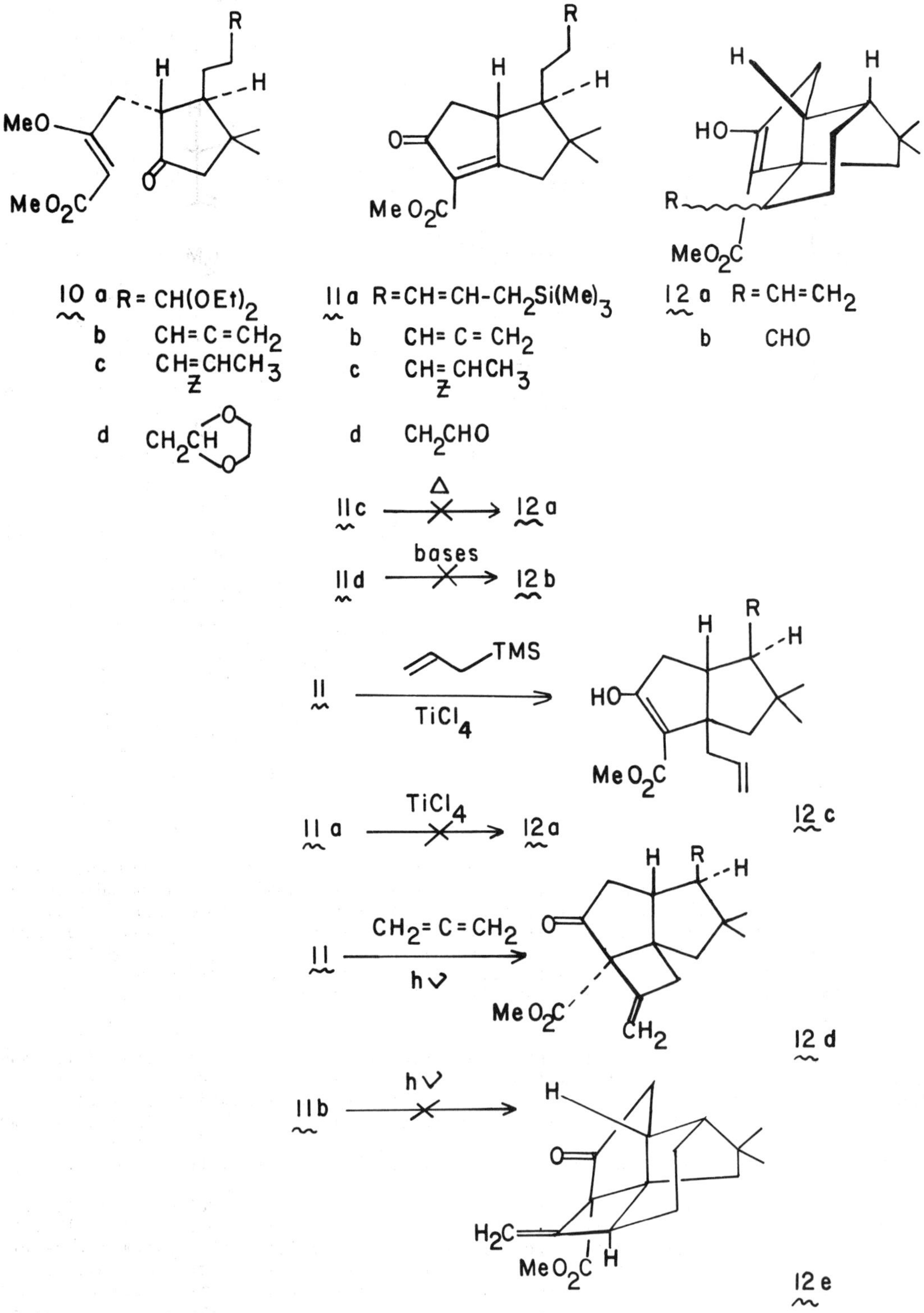

10 a R= CH(OEt)$_2$
  b   CH=C=CH$_2$
  c   CH=CHCH$_3$
           Z

  d   CH$_2$CH (dioxolane)

11 a R=CH=CH-CH$_2$Si(Me)$_3$
  b   CH= C = CH$_2$
  c   CH= CHCH$_3$
           Z

  d   CH$_2$CHO

12 a   R= CH=CH$_2$

  b   CHO

11c —Δ—×—→ 12a

11d —bases—×—→ 12b

11 —(CH$_2$=CH-CH$_2$-TMS / TiCl$_4$)—→ 12c

11a —TiCl$_4$—×—→ 12a

11 —(CH$_2$=C=CH$_2$ / hν)—→ 12d

11b —hν—×—→ 12e

bis ethyleneketal 18 through the action of ethylene glycol and tosyl acid in toluene under
reflux. With its keto functions thus protected, the vinyl group of compound 18 underwent
clean hydroboration. Oxidation with alkaline hydrogen peroxide provided alcohol 19 and,
thence, mesylate 20. The latter was transformed by the action of lithium bromide in acetone
into compound 21. Deketalization-aldolization of 21 afforded 14. The yield of 14 from 17
was ca. 42%. Alternatively, Finkelstein reaction of 20 afforded the iodo compound 22, which
reacted with dimethyl sodiomalonate to give 23. After deketalization and aldolization, 13
was in hand.

We have not yet fully elucidated the complex chemistry which ensues when compound 13 is treated with bases under a variety of conditions. Several structures await complete formulation.
However, it may be stated that under no conditions which we have yet devised could we detect
compound 15. An interesting result was obtained on treatment of 13 with D.B.U. in an attempt
to produce the elusive 15. Instead there resulted as the major product its isomer, 26. It
appears that this compound arises from 13 by way of its double bond isomers 24 and 25.

In a different attempt to reach compound 15, the reaction of 14 with dimethyl sodiomalonate
in tetrahydrofuran under reflux was investigated. There were obtained ca. 10-15% of 13 and
a 70% yield of still another isomer. The spectroscopic properties of this compound were not
at variance with its formulation as the long sought 15. Given the fact that we could not
obtain the alleged 15 from 13, the viability of its assignment depended on our ability to synthesize it via the other plausible mechanistic pathway, ie. from intermediate 16. This compound was obtained from a Mukaiyama-like reaction (13) of 14 with 27 (14) in the presence of
titanium tetrachloride, affording 16 in 65% yield.

Unfortunately, reaction of 16 with a variety of bases under an assortment of conditions failed to produce any detectable amount of product to which the structure 15 had tentatively

18 R = CH=CH$_2$
19 R = CH$_2$CH$_2$OH
20 R = CH$_2$CH$_2$OMs
21 R = CH$_2$CH$_2$Br
22 R = CH$_2$CH$_2$I
23 R = CH$_2$CH$_2$CH(CO$_2$Me)$_2$ $\longrightarrow$ 13

been assigned. The only observable reaction was a virtually instantaneous reversion of 16→14.

At this point, it seemed prudent to provide rigorous support for the assignments of structure and, particularly, stereochemistry to compound 16. This was accomplished by X-ray crystallographic means. Indeed, our formulations were fully corroborated. Given the extremely mild conditions of the Mukaiyama reaction by which 14 is cleanly converted to 16, the assignment of stereochemistry to the former seems equally valid.

Remaining unanswered was the structure of the major product of the reaction of 14 with dimethyl sodiomalonate. Its formulation as 15 seemed to be in particular jeopardy since the latter could not be obtained from either of its plausible mechanistic precursors, 13 or 16.

There is little advantage in recounting here the chemical developments by which the correct structure of this product was eventually deduced. Instead, we pass directly to the most definitive argument, i.e. that provided by X-ray crystallographic measurements, from which structure 28 emerges.

In gross terms it is recognized that 28 would arise from a compound like 16 wherein cyclization had occurred to the β-ketoester enolate rather than by way of the malonate. Of course, in compound 16 itself, no such choice exists since the bromoethyl group is, with respect to the B-ring, trans to the β-ketoester. For such a cycloalkylation to be possible, it is first necessary to epimerize the bromoethyl side chain from its more stable exo arrangement to the endo series. While in absolute terms it would be difficult to contemplate inversion in the stereochemistry of the bromoethyl group, its epimerization in relative terms can easily be envisioned by means of intermediates 14a and 14(endo). Facile enolization at the junction center had already been implicated in the conversion of 13→26 (vide supra). Protonation of 14a in the alternate sense would provide 14(endo). Michael addition of malonate to 14(endo) then produces 28a in which the β-ketoester enolate and the bromoethyl side chain are cis to the B-ring. Cycloalkylation would afford system 28.

Clearly, any equilibrium between 14 and the presumed 14(endo) must favor the former. In this case it would appear that the equilibrium is driven by the exothermic conversion of 14 (endo) to 28a, and thence to 28. Throughout all these events, the desired result, i.e.

$$15 \xleftarrow[\text{bases}]{} 13(\Delta^{2,3}) \xrightarrow{\text{DBU}} \left[24(\Delta^{3,4}) \longrightarrow 25(\Delta^{4,5}) \right] \longrightarrow 26 \ E = CO_2Me$$

cycloalkylation of 16→15 via deprotonation of the malonate center, is apparently non-competitive. Of course, we have not rigorously shown that 16 is in fact generated as a non-productive intermediate in the reaction of 14 with dimethyl sodiomalonate, but this would certainly seem likely. Moreover, as shown before, compound 16, independently prepared, failed to undergo cyclization to 15.

Finally in this regard, reaction of 16 with diazomethane in the presence of catalytic aluminum chloride afforded enol ether 29. Attempted cycloalkylation of this compound, in which reversion to 14 has been prevented, was also unsuccessful. While we have not yet sorted out the apparently complex products which are generated from the reaction of 29 with bases, no monomeric cycloalkylation product could be detected.

Synthesis of the key tricyclic compound

While the experiments described above had failed to reach the generalized tricyclic system 5, they did suggest some new lines of attack which might be productive. First, excellent access to compound 14 had been provided. Second, the feasibility of angular functionalization with formation of the cis fused bicyclo[3,3,0]octane system had been demonstrated (cf. compound 16). Given our repeated failures in producing compounds such as 15 from 13, 14, or 29, we turned our attention to a substrate of the type 30 as a prospect for cycloalkylation.

30 P = protecting group for ketone
     R = H or precursor for CH₂
     L = leaving group

Enolization of the cyclopentanone would be strictly prohibited (see protective arrangement P). The presense of a biasing group to promote regiospecificity in the introduction of the α-methylene group of 3 was optional (R = H or a precursor of CH₂). To maximize chances of successful cycloalkylation, the nucleophile would be of the more reactive monoester enolate variety. Due to the series of failures in the attempted synthesis of the tricyclic system 5, all considerations about providing for the stereochemistry of the carboxyl group or ensuring regiospecificity in the placement of the α-methylene function were subordinated to the central problem, i.e. that of constructing a relevant tricyclic system.

In view of the successful reaction of 14 + 27→16, we attempted a comparable reaction of 14 with 31, which was prepared according to the standard methodology of Rathke (15). Again, the Mukaiyama-like reaction (13) was quite successful. Much to our surprise there was obtained 32 in which the t-butyl rather than the t-butyldimethylsilyl had been cleaved. This was readily converted to the acid 33 by treatment with $(n-Bu)_4N^+F^-$. The overall yield of homogeneous 33 was 70%. For our purposes, however, crude diester 32 was subjected to acid-induced hydrolysis and decarboxylation(1 M HCl in dioxane; reflux). After esterification of the monoacid with diazomethane, ketoester 34 was obtained in 72-73% yield from 14.

We had accepted the loss of the potentially useful β-ketoester functionality because we had been unable to achieve the ketalization of compounds such as 33, its derived methyl ester, or earlier, 14 itself. Happily, 34 underwent smooth ketalization with ethylene glycol using tosyl acid in toluene under reflux.thereby providing 35. With a view toward improving the quality of the leaving group, compound 35 was subjected to a Finkelstein reaction with sodium iodide in acetone, affording compound 36 in 93-94% yield from 34.

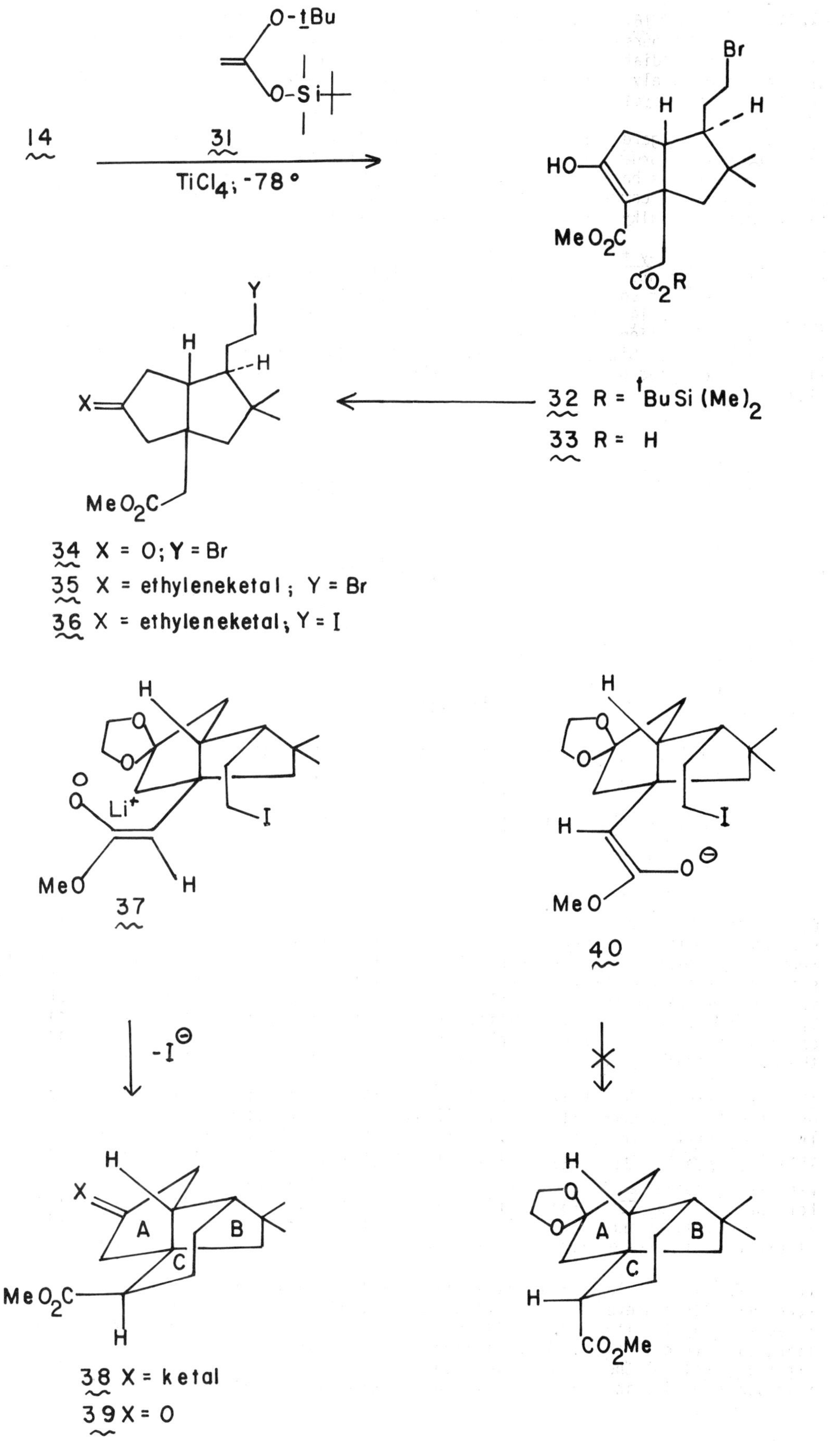

$$14 \xrightarrow[\text{TiCl}_4; -78°]{31} $$

where 31 is the silyl ketene acetal with O-tBu and O-Si groups.

32 R = ^{t}BuSi(Me)$_2$

33 R = H

34 X = O; Y = Br

35 X = ethyleneketal; Y = Br

36 X = ethyleneketal; Y = I

37

40

$-I^{\ominus}$

38 X = ketal

39 X = O

The stage for the final attempt at system 5 was now fully set. The iodo compound 36 was subjected to the action of lithium hexamethyldisilazide in THF at -78°. The resultant anion, 37, after warming to -23° and addition of ca. 10% hexamethyphosphorous triamide, was brought to room temperature. There was then isolated, in 55-60% yield, the long sought tricyclic system 38. Deketalization of compound 38 (acetone, tosyl acid, room temperature) afforded the tricyclic ketoester 39.

Surprisingly, the pmr spectrum of 39 (16) indicated that it was the epimer containing the much desired axial carbomethoxy grouping! The apparently exclusive formation of 38 (and, thence, 39) under these conditions of kinetic control can be rationalized in terms of a strong rotameric preference for conformation 37 relative to 40. In the cycloalkylation reaction, it is only in these conformers that the stereoelectronic requirements for backside displacement can be met. The cycloalkylation product of 37 is such that the carbomethoxyl is syn to the A ring (i.e. axial to a chair-like C ring), while from 40 it would emerge syn to the B ring (i.e. equatorial to a chair-like C ring). Much to our delight, the former situation prevailed in practice.

Completion of the total synthesis

To reach quadrone from compound 39, there remained only the α-hydroxymethylation of the ketone in the required sense. Several preliminary experiments designed to ascertain the direction of substitution about the cyclopentanone all pointed to the expected but unwanted result, i.e. that the reactivity is manifested away from the neopentyl center. Accordingly, a "blocking" strategy would be necessary.

Reaction of 39 with Bredereck's reagent (17) smoothly afforded the vinylogous amide 41. The

42 X = CHN (Me)$_2$
1 X = H (quadrone)

latter reacted with lithium di-isopropylamide and formaldehyde to spontaneously afford a lactone which we formulated as 42, i.e. α-dimethylaminomethylene quadrone. Unfortunately, a variety of experiments designed to retrieve quadrone from its "protected" derivative 42 were unsuccessful. While we shall not describe these attempts (since they did not lead to well defined products), they did tend to indicate that quadrone is not easily isolated from harshly acidic or basic reaction conditions. A deblocking maneuver which could be carried out under the mildest possible circumstances appeared to offer the best chance for success in dealing with the terminal compounds in this series.

We again exploited the natural sense of enolization of the ketone function. Selenylation (18) of compound 39 afforded 43, and thence, by oxidative elimination, enone 44. Of course, enolization of the enone in the γ sense is strictly prohibited in this case by geometric constraints. Accordingly, it was not surprising to find that reaction of 44 with lithium di-isopropylamide followed by reaction of the resultant enolate with formaldehyde produced the α'-hydroxymethylenone 45. It was expected and found that catalytic hydrogenation of the double bond (palladium on carbon; methanol-ethyl acetate) led to the cis fused A:B system in the form of compound 46. However, unlike the situation with 41, this compound showed no tendency to lactonize to quadrone. Again (cf 42), attempted alkaline hydrolysis of 46 was unrewarding. No clearly definable products could be isolated.

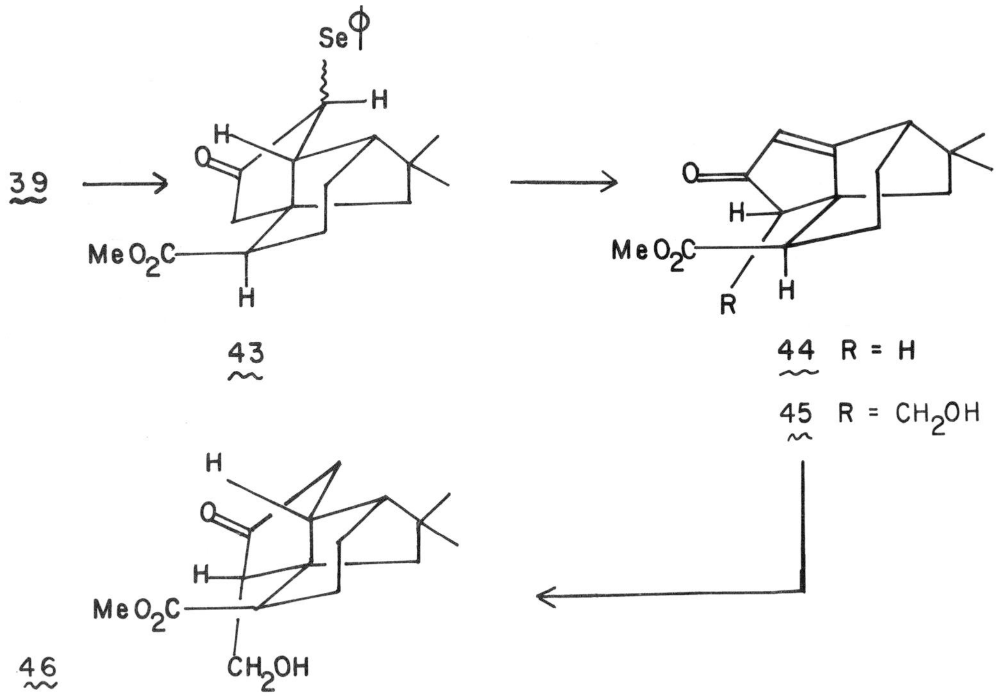

39 →

43

44 R = H

45 R = CH₂OH

46

In keeping with the desirability of avoiding harsh hydrolytic conditions in the terminal steps of the synthesis, we returned to ketoester 39. Alkaline hydrolysis afforded the acid 47. This was converted by selenylation and oxidative elimination (18) to the enone-acid 48 (78% from 39). The latter reacted with lithium di-isopropylamide (3 eq) followed by formaldehyde to provide compound 49 in 68% yield. Catalytic hydrogenation, as before, afforded keto-acid 50, mp 153-155°, which again showed no inclination toward spontaneous lactonization to quadrone. However, reaction of 50 with tosyl acid in benzene at 40-50° did afford another desired goal, the α-methyleneacid 3, mp 177-179° (94% yield).

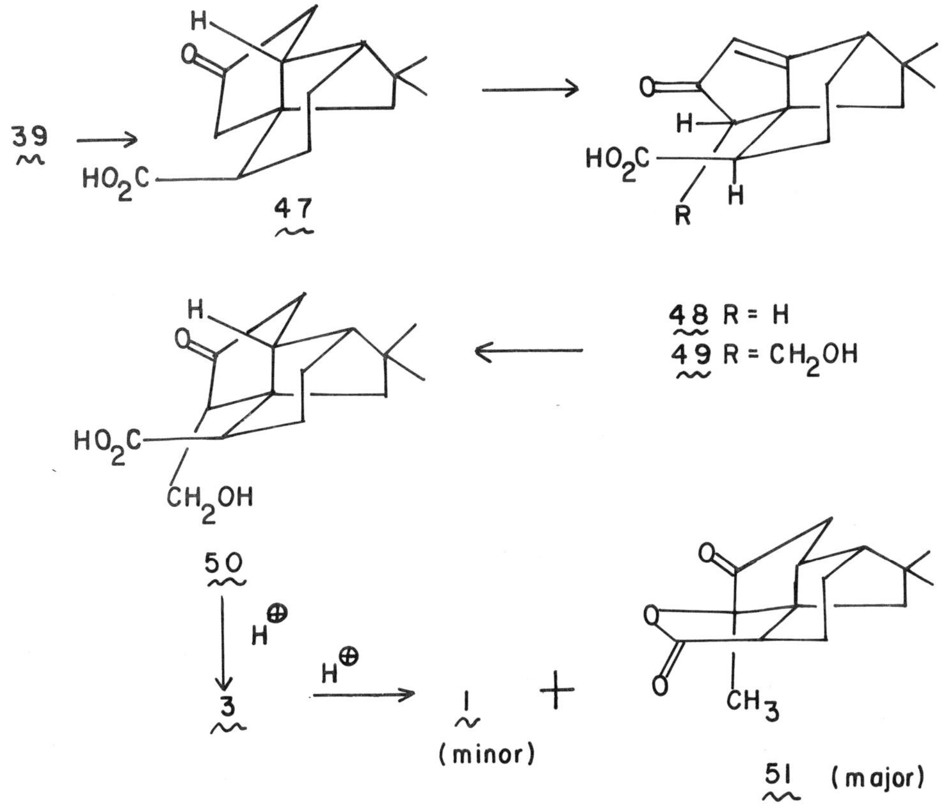

39 →

47

48 R = H
49 R = CH₂OH

50

3

1 (minor) + 51 (major)

With compound 3 finally secured, the remaining transformation in quadrone's total synthesis should have been trivial. This was not to be the case. Compound 3 showed no tendency toward spontaneous lactonization to quadrone. Reaction of 3 with tosyl acid in benzene under reflux did indeed afford dl-quadrone, but only as the minor product. The major product was 51, a substance we have termed isoquadrone.

Although the total synthesis of quadrone was finished, we naturally desired to achieve a more satisfactory ending to this long endeavor. We presumed that the formation of 51 from 3 involved the intermediacy of enol 52. From this diene, protonation of the exocyclic methylene in the expected sense would generate cation 53 and, thence, isoquadrone. Without access to an enol such as 52, the formation of the wrong lactone would be kinetically unlikely. With this in mind, we examined the acid catalyzed cyclization of enone-acid 49, in which enolization is prohibited. Reaction of 49 with tosyl acid in benzene afforded the expected cross-conjugated dienone-acid 54, as well as a neutral product whose spectral properties clearly revealed it to be lactone 55. This disaster naturally must have evolved from conjugate addition of the carboxyl function to the wrong double bond.

We returned to the enone-acid 3 with the hope of finding different conditions which would give a favorable ratio of quadrone to isoquadrone. None attempted were very promising. However, a serendipitous observation was made. T.L.C. examination of the residue formed by melting the α-methylene acid 3 indicated the presence of quadrone without its isomer 51. Moreover, the same result was achieved with the hydroxymethyl precursor 50. T.L.C. studies during the course of this transformation clearly indicated that the sequence involved 50→ 3→1. There was no positive evidence for the direct formation of quadrone from 50, although this cannot be ruled out.

1 = quadrone

On a "preparative" scale (*ca.* 50mg), this result translated to a 51% yield of homogeneous quadrone, mp 140-142°, without recycling of acid 3 (which persists in significant quantity). The chromatographic properties of dl-quadrone, so synthesized, are identical with those of an authentic sample kindly provided by the National Cancer Institute. More decisively, the nmr ($CDCl_3$,270 MHz), infrared ($CHCl_3$), and mass spectra of synthetic and natural quadrone were superimposable. The total synthesis of quadrone was complete.

Acknowledgements - This research was supported by PHS Grant CA-12107. We also acknowledge Fellowships from the Chaim Weizmann Foundation and from the National Cancer Institute to R.C.G. and from the Andrew Mellon Foundation to K.V.. We gratefully acknowledge the National Cancer Institute for providing us with a sample of authentic quadrone.

References

1. R.L. Ranieri and G.J. Calton, Tetrahedron Lett., 499 (1979).
2. G.J. Calton, R.L. Ranieri,and M.A. Espenshade, J. Antibiot., 31, 38 (1978).
3. S. Danishefsky, M. Hirama, K. Gombatz, T. Harayama, E. Berman, and P. Schuda, J.Am.Chem. Soc., 101, 6536 (1978).
4. S. Danishefsky, M. Hirama, K. Gombatz, T. Harayama, E. Berman, and P. Schuda, J.Am.Chem. Soc., 101,7020 (1979).
5. S.M. Kupchan, M.A. Eakin, and A.M. Thomas, J. Med. Chem., 14, 1147 (1971).
6. R.K. Boeckman, Jr. J. Org. Chem., 38, 4450 (1973).
7. R.V. Stevens, R.E. Cherpeck, B.L. Harrison, J. Lai, and R. Lapalme, J.Am.Chem.Soc., 98, 6317 (1976). For a recent, more convenient synthesis of enone 8, see P.D. Magnus and M.S. Nobbs, Synth. Commun., 10 (4), 273 (1980).
8. S.M. Weinreb and J. Auerbach, J.Am.Chem.Soc., 97, 2503 (1975).
9. For a recent use of compound 9 as an alkylating agent see G. Stork, D.F. Taber, and M. Marx, Tetrahedron Lett., 2445 (1978).
10. A. Hosomi and H. Sakurai, J.Am.Chem.Soc., 99, 1673 (1977).
11. P.E. Eaton, Tetrahedron Lett., 3695 (1964).
12. The only reference found for a compound in this iodo series (ethyl 4-iodo-3-ethoxy-2-butenoate) involved its formation during a synthetic study of halogenated allenyl esters (J. Tendil, M. Verny, and R. Vessiere, Bull. Soc. Chim. Fr., 565 (1977)). The route we used (Finkelstein reaction with bromide 9, *cf.* ref. 8) was, for us, the most expeditious one. Thus formed, the iodide was used without purification in the reaction described in the text.
13. K. Saigo, M. Osaki, and T. Mukaiyama, Chem. Lett., 163 (1976).
14. Y. Kuo, F. Chem, and C. Ainsworth, Chem. Commun., 137 (1971).
15. M.W. Rathke and D.F. Sullivan, Synth. Comm., 3, 67 (1973).
16. 600 MHz pmr data for ketoester 39 is as follows: ($CDCl_3$) δ1.17(s, 3H), 1.19(s, 3H), 1.76-2.0(m, 5H), 1.58(d,J=14.8 Hz,1H), 1.76(d, J=14.8 Hz, 1H), 2.24(d, J=17.7 Hz, 1H), 2.36 (d, J=17.7 Hz, 1H), 2.48(d, J=10.2 Hz, 2H), 2.76(d, J=7.5 Hz, 1H), 2.88(t, J=10.2 Hz, 1H), 3.68(s, 3H). Compound 38 had identical signals at δ2.55 and 2.65 corresponding to those at δ2.76 and 2.88, respectively, in compound 39.
17. H. Bredereck, G. Simchen, S. Rebsdat, W. Kantlehner, P. Horn, R. Wahl, H. Hoffmann, and R. Grieshaber, Chem. Ber., 107, 41 (1968).
18. K.B. Sharpless, R.F. Lauer, and A.Y. Teranishi, J.Am.Chem.Soc., 95, 6137 (1973).

BIOMIMETIC PRENYLATION REACTIONS

M. Julia

Laboratoire de Chimie de l'Ecole Normale Supérieure, 24 rue Lhomond,
75231, Paris Cedex 05, France

Abstract - The elucidation of the biosynthesis of squalene raises a number of synthetic problems. Two aspects of this series of events will be discussed. The head to tail attachment of dimethylallyl and isopentenyl synthons has been investigated and conditions found in which high yields (at small conversions) can easily be obtained. Limonene and some of its derivatives have been prenylated in a similar way, leading to α-bisabolol and delobanone. The 1.2' attachment of two dimethylallyl synthons to each other could be realized and this led to the investigation of a biosynthetic proposal for presqualene alcohol ring closure. Proper choice of the leaving group in 1.3 elimination reactions will be discussed and the results of model experiments described. The information obtained could be used for a synthesis of chrysanthemyl alcohol.

INTRODUCTION

In the large family of terpene compounds Nature presents us with a variety of problems associated with the linking together of so called isoprene units. In the recent past, considerable information has been accumulated on the mechanism of these condensation reactions, including their stereochemistry (Ref.1). We shall be concerned with three of these reactions :
a) the head to tail attachment of isoprene units (the growth reaction) b) the non head to tail attachment found in non regular terpenes such as presqualene alcohol (PSA), lavandulol etc ... c) the ring closure in PSA.
As regards the head to tail attachment, Cornforth and Popjack using suitably chirally labelled starting material were able to show that in the formation of (E)-GPP loss of hydrogen and attack on the terminal sp$_2$ carbon occur on the same side of the double bond of isopentenyl pyrophosphate (IPP). To account for this fact two suggestions have been put forward (Fig. 1)

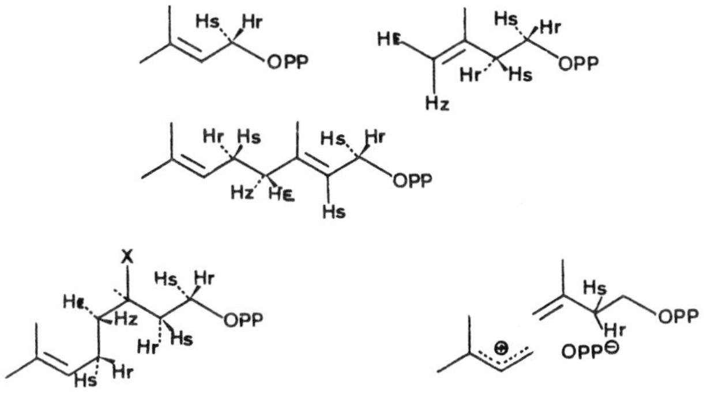

Fig. 1.

In one (Ref. 2) it is assumed that a suitable nucleophilic group X in the enzyme becomes attached to C$_3$ of the IPP residue and <u>anti</u> to the incoming electrophilic prenyl residue. This would be followed by <u>anti</u> elimination of this group X with H$_R$ on C$_2$.

On the other hand Poulter and Rilling (Ref. 3) present evidence that ionisation of dimethylallyl pyrophosphate (DMAP) precedes attack and suggest that, since the "attacking" DMAP has to be stacked in the active site of the enzyme directly underneath (on the <u>si</u> side of) the IPP, the pyrophosphate anion just separated from the prenyl carbocation would be ideally located to remove the pro R proton.

HEAD TO TAIL (1.4') AND TO MIDDLE (1.2') ATTACHMENT OF ISOPRENE SYNTHONS

It might seem odd that so little success has been achieved so far in the biomimetic attachment of isoprene units. Alkylation of olefins with carbocations and cationic polymerisation are well known. Intramolecular olefin alkylation is fairly common and is usually called neighbouring group participation by a carbon-carbon double bond. An early well authenticated case of olefin alkylation is that of W.S. Johnson and R.A. Bell (Ref. 4) : solvolysis of benzhydryl chloride in the presence of 2.2-diphenyl 4-methyl 4-butenoïc acid. Cyclohexene could be benzylated by heating with benzyl tosylate (Ref. 4a). Very recently a number of olefin prenylations have been accomplished with prenyl halides and Lewis acids.

Our first attempts at generating a dimethylallyl electrophilic species suitable for a prenylation of isopentenyl derivatives were carried out with dimethylallyl acetate (DMAA). As a catalyst for the condensation we first tried lithium perchlorate with an excess [stoichiometric proportions of reagents led to substantial reaction of DMAA with itself (see below)] of isopentenyl acetate. The lithium cation has recently been recognized as an efficient Lewis acid catalyst for a variety of reactions (Ref. 5) (Fig.2). It soon turned out (Ref.6) that it

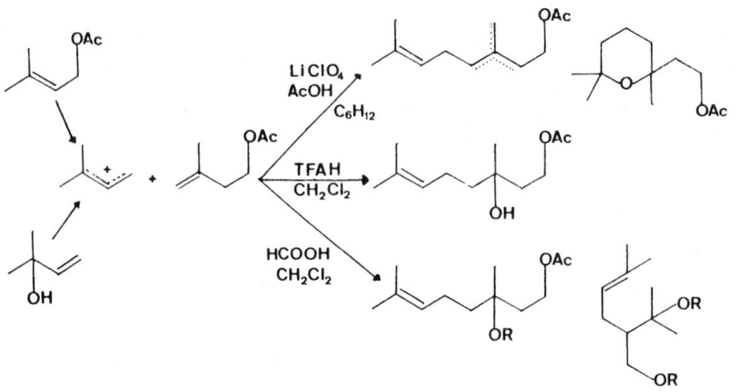

Fig. 2.

can bring about the attachment of these synthons and produce a mixture of the expected C_{10} diene esters with the monoterpene skeleton. A stoichiometric amount was found necessary. Acetic acid while unable by itself to achieve the condensation increased the efficiency of lithium perchlorate. A further improvement was observed on dilution with cyclohexane (conditions A) when almost quantitative yields of C_{10} dienesters were produced at low conversions. Others solvents led to inferior results. A number of other Lewis acids were tried, magnesium perchlorate proved very efficient. Some Brønsted acids (H_2SO_4, $HClO_4$, CF_3COOH) also induced the condensation but the yields were low and cyclisation products of the C_{10} glycol began to appear. In the lithium perchlorate catalysed reaction, geranyl acetate and the methylene acetate were minor among the isomeric dienesters, the main compounds were the (E+Z) β,γ-unsaturated esters. This is in agreement with what has long been known in the deprotonation of carbocations bearing an oxygen function in the γ-position as shown in the dehydration of -1.3 diols (Ref. 7) (Fig.3). Assistance in proton removal by the oxygen atom present in the molecule has been held responsible for this orientation. This chemical fact would seem to be a further reason to have a nucleophilic group X take part in the biosynthesis. Whatever that may be, the problem of the <u>direction</u> of elimination in the adduct cannot be ignored.

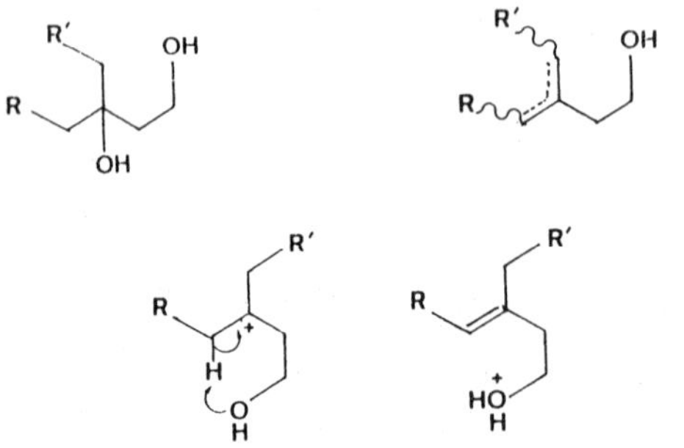

Fig. 3.

Other means of generating a prenyl cation were next tried, particularly the treatment of the readily available α-dimethylallyl alcohol (DMVC) with acids. We benefited from the extensive investigations of W.S. Johnson and his group (Ref. 8) on the cyclisation of polyenic allylic alcohols triggered by various acids.

Another reason to use an isoprenoïd alcohol in acid medium was that a molecule of water would be produced and, in previous experiments with DMAA and aqueous Brønsted acids, derivatives of the C_{10} glycol had been formed (see above). This of course reminds one of the role assigned to the X nucleophile in the biosynthetic scheme. Among the various systems tried, trifluoroacetic acid (TFA) in methylene chloride (conditions B) brought about the formation of the monoterpene glycol monoacetate. The conversion had to be kept low (14%) to get high yields (80%) (Fig.2).

A third set of reaction conditions (C) was investigated in some detail. Neat formic acid was not very good but, on adjunction with dichloromethane, reasonably good results were obtained. The monoterpene C_{10} glycol was formed on about 50% yield as a mixture of mono + diformate together with the corresponding C_{10} esters formed by 1.2' attack (see below). An improvement was observed with dichloromethane saturated with concentrated sulphuric acid when near quantitative yields (relative to the IPA consumed) were obtained (Fig.2).

The products obtained above had to be transformed into geranyl derivatives. The unsaturated diene acetates were hydrolysed and the mixture of isomeric alcohols oxidized with pyridinium chlorochromate. The double bonds were then isomerized to the α position under basic conditions (Ref. 9). The C_{10} glycol could also be oxidised to the corresponding hydroxyaldehyde which in turn was smoothly dehydrated to citral.

It had been mentioned above that the third set of conditions found to lead to an efficient head to tail attachment of isoprene units also led to 1.2' attack of a prenyl residue on another. This was investigated in more detail as : it could lead to an efficient access to the interesting lavandulol family and it has been suggested as an important first step on the way leading from farnesyl pyrophosphate to squalene. It was indeed found that DMAA undergoes self condensation under conditions A to give lavandulyl acetate in a yield of 50% (glc), 30% (isolated) (Fig.4).

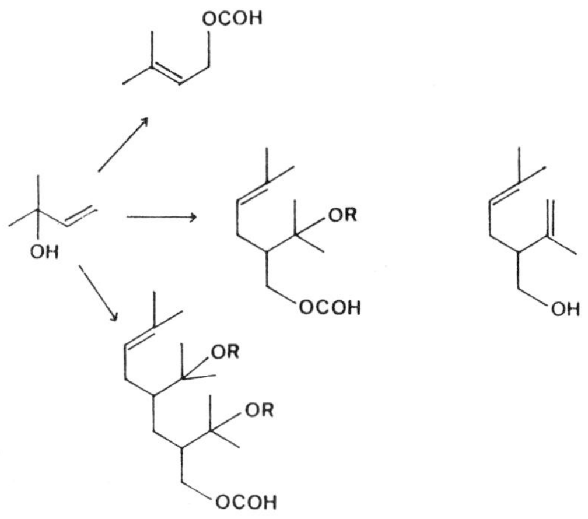

Fig. 4.

If DMAA is now treated with DMVC in the presence of TFA the branched C_{10} glycol monoacetate is formed in a yield of 63%. Formic acid in dichloromethane efficiently converted DMVC into 70% prenyl formate + 20% C_{10} glycol. This is easily separated and dehydrated, giving lavandulol in over 80% yield. An interesting byproduct of this investigation proved to be a C_{15} triol obviously formed by further attack of a prenyl electrophile on the C_{10} glycol. The trisubstituted double bond in this glycol so appears to be at least as easily prenylated as prenyl formate itself.

The interesting question arose as to what influence, if any, the acetoxy group in IPA has on the condensation. 2-methyl-1-heptene was therefore compared with IPA under the prenylation conditions. Under conditions B and C the olefin is converted into tertiary esters with formation of no more than a trace of condensation product. This could point to some participation of the carbonyl group of the ester which might favour prenylation over protonation of the double bond. Isopentenol itself however behaved very much like its acetate although it is not converted into its trifluoroacetate under the reaction conditions. Experiments are underway to get information on this point.

The next step in the investigation was of course to bring about a similar head to tail attachment between a geraniol derivative and isopentenyl acetate to gain access to the C_{15} family.
Geranyl acetate with IPA under conditions (A) gave the expected mixture of the C_{15} triene-
acetate but the yield was very low (5%) at 45% conversion of IPA and 55% conversion of
GA (Fig.5)

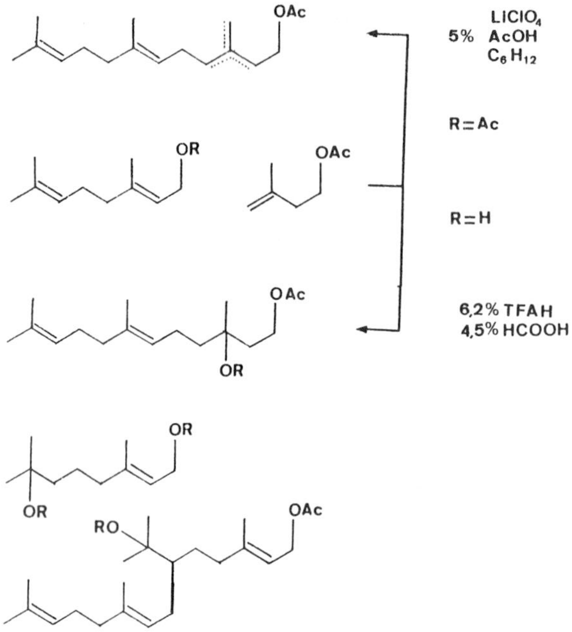

Fig. 5.

Under conditions B geraniol gave the expected C_{15} glycol monoacetate (6%) together with   terpineol
and 7-hydroxy dihydrogeraniol. Conditions C also gave the C_{15} glycol (4,5%).

STRUCTURAL AND SYNTHETIC STUDIES IN THE BISABOLOL SERIES

Another possible application of the prenylation reaction was the condensation of the prenyl
"cation" with an appropriate C_{10} target. Starting from nerolidol an interesting biosynthetic
family of compounds runs through α-bisabolol, β-bisabolol, α-alaskene and finally α-cedrene
or related compounds (Ref. 10). Looking at this scheme it seems an obvious thing to try to
produce the "α-bisabolyl cation" by electrophilic attack of a prenyl cation equivalent on the
extracyclic double bond of limonene (Fig.6).

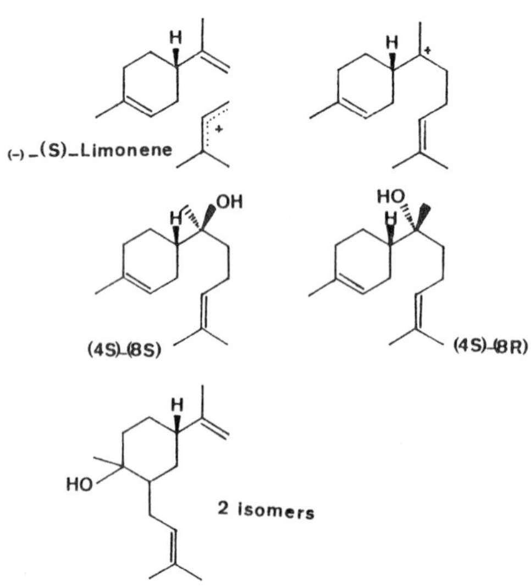

Fig. 6.

However difficulties were to be expected concerning the propensity of 1.1-disubstituted double
bonds to add carboxylic acids under the conditions used to generate the desired "cation", and
the presence of another double bond in the molecule of limonene. Perusal of the literature
showed that, according to the electrophilic reagent used, one or the other of the double bonds
was mainly attacked . The advantages of this route would be of course simplicity and direct-
ness. Moreover the ready availability of the enantiomers of limonene would make it possible to
conserve stereochemistry at C_4 of the target compound instead of producing a totally racemic
mixture.

The situation concerning α-bisabolol was rather confused in that the (S)-configuration has
been proposed for carbon atom 4 of (-)-α-bisabolol by Knöll and Tamm (Ref. 11) and confirmed
by correlation with (-)-β-bisabolene (Ref. 12). However two recent syntheses (Ref. 13, 14a &
14b) have led to opposite conclusions as to the configuration of carbon atom-8.
Authentic (-)-α-bisabolol was isolated from chamomile oil (german extra) [Kindly provided by
Dr. Willis of Fritzsche, Dodge & Olcott (New-York)] . It showed one peak on a capillary glc
column and one signal at 1.11 ppm in the NMR (250 MHz). A commercial sample (Dragosantol
of the Dragoco Co) showed two peaks in the glc and two methyl signals in the NMR at 1.11 and
1.14 ppm. This is in agreement with literature data (Ref. 13 & 14) for α-bisabolol and its
C_8 epimer.

The main question was the relative configuration about the two chiral centers. This was solved
by an X-ray structure determination carried out on the nicely cristalline p-phenylazophenyl-
urethane (Ref. 15) of natural (-)-α-bisabolol. This definitely showed (Ref. 16) the stereochemistry to be RR or SS for α-bisabolol. Therefore (-)-α-bisabolol must have the (4S,8S)
absolute configuration.

Attempted prenylation of limonene under conditions A and B above gave practically no condensation product but α-terpineol. However in the presence of formic acid in dichloromethane,
which makes it possible to work in a homogeneous medium, DMVC and limonene reacted to give
after hydrolysis a mixture of : a) α-terpineol   b) two isomeric C_{15} alcohols (A + B) resulting from attack of the ring double bond and c) two isomeric C_{15} alcohols (C + D) similarly
formed with the extracyclic double bond. A and B were easily separated by chromatography on a
column of silica gel and identified by their spectral properties. C and D gave on glc with
a capillary column the same two peaks as Dragosantol, one of which coincided with that of
α-bisabolol. This mixture of alcohols (C + D) from (-) (S)-limonene was p-nitrobenzoylated and
the mixture of esters separated by hplc. After reduction with lithium aluminium hydride, two
alcohols were obtained, one of which proved identical with (-)-α-bisabolol thus confirming the
(4S, 8S) configuration. The other alcohol is the epimer at C_8 i.e. (4S, 8R). Knowledge of the
stereochemistry of α-bisabolol enables one to draw interesting comparisons between cyclisa-
tions in the mono (Ref. 18) and sesquiterpene series.

Very recently A.G. Gonzalez and his colleagues (Ref. 19) have used a similar technique to
synthetize 8-desoxy isocaespitol, a halogenated sesquiterpene isolated from the marine alga
Laurencia caespitosa. The terminal bromohydrin obtained by N-bromosuccinimide treatment of
commercial farnesyl acetate was treated 48h. at room temperature in acetic acid/acetic anhydride containing lithium perchlorate (hydrated) to give a tetrahydropyran derivative, addition of bromine chloride to which gave the natural product.

Regioselective prenylation of the extracyclic double bond of (+)-(S)-carvone required more
vigourous conditions than that of limonene. Delobanone (Ref. 21) was obtained (major isomer)
together with its C_8 epimer (Ref. 20). Reduction of the carbonyl group afforded (+)-α-bisa-
bolol (major) and its C_8 epimer.

Similar treatment of (-)-(S)-perillaldehyde or (-)-(S)-methyl perillate gave a separable mixture of hydroxyaldehydes, the major isomer of which was converted to (-)-α-bisabolol (Fig. 7).

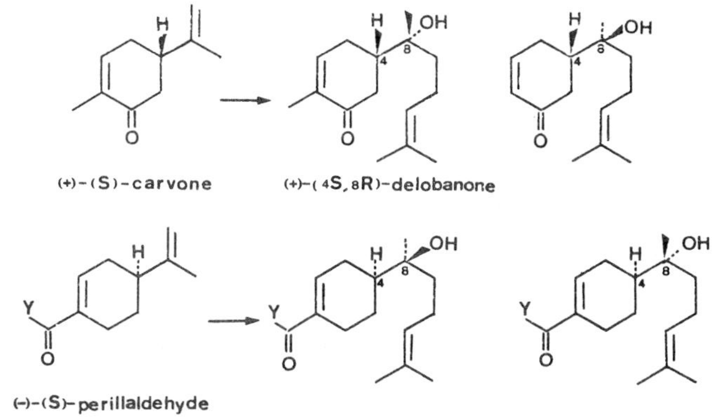

(+)-(S)-carvone   (+)-(4S,8R)-delobanone

(-)-(S)-perillaldehyde

Fig. 7.

Prenylation of camphene similarly led to iso-santalene (Ref. 40) (Fig.8).

Fig. 8.

BIOMIMETIC SYNTHESIS OF CHRYSANTHEMYL ALCOHOL

The discovery of presqualene alcohol (Ref. 22) as an intermediate on the way from two molecules of farnesyl pyrophosphate to squalene raises a number of problems connected with the
closing as well as the opening of the cyclopropane ring. Stereochemical details have been
elucidated. In the ring closing reaction one terminal -CH$_2$OPP is intact whereas in the other
removal of the H$_S$ takes place (Ref. 23) (Fig. 9).

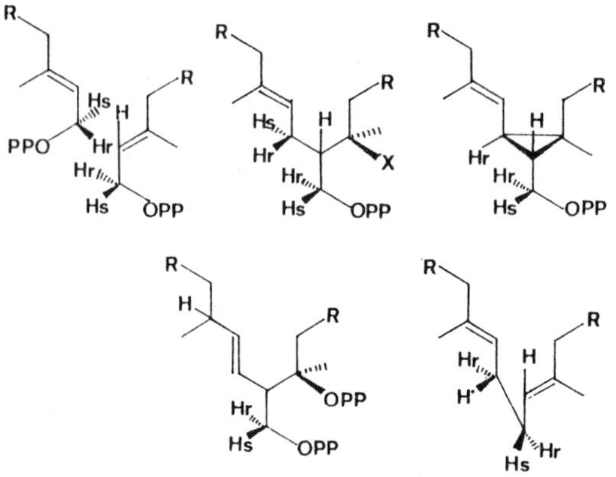

Fig. 9.

In order to account for these facts several hypotheses (Réf. 24, 25, 26 & 27) have been put
forward. In one of them, electrophilic attack by one moiety would take place on C$_2$ of the
other ; a situation obtains which is very similar to that which we have encountered above
in the lavandulol series. A nucleophilic X group is supposed to neutralise the positive
charge on C$_3$. Base promoted 1.3-elimination would then produce the cyclopropane ring (Ref.24).
The question is how could this be persuaded to take place instead of the more familiar
1.2-elimination. The strain in the cyclopropane ring is not very much larger than in a
double bond (27 and 22 Kcal/mole) (Ref. 28). Further the proton to be removed is allylic.
What is needed therefore is a leaving group that would not go off normally, even in the
Hofmann way, but only with the advantage of the allylic activation.

Literature date indeed show that the relative rates of elimination of the various leaving
groups very much depend on the substrate : k_{Cl}/k_{PhS} is about 10^5 for 2-hexyl derivatives
but less than 70 for β-carbonyl derivatives. k_{PhS}/k_{OAc} is about 0.01 for the former and 10
for the latter ! Considerable information could be obtained from the work of C.J.M. Stirling
(Ref. 29) and his group who investigated the behaviour of many weak leaving group β to an
electron withdrawing group. Onium ions which are considered as not very good   (Réf. 30 & 31)
leaving groups then become outstanding.

The literature in fact contains examples of this very situation. More than 40 years ago
Ingold and Rogers (Ref. 32) obtained by decomposition of 3.3-dicarbethoxy-4-phenyl-butyl trimethyl ammonium hydroxide a compound which was recognised 20 years later by Weinstock
(Ref. 33) and Rogers (Ref. 34) to be a cyclopropane derivative formed by displacement of the
onium leaving group by an α-carbethoxy carbanion. C. Bumgardner (Ref. 35) obtained phenyl-
cyclopropane in 80% yield when treating 3-phenyl propyl-trimethyl ammonium hydroxide with
sodamide in liquid ammonia. He made interesting observations on the proportions of β and α
eliminations with various leaving groups.

For model experiments a series of 2-methyl 4-phenyl 2-butyl derivatives were prepared with
a variety of leaving groups and submitted to elimination conditions (Table I).

TABLE I.

Z	Conditions	% Recov.	yield % elim.	7 %	8 %	9 %
$\overset{+}{S}Me_2$	A	0	75	13	87	0
	B	0	38	7	83	0
	C	0	94	7	83	0
	D	0	25	20	80	0
$\overset{+}{N}Me_2Et$	C	68	24	10	90	0
	D			0	100	0
SOPh	A,B	98				
	C	51	15	27	73	0
	D	0	66	26	67	7
SO_2-Ph	A,B	90-100				
	D	0	4		33	66
S-Me	D	28	63	32	68	
S-CH$_2$-Ph	D	0	80	20	80	
SC$_{12}$H$_{25}$	D	100				
S-tBu	D	83	25			99
	E	50	50			99
	F	50	48			99
S-Ph	A,B,C	95				
	D	8	87	0	0	99
	E	70	30	0	0	99
	F	0	90	0	0	99
	G	65	35	86	14	

A : LDA, THF, 3h, -78° ; B : id. +HMPA ; C : NaNH$_2$, NH$_3$, 3h ; D : BuLi TMEDA, hex. 24h, 19° ; E : BuLi, tBuOK, hex. 48h, 19° ; F : LiNEt$_2$, Et O, HMPA, 20h, 19° ; G : tBuOK, DMSO, 60h, 19°.

With the dimethylethylammonium or dimethyl sulfonium groups no cyclo propane but only olefins could be detected. This stands in contrast with Bumgardner results but it must be recalled that leaving groups go off much more easily from tertiary than primary carbon atoms. The benzene sulphinate ester was attacked predominantly at sulphur. The phenylsulphoxide gave a small proportion of 1.3 and much 1.2-elimination. The phenylsulphone gave very little elimination but 1.3-elimination predominated over 1.2.

Various sulphides were next tried. The methyl and benzyl sulphides led to 1.2 elimination whereas the lauryl sulphide was very inert. This is in good agreement with recent results by Biellmann (Ref. 36) on the α',β-elimination of aliphatic sulphides.

The phenylsulphide underwent elimination when treated with a strong base such as butyl lithium (TMEDA) in hexane ; 1.1-dimethyl 2-phenyl cyclopropane was however practically the only product formed under those conditions ; so 1.3 elimination indeed has taken place.

The t-butyl sulphide underwent elimination under more severe conditions (n-BuLi/t-BuOK) (Ref. 37) to yield dimethyl phenyl cyclopropane.

Information about the cyclisation of the phenyl sulphide was obtained by quenching the reaction mixture with D_2O or carbon dioxide. It turned out that metalation is very rapid on the phenyl thio ring (mainly ortho to the sulphur atom) and in the benzylic position. Quenching after 10h at room temperature indicated metalation mainly meta and para to the sulphur in the starting material. Apparently then the ortho metalated product undergoes ring closure whereas the m and p isomers are inert.

The corresponding chrysanthemyl precursor was next synthesized (Fig. 10). With the now readily available lavandulyl diol, difficulties were experienced in converting the tertiary hydroxyl to a phenyl thio group owing the protonation of the trisubstituted double bond with participation of one of the oxygen atoms.

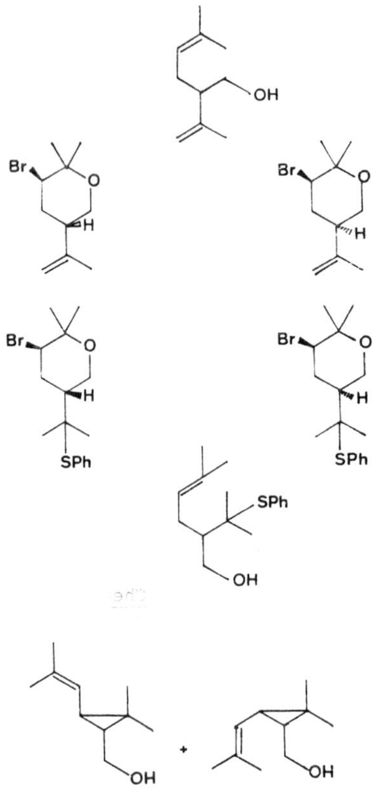

Fig. 10.

This unwanted participation of the double bond in the C_{10} glycol was used to put it out of the way in a reversible manner. Bromination of lavandulol with N-bromosuccinimide in wet ether led to the bromotetrahydropyranylether (2 epimers) in which now both the Ritter reaction and the perchloric acid catalyzed addition of phenylthiol could be carried out efficiently. After the appropriate modifications, reductive elimination regenerated the primary alcohol group and the double bond and gave 2-prenyl 3-methyl 1-butanol with an amino or phenyl thio group in the 3-position.

When the phenyl thio alcohol was treated with BuLi/TMEDA in hexane, about 50% of the starting material disappeared in 24h at room temperature and two new compounds were formed which were easily isolated by chromatography on silicagel. They were identified as cis and trans chrysanthemyl alcohol (20/80) by glc on a capillary column, ^{1}H NMR, ^{13}C NMR and glc-mass spectrum. Lavandulol could hardly be detected. The main component (trans) of the mixture was isolated by hplc, after conversion to the 3.5-dinitro benzoates, as a crystalline solid m.p. 105.5° alone or admixed with an authentic sample.

It is remarkable that the reaction stops at 50% conversion . Quenching with D_2O shows that the "starting material" is metalated in the aromatic ring ortho to the sulphur. This carbanion thus appears to be inert, unlike what has been observed in the model experiments.

This cyclopropane synthesis might be of use in the terpene field. A simple conversion of limonene into 2-carene has been carried out in two steps (Réf. 39) (Fig. 11).

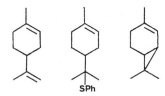

Fig. 11.

It thus proved possible to carry out in the laboratory a reaction similar to the scheme proposed ten years ago by Rilling and Poulter (Ref. 24) for the biosynthesis of presqualene alcohol.Obviously several problems require investigation. If this is the real biosynthetic route, how does the living cell carry out the cyclisation step ? How is the stereochemistry controlled ? This problem might be related to the one alluded to above concerning the direction of elimination of HX in the head to tail adduct in the prenyl transferase reaction. In the condensation of the two farnesyl units how is the attack directed on C_2' instead of C_6' or C_{10}'?

Acknowledgement - The results reported here are due to the skill and dedication of the people who worked together on this programme. Drs. C. Perez and L. Saussine for the first chapter and Dr. J.D. Fourneron, Dr. L.M. Harwood and Mr. D. Babin for the second and third chapters. Their contribution is gratefully acknowledged. Thanks are also due to the C.N.R.S. for generous support.

REFERENCES

1. J.W. Cornforth, Quart. Rev. 23, 125 (1969) ; Chem Soc. Rev. 2, 1 (1973) ; Tetrahedron 30 ; 1515 (1974) ; Angew. Chem. Internat. Edit., 7, 903 (1968) ; K.H. Overton, Chem. Soc. Rev. 8, 447 (1979).
2. J.W. Cornforth, R.H. Cornforth, G. Popjack & L. Yengoyan , J. Biol. Chem. 241, 3970 (1966); G. Popjack and J.W. Cornforth, Biochem. J. 101, 553 (1966) ; J.W. Cornforth, Angew. Chem. 80, 977 (1968).
3. C.D. Poulter & H.C. Rilling, Acc. Chem. Res. 11, 307 (1978).
4. W.S. Johnson & R.A. Bell, Tetrahedron Letters, 27 (1960 12).
4a. D. Klamann, P. Weyerstahl & M. Fligge, Angew. Chem. 77, 1137 (1965).
5. A. Thiry, M.E. Degeorges & E. Charles, (Progil) Fr. Pat. 1, 271, 563 (1960); B. Rickborn & R.M. Gerkin, J. Amer. Chem. Soc. 93, 1693 (1971).
6. a) M. Julia, C. Perez & L. Saussine, J. Chem. Res. (S) 268 (1978) (M) 3401 (1978) ; b) M. Julia & L. Saussine, ibid., (S) 269 (1978) (M) 3420 (1978) ; c) M. Julia, C. Perez & L. Saussine, ibid., (S) 311 (1978) (M) 3877 (1978).
7. A. St. Pfau & P.A. Plattner, Helv. Chim. Acta 15, 1250 (1932) ; R.T. Arnold, ibid, 32, 124 (1949).
8. W.S. Johnson, Bioorganic Chemistry 5, 51 (1976) ; Angew. Chem. 88, 33 (1976) ; Angew. Chem. Int. Ed. 15, 9 (1976).
9. G. Cardillo, M. Contento & S. Sandri, Tetrahedron Letters, 2215 (1974).
10. N.H. Anderson, Y. Ohta & D.D. Syrdal, Biorganic Chemistry, vol. II Substrate Behaviour Chap. I, p.1 ; van Tamelen Ed. 1978 Academic Press, New York.
11. W. Knöll & C. Tamm, Helv. 58, 1162 (1975).
12. I. Yosioka, T. Nishino, T. Tani & I. Kitagawa, Yakagaku Zasshi 96, 1229 (1976) ; Chem. Abstr. 85, 198083 a (1976).
13. A. Kergomard & H. Veschambre, Tetrahedron 33, 2215 (1977).
14. a) M.A. Schwartz & G.C. Swanson, J. Org. Chem. 44, 953 (1979).
    b) T. Iwashita, T. Kusumi & H. Kakisawa, Chem. Letters, 947 (1979).
15. K.G. O'Brien, A.R. Penfold & R.L. Werner, Aust. J. Chem. 6, 166 (1953).
16. T. Prangé, D. Babin, J.D. Fourneron, M. Julia, Comptes Rendus Ac. Sci.Série C, 289, 383 (1979).
17. D. Babin, J.D. Fourneron & M. Julia, Tetrahedron, in the press.
18. S. Godtfredsen, J.P. Ulbrecht & D. Arigoni, Chimia (Switz) 31, 62 (1977).

19. A.G. Gonzalez, J.D. Martin, C. Perez, M.A. Ramirez & F. Ravelo, Tetrahedron Letters 187 (1980).
20. L.M. Harwood & M. Julia, Tetrahedron Letters, 21, 1743 (1980)
21. K. Takeda, K. Sakurawi & H. Ishii, Tetrahedron 27, 6049 (1971).
22. H.C. Rilling & W.W. Epstein, J. Amer. Chem. Soc. 91, 1041 (1969).
23. D. Popjack, H.J. Ngau & W.S.,Agnew. Bioorgan. Chem. 4, 279 (1975).
24. H.C. Rilling, C.D. Poulter, W.W. Epstein & B. Larsen, J. Amer. Chem. Soc. 93, 1783 (1971).
25. E.E. van Tamelen & M.A. Schwartz, ibid 93, 1780 (1971).
26. T. Cohen, G. Herman, T.M. Chapman & D. Kuhn, J. Amer. Chem. Soc. 96, 5627 (1974).
27. B.M. Trost & W.G. Biddlecom, J. Org. Chem. 38, 3438 (1973).
28. J.D. Roberts & M.C. Caserio, Basic Principles of Organic Chemistry, p. 112 WA Benjamin New York (1964).
29. C.J.M. Stirling, Acc. Chem. Res. 12, 198 (1979).
30. C.K. Ingold, Structure & Mechanism in Organic Chemistry, 2nd Ed., p. 663-668, Cornell University press, Ithaca, New York (1969).
31. A.F. Cockerill & R.G. Harrison in : The Chemistry of double bonded functional group ; S. Patai Ed. I, p. 149, Wiley,New York (1977).
32. C.K. Ingold & M.A.T. Rogers, J. Chem. Soc. 722 (1935).
33. J. Weinstock, J. Org. Chem. 21, 540 (1956).
34. M.A.T. Rogers, J. Org. Chem. 22, 350 (1957).
35. C.L. Bumgardner, J. Amer. Chem. Soc. 83, 4420, 4423 (1961) ; J. Amer. Chem. Soc. 29, 767 (1964) ; J. Amer. Chem. Soc. 88, 5518 (1966) ;
    C.L. Bumgardner, J.R. Lever & S.T. Purrington, J. Org. Chem. 45, 748 (1980) ;
    C.L. Bumgardner, Chem. Comm. 374 (1965) ; see also R. Baker & M.J. Spillett, J. Chem. Soc., 581 (B 1969) ; M. Apparu & M. Barrelle, Tetrahedron 34, 1691 (1978) ; S. Martel & C. Huynh, Bull. Soc. Chim. Fr., 985 (1967); M. Julia & A. Guy Rouault, ibid, 1441
36. J.F. Biellmann, H. d'Orchymont & L. Schmitt, J. Amer. Soc. 101, 3283 (1979).
37. M. Schlosser & J. Hartmann, Angew. Chem. 85, 544 (1973) ; Angew. Chem. Int. Ed. 12, 508 (1973).
38. D. Babin, J.D. Fourneron & M. Julia, Bull. Soc. Chim. Fr., submitted for publication
39. D. Babin, J.D. Fourneron, L.M. Harwood & M. Julia, unpublished results.
40. W. Rojahn, W. Bruhn & E. Klein, Tetrahedron 34, 1547 (1978).

ASPECTS OF CARBOCYCLIC RING FORMATION AND MODIFICATION

H. Nozaki, H. Yamamoto, K. Oshima, K. Utimoto and T. Hiyama

Department of Industrial Chemistry, Kyoto University, Kyoto 606, Japan

Abstract - Biomimetic cyclization of diethyl neryl phosphate and of (Z)-$Me_2C=C(OSiMe_3)CH_2CH_2CMe=CHCH_2OAc$ has been effected by means of reagents containing aluminum as the key atom to afford limonene and karahanaenone, respectively, with high selectivities. The synthetically useful reactions are rationalized by assuming the "combined acid-base attack" of the reagents and intimate ion-pair character of the reacting species. Humulene skeleton has been prepared by cyclizing $MeC^{\ominus}(COOMe)COCH_2CH_2CMe=CHCH_2CH_2CMe=CH=CH_2 \cdot Pd(OAc)L_n$ species stereospecifically. Intramolecular Friedel-Crafts reaction of $Me_3SiC\equiv C(CH_2)_nCOCl$ gives cycloalk-2-ynones for $n \geq 8$, while 3-chloro-2-trimethylsilylcycloalk-2-enones for $2 \leq n \leq 5$. In contrast, intramolecular crossed aldol reactions mediated by aluminum enolates suffer from certain limitations. Finally, methods of cyclopentenone ring annulation are discussed in connection with novel syntheses of muscopyridine and nootkatone.

INTRODUCTION

This paper involves (1) biogenetic type terpenoid monocyclization, where various types of organoaluminum, amphoteric reagents play key roles in highly selective reactions, (2) medium and large ring cyclization utilizing weak interaction between both termini of a chain and the reagent, and (3) ring enlargement through carbenoid reactions and cyclopentenone annulation. Synthetically useful reactions between the functional groups located at the ends of an aliphatic chain would be the main subject and a few cases of electrocyclic reactions will be discussed only briefly in the last chapter.

BIOMIMETIC MONOCYCLIZATION OF TERPENOIDS

The eq 1 shows the start of an inquiry into synthetic applications of certain organoaluminum reagents, an account of which has already appeared (ref. 1). Notably, the reaction of neryl phosphate with trimethylaluminum gave 4-t-butyl-1-methylcyclohexene as a cyclization-alkylation product, whereas geranyl isomer produced only methylated geranyl and linalyl

$$ (1) $$

R	X	Solvent	Y%	Y%	Y%
Me	Me	CH_2Cl_2 below 20°, 4 hr	68		
Me	PhO	CH_2Cl_2 (besides 8-10% PhO-neryl)		59	10
^iBu	-O-	CH_2Cl_2		66	9
		THF		50	30
Me	PhNH	CH_2Cl_2		49	13
^iBu	Me	CH_2Cl_2 or hexane		50	<0.5

241

hydrocarbons in roughly 9:1 ratio. No methylated neryl product has been isolated in both reactions. Treatment of diethyl geranyl phosphate or the neryl isomer with dimethylaluminum phenoxide in h e x a n e afforded geranyl phenyl ether and neryl phenyl ether, respectively, in a strictly stereospecific manner. The etherification in hexane, which is a poor solvent and the whole forms a white suspension, proceeds without affecting the double bond at all. No allylic migration, no (E)-(Z) transformation, and no cyclization have been observed. In d i c h l o r o m e t h a n e, however, the clear, homogeneous reaction system produced a mixture of limonene and terpinolene as shown in the Table below eq 1. The limonene/terpinolene ratios are dependent on the reagents and solvents. Eventually, more than 99% selectivity of limonene has been attained by utilizing a sterically hindered aryloxide (last row), although the yield is only moderate. Where X = -O-, the reagents are R$_2$Al-O-AlR$_2$. Tetraisobutyldialuminoxane (TIBAO) is a unique reagent in the sense that diethyl geranyl phosphate also can be cyclized in THF into a mixture of limonene and terpinolene (ca. 5:3 ratio) in 75% yield by the action of this particular reagent.

Extension of the cyclization technique to farnesyl phosphate (eq 2) yielded the cyclohexene derivatives, α- and β-bisabolene, exclusively. No products arising from participation of the farthest C=C have been isolated. Control of the β/α ratio of bisabolene has not been successful. Eq 3 shows the preparation of a spiro-type sesquiterpene, chamigrene.

Reagent/Solvent	Y%	β/α
PhOAlMe$_2$/CH$_2$Cl$_2$	95	64:36
(^{i}Bu$_2$Al)$_2$O/CH$_2$Cl$_2$	84	68:32
/THF	69	48:52
2,6-^{t}Bu -4-MeC$_6$H$_2$O-Al^{i}Bu$_2$/CH$_2$Cl$_2$	73	84:16

(2)

Y72% (no β-product)
(^{i}Bu$_2$Al)$_2$O/hexane
-78° 10 hr, 20° 3 hr

(3)

(4)

MeAl(OCOCF$_3$)$_2$/CH$_2$Cl$_2$ at r.t.

7:3

the cis isomer

12 hr, Y62%

cis + trans
66%   34%

Et$_2$AlOAr/CH$_2$Cl$_2$
at r.t., Ar =

(5)

X	Y%
Cl	58
Br	51

(6)

quantitative

TiX$_4$:PhNHMe (1:1)/CH$_2$Cl$_2$ at -23°, 1 hr

Karahanaenone is a terpene ketone of 7 membered ring as shown in eq 4. The biomimetic cyclization of neryl group must involve the addition to the terminal C=C occurring in the anti-Markovnikov sense. Introduction of trimethylsilyloxy group on this olefinic carbon should stabilize the cation ($^\oplus$C-OSiMe$_3$), which is formed upon cyclization of the allylic cation. The initial anticipation has led to a novel type of karahanaenone synthesis, but the observation with the geranyl isomer has turned out to be incompatible with the supposition of "free" allylic carbocation intermediacy. The latter reacted much more sluggishly to afford a mixture in lower yields.

Recent observation of Itoh in this laboratory, as shown in eq 5, provides a means of cyclohexanone annulation. Cyclization of trans-disubstituted cyclohexene substrate proceeded smoothly in the presence of sterically hindered aryloxide to provide the stereomerically homogeneous cis-octalone product. The reagent in eq 4 turned out to be less effective in this case. As given under the formulas, the cis-disubstituted cyclohexene required much longer reaction time and afforded a mixture in lower yield again.

Inspection of the molecular models indicates that coordination of R$_2$AlX reagent on the trans-substrate produces an adduct whose structure is close to the intermediary zwitterion shown in eq 5 (right). Smooth nucleophilic attack on silicon by the leaving group + reagent complex [O=C(Me)-O-AlR$_2$-X]$^\ominus$ is expected only in this case. Such kind of <u>combined acid-base attack</u> is commonly observed in reactions mediated by the reagents having <u>aluminum as the key atom</u> (ref. 2). In other words, R$_2$AlX reagent coordinates the leaving group as a Lewis acid to accelerate the ionization, but this is not the decisive step in many cases. Nucleophilic attack of [Lv-AlR$_2$-X]$^\ominus$ complex on the resultant carbocation part in tight ion-pair stage is responsible to determine the reaction rate and product distribution.

Eq 6 shows another methodology of cyclization of nerol itself. The reagent is prepared from titanium tetrahalide and N-methylaniline in 1:1 ratio and the ring closure is accompanied by halogenation. Treatment of geraniol yielded the corresponding allylic halides quantitatively. Such reactions do not take place in the absence of equimolar amine.

Seven membered ring closure is attained in eq 7. Note that the difference in the substrate structure as compared with eq 6 is only the mode of substitution of the distant C=C bond. If R = Cl, the main product is the dichloride and the combined yield is fairly good. Synthetic application of the cyclization is given in eq 8 indicating a novel route to nezukone. The barrier in the whole sequence existed after cyclization. Every attempt of nucleophilic substitution to convert the C-Cl bond into C-O bond failed to succeed. The reactions shown in the scheme yielded the desired product (ref. 3).

R	X	Y%
H	Cl	65
H	Br	56
Me	Cl	66
Cl	Cl	18 besides

(7)

(8)

$$TiCl_4-PhNHMe \quad / \quad CH_2Cl_2$$

1. Mg, 2. n-Bu$_3$SnCl

$$CrO_3 2C_5H_5N$$

X	R	reagent	Y%
SiMe$_3$	H	TiCl$_4$-PhNHMe/CH$_2$Cl$_2$	79
SiMe$_3$	Ac	MeAl(OCOCF$_3$)$_2$	29
SnBu$_3$	H	TiCl$_4$-PhNHMe/CH$_2$Cl$_2$	73
SnBu$_3$	Ac	MeAl(OCOCF$_3$)$_2$	51

(9)

Terminally silylated or stannylated nerol derivatives as shown in eq 9 give 100% selectivity of limonene, of course. The titanium halide-amine complexes are effective cyclization reagents and unique halogenating ones.

Attempted application of these techniques to the cyclization of farnesol derivatives failed to afford 10 or 11 membered ring products at all, although the initial transformation of farnesyl pyrophosphate into humulene accounts for the biogenetic origin of bicyclic sesqui-terpenes, reportedly. Extensive modification of farnesol molecule into the acyclic keto ester in eq 10 enabled us to mimic the biogenetic eleven membered ring formation.

Humulene has an eleven membered ring with three incorporated C=C bonds all in (E) configu-ration. One can prepare the acyclic keto ester in (E,E) configuration by standard tech-nique, which is subjected to the key cyclization reaction proceeding between the sodium enolate moiety and pi-allyl palladium complex structure both being located at the ends of a carbon chain. This unique reaction as discovered by J. Tsuji and later extensively studied by B. M. Trost (ref. 4) proceeded under strict retention of configuration of the allylic C=C bond as shown in eq 10, rather surprisingly (ref. 1).

Introduction of the third double bond in the step d or the step f in eq 10 occurred 100% stereoselectively thanks to the efficacy of the amphoteric organoaluminum reagents (ref. 1 and 2).

a: $^{\ominus}$CH$_2$COC$^{\ominus}$MeCOOMe, b: NaH, Pd/PPh$_3$/PPh$_2$(CH$_2$)$_3$PPh$_2$, Y45% for (E,E)-isomer, Y59% for (E,Z)-isomer (m. 68°), c: (1) LiAlH$_4$, (2) TsCl, (3) K-OBu^t, d: Et$_2$AlNPhMe producing the primary alcohol Y = OH, which was transformed into humulene (Y = H) upon treat-ment with (1) Me$_2$S/NCS/Et$_3$N, (2) H$_2$NNHTs, (3) LiAlH$_4$, e: (1) LiAlH$_4$, (2) TsCl, (3) LiAlH$_4$, (4) MsCl, f: Me$_2$AlOPh (Y83%), aq AcMe (Y ca. 70%) producing humulene directly.

MEDIUM AND LARGE RING FORMATION

Eq 10 above shows that the entropy barrier in medium ring cyclization could be overcome by setting interactions between functional groups at both termini and the reagent(s) involved. A working hypothesis, not proved mechanism, has been given in eq 11. This indicates the possibility of an aluminum enolate being produced by zinc reduction of α-bromo ketones or esters in the presence of diethylaluminum chloride, for example. Actually, this kind of technique has provided a means of transforming α-bromo ketones into the corresponding enolates regioselectively in a predictable way, which are combined with the coexisting carbonyl component irreversibly to provide crossed aldol products in good yields under mild reaction conditions (ref. 1).

Intramolecular Reformatsky type reaction in eq 12 gives large ring lactones and the technique has been extended to the preparation of 10 membered ring of unsaturated lactone by J. Tsuji (ref. 5) which has given almost comparable yield. Wittig-Horner type products, R^{1}R^{2}C=CHCOOR and R^{1}R^{2}C=CClCOOR, and not Reformatsky beta-hydroxy esters are produced in the reaction of dibromoacetates or trichloroacetates with carbonyl components R^{1}COR^2. The intermediacy of dicarbanion species is postulated (ref. 6).

Aluminum enolates provide another possibility in synthetic methodology (ref. 7) and, as far as the cyclization is concerned, J. Tsuji et al. (ref. 8) have applied intramolecular aldolic reaction of 2,15-hexadecanedione to muscone synthesis.

$$\text{(11)}$$

$$\text{(12)}$$

$$\text{(13)}$$

$$\text{(14)}$$

Michael addition of R₃Al and the resulting enolate reaction have been described (ref. 9). Eq 13 and 14 show novel cyclization technique by means of Me₂AlSPh, in which PhS group is the active nucleophile, of course (ref. 10).

All these observations appeared to support the expectation that aluminum enolate-mediated procedures could provide novel access to medium ring. Attempted 11 membered ring closure, however, failed to succeed upon repeated experiments with the (E,E) isomer of a bromo keto aldehyde, Me₂CBrCOCH₂CH₂CMe=CHCH₂CH₂CMe=CHCHO (Zn, Et₂AlCl), and with a keto aldehyde, CH₂=CHCO(CH₂)₈CHO (Me₂AlSPh).

Utimoto has been able to find an entrance to medium and large rings as shown in eq 15 (ref. 11, 12), which consists in the application of the Calas reaction to cyclization. The route is characterized by the ready availability of the starting

$$\text{(15)}$$

a: KNH-(CH₂)₃NH₂   b: EtMgBr, Me₃SiCl
c: Jones' reagent, (COCl)₂   d: AlCl₃

$$\text{(16)}$$

$$\text{(17)}$$

material. An internal acetylenic alcohol is isomerized in step a to the terminal one, which is easily transformed into the required terminally trimethylsilylethynyl-substituted alkanoyl chloride. A solution of 1 mmol of the acid chloride in 25 ml of CH_2Cl_2 is added to a suspension of 3 mmol of $AlCl_3$ in 200 ml. of refluxing CH_2Cl_2 under high dilution condition within 2.5 hr. The isolated yields of cycloalkynones (ring size or n + 3 in parentheses) are 46% (11), 56% (13), 77% (15) and are not optimized yet.

The cycloalkynones are easily transformed into 1,3-cycloalkanediones, pyrazolophanes, and isoxazolophanes. Cyclopentadec-2-ynone is readily derived into muscone by simple treatment with Me_2CuLi and the subsequent hydrogenation (PtO_2) of the resulting mixture of 3-methylcyclopentadec-2- and 3-enones.

Optically active muscone is synthesized in eq 17 with (R)-(+)-citronellol as a chiral pool. Ozonolysis of C=C linkage, reduction (X = OH), iodination (X = I), reaction with R-C≡C-MgX, KAPA treatment, trimethylsilylation, oxidation of the terminal CH_2OH group into COOH, and final chlorination give the acid chloride required for ring closure, which has proceeded to afford 52% yield of the acetylenic ketone. Hydrogenation gives (R)-muscone quantitatively.

One might worry about possible drop of e.e. during the KAPA treatment, but this is not the case as Utimoto has shown. One can prepare 3-D-citronellol from regular citral by reduction with $Fe(CO)_5$/base/MeOD/D_2O (Noyori's procedure), aqueous work-up, and carbonyl reduction. This is transformed into the internal acetylenic alcohol and subjected to the KAPA treatment. The D-content before and after isomerization remained the same value, 97%. This excludes the possibility of the super-base attack on the methine proton of the methylated carbon.

(R)-Muscopyridine has been synthesized by a route of eq 18. The chiral pool is pinane as prepared from beta-pinene according to the H. C. Brown's procedure (ca. 90% e.e.). Pyrolysis gives the diene indicated in the scheme. Ozonolysis of the triply substituted C=C linkage is followed by reduction to CH_2OH and tosylation. Condensation with THP-O-$(CH_2)_4$MgBr produces the lower part of the molecule. Hydroboration of the vinyl group and oxidation to aldehyde are followed by condensation with 1-pentynyllithium. Treatment with KAPA, silylation, oxidation with Jones reagent, and final chlorination [(COCl)$_2$] give the desired acid chloride. Unexpectedly, the cyclization proceeded with 82% isolated yield of the acetylenic ketone. Maybe the participation of the keto carbonyl group is important in this cyclization. Following steps   require no explanation.

Incidentally the methylene group between the pyridine ring and carbonyl group of the beta-keto derivative of muscopyridine in eq 18 exhibited a typical AB quartet. The "benzylic" protons of muscopyridine showed signals much more complex but the methylene protons are evidently not equivalent to each other. In the absence of the beta-methyl group the

(18)

(19)

alpha-methylene protons of such [m](2,6)pyridinophane are equivalent even with smaller number of m, for example m = 7 (ref. 13). The remarkable effect of the methyl substituent on
the conformation of the aliphatic chain is the subject of current studies.

Eq 19 shows that the intramolecular Friedel-Crafts cyclization proceeds also with shorter
chain-length substrates to afford 3-chloro-2-trimethylsilylcycloalk-2-enones, whose yields
(the ring-sizes or n + 3 in parentheses) are as follows: 31% (5), 96% (6), 65% (7), and
40% (8). Regrettably, no isolable products have been obtained in cases of 9 and 10 membered rings being expected. Obviously, there might be some interactions between both
terminal functional groups as well as the reagent, aluminum chloride, which must help to
lower the entropy barrier of the cyclization. Furthermore, the cyclopentenone formation
in eq 19, at least, requires the disappearance of the linear arrangement of Si-C≡C-CH$_2$
moiety before the ring formation. Otherwise, the carbonyl carbon and silyl-substituted
carbon can not approach to each other to form a new C-C bond. More work should be done
to understand the role of the silicon group thoroughly.

CARBENOID REACTIONS AND CYCLOPENTENONE ANNULATION

Another route to dl-muscopyridine is shown in eq 20 (ref. 14). The key transformation of
enone into dione involving the steps d through g has been mentioned by A. S. Dreiding (ref.
15) and the cyclopentenone annulation on cycloalkanones mediated by propargyl alcohol by
R. A. Raphael (ref. 16) and by A. S. Dreiding (ref. 17). We were able to extend the
technique to the cycloalk-2-enone system and to observe the electrocyclic ring closure
(in the brackets) of OH-substituted heptatrienyl cation into the 3-ethenylcyclopent-2-
enone system. The required methyl group of muscopyridine is introduced smoothly by 1,6-
addition to the bicyclic dienone.

a: LiC≡CCH$_2$OLi, b: (HO)$_2$SO$_2$-MeOH (1:2), -20° to r.t. 12.5 hr, Y65%, c: 2MeMgI,
0.12 CuCl, 0°, 1.5 hr. Y74%, d: NaBH$_4$/MeOH aq, e: MCPBA/CH$_2$Cl$_2$, f: TsCl/C$_5$H$_5$N,
g: CaCO$_3$ aq/dioxane refl. (Y71% for d through g), h: NH$_2$OH.HCl/EtOH 150-160°.

Incidentally, this type of cyclopentenone annulation has been shown to furnish the nootkatone skeleton with unique selectivity, as shown in eq 21 (ref. 18). The step C must involve
conrotatory cyclization of -CMe=C-⊕C(OH)-C=CHMe group and the resulting cyclopentenone
should (and actually did) contain cis-stereochemistry of the vic-dimethyl substituents. The
methoxycarbonyl group can have both alpha- and beta-configuration and the resulting product
is a mixture of both dl-epimers. Conversion of this moiety into acetyl and then into isopropenyl group requires the protection of cyclopentanone keto group. The resulting product
has been found to be stereochemically homogeneous and to possess the desired alpha-con-
figuration with respect to the isopropenyl group. The epimerization is ascribed to the
strong basicity of the Wittig reagent in the reaction with the methyl ketone intermediate.

The ring expansion of hydrindanone system into octalone has proceeded 100% regioselectively to involve the exclusive migration of the olefinic carbon (ref. 19). The final step is, of course, the shift of C=C bond to be conjugated with C=O group and is easily performed.

The one carbon ring expansion in eq 21 is an efficient method of muscone synthesis starting from cyclotetradecanone as outlined in eq 22 (ref. 19). Cyclotetradecanone is available from 12 membered ring diester *via* the acyloin disilyl ether whose electrocyclic rearrangement produces 14 membered ring under two carbon ring expansion (ref. 20).

The dibromomethane-adduct of 2-methylcyclotetradecanone (ref. 21) is treated with n-BuLi in step g to afford a beta-oxido carbenoid whose rearrangement under LiBr removal proceeds with nearly exclusive regioselectivity under migration of the methylated carbon to afford muscone, if one chooses appropriate sets of reaction solvent, temperature, and period (ref. 19).

(21)

a: PhH at 160-180°, AcOH aq or KF/MeOH, b: HC≡CCHMeOH, KOH aq 40°, c: (HO)₂SO₂ and MeOH (1:1), 50° 30 min. Y49-60%, d: NaBH₄-CeCl₃·7aq, e: KOH, MeOH aq, 80° 30 min., f: excess MeLi/ether at 0° 2 hr. Y97% for d through f, g: Ph₃P=CH₂, DMSO at r.t., h: C₅H₅N.HCl.CrO₃/CH₂Cl₂, 1.5 hr. Y80% for g & h, stereopurity over 88%, i: LiCHBr₂/THF -78° to r.t. 3 hr. Y84%, j: 2BuLi/THF -95° to r.t. 1.6 hr. Y83%, k: 5% (HO)₂SO₂ aq/THF 1:1, r.t. 4 hr. Y74%. stereopurity >93%.

(22)

a: Na, Me₃SiCl, xylene, b: HCl aq, c: P(OEt)₃/refl. PhH, c: Zn/AcOH-HCl, e: MeI, LiNⁱPr₂, f: LiCHBr₂ from CH₂Br₂ + LDA or LiTMP (2,2,6,6-tetramethyl-piperidide), g: 2BuLi/ether, -78° Y79%.

Another route to muscone from cyclotetradecanone also utilizes lithium carbenoid reaction as shown in eq 22 (ref. 22). Dichlorocarbene-adduct of enol ether is treated with 2 equiv of MeLi. The reaction should involve initial deprotonation, chloride loss, and readdition of another equiv of MeLi. Doering-Skattebøl cleavage of the resulting lithium carbenoid in brackets produces the one-carbon ring expanded enone under concomitant methylation (ref. 23).

$$(23)$$

a: $HC(OEt)_3$, TsOH, b: $CHCl_3$, NaOH aq, $n-C_{16}H_{33}NMe_3^+Cl^-$, c: 2MeLi, THF-HMPA

(5:1), Y69%

$$(24)$$

a: $LiCCl_2CH=CH_2$ ($CHCl_2CH=CH_2$, LiN^iPr_2), Y66%, b: CF_3COOH, Y90%, c: $CH_2=CHOEt$

+ C_5H_5NHOTs, d: $CHCl_3$, NaOH, $n-C_{16}H_{33}NMe_3^+Cl^-$, e: EtOH, C_5H_5NHOTs, Y77%

overall for step c through e, f: CF_3COOH refl. 4 days, Y76%.

Further methodologies of cyclopentenone annulation are summarized in eq 24. The vinyl-
substituted lithium carbenoid (ref. 19) reacts *in situ* with cyclododecanone and the subsequent acid-treatment of the resulting adduct produces 2,3-decamethylenecyclopent-2-enone
(ref. 24). Eschenmoser's classical muscone synthesis therefrom would need not be mentioned
(ref. 25). Similar chloro-substituted pentadienyl cation cyclization followed by hydrolysis
and deprotonation also affords cyclopentenone-annulated products and is based upon
dichlorocarbene adducts of appropriate allylic alcohols (ref. 26).

Acknowledgment — Thanks are given to valuable contributions of enthusiastic
students of this research group. Financial supports by the Ministry of Education, Science,
and Culture, Japanese Government, through Research Grants (203014, 303023, 403022, 311702,
411102, 247077) are gratefully appreciated.

REFERENCES AND NOTES

1. H. Yamamoto and H. Nozaki, *Angew. Chem. Intern. Ed. Engl.*, *17*, 169 (1978) and refs.
   cited there; see also Y. Kitagawa, S. Hashimoto, S. Iemura, H. Yamamoto, and H. Nozaki,
   J. Am. Chem. Soc., *98*, 5030 (1976).
2. K. Oshima, H. Yamamoto, and H. Nozaki, *Proceedings of the First International Kyoto
   Conference of Organic Chemistry (December 1979)*, to be published by Kodansha, Tokyo,
   1980.
3. T. Saito, A. Itoh, K. Oshima, and H. Nozaki, *Tetrahedron Lett.* 1979, 3519.
4. B. M. Trost, *Tetrahedron*, *33*, 2615 (1977).
5. J. Tsuji and T. Mandai, *Tetrahedron Lett.*, *1978*, 1817.
6. K. Takai, Y. Hotta, K. Oshima, and H. Nozaki, *Tetrahedron Lett.*, *1978*, 2417; *Bull. Chem.
   Soc. Jpn. in press*.
7. H. Nozaki, K. Oshima, K. Takai, and S. Ozawa, *Chem. Lett.*, *1979*, 379.
8. J. Tsuji et al., *Tetrahedron Lett.*, *1979*, 2257; *Bull. Chem. Soc. Jpn.*, *53*, 1417 (1980).
9. E. A. Jeffery, A. Meisters, and T. Mole, *J. Organometal. Chem.*, *74*, 373 (1974).
10. A. Itoh, S. Ozawa, K. Oshima, and H. Nozaki, *Tetrahedron Lett.*, *1980*, 361.
    See, however, E. C. Ashby and S. A. Noding, *J. Org. Chem.*, *44*, 4792 (1979).
11. K. Utimoto, M. Tanaka, M. Kitai, and H. Nozaki, *Tetrahedron Lett.*, *1978*, 2301.
12. K. Utimoto, M. Tanaka, and H. Nozaki, *ACS/CSJ Chemical Congress, Symposium of Synthesis
    and Chemistry of Acetylenic Compounds*, April, 1979, Hawaii.

13. S. Fujita and H. Nozaki, *Bull. Chem. Soc. Jpn.*, *44*, 2827 (1971); *Yuki-gosei-kagaku*, *30*, 679 (1972).
14. H. Saimoto, T. Hiyama, and H. Nozaki, *Chem. Soc. Jpn. 41st Spring National Meeting, April 1980*, 1R04, Osaka.
15. R. W. Gray and A. S. Dreiding, *Helv. Chim. Acta*, *60*, 1969 (1977).
16. R. A. Raphael et al., *J. Chem. Soc.*, *1953*, 2247; *J. Chem. Soc. Perkin Trans. 1*, *1976*, 410.
17. M. Karpf and A. S. Dreiding, *Helv. Chim. Acta*, *59*, 1226 (1976).
18. T. Hiyama, M. Shinoda, and H. Nozaki, *J. Am. Chem. Soc.*, *101*, 1599 (1979); T. Hiyama, M. Shinoda, and H. Nozaki, *Tetrahedron Lett.*, *1979*, 3529; cf. also R. M. Jacobson et al., *J. Org. Chem.*, *45*, 395 (1980). P. Magnus et al., *ibid.*, *45*, 1046 (1980)
19. H. Taguchi, H. Yamamoto, and H. Nozaki, *Bull. Chem. Soc. Jpn.*, *50*, 1592 (1977).
20. T. Mori, I. Nakahara, and H. Nozaki, *Canad. J. Chem.*, *47*, 3266 (1969).
21. H. Taguchi, H. Yamamoto, and H. Nozaki, *Bull. Chem. Soc. Jpn.*, *50*, 1588 (1977).
22. T. Hiyama, T. Mishima, K. Kitatani, and H. Nozaki, *Tetrahedron Lett.*, *1974*, 3297. Cyclic enol ether-dibromocarbene adducts react differently under initial Br-Li exchange with alkyllithium. See T. Hiyama, A. Kanakura, Hajime Yamamoto, and H. Nozaki, *ibid.*, *1978*, 3047.
24. T. Hiyama, M. Shinoda, and H. Nozaki, *Tetrahedron Lett.*, *1978*, 771.
25. For recent muscone syntheses, see G. Ohloff et al., *Helv. Chim. Acta*, *62*, 2655, 2673 (1979).
26. T. Hiyama, M. Tsukanaka, and H. Nozaki, *J. Am. Chem. Soc.*, *96*, 3713 (1974).

A TOTAL SYNTHESIS OF
RACEMIC ERIOLANIN

R. H. Schlessinger

Department of Chemistry, University of Rochester, Rochester, New York 14627, USA

The sesquiterpene lactones eriolanin (1), eriolangin (2), and ivangulin (3) constitute the membership of an unusual class of natural products classified as 1,10-*seco*-eudesmanolides. Eriolanin and eriolangin were isolated from the chloroform extracts of the plant *Eriophyllum lanatum* Forbes (Compositae) by Kupchan and coworkers.[1] Both natural products exhibit significant *in vivo* activity against P-388 leukemia in mice and *in vitro* activity towards cell cultures derived from human carcinoma of the nasopharynx (KB).[2] The structural elucidation of compounds 1 and 2 required an X-ray analysis of a mixed crystal of their allylic aldehyde analogues, together with nmr, IR, and mass spectral measurements. These studies demonstrate that 1 and 2 contain three contiguous chiral centers on a cyclohexene ring, together with a fourth chiral center located on an acyclic side chain.

Ivangulin was isolated by Herz from the plant *Iva angustifolia* Natl. (section Lineabractea).[3] A gross structural assignment was given by Herz and, some twelve years later, confirmed *via* total synthesis by Grieco[4] — this group has also synthesized both eriolanin and eriolangin.[5]

Not atypically, the structural determination of all three natural products, by modern means, resulted in very little chemical information concerning these systems; however, Grieco's synthetic work on these compounds was available to us before we embarked on our own effort. Of particular interest to us was Grieco's finding that neither 1 nor 2 show a pronounced tendency to become aromatic, a possibility which appeared to us distinctly viable in the absence of chemical data. Since 1 and 2 do not show this tendency, we were free to concentrate on the problems associated with stereochemistry — in particular, finding a tractable means of creating the three chiral centers on the cyclohexene ring in the correct orientation relative to the center on the acyclic side chain.

1, R = H
2, R = Me

3

It occurred to us that ring-opening of a bicyclooctane system such as 4 would lead to compound 5 — a substance bearing sufficient stereochemistry and functionality to permit its reformulation into the bicyclic lactone, 6, which bears a strong resemblance to the target natural products.

The construction of a bicyclooctene analogue of 4, compound 7, appeared to be feasible *via* a Diels-Alder reaction between a crotonate ester and an appropriate cyclohexadiene derivative. However, when methyl crotonate was reacted with the diene 8, a difficult to separate mixture of all possible isomers of 7 was formed.[6] In spite of very considerable effort, this reaction could not be rendered synthetically useful. An alternative construction of a suitable bicyclooctanone was then attempted using a sequential series of 1,4-addition reactions. Reusch discovered that kinetic enolates derived from various cyclohexenone systems, when reacted with electron deficient olefins, such as acrylate esters, yield bicyclooctanones.[7] In such cases where meaningful stereochemistry accrues, Reusch found that isomerically pure products are formed and that these substances reflect *endo*-Diels Alder stereochemistry.[7] The salient question with respect to our synthetic problem, was whether or not the kinetic enolate derived from a vinylogous ester like 9 would undergo the same type of sequential Michael addition reactions with methyl crotonate to afford the bicyclooctanone 10.

Vinylogous ester 9 was uneventfully prepared from 3,4,5-trimethoxybenzoic acid by: reduction with lithium in liquid ammonia,[8] reduction of the acid residue with lithium aluminum hydride, protection of the resulting hydroxymethyl group with chloromethyl methyl ether, hydrolysis of the methyl vinylogous ester into its corresponding β-diketone with aqueous potassium hydroxide, and finally, reaction of the β-diketone with hexamethyldisilizane to afford 9.[9] Kinetic deprotonation of 9 in THF solution at -78° with lithium diisopropylamide (LDA),[10] followed by reaction of the resulting enolate with methyl crotonate, smoothly formed a single compound, 10, in 70% yield.[11] The structure of this substance readily followed from its spectra.

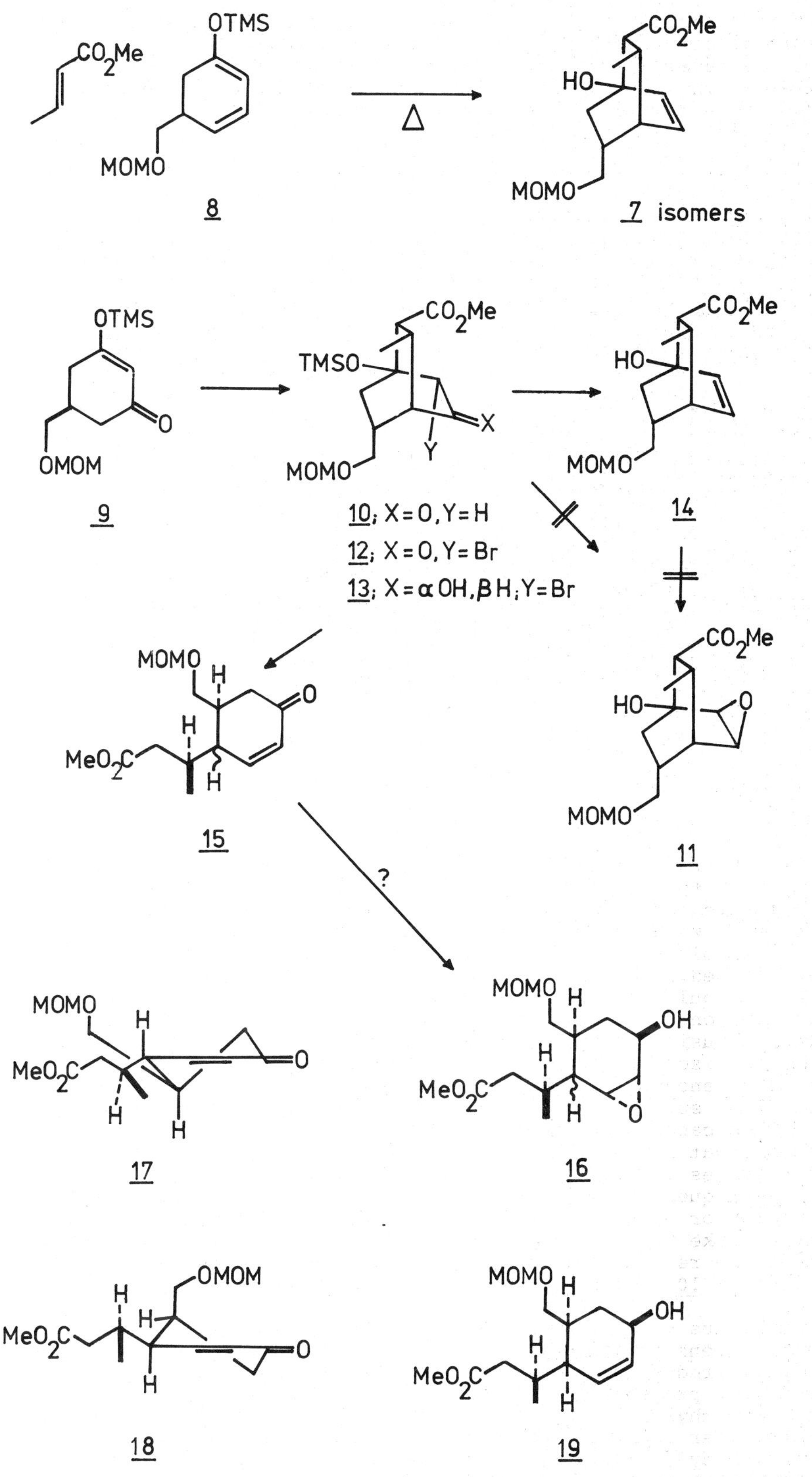

OTMS
CO_2Me
MOMO
8
Δ
HO
7 isomers
9
TMSO
X
Y
10; X = O, Y = H
12; X = O, Y = Br
13; X = α OH, β H; Y = Br
14
15
MeO2C
O
H
11
?
OH
16
17
18
OMOM
19

We then commenced a series of reactions which we hoped would
result in the reconstitution of 10 into the epoxide 11. To
this end, we deprotonated 10 with LDA in THF and treated the
resulting enolate with elemental bromine to obtain the α-
oriented, α-substituted bromoketone 12. A variety of reducing
agents, when reacted with 12, gave only the *cis*-bromohydrin
13 — we have as yet been unable to prepare the *trans*-bromo-
hydrin, an obvious precursor to the epoxide 11. In the hope
that the olefin 14 might serve as an intermediate leading to
11, we treated 13 with zinc dust in ethanol, and after acidification, obtained the bicyclooctene alcohol 14 in excellent
yield. Much to our great disappointment, this substance, as
well as alcohol protected forms thereof, proved completely
resistant to epoxidation by a variety of means. We had no
choice, therefore, but to ring-open 14 to obtain the cyclohexenone 15, and to attempt to utilize this enone system as a
synthetic intermediate. Our reluctance to employ 15 in this
manner stemmed from previous experience which suggested conversion of it into the *trans*-β-hydroxy-α-epoxide 16 could be
accomplished, stereospecifically, but not without difficulty.[12]
Ring-opening of 14 was carried out using a catalytic amount of
potassium *t*-butoxide in *t*-butanol for less than three minutes
at 22°.[13] From spectral data, it was clear that 15 was a
single compound. However, the stereochemistry of attachment
of the ester side chain to the γ-carbon of the cyclohexenone
was ambiguous even at 400 MHz in the proton resonance spectra.
Clearly, while this center is not present as sp^3-carbon in the
target natural product, its relative geometry would have great
influence on the upcoming synthetic steps.

Compound 15 presented to us two conformational choices, 17
and 18 depicted in what we currently believe to be their minimum energy configurations. Barring unforseen solvent or metal-
ion effects, we anticipated that hydride reduction of 17 would
result in the formation of epimeric allylic alcohols;[12] whereas, 18 would predominately form the β-allylic alcohol.[14] Reduction of 15 with sodium borohydride in ethanol, to our surprise, led to a single allylic alcohol in greater than 95%
yield.[15] This suggested that 15 is best formulated as 18, and
that the alcohol obtained would be the β-allylic alcohol 19.
We then treated 15 with basic hydrogen peroxide, fully expect-
ing to form the α-epoxy ketone 20, assuming 18 to be the op-
erative geometry. Indeed, this reaction led to a single epoxy
ketone. We then sought to interrelate 19 and 20 by Henbest
epoxidation of the former, a reaction known to yield *cis*-α-
hydroxy epoxides.[16] Subsequent oxidation of this material
should result in formation of the β-epoxy ketone 21, a substance that should clearly differ from the epoxy ketone 20.
The reaction sequence was carried out on 19 and the two epoxy
ketones carefully compared. They were found to be identical.

From the above, we obviously were unable to obtain definitive
structural information on the enone 15, but on the bright side,
we had found that stereospecific oxidation and reduction reactions could be performed on it. Thus, we were reduced to
a 50% chance of selecting the correct *trans*-epoxy alcohol for
use in further synthetic manipulation. To this end, we selected the allylic alcohol 19 (which must now be viewed with
unknown stereochemistry at the point of attachment of the
ester side chain to the cyclohexenol ring) and derivatized the
alcohol with *t*-butyl dimethylsilyl chloride, followed by treatment of the resulting silyl ether with aqueous N-bromosuccinimide
and potassium carbonate to obtain the *trans* epoxy-α-oxy system
22.[17] Parenthetically, desilylation of 22 and oxidation gave
an epoxy ketone, which we believe to be 20, and which is different spectrally from epoxy ketone 21.

We then decided to chain-lengthen 22 and the highest yield
method found involved: diisobutylaluminum hydride reduction
of the ester to the primary alcohol; mesylation of the alcohol; conversion of the primary mesylate into the iodide with
sodium iodide; and finally, displacement of the iodide with

$\underline{27}$; X = O; Y = Z = Ac

$\underline{28}$; X = H,OH; Y = Z = H

$\underline{29}$; X = H,OTBS; Y = TMS; Z = TBS

$\underline{24}$; X = H,OMOM; Y = H; Z = CH$_2$

$\underline{25}$; X = H,OMOM; Y = H; Z = H,OH

$\underline{26}$; X = O; Y = Ac; Z = H,OAc

$\underline{30}$; X = OTBS

divinyl copper lithium, thereby affording 23.

It was our intention to regiospecifically introduce the lactone residue at this point, utilizing dilithioacetate ring-opening of the epoxide residue of 23, according to methodology developed by Danishefsky.[18] Hence, 23 was desilylated with triethylammonium hydrofluoride, and the resulting epoxy alcohol treated with an excess of dilithioacetate in THF containing hexamethylphosphoramide. Acidification of the reaction mixture afforded an excellent yield of the *cis*-lactone 24. Ozonolysis of 24, followed by sodium borohydride reductive workup gave the lactone diol 25.

Introduction of the double bond present in the six-membered ring constituted the next phase of our work. Acylation of the alcohol residues of 25 with acetic anhydride in pyridine containing 4-dimethylaminopyridine (DMAP), followed by removal of the methoxymethyloxy protecting group using boron trifluoride etherate and ethanedithiol in methylene chloride solution and subsequent oxidation of the resulting primary alcohol with pyridinium chlorochromate afforded a very good overall yield of the aldehyde 26. When crude samples of 26 were treated with phenylselenylchloride in ethylacetate, we obtained, in 40% yield,[19] a single substance easily identified as the unsaturated aldehyde 27.[20] This material was immediately reduced with sodium borohydride and the acetate residues removed with potassium carbonate in dry methanol to afford the triol 28.

Remaining were the problems of methyleneation of the lactone ring, and esterification of the secondary alcohol. To accomplish this, we needed to conveniently differentiate the primary and secondary alcohols in 28. This was readily accomplished by treatment of the triol, first with *t*-butyldimethylsilylchloride in DMF and, second, with trimethylsilylchloride in pyridine affording, thereby, 29. Methylenation of 29 was then commenced by carbonation of the lactone, using lithium diisopropylamide to secure the lactone enolate, followed by reaction with gasseous carbon dioxide. The intermediate lactone carboxylic acid, without characterization, was treated with a mixture of diethyl amine and 30% formalin solution and then worked up with a mixture of acetic acid, water, and methanol.[21] This sequence afforded the α-methylene lactone 30, with the secondary alcohol free, in good overall yield. Esterification of 30 with the anhydride of methacrylic acid in pyridine containing DMAP, followed by treatment of the resulting lactone ester with methanolic hydrochloride acid, gave a single crystalline substance mp. 114-115°. To our immense pleasure, this substance proved absolutely identical, with the exception of optical rotation, to a generous sample of naturally occurring eriolanin (2) kindly provided by Professor Paul Grieco.[22]

I would like to acknowledge the experimental and intellectual excellence of my coworker on this project, Mr. Michael Roberts, an extraordinary graduate student who, during his three and a half years at Rochester, has completed not only the total synthesis of eriolanin, but also the total synthesis of the pseudoguaianolide, helenalin. We both acknowledge support of this work by the National Cancer Institute division of the National Institutes of Health.

REFERENCES AND FOOTNOTES

1. a) S. M. Kupchan, R. L. Baxter, C. K. Chiang, C. J. Gilmore, and R. F. Bryan, J. Chem. Soc., Chem. Commun., 842 (1973); b) R. F. Bryan and C. J. Gilmore, Acta, Crystallogr., Sect. B, 31, 2213 (1975).

2. P. A. Grieco, J. A. Noguez, Y. Masaki, K. Hiroi, M. Nishizawa, A. Rosowsky, S. Oppenheim, and H. Lazarus, J. Med. Chem., 20, 71 (1977), and references cited therein.

3. W. Herz, Y. Sumi, V. Sundarsanam, and D. Raulais, J. Org. Chem., 32, 3658 (1967).

4. P. A. Grieco, T. Oguri, C. L. J. Wang, and E. Williams, ibid., 42, 4113 (1977).

5. P. A. Grieco, T. Oguri, S. Gilman, and G. T. DeTitta, J. Am. Chem. Soc., 100, 1616 (1978).

6. Our experience is not unique; for additional examples, see T. Ibuka, Y. Mori, T. Aoyama, and Y. Inubushi, Chem. Pharm. Bull., 26, 456 (1978) and references cited therein.

7. K. B. White and W. Reusch, Tetrahedron, 34, 2439 (1978) and references cited therein.

8. M. E. Kuehne and B. F. Lambert, J. Am. Chem. Soc., 81, 4278 (1959).

9. S. Torkelson and C. Ainsworth, Syn., 722 (1976).

10. G. Stork and R. L. Danheiser, J. Org. Chem., 38, 1775 (1973).

11. The remaining 30% of the reaction mixture is polymeric in nature — compound 10 is the sole identifiable product. It should be noted that structure 10 represents a product reflecting thermodynamic control; a function in this regard, both of steric interactions and of gegen-ion transport.

12. For a similar example, see G. R. Kieczykowski, M. L. Quesada, and R. H. Schlessinger, J. Am. Chem. Soc., 102, 782 (1980).

13. For a similar ring-opening reaction, see W. C. Still and M.-Y. Tsai, ibid., 102, 3654 (1980) and references cited therein.

14. J. E. Baldwin, J. Chem. Soc., Chem. Commun., 736 (1976).

15. This same allylic alcohol also is obtained with lithium tri-t-butoxyaluminum hydride in THF solution.

16. H. B. Henbest and R. A. L. Wilson, J. Chem. Soc., 1958 (1957).

17. We have found this method of preparing trans α-epoxy alcohol derivatives which is based on work by R. A. B. Bannard and L. R. Hawkins, Can. J. Chem., 36, 1241 (1958), to be more stereoselective, in some cases, than the method described by C. G. Chardarian and C. H. Heathcock, Syn. Commun., 6, 277 (1976).

18. S. Danishefsky, M. Tsai, and T. Kitahara, J. Org. Chem., 42, 394 (1977).

19. This is the lowest yield reaction in the entire sequence; most other reactions proceeding in between 80 and 90%

yield.

20. The formation of the aldehyde is accomplished using 1.1
     eq. of pyridinium chlorochromate. Isolation of the pro-
     duct using a combination of aqueous Na_2SO_3/6NHCl followed
     by extraction with $CHCl_3$ results in a pale green oil
     which is submitted immediately to a freshly prepared solu-
     tion of PhSeCl in EtOAc. These conditions preclude the
     formation of a selenoxide prior to elimination. A ther-
     mally induced dehydroselenylation is believed to be re-
     sponsible for this unique result.

21. This very useful methylenation sequence has been de-
     scribed, in part, by W. H. Parker and F. J. Johnson,
     J. Org. Chem., 38, 2489 (1973).

22. We thank Professor Grieco for this sample of the natural
     product.

TXA$_2$ STRUCTURE. THE SYNTHESIS
OF ALTERNATIVES

R. C. Kelly

Experimental Chemistry Research I, The Upjohn Company, Kalamazoo, Michigan, USA

Abstract - The ene epoxide 4 was suggested to us by Prof. J. E. Baldwin as an alternate possibility for the structure of thromboxane A$_2$ (TXA$_2$). The synthesis of the four geometrical isomers of 4 is described. Similarly to TXA$_2$ these molecules are extremely labile but they are not converted chemically to the same products as TXA$_2$ nor do they possess the TXA$_2$ profile of bioactivity.

Today I am going to talk to you about the structure of TXA$_2$ and present our work on the synthesis of alternatives to the currently accepted structure. (Scheme 1) Let us begin by reviewing some of the current dogma in this area. As many of you are aware thromboxane A$_2$ (TXA$_2$) is produced biologically from prostaglandin H$_2$ (PGH$_2$). This same precursor leads also to the now classical prostaglandins (i.e. PGE$_2$, PGF$_2\alpha$, PGD$_2$, etc.) and to the more recently discovered PGI$_2$.

One of the major facinations in the prostaglandin field is the discovery of the Yin Yang pair of PGI$_2$ and TXA$_2$. These molecules both originating from the same precursor are created by the body in a delicate balance of opposing biological functions. Thus, the blood vessels convert PGH$_2$ to PGI$_2$ and PGI$_2$ prevents platelets from aggregating and adhering to blood vessel walls. Further, it causes smooth muscles, particularly blood vessels to relax. On the other hand, the platelets convert the same PGH$_2$ to TXA$_2$, which causes the platelets to aggregate and to adhere to blood vessels and causes blood vessels to contract. Thus, it is clear that this area is rich in opportunities for the pharmaceutical chemist. PGI$_2$ itself and analogs or congeners have already found use or have been suggested as useful in coronary by-pass operations, hemodialysis, peripheral vascular disease, angina and myocardial infarct. Similarly, since PGI$_2$ and TXA$_2$ have opposing actions and PGH$_2$ is converted by platelets to TXA$_2$, it is anticipated that a TXA$_2$ synthetase-inhibitor or antagonist would possess many of the desirable properties of a PGI$_2$ agonist.

To this point I have emphasized the differences between PGI$_2$ and TXA$_2$. However, in some important aspects they are very similar. For example, they are both very unstable, having half lives of less than a couple of minutes in the aqueous media of their generation. Further, they are both highly active and only extremely small amounts are required to produce biological effects. These factors led to great challenges in their structural elucidation. In each case one important technique was the isolation of stable breakdown products and derivatives from which the original structure was reconstructed. In the case of PGI$_2$ the structure was then conclusively proven by synthesis and comparison of the biological activities of natural and synthetic materials (Scheme 2).

With TXA$_2$ the story is a little more complicated. Let's look at it in closer detail. In 1975, Hamburg, Svensson and Samuelsson proposed a structure for TXA$_2$ based on its conversion to TXB$_2$ in water, to TXB$_2$ methyl acetal (1a) in methanol and to azide 1b in aqueous sodium azide. The mass spectral data on these derivatives is quite convincing. In addition, TXB$_2$ and its methyl acetal 1a have been synthesized by several groups further confirming their structure. (Scheme 3) In another set of experiments Samuelsson and co-workers converted ^{18}O labeled PGH$_2$ to labeled TXB$_2$. They were able to show from mass spectral fragmentations that all 3 PGH$_2$ ^{18}O's were incorporated into the TXB$_2$ as shown. This means that the ring oxygen of TXB$_2$ comes not from solvent but from the C-11 oxygen of PGH$_2$. Also, when TXA$_2$ was treated with deutero methanol the TXB$_2$ methyl acetal 1a was found to contain no deuterium. This later piece of data was used to exclude the enol ether 2, the counterpart of the enol ether PGI$_2$. There is to my knowledge no spectral data to date on TXA$_2$ itself. (Scheme 4) Thus, from the data in hand Samuelsson and co-workers formulated structure 3 for TXA$_2$. Structure 3 certainly is consistent with all the data. But is it the only consistent structure? While visiting Upjohn some time ago Prof. Jack E. Baldwin, now of Oxford University, suggested for our consideration the alternative TXA$_2$ structure ene-epoxide 4. This suggestion sat dormant in our laboratories for quite some time because

Scheme I

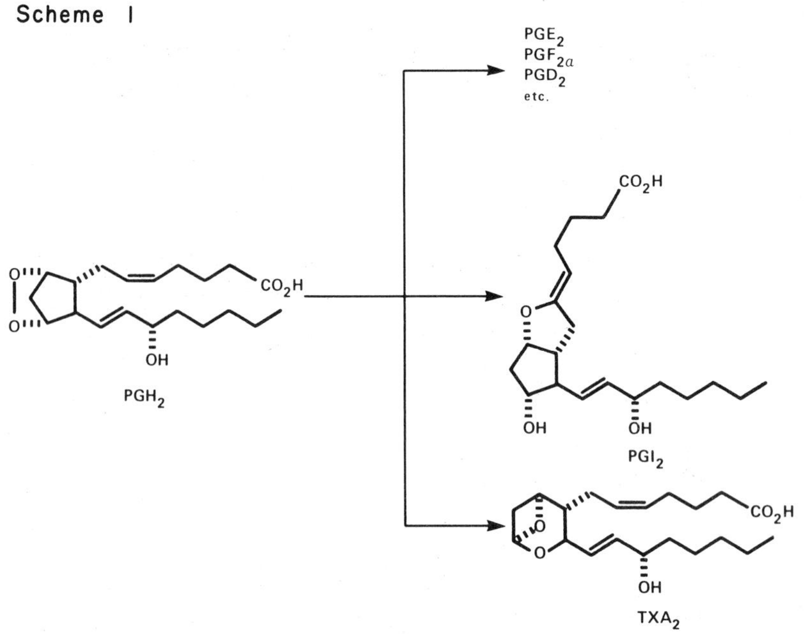

Scheme 2

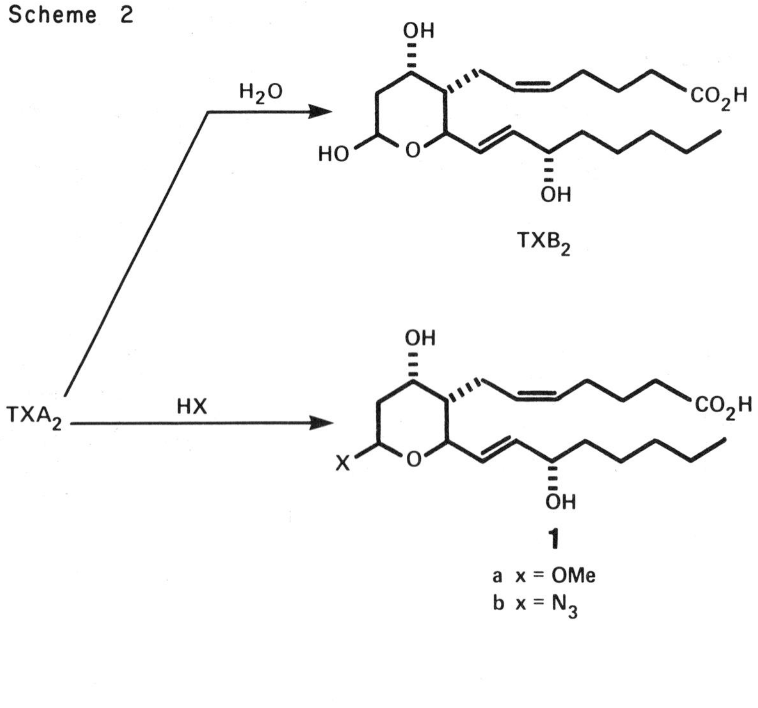

Scheme 3

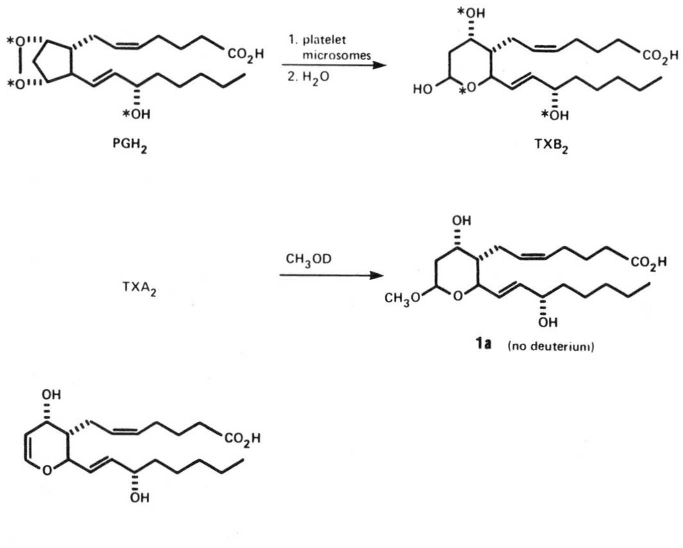

PGH$_2$

1. platelet microsomes
2. H$_2$O

TXB$_2$

TXA$_2$

CH$_3$OD

1a (no deuterium)

2

M. Hamburg, J. Svensson and B. Samuelsson, Proc. Nat. Acad. Sci, USA 72, 2994 (1975).

Scheme 4

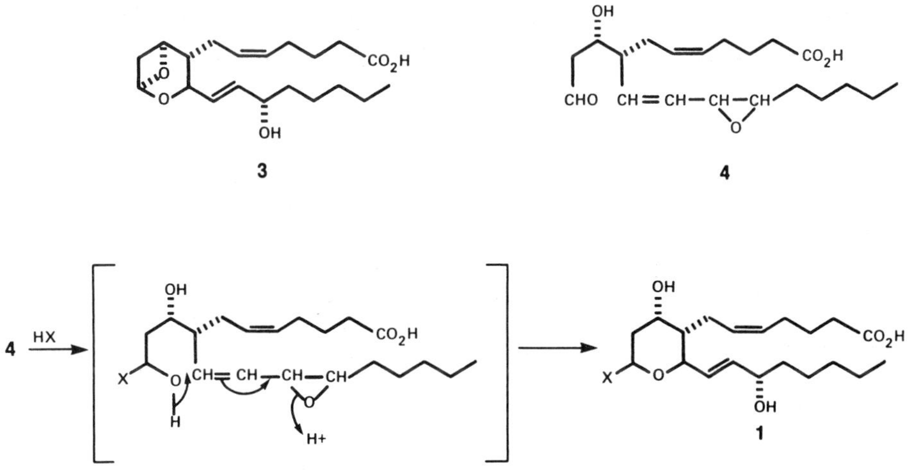

3

4

4 HX

5

1

we initially felt an allylic epoxide would not be sufficient reactive with MeOH or water
for such a structure to be consistent with TXA$_2$. As time past we encountered both in the
literature and by personal experience the high reactivity of certain allylic epoxides and
began to seriously reconsider structure 4.

In order to meet requirements found earlier by Prof. Samuelsson for the incorporation of
three labeled oxygens and the formation of derivatives such as TXB$_2$ methyl acetal from TXA$_2$
directly without passing through TXB$_2$, one needs to assume that the aldehyde moiety is attacked by nucleophile X and that it is the oxygen of the aldehyde which attacks the allylic
epoxide. While we were unaware of any close precedent for this postulated reaction, we did
not see any bar to it, a priori. In fact, at one point we started to believe that structure 4 represented a reasonable alternative proposal to the structure of TXA$_2$, one that was
amenable to test by total synthesis. During the rest of my talk I will be describing the
total synthesis and the biological properties of the sterioisomers represented by structure
4. (Scheme 5)

As soon as one takes structure 4 seriously it is quickly apparent that there are four
possible geometrical isomers, namely two isomers with a 12,13 trans (PG numbering) or 12,
13-cis double bond and two isomers with a 14,15-trans or 14,15-cis epoxide. To further
clarify the epoxide stereochemistry the 15 position in all cases possesses the 15(S)-con-
figuration but in one pair, 4a and 4c, the 14 configuration is (S) and in 4b and 4d the 14
configuration is (R).

In addressing the synthesis of these molecules one must consider two very formidable
obstacles, instability and stereochemistry. Let us focus first on the problem of instability. If any of these isomers is TXA$_2$ itself, then the known half-life in pH 7.5 H$_2$O
is about 30 secs. If on the other hand, none is TXA$_2$, still the structure itself tells us
that the β-hydroxy aldehyde will not stand stronger than very mild base and the allylic
epoxide only mild acid. Our problem then is to release this delicate balance of functionality under the mildest possible conditions. (Scheme 6)

Our strategy was to generate and protect the aldehyde function at an early stage, then form
the epoxide, and finally release the aldehyde generating the desired final products.
Clearly, we needed at this point an aldehyde protecting group removable under the mildest
conditions possible. While a number of aldehyde protecting groups were tried what eventually worked is the orthonitrophenylethyleneglycol acetal shown here. This marvelous group
may be removed merely by photolysis at 350 nm in almost any solvent, even neutral aprotic
ones. The reaction proceeds as shown on the slide by transfer of an oxygen from the nitro
group to the benzylic position forming a hemi acetal which falls apart to the aldehyde and
the α-hydroxyketone 10. There are problems associated with this group, namely, difficulties in its introduction and with creation of two new chiral centers resulting in problems
with separation of isomers which will be discussed later.

Our next great obstacle in the synthesis (Scheme 7) was to control the stereochemistry at
all six centers while generating the requisite functionality. On this slide is shown our
plan of attack. Our intention was to use the chemistry generated by our co-workers at
Upjohn, Schneider and Morge, as shown at the top of this slide to create the acetoxy aldehyde and after proper protecting group manipulation to convert the 2-ene-1,4-diol moiety of
12 to a 1-ene-3,4-diol equivalent which could then be converted to the allylic epoxide 14.
While the basic foundations of our retrosynthetic analysis held, the details of the super
structure were altered by the actual course of events. (Scheme 8)

Immediately focus your attention on structure 15. You will notice two things. First we
chose to start with the 9- and 15-hydroxyls protected, not with acetates as on the previous
slide, but with t-butyl dimethyl silyl groups so as to be able to differentiate these positions from the acetoxy group generated at position 12 in the lead tetraacetate reaction.
Second, note that the stereochemistry at 15 was switched from 15(S) in the previous slide
to 15(R) in this slide. We in fact originally carried out all the subsequent chemistry in
the 15(S) series but because of the course led by that chemistry we generated an inversion
at C-15. This point will become clearer as we proceed. Returning to derivative 15, we had
expected the lead tetraacetate reaction to produce compound 16 and 16 only on the basis of
the earlier precedent. Thus we were quite surprised when after treating 15 with lead
tetraacetate in toluene we isolated a 1:1 mixture of two compounds as evidenced by TLC.
This seemingly difficulty turned out to be a blessing in disguise. After careful chromatographic separation of the mixture and mass spectral, proton and ^{13}C NMR spectral analysis
of the pure compounds, we were able to assign most of the structural details shown by
structures 16 and 17. The stereochemistry of the acetoxy at C-14 in compound 17 only
became clear at a later point but for clarity I show it here already defined. As you can
see merely by switching the 9,15-protecting groups from acetate to t-butyldimethylsilyl we
went from a reaction giving only 12 substitution to one giving both 12 and 14 substitution.
Why this is so we do not know but we were delighted because we were able to obtain directly
in 40% yield after chromatography compound 17, a material containing a 1,3-rearranged
structure we had planned to produce by additional chemistry. At this point we decided to

Scheme 5

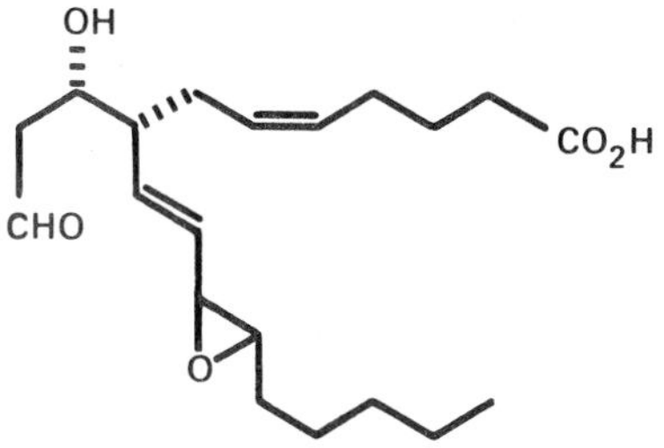

4a

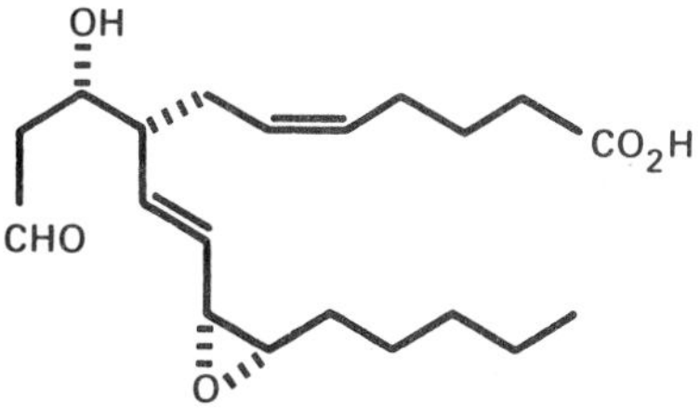

4b

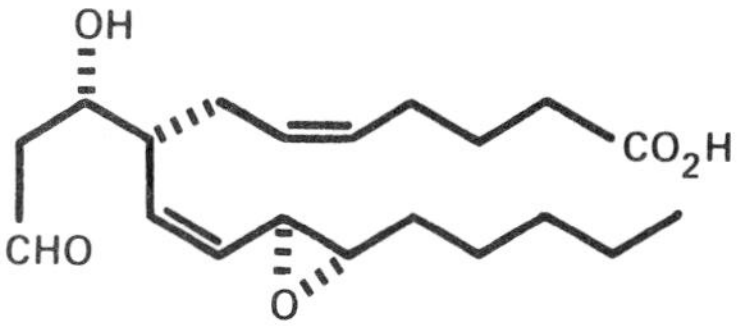

4c

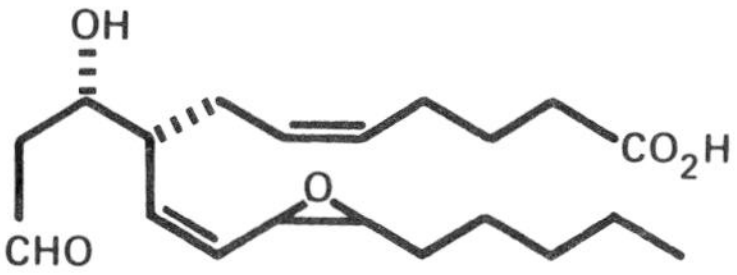

4d

Scheme 6

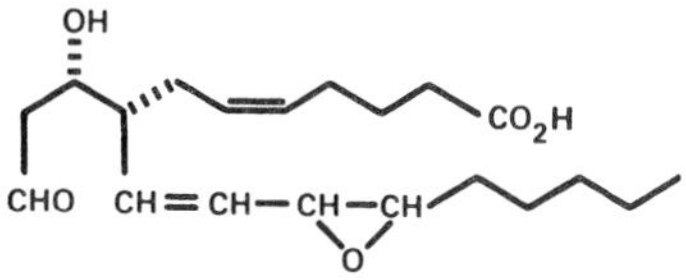

4

unstable to
acid and base

RCH $=$ X $\xrightarrow[\text{aprotic solvent}]{\text{Preferably neutral}}$ RCHO

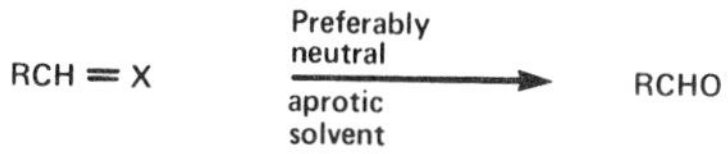

6   **7**

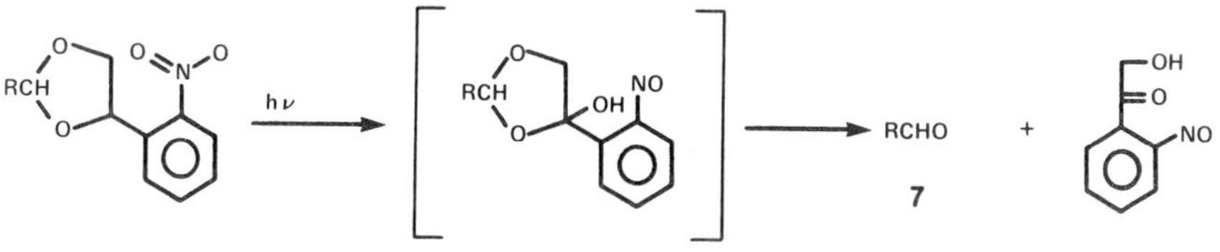

8   **9**   **10**

Scheme 7

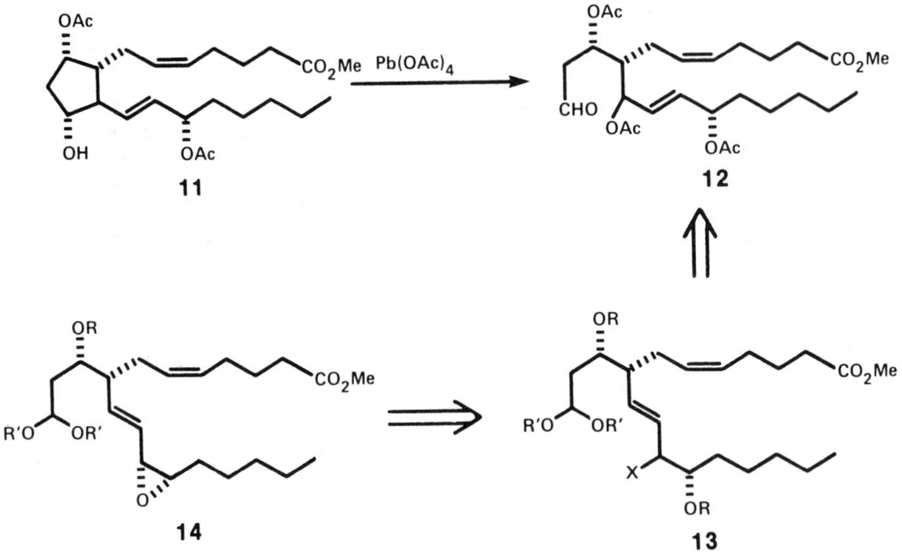

11   →   12

14   ⟹   13

W.P. Schneider and R.A. Morge, Tetrahedron Lett., 3283 (1976)

Scheme 8

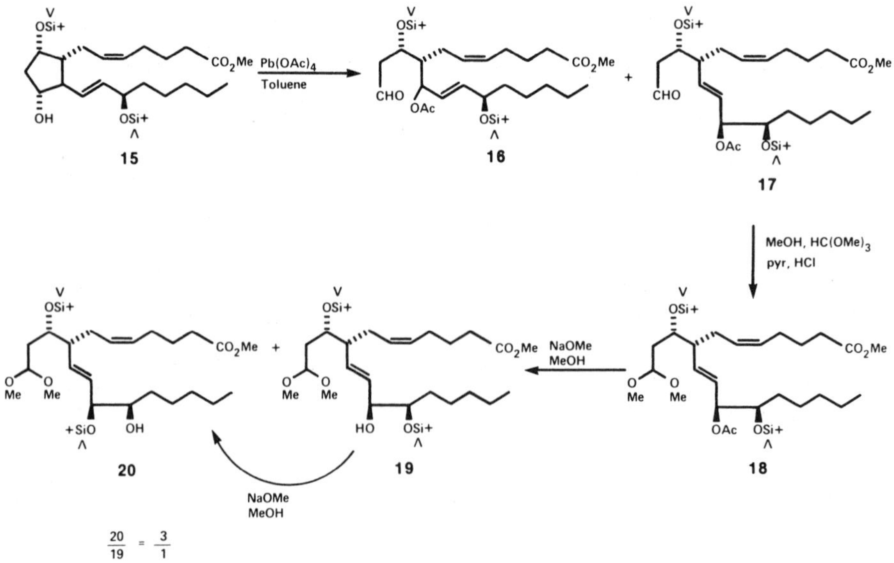

15   →   16   +   17

18   →   19   →   20

$$\frac{20}{19} = \frac{3}{1}$$

accept our good fortune and run with it. We left compound 16 behind and carried out the further chemistry with 17 directly. The β-substituted aldehyde structure 17 is quite unstable so it was converted immediately after purification to the dimethyl acetal 18.

Our plan at this point was to saponify 18 to 19 and to convert the free 14 hydroxyl of 19 to a leaving group. This then would be available for displacement by the liberated 15-OH to produce the desired epoxide. Had our plans proceeded along these lines then the 15(S) starting material would have been the correct one and not the 15(R) we are showing here. The slide of course shows what changed our plans. When 18 was treated with a trace of sodium methoxide in methanol the major product was not the expected product 19 but its isomer 20 where the t-butyl dimethyl silyl group that was at the 15 position has hopped to 14. The assignment of these structures turned out to be simple when approached in the right way. First each of 19 and 20 was converted to its TMS derivative and then subjected to mass spectrometry. In the mass spectrometer a major cleavage is seen between vicinal oxygen so that 19 loses the tail 6 carbons with a t-butyldimethylsiloxy attached and 20 looses it with a trimethylsilyloxy. While there is precedent for such migration of silyls between vicinal hydroxyls, they generally occur under much stronger base conditions. Added to the surprise of the migration was the fact that the ratio was not 1:1. In this case and all the others which we have done the silyl attached to the allylic hydroxyl always predominates. At this point we do not know whether it is steric or electronic factors which influence this migratory ratio.

It is this migration of silyls which changed our synthetic plan from the originally intended substitution of a leaving group at C-14 to one at C-15. In many ways this rearrangement made our job easier because a suitable leaving group in the allylic position might have been more difficult to find.

One more point should be mentioned, about the migration, namely, that it is a true equilibrium. Thus one can start with a pure sample of either isomer and treat it with dilute sodium methoxide in methanol and obtain the identical mixture. In the case at hand this is not so important because we already had the greater amount of the desired isomer but later on in our story we're going to want the less abundant isomer. That was obtainable by multiple equilibrations and chromatographies. (Scheme 9)

Continuing now with this series, alcohol 20 was converted to its mesylate, compound 21. Finally, we are ready in compound 21 to exchange the dimethyl acetal protecting group for the o-nitro phenyl ethylene glycol group. We could have hardly done so sooner because as indicated on the slide introduction of the o-nitro phenyl ethylene glycol group creates two new chiral atoms and thus produces 4 diastereomers. This would have created insurmountable problems at an earlier stage. At this particular stage little trouble was produced. All four compounds were separated chromatographically and their structures unambiguously assigned by proton NMR. In this particular case the pure diastereomers were carried to final products but later on when material was more dear, mixtures of the diastereomers were used.

The introduction of the o-nitrophenylethyleneglycol acetal proceded in all cases in 60-80% yield. We feel particularly fortunate in this since attempts to introduce it to the β-substituted free aldehyde failed entirely. We believe that the introduction of this and other acetal functions into labile aldehydes might be facilitated by exchange with the more easily introduced dimethyl acetal.

Treatment of the bis silyl mesylate 22 with tetrabutylammonium fluoride gave the hydroxy epoxide 23. Proton NMR data on this material defined the C-14 stereochemistry for us for the first time. The epoxide protons show up in an unobscurred region and show a coupling of 2 Hz characteristic of a trans epoxide. (Scheme 10)

To complete the synthesis 23 was hydrolyzed with sodium hydroxide in ethanol and the resultant sodium salt 24 photolyzed to the desired product 4a. We know it's 4a because we can convert it to its methyl ester 25 as shown and compare it with 25 prepared directly from 23 by photolysis in CDCl$_3$. The next slide (Scheme 11) shows an NMR of this material.

You will note the signal for the aldehyde proton at 9.8 δ, the signal for the C-12 vinyl proton at 5.9 δ with a 16 Hz coupling characteristic of a trans double bond and the signal for the C-14 epoxide proton at 3.3 δ with a small coupling of 2 Hz indicative of a trans epoxide. Thus we have achieved the synthesis of the first of our four isomers. Before telling you about its chemistry or biology I'll quickly outline the synthesis of the remaining three. (Scheme 12)

Our next goal was the synthesis of the trans-alkene-cis-epoxide. This was accomplished by oxidation of the C-14 alcohol 19 to the ketone 26 followed by sodium borohydride reduction back to a mixture of alcohols containing the original 19 and a new alcohol 27. The threo alcohol 27 predominated by more than 2:1. Alcohol 27 was then converted in the manner previously described for 19 to the trans-alkene cis-epoxide 4b. (Scheme 13)

Scheme 9

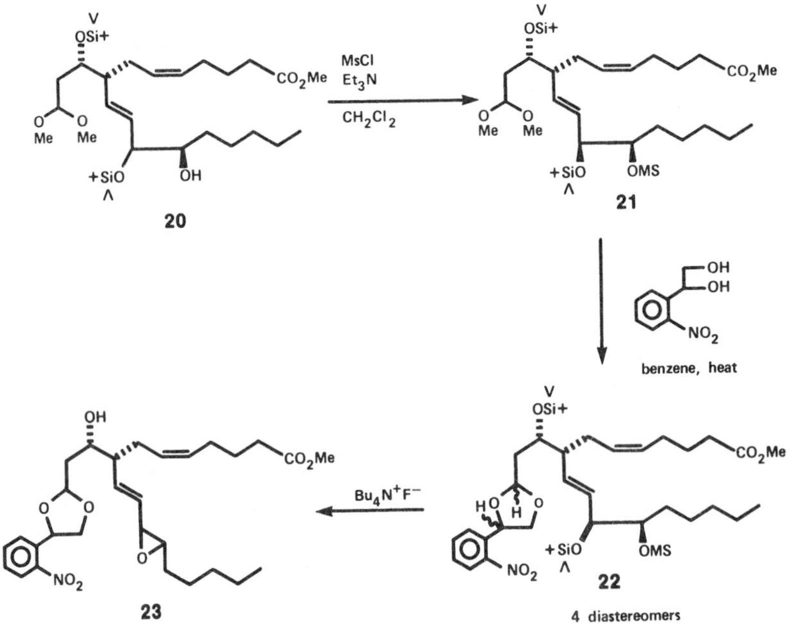

Scheme 10

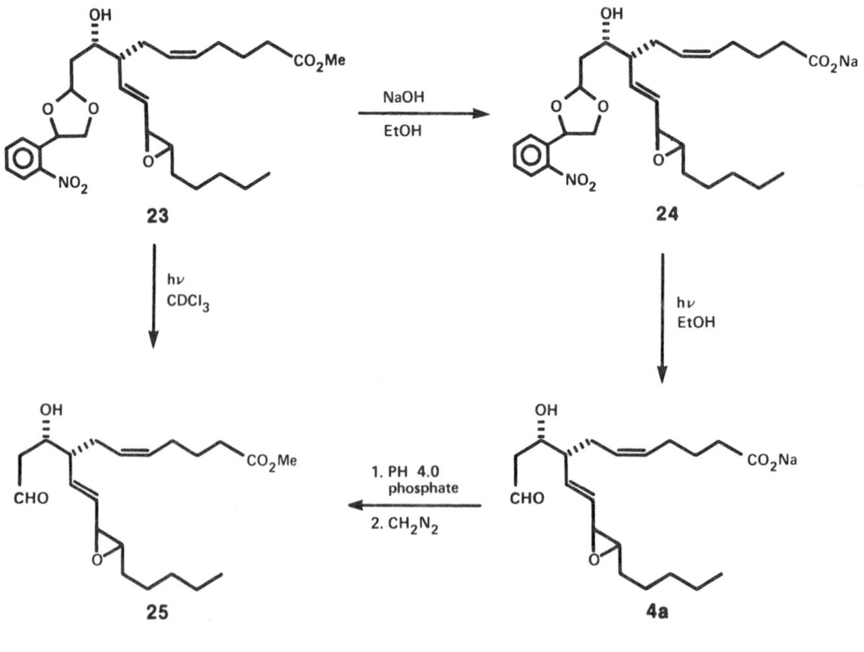

Scheme 11

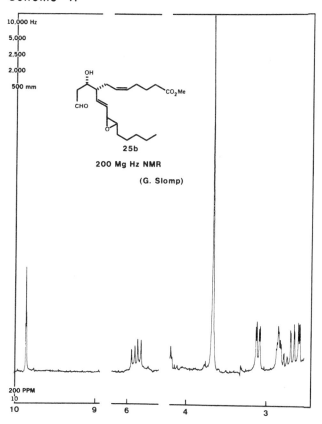

25b

200 Mg Hz NMR

(G. Slomp)

Scheme 12

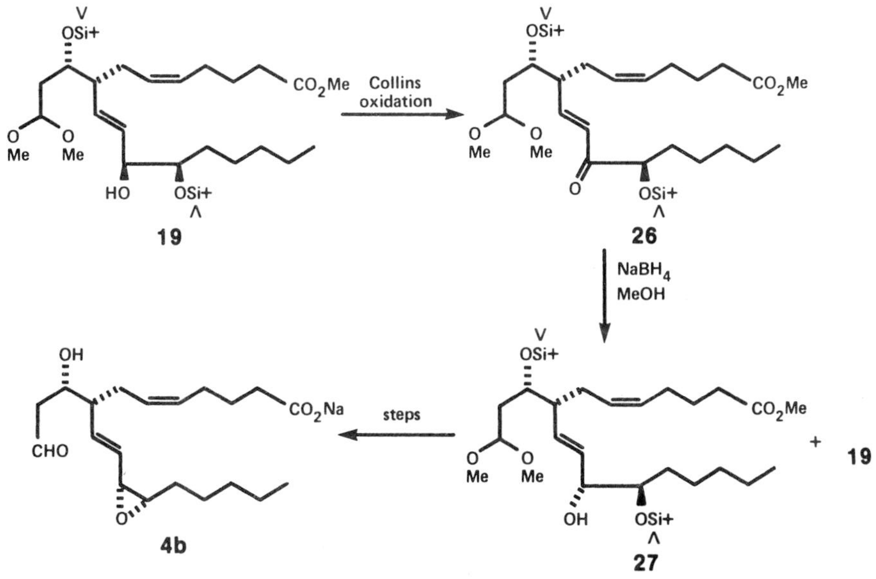

Our next goal was the synthesis of the cis-alkene epoxides. This was accomplished by the photochemical isomerization of trans enone 26 to the cis enone 28. Whereas reduction of trans ketone 26 with sodium borohydride had been rapid and quite clean, the reduction of the cis enone 28 was slow and gave a complex mixture. The mixture was completely separated chromatographically and the structures assigned on the basis of mass spectral and proton and ^{13}C NMR data.

We believe the ketone of the cis enone is considerably more hindered than its trans counterpart allowing two previously unseen reactions to occur. In the cis enone case 1,4 reduction now occurs as evidenced by isolation of dihydro product 29. Also, during the longer reduction times required in this case the sodium borohydride was sufficiently basic to catalyze the silyl migration. There is thus found the threo pair of silyl migrated isomers 30a and b and the erythro pair 31a and b. (Scheme 14)

Isomer 30a was rearranged and combined with 30b from the previous chromatography and 30b converted in the manner shown earlier to the cis-alkene-cis-epoxide 4d. (Scheme 15)

Finally, 31b was prepared and converted in the same way to 4c. Thus, we have prepared all four isomers.

The burning question now is; what are these materials like chemically and biologically? Are any of these TXA$_2$. The answer is they are very interesting but none is TXA$_2$. (Scheme 16)

All of these compounds are extremely labile both to acid and base. They are hydrolyzed by neutral water in a few minutes and react with anhydrous dry methanol in a matter of a few hours. They are thus very close in reactivity to that reported for TXA$_2$. However, they do not react as described for TXA$_2$. In water they hydrolyze by 1,2 and 1,4 opening of the allylic epoxide producing the non TXA$_2$ derived compound 36 and material which by TLC and GC mass spec we assign as TXB$_2$ or its C-12 isomer (compound 35). In methanol no TXB$_2$ methyl acetal is formed only the 1,4 and 1,2 opening products 33 and 34. (Scheme 17)

What do these materials show biologically? On the rat aorta the trans alkenes 4a and 4b are spasmogenic at 100-200 ng/ml or about 10-50 times less active than TXA$_2$. The cis alkenes 4c and 4d show no activity in this screen.

On washed platelets none of these compounds promotes platelet aggregation at 100 times the doses at which TXA$_2$ causes platelet aggregation. Thus none of these compounds is TXA$_2$.

What then have we accomplished? We have completed the synthesis of several highly sensitive complex molecules. We have removed from consideration alternate TXA$_2$ structures which I at least felt had reasonable possibilities of being TXA$_2$. We had the delight of carrying out what for us was exciting and challenging chemistry. I hope that my re-telling of the story has been as pleasurable for you as it has for me.

Before relinquishing the podium, I would like to acknowledge my co-workers on this project. I cannot give enough praise to Ilse Schletter who carried out single handed most of the chemistry you have seen today. I also wish to express deep appreciation to Bob Gorman and Jim Aiken of our Experimental Biology unit at Upjohn who set up and ran the tests for these materials seconds after we carried out their preparation. I also wish to thank the individuals of Upjohn Physical and Analytical Chemistry Unit, Dick Wnuk for mass spectra and George Slomp, Steve Miszak, and Terry Scahill for running 100 and 200 MgHz NMR spectra. Without their aid we would not have been able to successfully complete this project.

Scheme 13

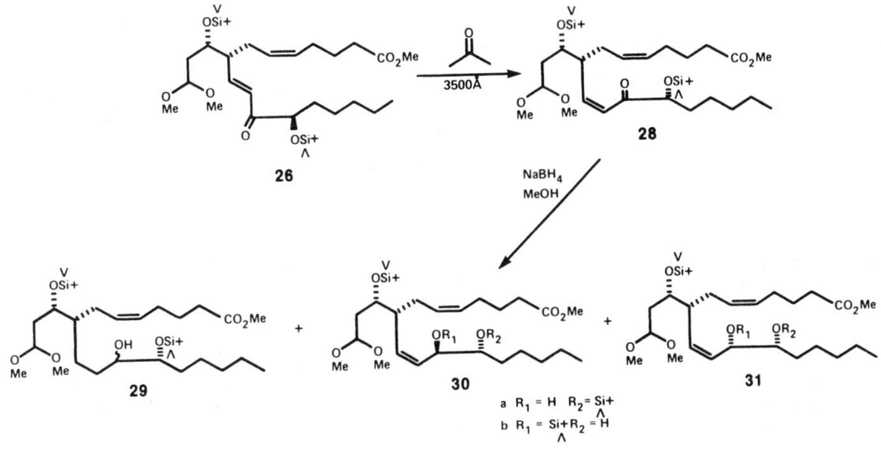

Scheme 14

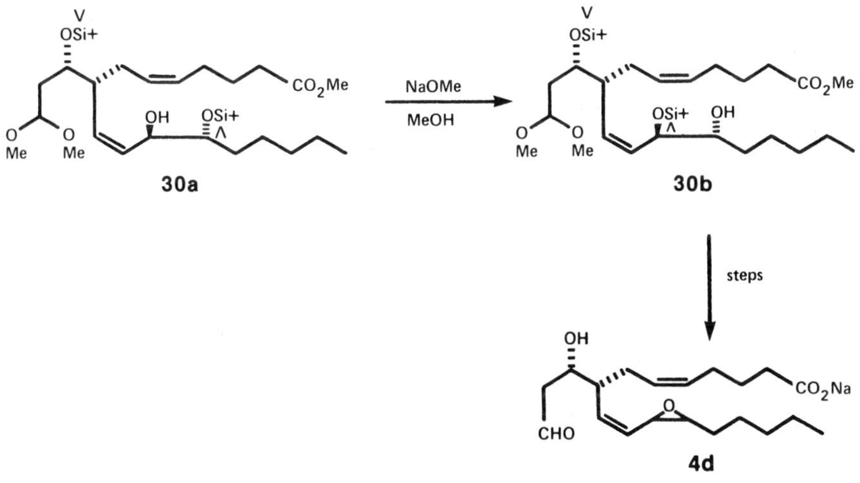

steps

Scheme 15

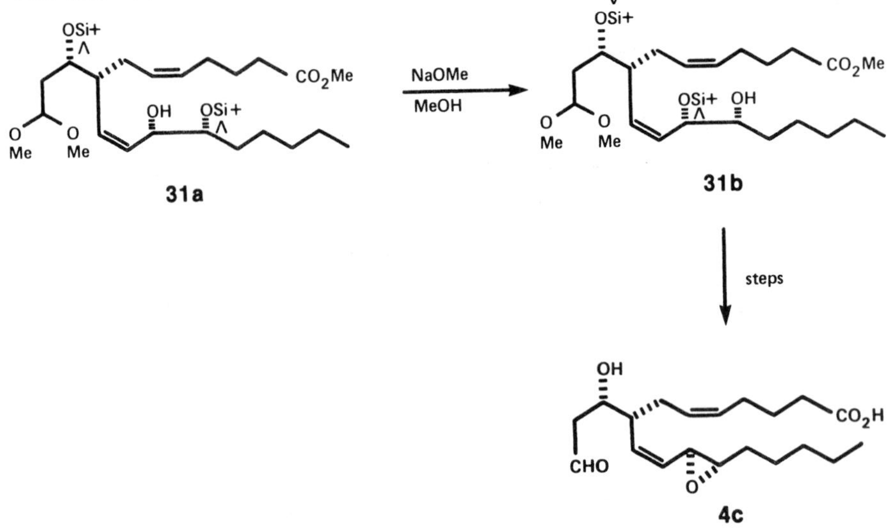

Scheme 16

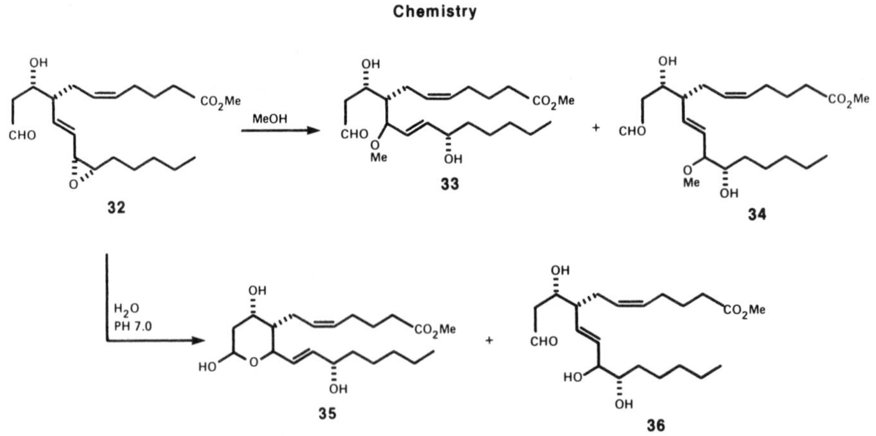

Chemistry

Scheme 17

BIOLOGICAL ACTIVITIES

A. Rat aorta

Compounds **4a** and **4b** are weak agonists on rat aorta
ED$_{50}$ ~ 100-200 ng/ml

Compounds **4c** and **4d** show no activity on rat aorta
at 10 μg/ml.

B. Washed Human Platelets

None of the compounds **4a-d** causes platelet aggregation.

CONCLUSION: None of compounds **4a-d** is TXA$_2$.

NEW METHODOLOGIES RELATED TO PROSTAGLANDIN SYNTHESIS

R. Noyori

Department of Chemistry, Nagoya University, Chikusa, Nagoya 464, Japan

Abstract—Newly developed tools for prostaglandin synthesis are
described. (1) Transition metal catalyzed reactions of strained oxygen
ring systems provide useful methods for site-specific oxygenation of unsaturated carbon skeletons. Palladium(0) complexes catalyze the
isomerization of α,β-epoxy ketones to β-diketones under mild conditions.
The reaction pathway of the palladium(0) catalyzed reaction of 1,3-diene
1,2-epoxides is highly dependent on the structure and substitution pattern
of the substrates; epoxides derived from flexible dienes isomerize to
dienol isomers, whereas cyclic diene epoxides are transformed to β,γ-
unsaturated ketones selectively. 1,3-Diene 1,4-epiperoxides give 4-
hydroxy-2-alkenones or syn diepoxides depending on the nature of the
catalysts. (2) New procedures for vicinal carbon group introduction to
α,β-unsaturated ketones have been elaborated on the basis of stoichiometric use of organocopper reagents. (3) A highly efficient asymmetric
reduction of carbonyl functions using an aluminum hydride reagent
modified by axially dissymmetric binaphthol ligand has been applied to
the generation of the 11R and 15S configuration of prostaglandins. Utility
of such methodologies is illustrated by the straightforward synthesis of
certain prostaglandin derivatives.

Synthesis of prostaglandins has occupied one of the most important places in synthetic
organic chemistry since the structural elucidation in the early 1960s. Although numerous
useful tools for such purpose have been devised, efforts are still to be continued to find new
synthetic methods and reactions which realize efficient entries, particularly in view of the
significant biological properties of newly discovered prostacyclins and thromboxanes.
Described herein are novel methods which accomplish selective transformation of oxygen
functional groups and direct construction of the basic carbon frameworks.

Transition metal catalyzed reactions of epoxides and 1,4-epiperoxides

Eminent nucleophilic property of low-valent transition metal species allows selective
skeletal changes of certain epoxides and epiperoxides, which provide versatile tools for
site-specific oxygenation of unsaturated carbon skeletons.

Palladium(0) catalyzed isomerization of α,β-epoxy ketones to β-diketones.

Under the
influence of catalytic amounts of tetrakis(triphenylphosphine)palladium(0) and 1,2-bis-
(diphenylphosphino)ethane (dpe), α,β-epoxy ketones (1) are converted to β-diketones (2) (1).
The isomerization is achievable under entirely neutral and aprotic conditions. In order to
obtain reasonable reaction rate and to avoid palladium metal precipitation so that the
catalyst gives a high turnover, addition of the bidentate auxiliary ligand, dpe, is necessary.
Some examples of the catalytic reaction are given in Table 1. A 14-electron, bent di-
coordinate palladium(0) species is a selected candidate for the actual active catalyst that
causes nucleophilic ring opening of the epoxides.

TABLE 1. Palladium(0) catalyzed isomerization of α,β-epoxy ketones to β-diketones[a]

Epoxy ketone	Product	% Yield
trans-3,4-Epoxy-2-pentanone	2,4-Pentanedione	81
trans-3,4-Epoxy-5-methyl-2-hexanone	5-Methyl-2,4-hexanedione	80
trans-4,5-Epoxy-6-tridecanone	4,6-Tridecanedione	90
trans-1,2-Epoxy-1-phenyl-3-nonanone	1-Phenyl-1,3-nonanedione	82
trans-2,3-Epoxy-1,3-diphenyl-1-propanone	1,3-Diphenyl-1,3-propanedione	84
trans-2,3-Epoxycyclododecanone	1,3-Cyclododecanedione	54
2,3-Epoxycyclooctanone	1,3-Cyclooctanedione	52
2,3-Epoxycycloheptanone	1,3-Cycloheptanedione	60
2,3-Epoxycyclohexanone	1,3-Cyclohexanedione	62
5-Isopropyl-2-methyl-2,3-epoxycyclohexanone	5-Isopropyl-2-methyl-1,3-cyclohexanedione	18
2,3-Epoxycyclopentanone	1,3-Cyclopentanedione	94

[a] Reaction was carried out using $Pd[P(C_6H_5)_3]_4$ and 1,2-bis(diphenylphosphino)-ethane (1.5—10 mol %) in toluene at 80—140 °C under argon.

Palladium(0) catalyzed isomerization of 1,3-diene 1,2-epoxides. 1,3-Diene epoxides, when exposed to tetrakis(triphenylphosphine)palladium(0) and some added triphenylphosphine, undergo three types of isomerizations, depending on the structure and substitution pattern of the substrates (2). Epoxides derived from flexible 1,3-dienes produce the dienol isomers via the metal mediated hydrogen transfer from the C-2 or C-4 (1,3-diene numbering) alkyl groups. By contrast, epoxy substrates derived from 1,3-dienes possessing ordinary ring size afford the corresponding β,γ-unsaturated ketones. No trace of α,β-unsaturated isomers have been detected. Examples of such reactions are shown in Table 2. The divergence of the reaction course would result mainly from the difference in ground-state geometry of epoxy compounds. A molecular orbital geometrical optimization

TABLE 2. Palladium(0) catalyzed reaction of 1,3-diene epoxides[a]

Epoxide	Solvent	Temp, °C (Time, h)	Product (% yield)
	Ether	60 (24)	(70)
	THF	60 (24)	(81)
	THF	50 (24)	(90)
	THF	60 (25)	(39)
	THF	110 (24)	(95)
	Benzene	130 (36)	(81)
	Benzene	120 (24)	(13) and (79)
	THF	120 (22)	(74)
	None[b]	120—140	(56)
	Dichloromethane	25 (3)	(94)
	Dichloromethane	0—10 (2)[c]	(88)

[a] Reaction was carried out under argon atmosphere using 0.1—1 mol % of tetrakis-(triphenylphosphine)palladium(0) as catalyst. In most cases some triphenyl-phosphine (1—2 equiv to the catalyst) was added. [b] Vapor-phase reaction. [c] Only 0.00013 mol % of palladium(0) catalyst was used. No triphenylphosphine was added.

suggests that the olefinic bond of 1,3-cyclohexadiene epoxide is conjugated considerably with the epoxy C—O bond but not for 1,3-butadiene epoxide (a flexible substrate model). The epoxide—enone conversion is considered to be initiated by nucleophilic attack of palladium atom on C-2, whereas the C-4 attack would lead to the dienols. The C-2 alkylated substrates prefer the reaction at C-4 for steric reasons. The epoxide—enone transformation allows ready preparation of 4-hydroxy-2-cyclopentenone starting from cyclopentadiene.

Transition metal catalyzed reaction of 1,4-epiperoxides. 1,4-Epiperoxides derived by cycloaddition of 1,3-dienes and singlet oxygen isomerize readily in the presence of metal catalysts. The reaction pathway is highly dependent on the electronic nature of the catalysts employed. Examples are illustrated in Table 3. Electron-donating tetrakis(triphenylphosphine)palladium(0) promotes the rearrangement of the epiperoxides to 4-hydroxy-2-alkenones. On the other hand, under the influence of palladium(II) chloride or a lithium salt, which acts as an electron-pair acceptor, syn diepoxides have been obtained. Cyclopentadiene 1,4-epiperoxide gives an open-chain epoxy enal under such conditions. Possible mechanisms for these reactions follow.

Organocopper conjugate addition to α,β-unsaturated ketones

The conjugate addition to enones based on the stoichiometric use of organometallic reagents has been developed. This method is useful for vicinal carbon group introduction.

TABLE 3. Palladium(0) catalyzed isomerization of 1,3-diene 1,4-epiperoxides[a]

1,4-Epiperoxide	Catalyst	Product (% yield)			
	Pd(PPh$_3$)$_4$		(54)[b]		(0)
	PdCl$_2$(PPh$_3$)$_2$		(0)		(62)[b,c]
	Pd(PPh$_3$)$_4$		(42)		(12)
	PdCl$_2$(PPh$_3$)$_2$		(0)		(88)
	LiClO$_4$				(79)[b]
	Pd(PPh$_3$)$_4$		(73)		(27)
	PdCl$_2$(PPh$_3$)$_2$		(0)		(78)
	LiClO$_4$		(0)		(67)
	Pd(PPh$_3$)$_4$		(51)[d]		
	PdCl$_2$(PPh$_3$)$_2$		(0)[e]		
	Pd(PPh$_3$)$_4$		(66)		
	Pd(PPh$_3$)$_4$		(74)		
	PdCl$_2$(PPh$_3$)$_2$		(0)[e]		

[a] All reactions were carried out in dichloromethane under argon atmosphere in a sealed glass tube using a catalyst (5—10 mol %), at 60—100 °C for 10—50 h. [b] This reaction was conducted at 4 °C. [c] THF was used as a solvent. [d] Benzene was used as a solvent. [e] Complete recovery of the starting epiperoxide.

Conjugate addition by stoichiometric use of organometallic reagents. Nucleophilic introduction of organic moieties to the β position of α,β-unsaturated ketones can be attained by using organocopper reagents formed from equimolar amounts of organolithium compounds and copper(I) iodide and 2 to 3 equivalents of tri-n-butylphosphine (3). The

$$-\overset{|}{C}=\overset{|}{C}-\overset{|}{\underset{O}{C}}- \quad \xrightarrow[\text{ether}]{\text{RLi}-\text{CuI}-2(n\text{-}C_4H_9)_3P} \quad -\overset{|}{\underset{R}{C}}-\overset{|}{\underset{H}{C}}-\overset{|}{\underset{O}{C}}-$$

TABLE 4. Organocopper conjugate addition to enones[a]

Enone substrate	Entering group	% Yield of product
5-Methyl-trans-3-hexen-2-one	n-Butyl	94
5-Methyl-trans-3-hexen-2-one	Isopropyl	95
5-Methyl-trans-3-hexen-2-one	t-Butyl	86
5-Methyl-trans-3-hexen-2-one	Phenyl	84
4-Methyl-3-penten-2-one	n-Butyl	88[b]
4-Methyl-3-penten-2-one	Isopropyl	99[b]
4-Methyl-3-penten-2-one	Phenyl	70[b]
4-Methyl-3-penten-2-one	Isopropenyl	66
2-Cyclopentenone	Methyl	73[b]
2-Cyclopentenone	n-Butyl	93
2-Cyclopentenone	Isopropyl	90
2-Cyclopentenone	t-Butyl	66
2-Cyclopentenone	Isopropenyl	98
2-Cyclohexenone	Methyl	95[b]
2-Cyclohexenone	n-Butyl	100
2-Cyclohexenone	Isopropyl	94
2-Cyclohexenone	t-Butyl	91
2-Cyclohexenone	Isopropenyl	100
3,5,5-Trimethyl-2-cyclohexenone	n-Butyl	86
3,5,5-Trimethyl-2-cyclohexenone	Isopropyl	100[b,c]
3,5,5-Trimethyl-2-cyclohexenone	Isopropenyl	100

[a] Reaction was run using a reagent prepared in situ by mixing copper(I) iodide, organolithium, and tri-n-butylphosphine in 1:1:2–3 mol ratio in ether (-78 °C for 0.5–1 h and then -40 °C for 1–3 h). [b] Reaction at 0 °C for 5–10 min. [c] Conversion was 44%.

reaction proceeds smoothly in ether at low temperature to give the conjugate addition products. When >2 equivalents of the tertiary phosphine ligand was present, the reaction system remains homogeneous throughout the reaction. Through this general method, methyl, primary alkyl, secondary alkyl, and tertiary alkyl groups can be introduced to enones. sp^2-Hybridized groups such as isopropenyl or phenyl group are equally employable. Examples are illustrated in Table 4.

Vicinal carbon group introduction via an organocopper/aldehyde joining process. The above described method allows generation of the non-equilibrating, regiochemically defined enolate 4 from enone 3. If this conjugate addition proceeds ideally, 4 is a sole strong nucleophile present in the reaction system. Therefore the reactive intermediate can be trapped efficiently by a stoichiometric amount of an aldehyde to give the aldol adduct 5. In certain cases, addition of boron trifluoride improves the product yield. Table 5 shows examples of the present method.

TABLE 5. Vicinal carbon group introduction to enones via an organocopper/ aldehyde joining process[a]

Enone	β-Entering group	Aldehyde trap	% Yield of product
2-Cyclohexenone	n-Butyl	Butanal	88[b]
2-Cyclohexenone	n-Butyl	Butanal	85
2-Cyclohexenone	n-Butyl	2-Methylpropanal	61[b]
2-Cyclohexenone	n-Butyl	2-Methylpropanal	94
2-Cyclohexenone	n-Butyl	Benzaldehyde	75[b]
2-Cyclohexenone	n-Butyl	Benzaldehyde	85
2-Cyclopentenone	n-Butyl	Formaldehyde[c]	66[b]
2-Cyclopentenone	n-Butyl	Butanal	85[b]
2-Cyclopentenone	n-Butyl	Benzaldehyde	83[b]
2-Cyclopentenone	n-Butyl	Acrolein	70[b]
2-Cyclopentenone	n-Butyl	Cinnamaldehyde	97[b]
2-Cyclopentenone	n-Butyl	6-Carbomethoxyhexanal	79[b]

[a] All reactions were carried out under argon atmosphere. Conjugate addition was performed by a similar method to that given in Table 4. The trapping reaction of an enolate with an aldehyde was done at -78—0 °C for 1—3 h. [b] A stoichiometric amount of BF_3 was added in the trapping reaction. [c] Excess amount was used.

Vicinal carbon group introduction via an organocopper/orthoester joining process. The enolate 4 generated from the enone 3 can be trapped by orthoesters to give 6. Dimethyl-formamide dimethyl acetal also serves as the trapping agent. Here addition of boron tri-fluoride or other Lewis acids such as aluminum trichloride or titanium tetrachloride is required to promote the α-alkylation. Several examples are given in Table 6. The α-alkylation is considered as a bimolecular displacement reaction between a boron tri-fluoride-activated orthoester (an oxonium ion) and a BF_2 enolate (4).

TABLE 6. Vicinal carbon group introduction to enones via an organocopper/
orthoester joining process[a]

Enone	β-Entering group	Orthoester	% Yield of product
2-Cyclohexenone	n-Butyl	Trimethyl orthoformate	63
2-Cyclohexenone	t-Butyl	Trimethyl orthoformate	66
2-Cyclohexenone	Isopropenyl	Trimethyl orthoformate	60
2-Cyclopentenone	n-Butyl	Trimethyl orthoformate	86
2-Cyclopentenone	Isopropenyl	Trimethyl orthoformate	80
2-Cyclopentenone	n-Butyl	Triethyl orthoacetate[b]	60[c]
2-Cyclopentenone	n-Butyl	Dimethylformamide dimethyl acetal[b]	58[d]

[a] The organocopper conjugate addition was carried out at -78 °C. The enolate trapping was performed using stoichiometric amounts of an orthoester and boron trifluoride at -95 to -78 °C. [b] Reaction at -40 °C to room temperature. [c] 3-n-Butyl-2-ethoxyethylidenecyclopentanone. [d] 3-n-Butyl-2-dimethylaminomethylene-cyclopentanone.

α-Alkoxyalkylation of α,β-unsaturated ketones. A convenient one-pot procedure has been elaborated for α-alkoxyalkylation of enones, 3 → 7. This method is based on the efficiency of the trimethylsilyl trifluoromethanesulfonate (TMSOTf) catalyzed conjugate addition of phenyl trimethylsilyl selenide to enones, the TMSOTf promoted reaction of enol silyl ethers and orthoester (4), and the propensity of selenoxide to undergo ready β-elimination. Several examples are listed in Table 7. When the α-alkoxyalkylated enones (7) are subjected to the organocopper conjugate addition, the corresponding vicinally functionalized ketones (8) are formed.

TABLE 7. α-Alkoxyalkylation of α,β-unsaturated ketones[a]

Enone	Acetal trap	% Yield of product
3-Buten-2-one	$C_6H_5CH(OCH_3)_2$	(57)
3-Buten-2-one	$CH(OC_2H_5)_3$	(53)
2-Cyclopentenone	$CH(OCH_3)_3$	(58)
2-Cyclohexenone	$C_6H_5CH(OCH_3)_2$	(71)
2-Cyclohexenone	$C_6H_5C(CH_3)(OCH_3)_2$	(30)
2-Cyclohexenone	$C_6H_5CH=CHCH(OCH_3)_2$	(83)
2-Cyclohexenone	$CH(OCH_3)_3$	(41)
2-Cyclohexenone	$CH(OC_2H_5)_3$	(76)

[a] All reactions were carried out in CH_2Cl_2 under argon atmosphere. Usually conjugate addition of phenyl trimethylsilyl selenide to an enone was conducted at -78 °C for 30 min using equimolar amounts of an enone and the silyl selenide and TMSOTf catalyst (2—4 mol %). The trapping of the resulting enol silyl ether with an acetal or an orthoester was performed at -40—-20 °C for 0.5—1 h. Removal of the seleno group was done by treatment with hydrogen peroxide (0 °C for 10 min and then 15—40 °C for 10 min).

Enantioselective reduction of carbonyl functions with BINAL-H

A chiral hydride reagent 9 (empirical formula, abbreviated to BINAL-H) formed in situ by mixing lithium aluminum hydride, a simple alcohol such as ethanol, and optically pure (S)- or (R)-2,2'-dihydroxy-1,1'-binaphthyl in THF exhibits exceptionally high enantioface-differentiating ability in reduction of prochiral carbonyl compounds (5). The reduction of certain aromatic ketones has resulted in virtually complete enantioselectivity. This asymmetric reduction is applicable to prostaglandin synthesis (6).

The α and β faces of the C-15 (prostaglandin numbering) carbonyl group of the synthetic intermediates of type 10 are diastereomeric each other but have substantial enantiomeric relationship. Consequently, reduction of these substrates with the chiral hydride reagent gives rise to a very high degree of stereoselection. For instance, the 11-hydroxyl protected substrate 10a or 10b has been reduced with (S)-9 at -100 to -78 °C to give the 15S alcohols, 11a and 11b with 99.4 and 99.5% selectivity. The reduction of the 11-unprotected derivative 10c with the same reagent gives solely 11c possessing the correct natural configuration. Similarly, reaction of the monocyclic enone 12 and (S)-9 has afforded the prostaglandin $F_{2\alpha}$ derivative 13 exclusively.

(S)-BINAL-H [(S)-(9)]

(10)   (11)

a:R=COCH₃   b:R=THP   c:R=H

(12)   (13)

In addition, BINAL-H has been found to be extremely useful for the preparation of chiral building blocks for the conjugate addition approach. Reduction of the iodovinyl ketone 14a with (S)-9 at -100 to -78 °C produces the prostaglandin ω chain 15a possessing 3S configuration in 97% ee (95% yield). The bromovinyl ketone 14b gives 15b in 96% ee. In addition, treatment of the five-membered diketone 16 with (S)-9 at -100 °C gives rise to the chiral hydroxy enone 17 in 94% ee.

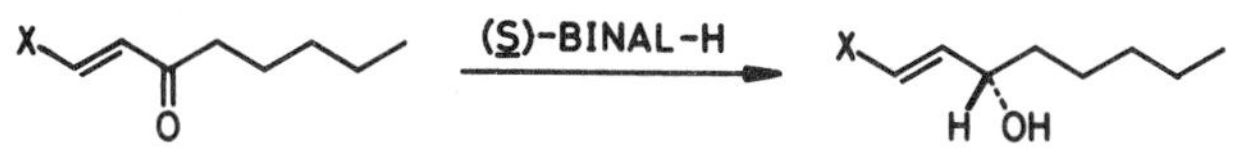

(14) → (15)

a: X = I   b: X = Br

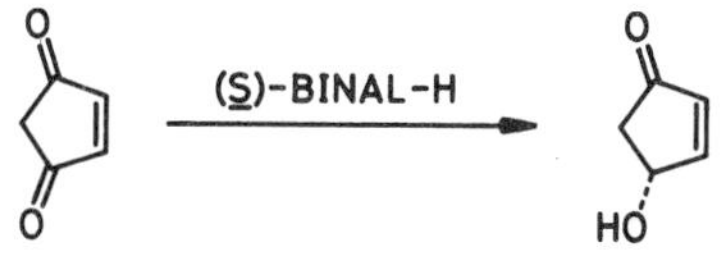

(16) → (17)

Synthesis of certain prostaglandin analogues

With a variety of new versatile synthetic tools in hand, highly convergent entries to prostaglandins and their analogues are accessible. Efficient conjugate addition routes starting from the chiral building blocks have been developed. Scheme I illustrates the synthesis of 7-hydroxyprostaglandin E_1 methyl ester (18), which exhibits potent inhibitory activity on blood platelet aggregation. Synthesis of a pyrazole prostacyclin 19 is shown in Scheme II. Such approaches possess great flexibility and find wide applicability.

Scheme I

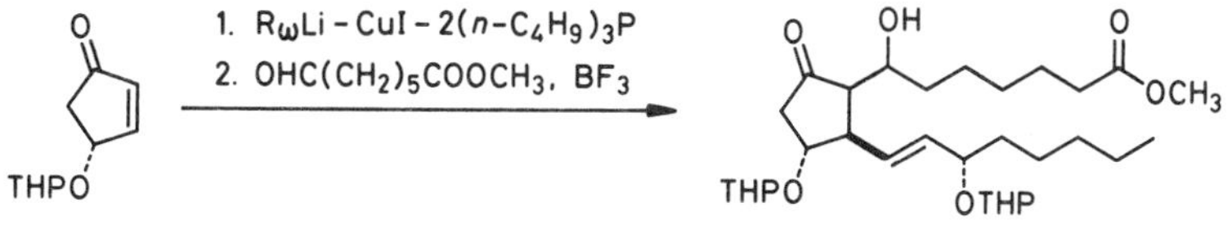

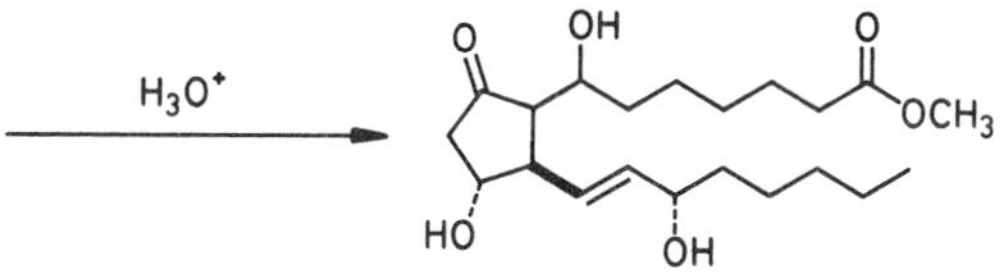

1. $R_\omega Li - CuI - 2(n-C_4H_9)_3P$
2. $OHC(CH_2)_5COOCH_3$, BF_3

H_3O^+

(18)

$R_\omega = $

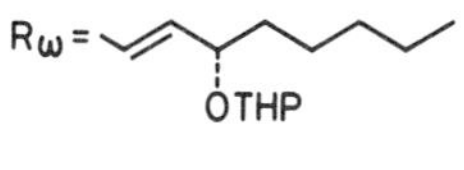

Scheme II

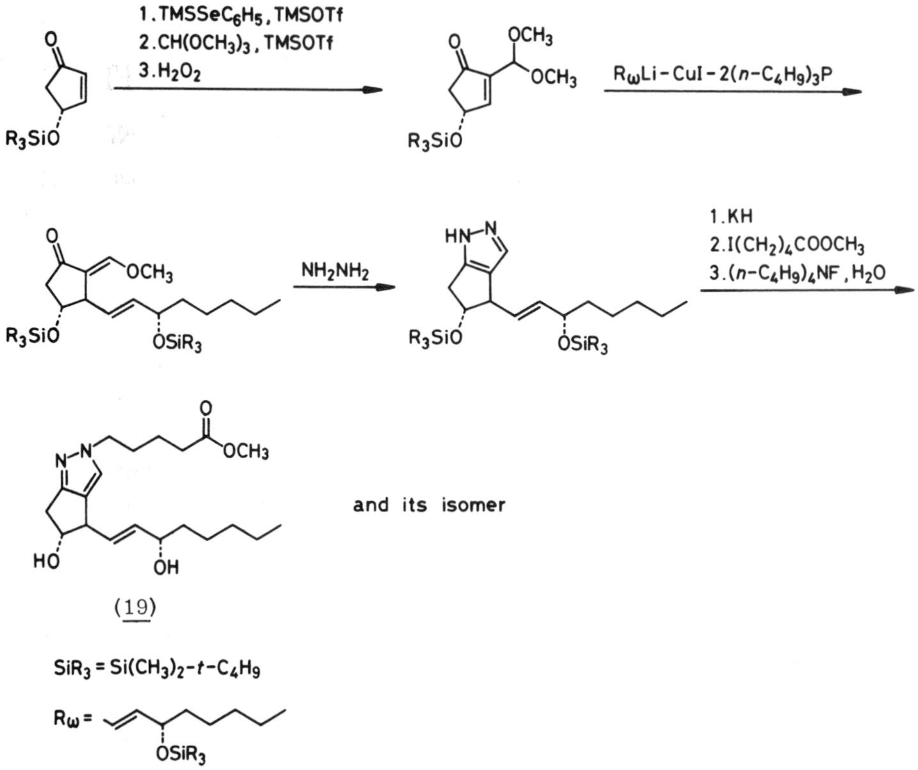

(19)

$SiR_3 = Si(CH_3)_2\text{-}t\text{-}C_4H_9$

$R\omega = $

References

1. M. Suzuki, A. Watanabe and R. Noyori, J. Am. Chem. Soc. 102, 2095-2096 (1980).
2. M. Suzuki, Y. Oda and R. Noyori, J. Am. Chem. Soc. 101, 1623-1625 (1979).
3. M. Suzuki, T. Suzuki, T. Kawagishi and R. Noyori, Tetrahedron Lett. 21, 1247-1250 (1980).
4. S. Murata, M. Suzuki and R. Noyori, J. Am. Chem. Soc. 102, 3248-3249 (1980).
5. R. Noyori, I. Tomino and Y. Tanimoto, J. Am. Chem. Soc. 101, 3129-3131 (1979).
6. R. Noyori, I. Tomino and M. Nishizawa, J. Am. Chem. Soc. 101, 5843-5844 (1979).

SYNTHETIC STUDIES IN THE ALKALOID FIELD

Cs. Szántay, L. Szabó, Gy. Kalaus, P. Győry, J. Sápi and K. Nógrádi

Institute for Organic Chemistry, Technical University, H-1521 Budapest, and
Central Research Institute of Chemistry, Budapest, Hungary

Abstract – The natural alkaloids (+)-vincamine (1),
(–)-eburnamonine (2) and the semisynthetic derivative
(+)-apovincaminic acid ethyl ester (3) are all established
pharmaceuticals used as cerebral vasodilators. In the course
of developing their industrial synthesis several unexpected
reactions and products were observed; these are presented in
this paper.

INTRODUCTION

(+)-Vincamine (1), an alkaloid first isolated from Vinca minor, has gained
wide application in recent years as a cerebral vasodilator. (–)-Eburnamon-
ine (2) is marketed with the same indication. Several syntheses of both have
been published (Ref. 1). A semisynthetic derivative of (+)-vincamine,
(+)-apovincaminic acid ethyl ester (3) is produced under the trade name
Cavinton by the pharmaceutical company Gedeon Richter, Hungary.

Fig. 1

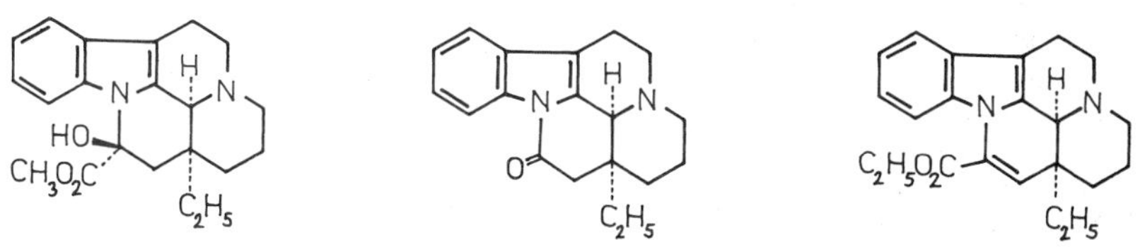

1   2   3

Some years ago a stereoselective total synthesis of (+)-vincamine, suitable
for industrial production was developed and published by our group (Ref.
2). In the course of this work and also of that directed towards an industrial synthesis of eburnamonine and Cavinton several unusual transformations
have been discovered, which, though not utilized in the main lines of the
large scale synthetic sequence, may be of general interest for organic
chemists. In the following we are going to discuss some of these reactions.

AN UNUSUAL ALKYLATION OF AN ENAMINE

In a traditional synthetic approach to vincamine-like alkaloids tetracyclic
lactams of type 4 are key intermediates (ref. 3, 4). They are obtained by
reacting enamines derived from butyric aldehyde with acrylic esters followed by condensation of the product (6) with tryptamine.

Fig. 2

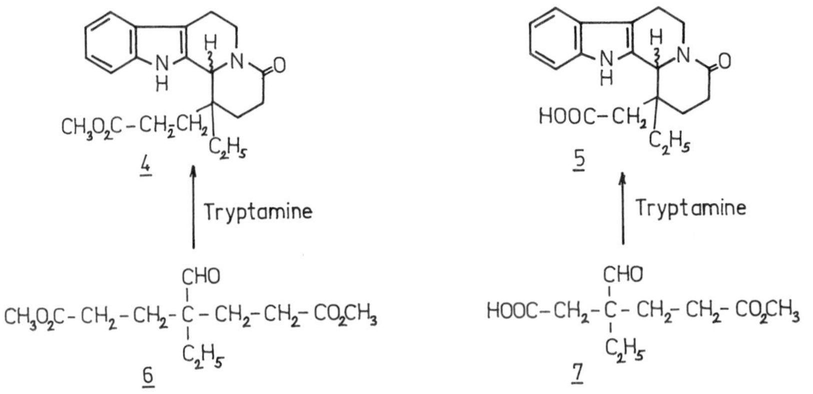

We had in mind to prepare according to this scheme the tetracyclic compound
5, which was intended as an intermediate in a sequence leading to eburnamonine. This required the synthesis of the ester-aldehyde 7 and as a model
first we studied the reaction of some enamines of butyraldehyde with
acrylic ester. It was known that the corresponding pyrrolidine enamine (8)
may be reacted, as required, with either one (Ref. 5) or two (Ref. 3)
moles of the acrylic ester.

We found however that taking the morpholine enamine 9, to our advantage,
the formation of the required mono-adduct (10) was exclusive.

Fig. 3

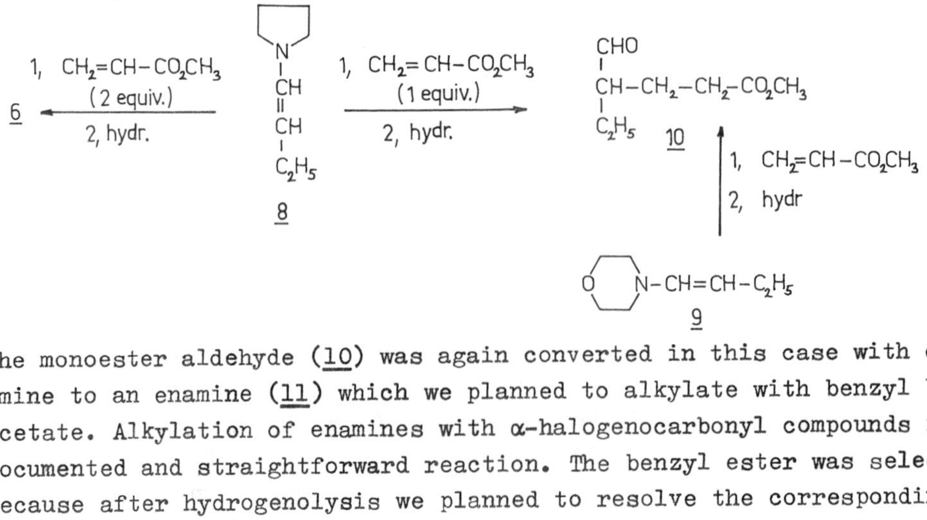

The monoester aldehyde (10) was again converted in this case with dibutyl-
amine to an enamine (11) which we planned to alkylate with benzyl bromoacetate. Alkylation of enamines with α-halogenocarbonyl compounds is a well
documented and straightforward reaction. The benzyl ester was selected
because after hydrogenolysis we planned to resolve the corresponding free
acid (7) and react the correct enantiomer with tryptamine. This would give

us an optically active intermediate which could then easily be transformed
to natural eburnamonine.

Fig. 4

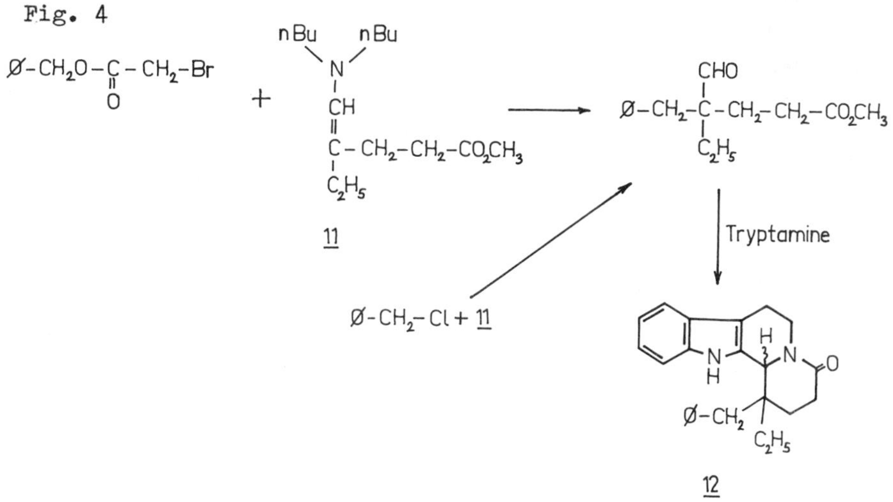

When in our first experiment the acid ester was reacted, without isolation,
with tryptamine, instead of the expected product (5) 12 was isolated (Ref.
6). The structure of 12 was not only established by the usual spectroscopic
techniques but also by an independent synthesis, namely by alkylating the
enamine 11 first with benzyl chloride and reacting the product with tryptamine. It has to be noted that only one product crystallized from the
reaction mixture in about 30% yield. This was found to be homogeneous by
TLC and in the NMR spectrum only one singlet for a benzylic CH_2 group
could be identified. This all indicated that only one of the two possible
diastereomers was isolated. Its detailed stereostructure has yet to be
established.

Though it is well known that esters of strong acids, such as the esters of
p-toluene sulfonic acid, are efficient alkylating agents, it was surprising
that in benzyl bromoacetate instead of the highly reactive α-carbon it is
the benzyl group which alkylates. This was a novel feature of enamine
chemistry.

REACTION OF ALKOXYCARBONYL TRYPTAMINE WITH AN ESTER ALDEHYDE

Syntheses of the above mentioned alkaloids are all based on tryptamine.
Guided by economic considerations we launched and carried to success a
project with the purpose of preparing tryptamine without catalytic hydrogenation. We were able to prepare in good yield (Ref. 7) α-ethoxycarbonyl
tryptamine (13), conversion of which to tryptamine by acidic hydrolysis
and decarboxylation had been reported before.

In connection with this project we looked at the reaction of our intermediate 13 with the aldehyde ester 6. After heating the partners in acetic
acid for 120 hours conversion was complete and from the resulting mixture,
along with 45% of isocarbostyryl, two new products could be isolated.

Fig. 5

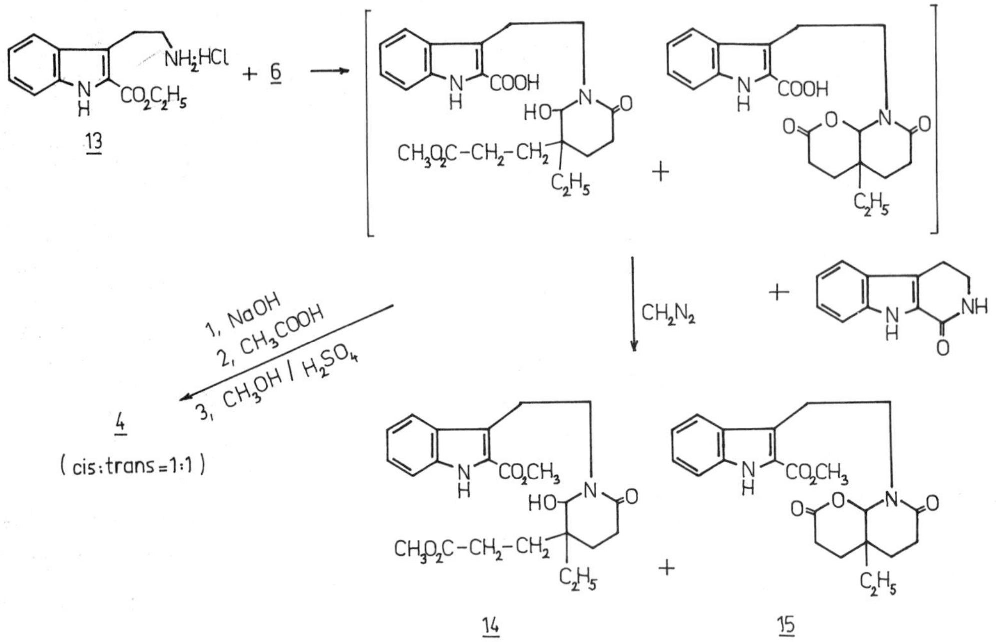

When the latter were treated, without separation, with diazomethane in
dioxane the amidocarbinol diesters 14 and 15 were obtained in 35% and 5%
yield respectively. The structure of the primary products was deduced from
that of the diesters.

On the other hand when the mixtrure of the intermediates was first hydrolysed with aqueous alkali, then cyclized by heating in acetic acid and
finally esterified with methanol - sulfuric acid then 4 was obtained in
60% yield as a 1:1 mixture of cis and trans stereoisomers. Note that 4 is
an intermediate of the already mentioned Kuehne synthesis.

ANOMALOUS NUCLEOPHILIC SUBSTITUTION AND OXIDATION

Compound 19 was intended to serve as an intermediate in the elaboration of
the eburnamonine ring system, since hydrolysis of the nitrile group and
cyclization to 2 were already known steps (Ref. 8).

We planned to prepare this nitrile (19) from a key intermediate in our
syntheses, that is from the enamine 16. In fact when 16 was reacted with
formaldehyde it gave in good yield the hydroxymethyl compound 17. Addition
was thus followed directly by reduction giving with total stereoselectivity
the trans compound. Exchange of the hydroxy function to chlorine afforded
the known trans-18 (Ref. 9). It was anticipated that reaction of 18 with
the cyanide anion would be a smooth SN2 process, since the chlorine atom
was associated with a primary carbon atom. Transformation of trans-18, both
in DMSO or DMF at 110° was however extremely slow, the starting material

Fig. 6

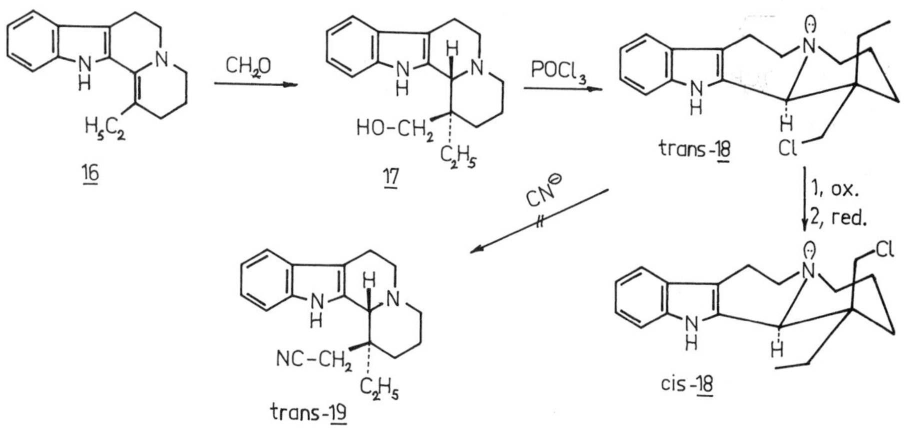

only disappeared after 120 hours and instead of the expected product (19) the pseudocyanide 20 was isolated, from which the nitrile group could be removed by reduction with borohydride anion to give 21.

Fig. 7

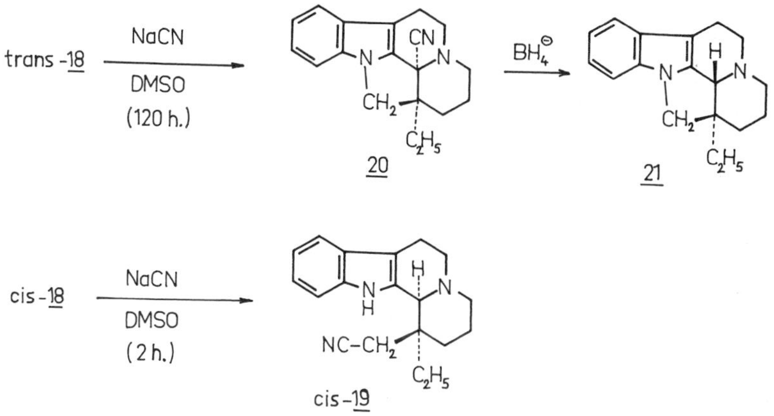

It can be assumed on one hand that the formation of an SN2 type transition state involving the cyanide anion was prevented by steric hindrance around the reaction center and on the other that the steric disposition of the chloromethyl group was favourable for ring closure. During prolonged heating also a carbon-nitrogen double bond was formed due to spontaneous oxidation and the electrophilic carbon terminal of this double bond was then attacked by the cyanide anion.

The above reasoning implies that with cis-18, characterized by a cis disposition of the ethyl group and the hydrogen at C-12b, and also having an axially oriented chloromethyl group, ring closure should be unfavoured whereas normal displacement should be unhindered.

In order to prepare cis-18 the trans compound was first oxidized with

mercury(II)acetate and then the unsaturated product was subjected to
catalytic hydrogenation. This afforded both cis-18 and trans-18 in a ratio
of 3:2.

In fact transformation of cis-18 was much faster than its trans stereoisomer and the required product (19) formed in low yield besides some
other compounds of yet unclarified structure. Thus at least in part the
cis-18 chloride behaved as expected.

The crowded environment around substituents attached to C-1 may rationalize
the unusual behaviour of 17 on oxidation.

Fig. 8

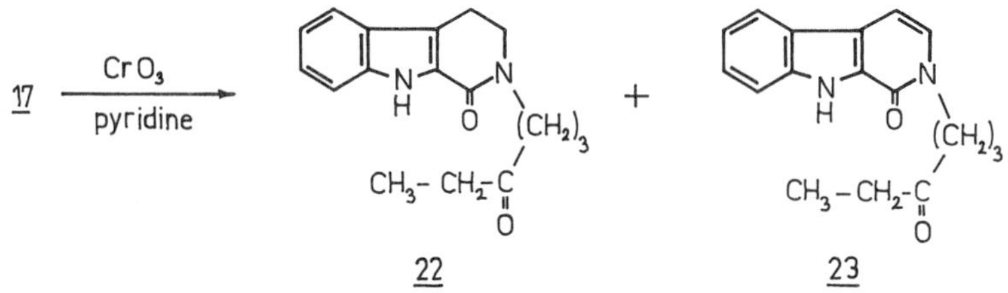

Oxidation of 17 was studied since the aldehyde derived therefrom was an
excellent intermediate in the synthesis of vinca alkaloids (Ref. 10).
Although oxidation of a primary hydroxyl group to an aldehyde is generally
easy, 17 with CrO_3 in pyridine or acetic acid failed to provide the
corresponding aldehyde, only products representing higher oxidation levels,
namely the dihydrocarbostyryl 22 and the carbostyryl 23, both derived by
ring cleavage were obtained in a ratio of 2:1.

REACTION OF ACRYLIC ACID DERIVATIVES WITH ENAMINES

In our successful industrial synthesis of vincamine mentioned in the
introduction we carried out the addition of α-acetoxyacrylic ester onto the
enamine 16. While investigating some other possibilities, among others,
α-halogenoacrylates were reacted with 16 in the hope that after reduction
of the product and substituting the halogen we would arrive at a hydroxy-
function. From there vincamine would be accessable by an oxidative transformation.

When 16 was reacted with chloroacrylonitrile or chloroacrylic ester the
primary adduct could not be trapped in either case since elimination of
hydrogen halide was so fast that only cyclization products of type 24 were
isolated. Reduction of the latter with borohydride anion led to products
in the trans series (Ref. 11).

Fig. 9

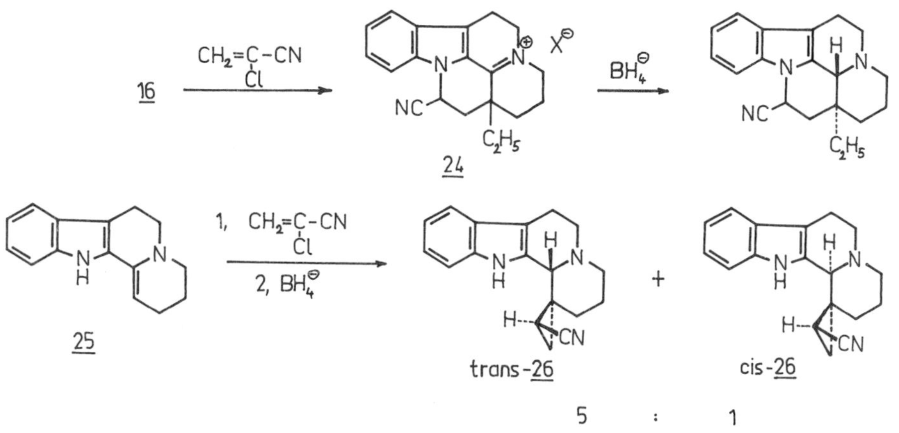

These results suggested that with α-chloroacrylonitrile the enamine 25 having no ethyl group at C-1 would also yield a pentacyclic compound analogous to 24. Contrary to our expectations, in the second alkylation step not the indole nitrogen was attacked, but a spiro compound containing a cyclopropane ring was formed. The product was isolated as its crystalline perchlorate salt which provided on reduction the nitriles cis-26 and trans-26.

Formation of a cyclopropane ring was rather surprising. One should envisage the sequence of events as follows. The carbanion arising in the first step abstracts the most acidic proton in the molecule, that is the one at the indole nitrogen. This N-anion attacks the electrophilic carbon atom adjacent to chlorine. This process should be a fast one, requiring small activation energy and therefore it is reasonable that with the 1-ethyl derivative the primary adduct could not be isolated only the pentacyclic product 24. In contrast with model 25, unsubstituted at C-1, it is the proton attached to this particular carbon atom which migrates and atom C-1 of the resulting enamine participates in the formation of the cyclo-propane ring.

The reaction was extended to the five-membered analogue of 25, i.e. to 27 which we prepared by the same method as 16.

Reaction of 27 with α-chloroacrylonitrile gave the cyclopropane 28 in about 50% yield. The reaction was stereoselective and the exo product was isolat-ed exclusively.

The outcome of the reduction of the iminium salt 28 depends on the condit-ions applied. Reduction with borohydride anion in methanol gave 29a and 29b in a ratio of 3:2. Conventional spectroscopic methods failed to provide a configurational assignment for the products, therefore an X-ray study was carried out (by A. Kálmán et al.) which gave unequivocal support to structure 29a.

Fig. 10

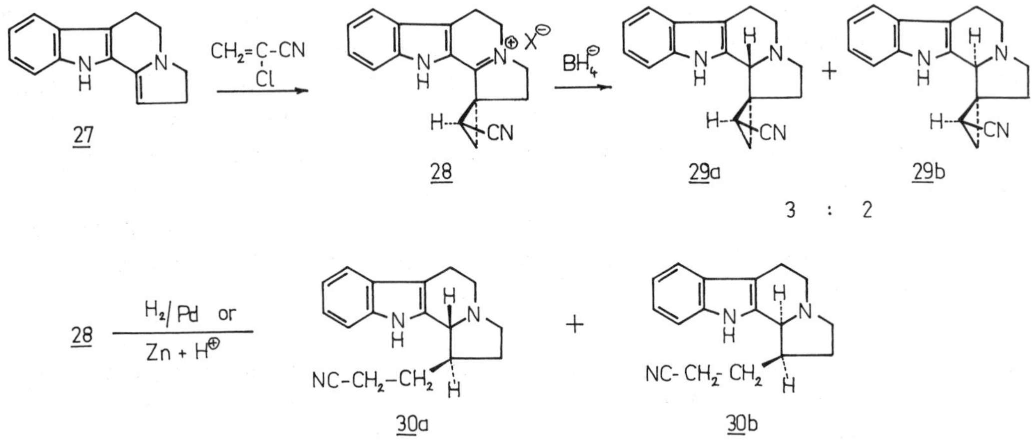

3 : 2

30a   30b

When 28 was hydrogenated over Pd/C or reduced with zinc in acetic acid then not only the double bond was saturated but also the cyclopropane ring was cleaved and in a ratio depending on conditions the trans and cis cyanoethyl compounds 30a and 30b were obtained.

Reaction of our enamines with bromoesters proceeded similarly and gave rise to cyclopropane carboxylates. We arrived at the same esters also by the hydrolysis of the nitriles with sulfuric acid in methanol. Hydrolysis of both nitriles (29a and 29b) gave a 1:1 ratio of epimers due to equilibrium epimerization at the center of chirality adjacent to the bridgehead nitrogen.

Reaction of enamines with α-halogenated acrylic acid derivatives appears to be a new route to cyclopropane carboxylic acid derivatives.

UTILIZATION OF THE UNWANTED EPIMER AND ENANTIOMER

In synthetic sequences leading to the alkaloids 1-3 (e.g. Ref. 3, 4) the acrylic ester derivative 31 was always accompanied by some of its trans epimer, which was of no value in the production of the end product.

Fig. 11

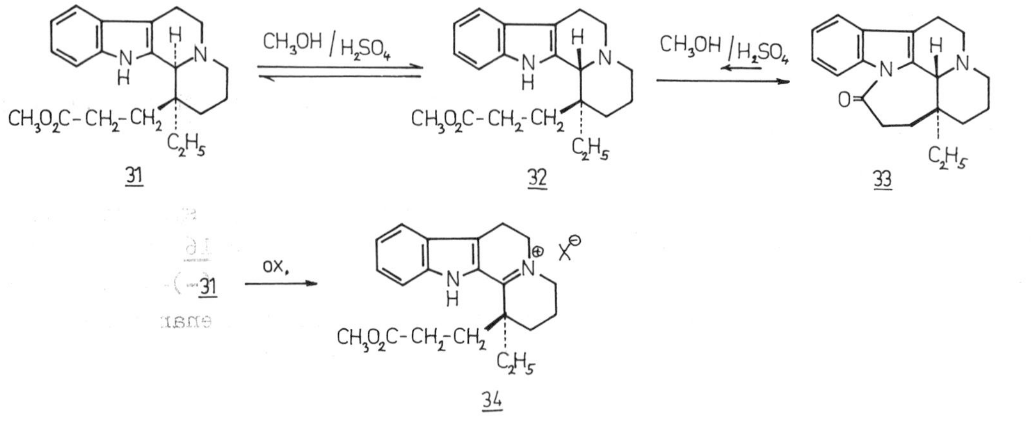

31   32   33

34

When $\underline{32}$ was boiled with methanolic sulfuric acid   , as observed previously also with the nitrile $\underline{29}$ and earlier with indoloquinolizines (Ref. 12),
epimerization took place at the bridgehead carbon adjacent to the nitrogen
and an equilibrium between the $\underline{cis}$ and $\underline{trans}$ epimers was established. The
process was however accompained by a slower cyclization of only the $\underline{trans}$
epimer to the lactam $\underline{33}$ and therefore on prolonged heating the latter
became predominant. No such cyclization could be observed with the $\underline{cis}$
compound ($\underline{31}$) and this can be explained, as with the pair $\underline{18}$, by steric
effects. Esterification of the corresponding nitriles using methanolic
sulfuric acid (Ref. 13) gave similar results: all of the three above
mentioned products ($\underline{31}$, $\underline{32}$ and $\underline{33}$) appeared.

Ester $\underline{31}$ can be conveniently resolved with the aid of dibenzoyl tartaric
acid and the recyclization of the unwanted enantiomer by its racemization
appeared to be a far more important research target than the utilization
of the minor epimer.

First oxidation to the iminium salt $\underline{34}$ with Hg(II)acetate or Pb(IV)acetate
was investigated. Optically active $\underline{34}$ was isolated as a perchlorate in 55%
and 57% yield, respectively. Re-reduction of the salt (Pd/C + H_2) gave the
epimers $\underline{31}$ and $\underline{32}$, which were separated by TLC on silicagel. The optical
rotation of $\underline{31}$ was identical with that of the starting material.

However, when the perchlorate salt of $\underline{31}$ was oxidized with sodium dichromate in acetic acid, the racemic salt $\underline{34}$ crystallized from the solvent in
53% yield, reduction of which yielded racemic $\underline{31}$.

On treating the optically active salt $\underline{34}$ under the same conditions with
sodium dichromate, no racemization was observed, $\underline{i.e.}$ bond breaking must
have occured during the oxidation process. All the above facts suggested
that during the oxidation of compound $\underline{31}$ a chromium containing radical was
formed in which the C_1-C_{12b} bond was homolytically broken causing racemization.

This unusual "Chromic effect" is being investigated using some other models
too.

ENANTIOSELECTIVE SYNTHESES

The best way to deal with unwanted enantiomers is to avoid their formation
by devising enantioselective procedures.

Earlier we reported the use of asymmetric induction in the synthesis of
vincamine (Ref. 2). When we reacted the prochiral enamine $\underline{16}$ with the readily accessible ester of α-acetoxyacrylic acid with natural (-)-menthol after
reduction, transesterification and oxidation the required enantiomer of
vincamine was obtained in about 40% enantiomeric excess.

We thought that it would be a practical alternative to use a member of the
natural "chiral pool" namely L-tryptophane as starting material, the more
so since natural amino acids had been used several times for the synthesis
of alkaloids (Ref. 14).

Fig. 12

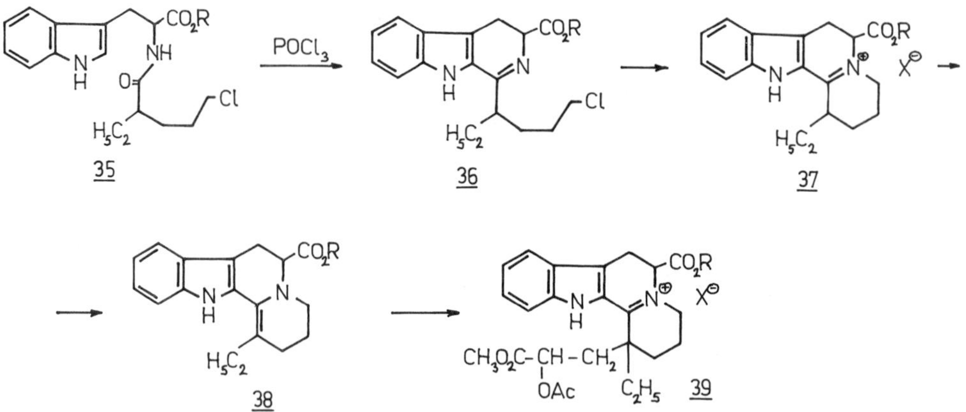

In order to reduce the tendency to racemization of its chiral center, the
isopropyl ester of L-tryptophane was prepared and allowed to react with
2-ethyl-5-chlorovaleryl chloride. In the acylation reaction, carried out
under mild conditions, the optically active amide 35 was isolated in a
yield of about 80%. Ring closure with phosphoryl chloride gave the β-car-
boline 36, isolated as the perchlorate. The base liberated from the perchlorate had high optical activity and was readily converted in dichloromethane, even at room temperature, to the tetracyclic iminium salt 37
showing high specific rotation.

As in our vincamine synthesis the enamine 38 liberated from the salt 37
reacted with the enol acetate of pyruvic acid methyl ester. The resulting
adduct 39, however, was optically inactive.

All of our attempts to avoid racemization failed. In the presence of a
weak base or of a small amount of a strong base no reaction took place or
only polymers were obtained.

In order to avoid racemization during addition a stronger electrophile
than acrylic esters or α-acetoxyacrylic esters, such as methylenemalonic
ester was selected as partner.

First model experiments with the achiral enamine 16 were carried out. To a
suspension of the iminium perchlorate derived from 16 in dichloromethane
less than one equivalent of triethylamine or a catalytic amount of potassium tert-butoxide and thereafter an excess of methylenemalonic ester was
added. After two days the mixture was worked up and the iminium salt of
the adduct was directly subjected to catalytic hydrogenation.

Fig. 13

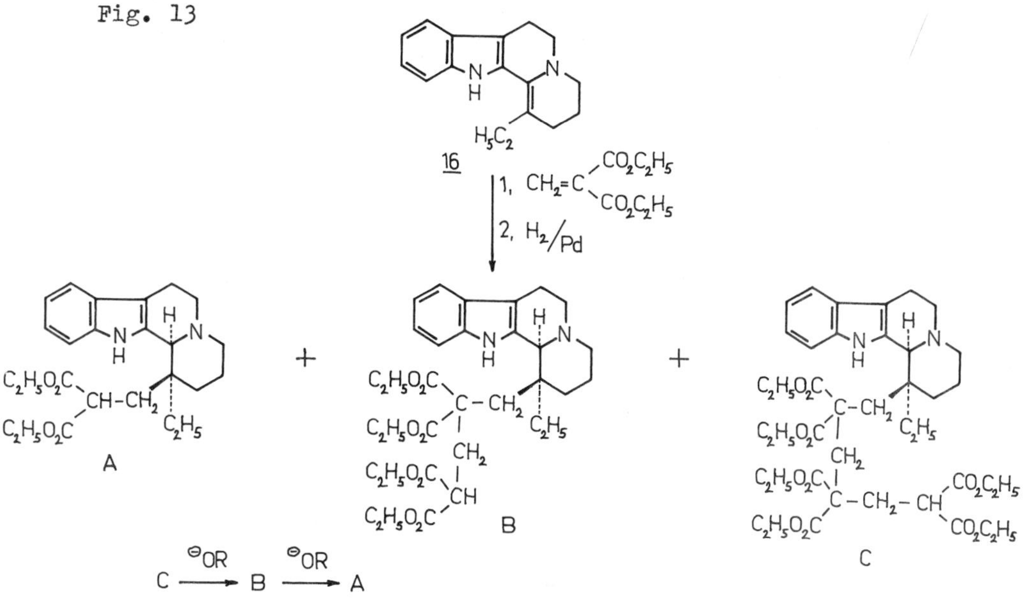

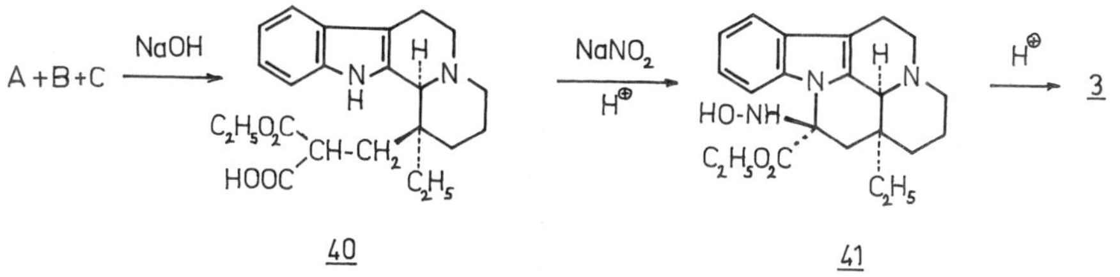

16

1, $CH_2=C\begin{smallmatrix}CO_2C_2H_5\\CO_2C_2H_5\end{smallmatrix}$

2, H_2/Pd

A   +   B   +   C

$C \xrightarrow{\ominus OR} B \xrightarrow{\ominus OR} A$

$A+B+C \xrightarrow{NaOH}$ 40 $\xrightarrow[H^{\oplus}]{NaNO_2}$ 41 $\xrightarrow{H^{\oplus}}$ 3

In 95% overall yield a three-component mixture was obtained, the products (A, B, and C) corresponded to the addition of one, two, and three molecules of malonic ester. Such multiple additions are not unprecedented (Ref. 15).

The reaction proceeded with remarkable stereoselectivity: all three products had, as the natural alkaloids 1 and 2, a cis configuration and less than 1% of a trans product could be isolated from the mother liquor.

Both the bis- (B) and the tris-adduct (C) could be converted to the mono-adduct (A) by treatment with ethanolic sodium ethoxide. The most practical solution however was the direct hydrolysis of the crude mixture with aqueous alkali, which gave in good yield 40, i.e. the half ester of the mono-adduct.

Treating the latter with sodium nitrite in acetic acid yielded the hydroxy-amino derivative 41 from which, when treated with ethanolic sulfuric acid racemic Cavinton (3) was obtained in excellent yield.

We also studied the possibility of converting 41 to vincamine. Although in the literature we did not find any example for an exchange of a hydroxy-amino group for a hydroxyl group, we supposed that this might be feasible in acidic media.

Fig. 14

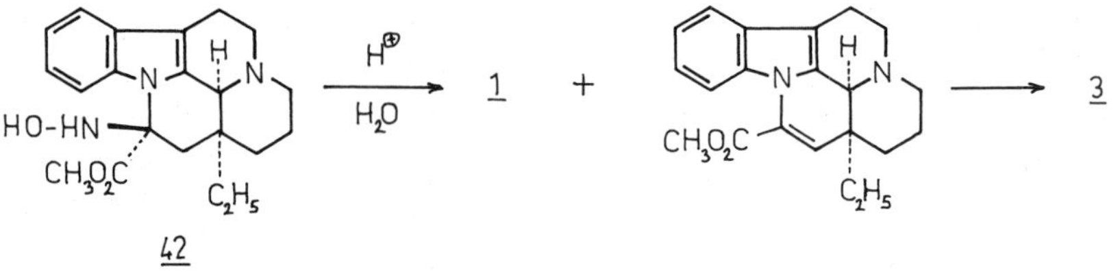

In fact on treatment with dilute sulfuric acid the methyl ester 42 gave
vincamine as the main product along with apovincaminic acid ester, which
can be easily converted to Cavinton. Product ratio can be modified by
altering reaction conditions. Despite the fact that the hydroxyl group is
introduced at a tertiary carbon atom, elimination can be suppressed.

After these encouraging model experiments we returned to work with the
optically active ester 37, which we reacted with methylenemalonic ester in
the presence of catalytic amounts of potassium tert-butoxide. The strong
electrophilic character of the ester permitted the retention of optical
activity, in other words racemization, as observed with α-acetoxyacrylic
ester, did not take place. Further transformation of the products is being
investigated.

We also studied the reaction of the optically active iminium salt 37 with
the highly reactive acrolein; this reaction had been utilized with the
enamine 16 by Buzás et al. (Ref. 16) in their synthesis of eburnamonine.

Fig. 15

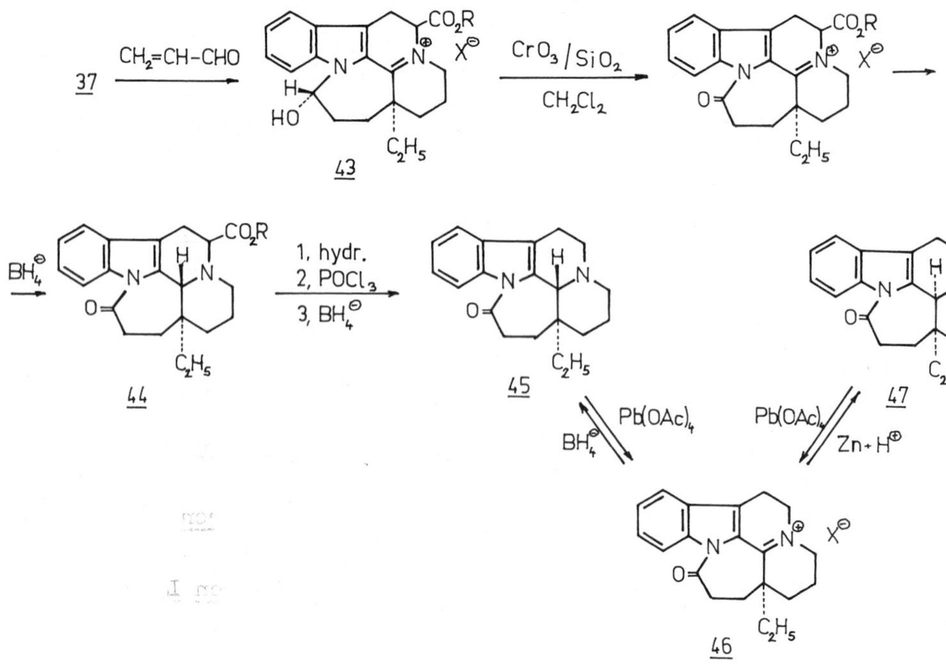

To a suspension of the iminium salt <u>37</u>, again in dichloromethane, first acrolein and then a catalytic amount of potassium tert-butoxide was added. When the mixture was worked up after 2 hours the optically active adduct <u>43</u> was obtained in 90% yield. This was then oxidized without purification with chromium(VI)oxide absorbed onto silica gel (Ref. 17) followed by reduction with borohydride anion. Thus after recrystallization <u>44</u> was obtained in 86% optical purity.

<u>44</u> was hydrolysed, treated with the method of Rapoport with $POCl_3$ and the product was reduced with borohydride anion. This gave us the lactam <u>45</u>, which we could correlate with (+)-vincamine in two ways. First we oxidized <u>47</u>, a known, optically active intermediate of the synthesis of vincamine, and the product, the iminium salt <u>46</u> was reduced with borohydride anion to yield (+)-<u>45</u>. When in turn <u>46</u> was reduced with zinc in acetic acid the product was (+)-<u>47</u>.

The asymmetric synthesis described here needs further improvement, but most importantly it can be stated that L-tryptophan, though its center of chirality was sacrificed in the course of the synthesis, assured the right absolute configuration in the end product.

REFERENCES

1. I.L. Herrmann, R.J. Cregge, J.E. Richman, G.R. Kieczykowski, S.N. Normandin, M.L. Rucsada, C.L. Semmelhack, A.J. Poss and R.H. Schlessinger, J. Amer. Chem. Soc., 101, 1540 (1979) and citations therein.
2. Cs. Szántay, L. Szabó, Gy. Kalaus, Tetrahedron, 33, 1803 (1977).
3. M.E. Kuehne, J. Amer. Chem. Soc., 86, 2946 (1964).
4. Roussel-Uclaf, Belg. Pat. 765.006.
5. G. Stork, A. Brizzolare, H. Landesman, J. Szmuszkovicz and R. Terrel, J. Amer. Chem. Soc., 85, 207 (1963)
6. Gy. Kalaus, P. Győry, L. Szabó and Cs. Szántay: J. Org. Chem. 43, 5017 (1978).
7. Cs. Szántay, L. Szabó and Gy. Kalaus, Synthesis, 1974, 354.
8. D. Cartier, J. Lévy and J. Le Men, Bull. soc. chim. France, 1976, 1691.
9. P. Pfäffli, W. Oppolzer, R. Wenger and H. Hanth, Helv. 58, 1131 (1975).
10. W. Oppolzer, H. Hanth, P. Pfäffli and R. Wenger, Helv. 60, 1801 (1977).
11. Gy. Kalaus, L. Szabó, Cs. Szántay, E. Kárpáty and L. Szporny, Arch. Pharm. 312, 312 (1979).
12. A.J. Gaskel and J.A. Joule, Tetrahedron, 23, 4053 (1967).
13. Gy. Kalaus, L. Szabó, P. Győry, É. Szentirmay and Cs. Szántay, Acta Chim. Hung. 101, 387 (1979).
14.a.) H. Rapoport, Stereoselective synthesis of natural products, p. 37 Excerpta Medica, Amsterdam-Oxford 1979;
   b.) S. Yamada, K. Murato and Takayuki Shioiri, Tetrahedron Letters 1976, 1605;
   c.) H.H. Wasserman, A.W. Tremper and J.S. Wu, Tetrahedron Letters 1979, 1089;

14.d.) L. Szabó, Gy. Kalaus, K. Nógrádi and Cs. Szántay, <u>Acta Chim. Hung.</u> <u>99</u>, 73 (1979).

15. K. Baum and A.M. Goest, <u>Synthesis</u>, <u>1979</u>, 311.

16. A. Buzás, C. Retourne, J.P. Jaquet and G. Lavielle, <u>Tetrahedron</u> <u>34</u>, 3001 (1978).

17. E. Santiello, F. Ponti and A. Manzocchi, <u>Synthesis</u> <u>1978</u>, 534.

CHIRALLY SELECTIVE SYNTHESIS OF NATURAL PRODUCTS. THE 10-HYDROXY ANALOGS OF DIHYDROQUININE

M. R. Uskoković and I. M. Kompis

Chemical Research Department, Hoffmann-La Roche Inc., Nutley, New Jersey 07110, USA

Abstract - A stereoselective synthesis of the 10R-hydroxy-7'-trifluorodihydrocinchonidine 5 possessing five chiral centers will be presented. Compound 5 is a metabolite of the specific supraventricular antiarrhythmic agent 6.

Several years ago, we developed a practical synthesis of quinine and quinidine (1), which was subsequently applied in the preparation of a large number of analogs with vinyl or ethyl side chains, and with different substituents in the aromatic rings. These analogs were tested as antimalarial and antiarrhythmic agents. The 7'-trifluoromethyldihydrocinchonine (Ro 20-7775/001) exhibited specific supraventricular antiarrhythmic activity.

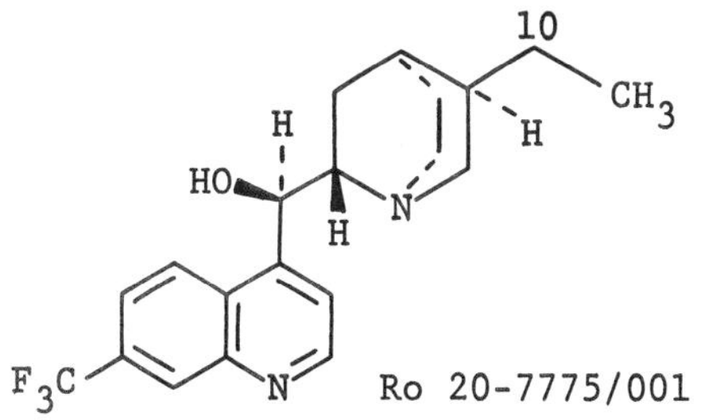

Ro 20-7775/001

The metabolism of this compound was investigated by M.S. Schwartz and his coworkers (2). Ro 20-7775/001 was incubated at 37°C and pH 7.4 with a 9000 x g supernatant fraction of rat liver and a NADPH generating system which would support cytochrome P-450 mediated drug oxidations. Five metabolites were formed enzymatically, two of them being epimers bearing a C-10 hydroxyl group:

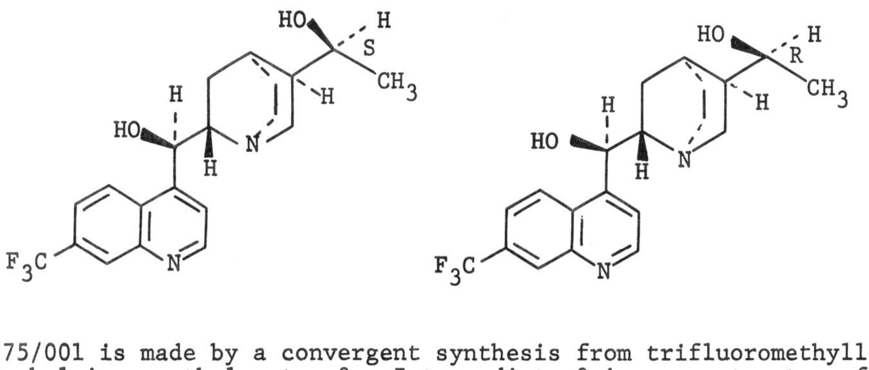

Ro 20-7775/001 is made by a convergent synthesis from trifluoromethyllepidine 1 and cincholoipon methyl ester 3. Intermediate 3 incorporates two of the three chiral centers of the quinuclidine ring. This synthesis is shown in the following Scheme.

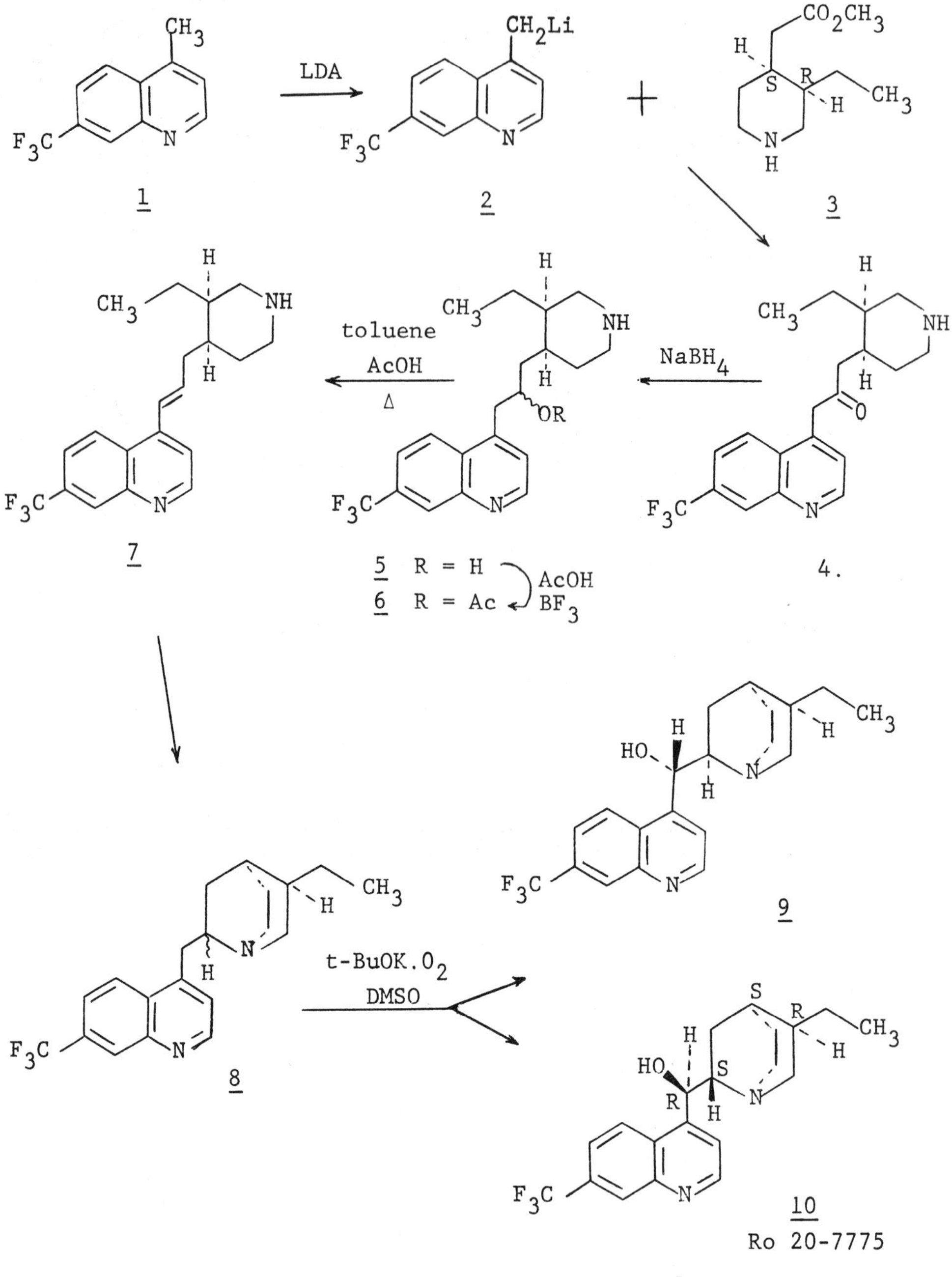

A similar synthetic approach to 10-hydroxy metabolites would require the quinuclidine precursors with three contigous chiral centers, as illustrated below.

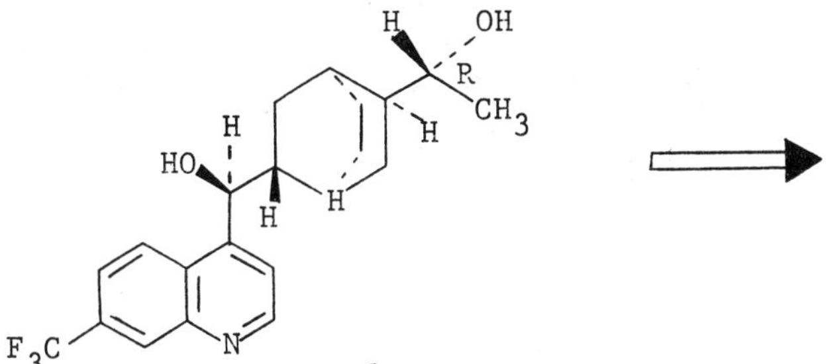

1

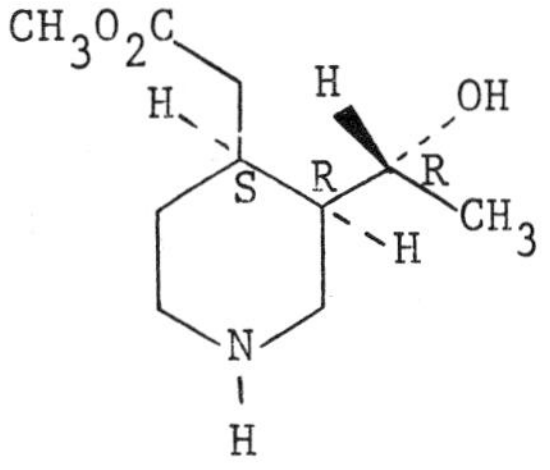

2

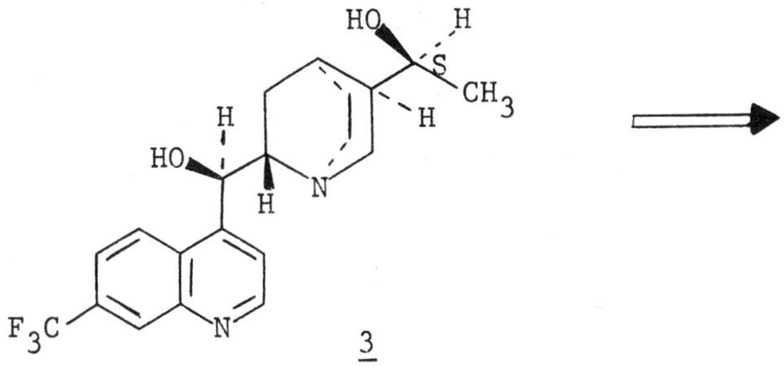

3

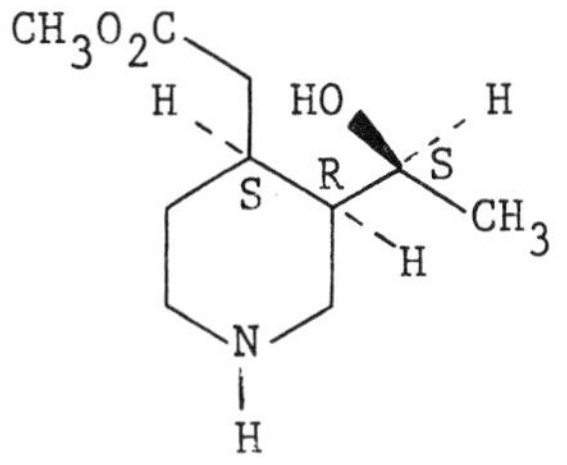

4

The syntheses of the S,R,R-quinuclidine precursor 2 and of the 10-hydroxy metabolite 1 have been completed and are illustrated in the following schemes.

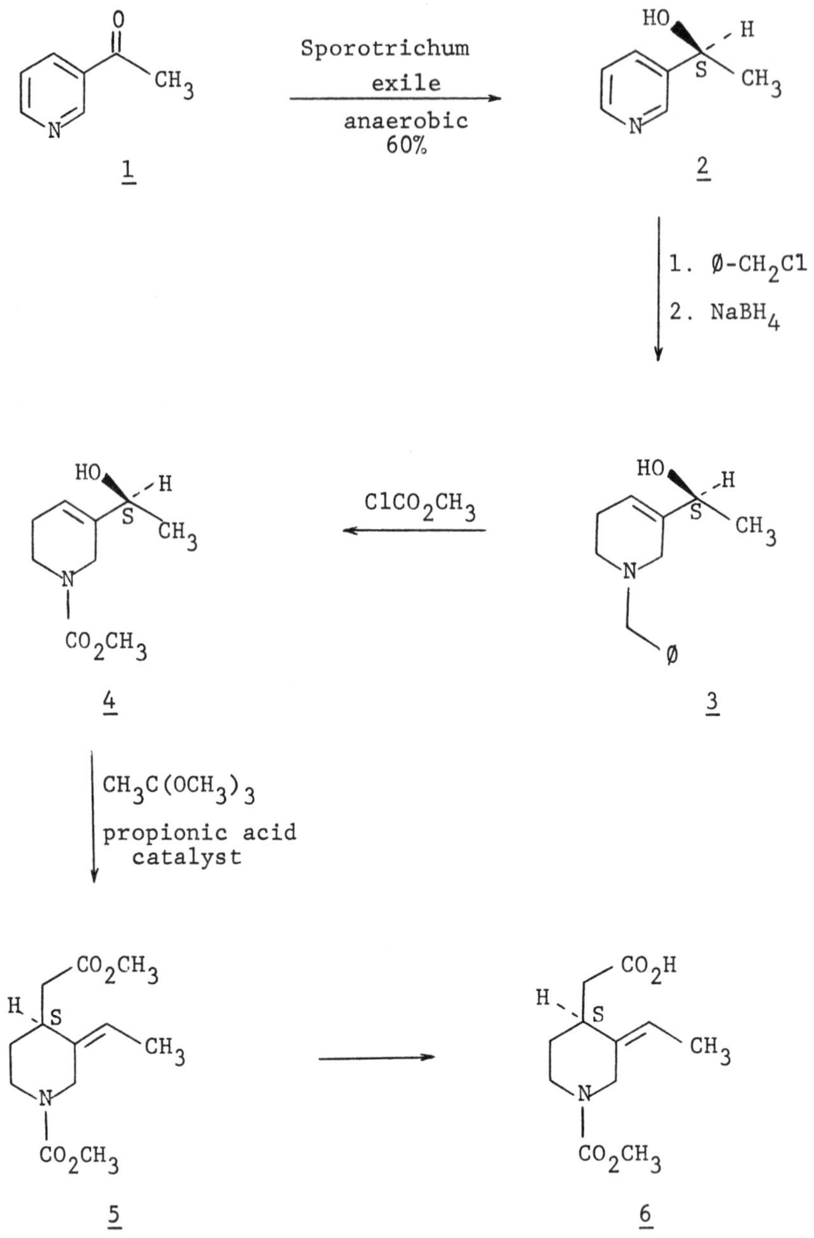

Sporotrichum
exile
anaerobic
60%

1   2

1. Ø-CH₂Cl
2. NaBH₄

ClCO₂CH₃

4   3

CH₃C(OCH₃)₃
propionic acid
catalyst

5   6

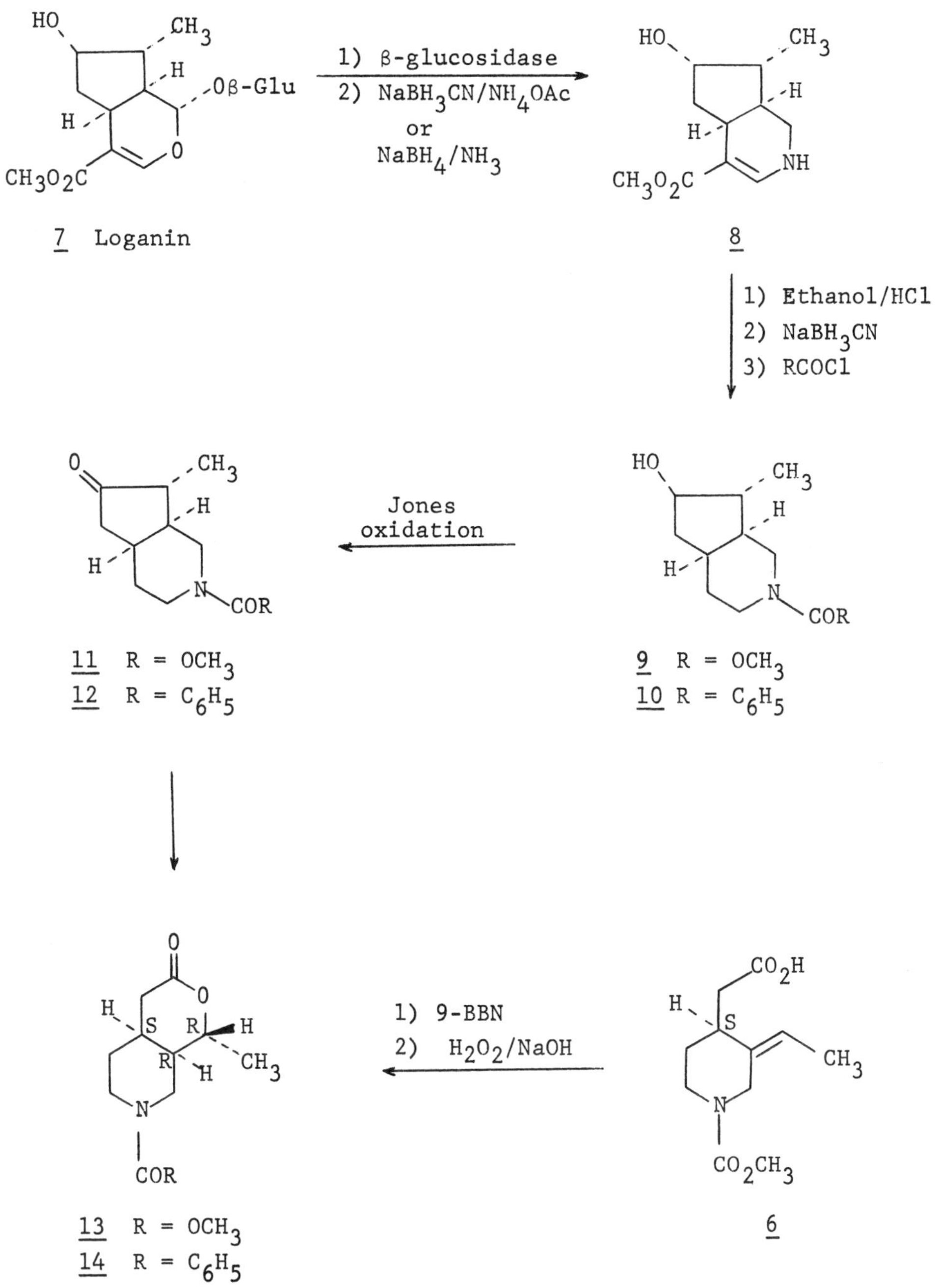

7 Loganin
1) β-glucosidase
2) $NaBH_3CN/NH_4OAc$
or
$NaBH_4/NH_3$
8
1) Ethanol/HCl
2) $NaBH_3CN$
3) RCOCl
Jones oxidation
11 R = OCH_3
12 R = C_6H_5
9 R = OCH_3
10 R = C_6H_5
1) 9-BBN
2) H_2O_2/NaOH
13 R = OCH_3
14 R = C_6H_5
6

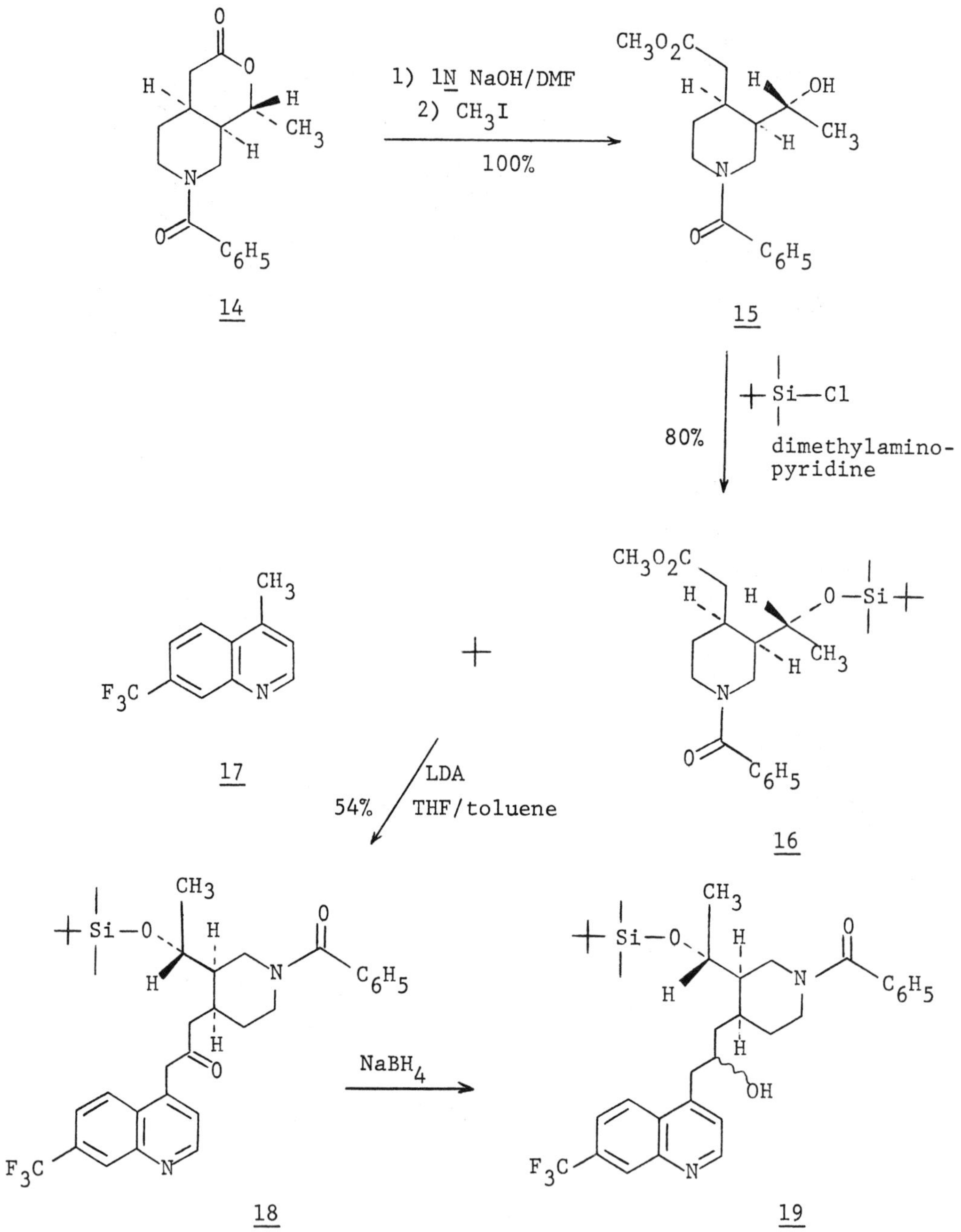
1) 1N NaOH/DMF
2) CH3I
100%
14
15
80%
dimethylamino-
pyridine
17
16
LDA
THF/toluene
54%
NaBH4
18
19

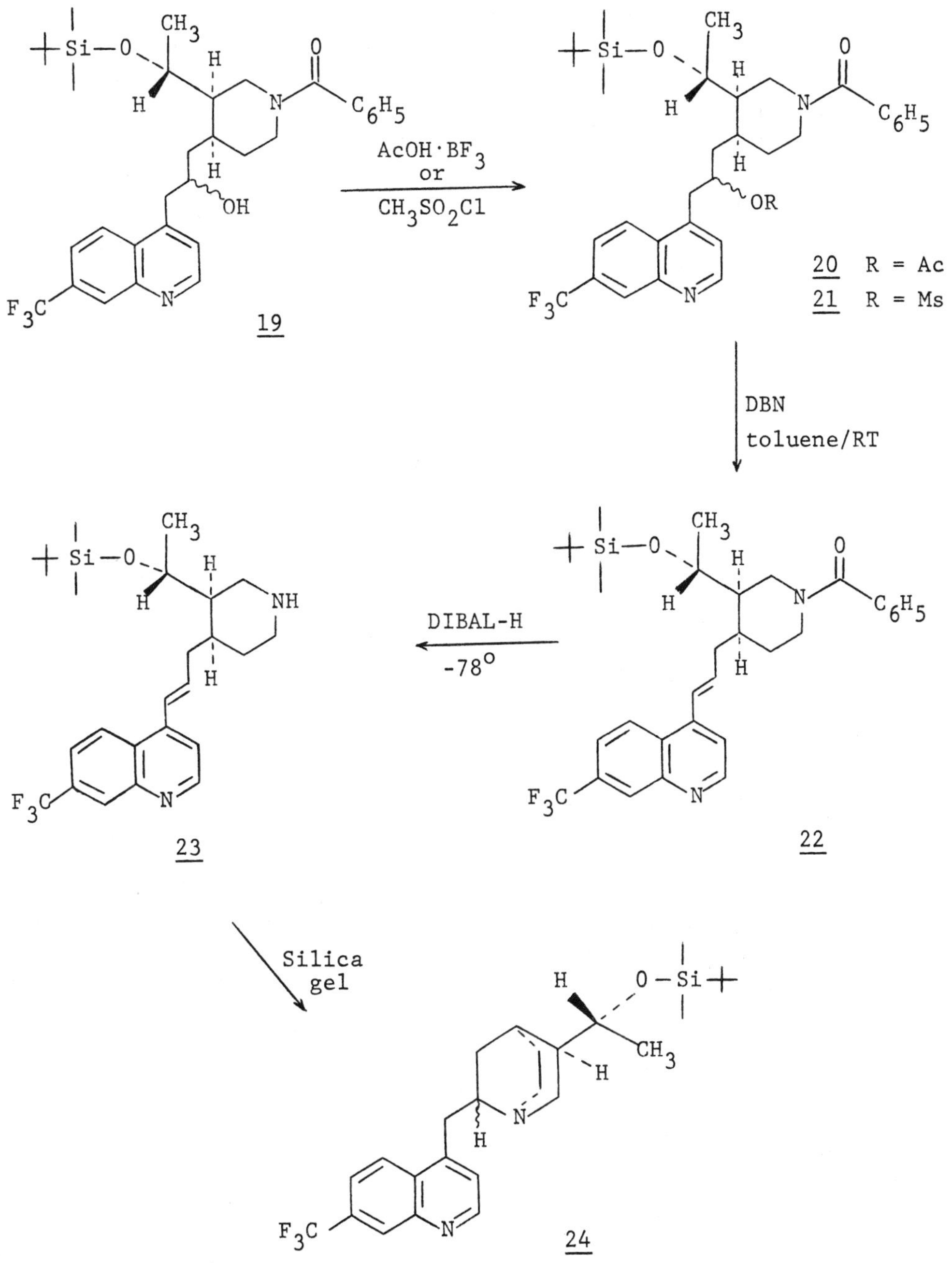
CH3
O
Si—O
H
N
C6H5
H
OH
F3C
N
19
AcOH·BF3
or
CH3SO2Cl
OR
20 R = Ac
21 R = Ms
DBN
toluene/RT
22
DIBAL-H
-78°
NH
23
Silica
gel
O—Si
CH3
24

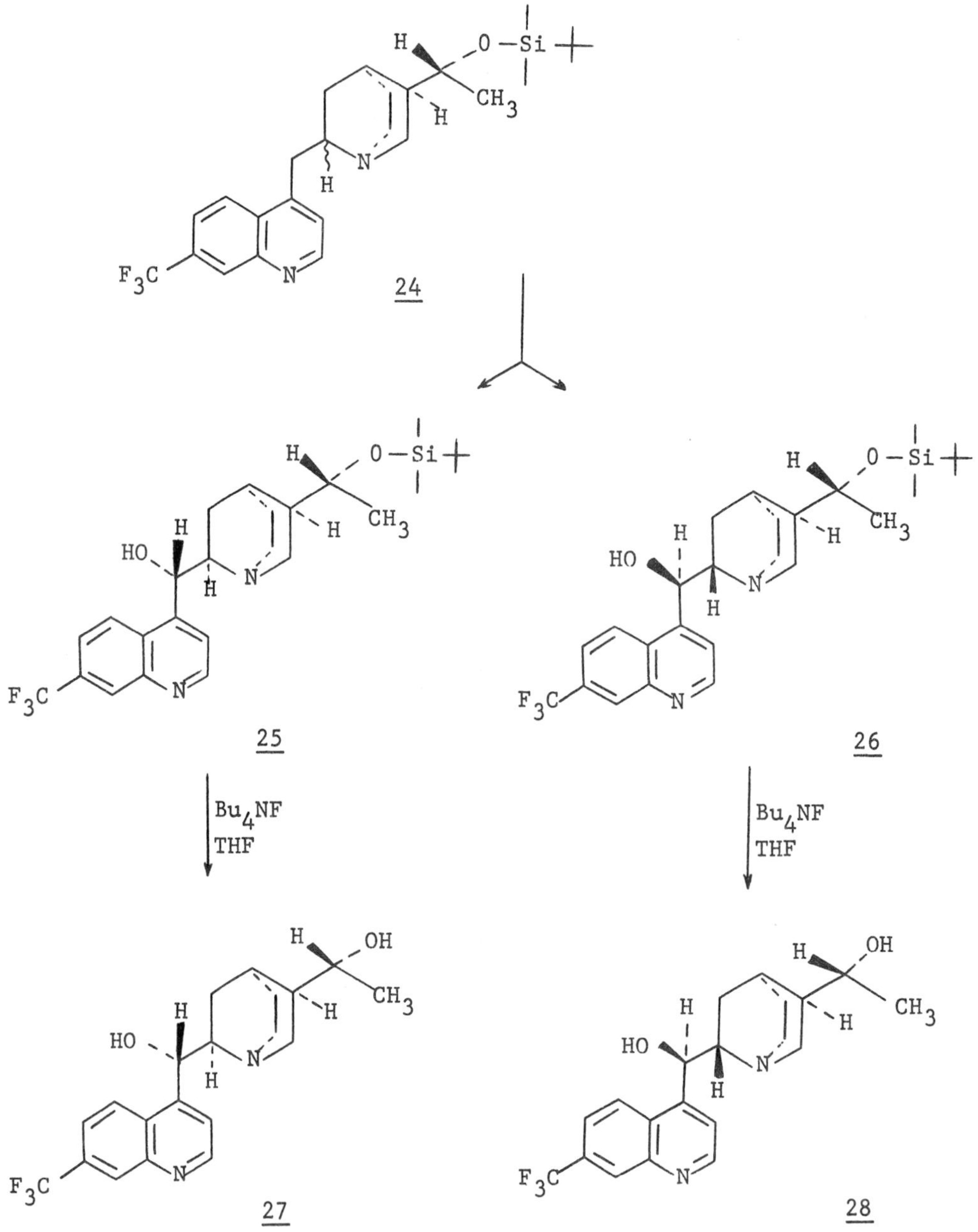
O—Si
H
CH3
H
N
H
F3C
N
24
HO
25
26
Bu4NF
THF
Bu4NF
THF
OH
27
28

AN APPROACH TO THE SYNTHESIS OF HIGHER-CARBON SUGARS

O. Achmatowicz

Warsaw Agricultural University, 02-528 Warsaw, Poland

Abstract: A method for the synthesis of higher-carbon aldoses and 2-ketoses has been developed in which the relative configuration of the side-chain and the pyranose ring results from that of the substrate. Optically active furan derivatives were obtained by acylation of furyllithium or 5-hydroxymethylfuryllithium with chiral hydroxyacids and reduction of the resulting ketone. Configuration of a new a-symmetric center in the epimeric alcohols was established by chemical or spectroscopic evidence. (1R,2S,3R) and (1S,2S,3R)-1-(2-furyl)-1,2,3-butantriols were transformed into corresponding pyranos-4-octuloses of the D and L-series, respectively. By reduction of the ketone group and trans-hydroxylation of the double bond in the D-isomer, 8-deoxyoctoses with gluco and galacto configuration of the pyranose ring were obtained.

INTRODUCTION

Carbohydrates containing more than six carbon atoms are often termed higher-carbon sugars (Ref. 1). To this class of monosaccharides belongs a wide variety of naturally occurring aldoses and ketoses with 7, 8, 9 and even 11 carbon-chain structures (Ref. 2). They have been isolated from various sources. For instance, N-acetylneuraminic acid ($\underset{\sim}{1}$) is found in glycoproteins and microbial polysaccharides (Ref. 3), whereas lincosamine ($\underset{\sim}{2}$) and hikosamine ($\underset{\sim}{3}$) constitute the sugar basic unit of the antibiotics lincomycine (Ref. 4) and hikizimycine (Ref. 5), respectively. 2-Ketoheptose ($\underset{\sim}{4}$) and 2-ketononose ($\underset{\sim}{5}$) were isolated in minute quantities from the avocado pear (Ref. 6).

Syntheses of rare sugars are usually accomplished by modification of   readily available
monosaccharides. In the case of higher-carbon sugars, it consists of elongating the carbon
chain by one or two appropriately functionalized carbon units and relating the configurations of the new chiral centers with those of the starting material. Recently a new approach to syntheses of higher-carbon sugars has been reported. It has been shown that the
condensation of dithiane, obtained from erythrose, with di-O-isopropylidene-α-D-galacto-
hexodialdo-1,5-pyranose (Ref. 7) or a galactose phosphorane with carbohydrate aldehydes
(Ref. 8) proceeds with formation of a C-C bond leading to a complex carbohydrate. However, in both cases monosaccharides are used as starting material. Stereoselective syntheses of higher-carbon sugars from non-sugar precursors have so far been described only
by one group. Kunieda et al. (Ref. 9) used telomers prepared in a free-radical telomeriza-
tion of vinylene carbonate in media of polyhalomethanes, as substrates for the synthesis of
aldosugars. From telomers with three vinylene carbonate units several racemic aldohepto-
ses and aldooctoses were obtained.

SYNTHESIS OF MONOSACCHARIDES FROM FURAN DERIVATIVES

We have developed a general method of a stereoselective synthesis of monosaccharides
from furan compounds. It consists in the following transformations. Furyl alcohol of the
type 6 (R=H, CH_3, CH_2OH, CH_2NHAc, CH_2NO_2) is treated with methanol and bromine at
-40°C to give dimethoxydihydroderivative 7, which, upon acid hydrolysis, yielded ulose 8.
Subsequent methylation with trimethyl orthoformate in the presence of Lewis acid (BF_3,
$AlCl_3$) afforded two glycosides 9 and 10 with preponderance of the α-anomer 9.

Reduction of the carbonyl group in 9 with metal hydride leads stereoselectively to alcohol
11 and, after inversion of configuration at C-4, to alcohol 12. Epoxidation of 11 and 12
with subsequent splitting of the oxirane ring or cis-hydroxylation of the double bond afforded methyl pyranosides of the respective sugars.

DEAD = $\begin{matrix} N-CO_2Et \\ \| \\ N-CO_2Et \end{matrix}$

TPP = $P(C_6H_5)_3$

CIS or TRANS HYDROXYLATION

MONOSACCHARIDES

To explore the scope of this approach and examine the stereoselectivity of all steps of the
synthesis, we have obtained by this method all aldopentoses (Ref. 10), 6-deoxyhexoses
(Ref. 11), some aldohexoses (Ref. 12) and 6-amino or 6-nitro-6-deoxyhexoses (Ref. 13).

This approach was applied and proved very efficient in the synthesis of antibiotic sugars:
cinerulose A (Ref. 14), amicetose (Ref. 15) and noviose (Ref. 16).

SYNTHESIS OF 2-KETOSES

Recently, we have shown that starting from the 2,5-disubstituted furan derivative 13, 2-ketohexoses could be obtained by a method analogous to the one described above.

Benzyl ether 13 formed diulose 14 in three steps. Sodium borohydride reduction of 14 afforded alcohol 15 and treatment of the latter with benzoic acid in the presence of diethyl azodicarboxylate-triphenylphosphine reagent with subsequent removal of benzoyl group led to alcohol 16 stereoisomeric at C-5.

Stereoselective cis or trans hydroxylation (via epoxide) of the double bond in both 15 and 16, followed by catalytic removal of the benzyl group, gave methyl pyranosides of racemic tagatose, sorbose, fructose and psicose (Ref. 17). All compounds gave consistent analytical and spectral (IR, NMR, MS) data and the structures of the obtained derivatives of the 2-ketoses were confirmed by direct comparison with samples prepared from natural sugars.

In a similar manner, 2-ketoheptoses, thus higher-carbon sugars, could be obtained. Condensation of butyl glyoxylate (18) with benzyl-furfuryl ether (17) gave with a high yield ester 19, which in four steps afforded dibenzyl ether 20. The latter was in turn converted into a mixture of diulosides 21 and 22 by the usual sequence of reactions. Since glycosidation with trimethyl orthoformate in the presence of Lewis acid was found to lead almost to the equilibrium mixture of the anomers (Ref. 11, 12), structure 21 was assigned to the major product on the basis of conformational analysis considerations.

The next three steps leading to the final 2-ketoheptose derivative proceeded with high stereoselectivity. Sodium borohydride reduction of 21 gave alcohol 23 which on epoxidation
with m-chloroperbenzoic acid afforded anhydro compound 24. Splitting of the oxirane ring
in 24 followed by catalytic hydrogenolysis of the benzyl groups yielded methyl α-DL-glu-
co-2-heptulopyranoside (Ref. 17). Its optically active form was obtained previously by the
action of the bacterium on perseitol (Ref. 18), naturally occuring heptitol. Other 2-keto-
heptoses could be easily synthesized by appropriate functionalization of diulose 21.

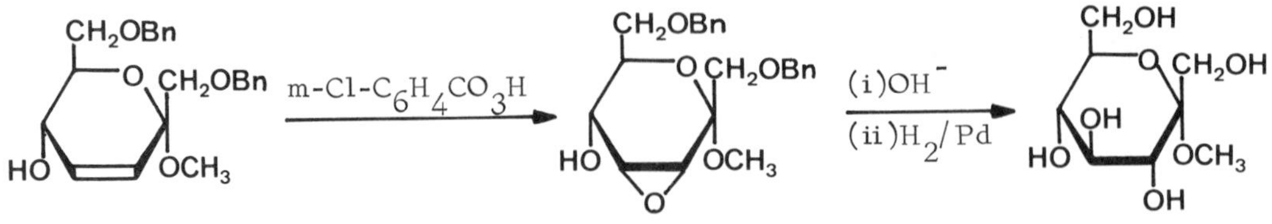

SYNTHESIS OF HIGHER-CARBON SUGARS FROM FURAN DERIVATIVES

Synthesis of the 2-ketoheptose did not require relating the configuration of the pyranose
ring with the chiral center(s) in the side chain. This problem could also be approached by
the present method. The transformation of the furan derivative into monosaccharide carried
out by us is summarized in the following equation:

The carbinol carbon atom in 26 becomes C-5 in the sugar 27 and in the case of enantiomerically pure 26 it defines configurational series D or L of the latter. Starting with furan
derivatives with 3, 4, etc. carbon side-chain one may expect to obtain heptoses, octoses,
etc., i.e. higher-carbon sugars. Moreover, the relative configuration of the pyranose
ring and that of C-6 in, for example, sugar 29 follows from the configuration of the side-
chain chiral centers in 28 and could be chosen at the first step of the synthesis.

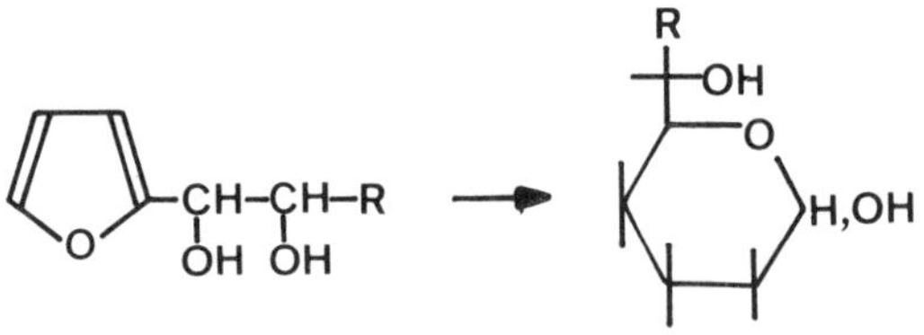

A prerequisite for the success of this synthetic scheme of higher-carbon sugars was the
retention of the configuration of the carbinol carbon atom in the furan compound in the
course of transformations leading to the pyran system. From the outset it was not certain
whether this condition could be met. A possibility of epimerization of the carbinol center
at the stage of hydrolysis of the dihydrofuran derivative 30 was indicated by the isolation
(3%) of the hydroxyketone 34 as a side product (Ref. 19). Its formation could be rationalized as cyclization and aromatization of the tautomer 33 of the ketoaldehyde 31, an intermediate in the formation of the ulose 32. Similar side-products were noticed during the
hydrolysis of other dihydrofurans (Ref. 19).

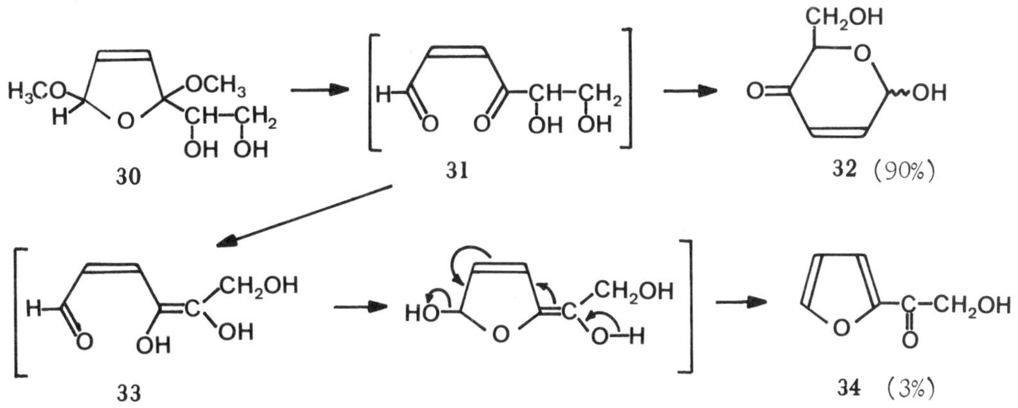
30
31
32 (90%)
33
34 (3%)

This crucial question was settled by preparation of the optically active methyl 2,3-dideoxy-α- and β-hex-2-enopyranosid-4-uloses (36 and 37) from methyl (R)-(-)-(2-furyl)glycolate 35. A comparison of their optical rotations (in the case of α-anomer for the 6-benzoate) with the literature data for methyl 6-O-benzoyl-α-D-glycero-hex-2-enopyranosid-4-ulose (Ref. 20) and methyl β-D-glycero-hex-2-enopyranosid-4-ulose (Ref. 21) obtained from D-glucose, demonstrated an enantiomeric relation of the respective compounds and indicated unchanged optical purity of the synthetic samples in relation to that of the starting material (Ref. 22).

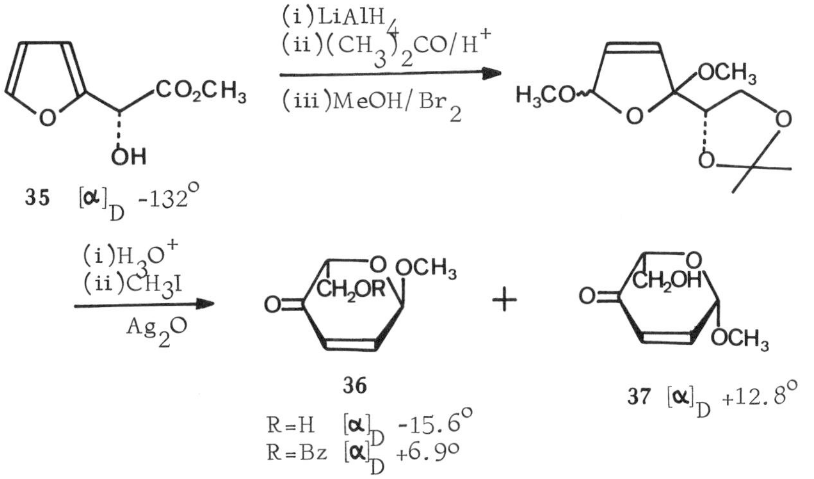
(i) $LiAlH_4$
(ii) $(CH_3)_2CO/H^+$
(iii) $MeOH/Br_2$
35 $[\alpha]_D$ -132°
(i) H_3O^+
(ii) CH_3I
Ag_2O
36
R=H $[\alpha]_D$ -15.6°
R=Bz $[\alpha]_D$ +6.9°
37 $[\alpha]_D$ +12.8°

Therefore, it could be concluded that the configuration of the carbinol asymmetric center in the starting ester 35 was not altered during the foregoing transformations. This point was clearly demonstrated by completion of the synthesis of methyl α-L- and α-D-glucopyranosides from methyl (R)-(-) and (S)-(+)-(2-furyl)glycolates, respectively, without any loss of optical activity (Ref. 22).

FURAN DERIVATIVES

For the synthesis of 8-deoxyoctoses or 8-deoxyaminooctoses and 2-ketononoses to which we have turned our attention, since amongst these types of sugars appear biologically active natural compounds (e.g. antibiotic sugar lincosamine, neuraminic acid, ulosonic acids) furan derivatives of the general formulas 38, 39 and 40 were required.

38 39

$$HOH_2C-\text{[furan]}-CH-CH-CH-CH_2$$
$$\phantom{HOH_2C-\text{[furan]}-}OH\ \ OH\ \ OH\ \ OH$$

40

These compounds should posess definite configuration and, preferably, be available in their optically active state. Therefore, for their preparation we have utilized as substrates readily accessible optically active hydroxyacids and the known reaction of lithium salts of carboxylic acids with organolithium compounds yielding respective ketones (Ref. 23).

(1R,2S,3R) and (1S,2S,3R)-1-(2-furyl)-1,2,3-butantriol

Lævorotatory (2S,3R)-2,3-dihydroxybutyric acid (41) was efficiently obtained by hydroxylation of crotonic acid and resolution of the product via brucine or quinidine salts (Ref. 24). Acylation of furyllithium with lithium (2S,3R)-2,3-O-isopropylidene-2,3-dihydroxybutyrate (42) gave an excellent yield of ketone 43. Hydride reduction of the carbonyl group in the latter yielded two epimeric (1R,2S,3R) and (1S,2S,3R)-1-(2-furyl)-1,2,3-butantriol derivatives 44 and 45, substrates for 8-deoxyoctoses synthesis.

The configuration of the new chiral center at C-1 in 44 and 45 was assigned by their degradation to the derivatives 46 and 47 of the (2-furyl)glycolic acid, the compound of the known absolute configuration (Ref. 25).

(1R,2R,3S) and (1S,2R,3S)-1-(5-hydroxymethyl-2-furyl)-1,2,3,4-butantetrol

Reaction of lithium salt of the (2R,3S)-threonic acid derivative 52 with dianion of furfuryl alcohol yielded ketone 53. Hydride reduction of the keto group in 53 gave two epimeric alcohols 54 and 55. Their (1R,2R,3S)-54 and (1S,2R,3S)-55 configuration was tentatively assigned by comparison of optical rotations with those of 45 and 44, respectively. Lithium salt 52 of the appropriately protected threonic acid used in acylation reaction was obtained from L-tartaric acid (48). A controlled hydrolysis of readily available diethyl 2,3-O-iso-propylidenetartrate (49) (Ref. 26) yielded monoester 50 in which carboxyl group was selectively reduced with sodium borohydride modified with tin tetrachloride (Ref. 27). Protection of the hydroxyl group in the resulting alcohol by treatment with dimethoxymethane in the presence of phosphorous pentoxide and hydrolysis of an ester group with 1 mole of lithium hydroxide afforded 52 (Ref. 28).

(1S,2S,3R) and (1R,2S,3R)-2-amino-1-(2-furyl)-1,3-butandiol

Hydrolysis with lithium hydroxide of the oxazoline 57 obtained in three steps from D-threo-nine 56 (Ref. 29) gave lithium salt 58 with (2S,3R) configuration, since overall transformation proceeded with one inversion at each asymmetric carbon atom (Ref. 29). Acylation of furyllithium with 58 afforded moderate yield of ketone 59. The latter, treated with lithium aluminum hydride, gave two products 60 and 61 in the ratio 3:1. They were separated by column chromatography. Their IR spectra showed that only the keto group had been reduced leaving the C=N bond unchanged. The (2S,3R) configuration in compounds 60 and 61 was evident from the method of preparation. In order to determine the configuration of the hydroxyl group at C-1 in both carbinols the oxazoline ring was cleaved by acid hydrolysis to give the N-benzoyl derivatives 62 and 63 which, in turn, were converted with 2,2-dimethoxypropane into the dioxanes 64 and 65, respectively. In the ^{1}H NMR spectrum of the iso-propylidene derivative of the major reduction product, coupling constants $J_{1,2}=1.3$ and $J_{2,3}$ =1.7 Hz were taken as evidence of the chair conformation (Ref. 30), and, consequently, its (1R,2S,3R) configuration. Therefore, the configuration of the minor product 61 must be (1S,2S,3R), consistent with predominance of the twist-boat conformation of its isopropylidene derivative 65, indicated by the coupling constants $J_{1,2}$=6.1 and $J_{2,3}$=3.7 Hz (Ref. 31).

SYNTHESIS OF 8-DEOXYOCTOSES

Application of our approach to the stereoselective synthesis of higher-carbon sugars from furan derivatives is demonstrated by the synthesis of methyl 6,7-O-isopropylidene-8--deoxy-α-D-threo-D-gluco (79) and D-galacto (82) octopyranosides.

Reaction of 2,3-O-isopropylidene-1-(2-furyl)-1,2,3-butantriol 44 and 45 with methanol and bromine at -40°C yielded dimethoxydihydro derivatives 66 and 71. Acidic hydrolysis of the dihydro compound or, even better, treatment with pyridinium p-toluenesulfonate (Ref. 32) in boiling acetone yielded uloses 67 and 72.

Uloses 67 and 72 could also be obtained directly from furan compounds 44 and 45, respectively, by oxidation with m-chloroperbenzoic acid, according to the Lefebvre procedure (Ref. 33).

Treatment of 67 with pyridinium p-toluenesulfonate in a methanol solution led to anhydro compound 68 which, in acidic medium, rearranged into isomeric derivative 69. Coupling constant $J_{5,6} \sim 0$ value in anhydrouelose 68 ^{1}H NMR spectrum confirmed the relative configuration at C-5 and C-6 (Ref. 34).

Anhydrosugars are convenient intermediates in numerous transformations of monosaccharides (Ref. 34). These obtained by us could also be utilized readily for such purposes.

Methylation of the uloses 67 and 72 with trimethyl orthoformate in the presence of boron trifluoride or with methyl iodide in the presence of silver oxide gave, in both cases, a mixture of α- and β-glycosides in a ratio approximately 2:1 (^{1}H NMR). Separation of methyl glycosides 73 and 74 and hydride reduction of carbonyl group in α-anomer 73 led to 2,3-unsaturated alcohol 75 as a major product, accompanied by isomeric alcohol 76.

The mixture of anomers 70, without any separation, was treated with lithium aluminum hydride and since hydride reduction of α-anomers proceeds stereoselectively (Ref. 35) alcohol 77 was separated, after acetylation, as the major product. Removal of the acetyl group and epoxidation with m-chloroperbenzoic acid introduced oxirane ring stereoselectively cis to the free hydroxyl group at C-4 (Ref. 36) resulting in the anhydro compound 78. Basic hydrolysis of 78 afforded methyl 6,7-O-isopropylidene-8-deoxy-α-D-threo-D-gluco-octopyranoside 79 (Ref. 31).

To achieve galacto configuration of the pyranose ring, appearing in licosamine (2), firstly an inversion of configuration at C-4 in 77 was performed. A reaction of 77 with benzoic acid in the presence of diethyl azodicarboxylate (DEAD) and triphenylphosphine (TPP) (Ref. 37) afforded smoothly benzoate 80. Epoxidation of 80 resulted in a mixture of epoxides. The major isomer was identified as the one with oxirane ring trans to the benzoyloxy group on the basis of coupling constants in the ^{1}H NMR spectrum (Ref. 38). Acidic hydrolysis of the oxirane ring in 81 and removal of benzoate group afforded methyl 6,7-O-isopropylidene-8-deoxy-α-D-threo-D-galacto-octopyranoside 82 (Ref. 31).

Syntheses aiming at other sugars with the use of furan derivatives, the preparation of which is described in this report, are in progress at present. It is worth mentioning that this method could be regarded as a procedure of four or five carbon extention of monosaccharide chain when suitably protected aldehydosugars instead of lithium salts of carboxylic acids are used in their reactions with furyllithium.

ACKNOWLEDGEMENTS

I am indebted to my coworkers, whose names are given in the references, for their contributions to the work described above. In particular I wish to thank Dr. Barbara Szechner for her contribution which have not yet been published.
The Polish Academy of Sciences is thanked for providing financial assistance (Grant MR--I-12).

REFERENCES

1. J.M.Webber, Adv. Carbohydr. Chem. 17, 15-63 (1962).
2. N.Sharon, Complex Carbohydrates, Addison-Wesley Publishing Company, Inc., London (1975).
3. L.Benzing-Nguyen and M.B.Perry, J.Org.Chem. 43, 551-554 (1978).
4. W.Schroeder, B.Banister and H.Hoeksema, J.Am.Chem.Soc. 89, 4448-4453 (1967).
5. R.L.Hamil and M.M.Hoehn, J.Antibiot.,Ser.A 17, 100-103 (1964).
6. Ref. 2, p. 13.
7. H.Paulsen, K.Roden, V.Sinnwell and W.Koebernick, Angew.Chem. 88, 477 (1976).
8. J.A.Secrist III and Shang-Ren Wu, J.Org.Chem. 44, 1434-1438 (1979).
9. T.Kunieda and T.Takizawa, Heterocycles 8, 661-694 (1977).
10. O.Achmatowicz Jr. and P.Bukowski, Can.J.Chem. 53, 2524-2529 (1975).
11. O.Achmatowicz Jr. and B.Szechner, Rocz.Chem. 49, 1715-1724 (1975); ibid 50, 729-736 (1976).
12. O.Achmatowicz Jr., R.Bielski, P.Bukowski, Rocz.Chem. 50, 1535-1543 (1976).
13. O.Achmatowicz Jr. and G.Grynkiewicz, Rocz.Chem. 50, 719-728 (1976).
14. O.Achmatowicz Jr. and B.Szechner, Bull.Acad.Polon.Sci.,Sér.sci.chim. 19, 309-311 (1971).
15. B.Szechner, unpublished results.
16. O.Achmatowicz Jr., G.Grynkiewicz and B.Szechner, Tetrahedron 32, 1051-1054 (1976).
17. O.Achmatowicz Jr. and M.H.Burzyńska, unpublished results.
18. G.Bertrand and G.Nitzberg, Compt.rend. 186, 1172-1175 (1928).
19. O.Achmatowicz Jr., P.Bukowski, B.Szechner, Z.Zwierzchowska and A.Zamojski, Tetrahedron 27, 1973-1996 (1971).
20. B.Fraser-Reid, A.McLean and E.W.Usherwood, J.Am.Chem.Soc. 91, 5392-5394 (1969).
21. B.T.Lawton, W.A.Szarek and J.K.N.Jones, Chem.Commun. 787-788 (1969); Carbohydr.Res. 15, 397-402 (1970).
22. O.Achmatowicz Jr. and R.Bielski, Carbohydr.Res. 55, 165-176 (1977).

23. M.J.Jorgensen, Org. React. 18, 1-97 (1970).
24. F.W.Bachelor and G.A.Miana, Can. J. Chem. 47, 4089-4091 (1969).
25. O.Achmatowicz Jr. and P.Bukowski, Bull. Acad. Polon. Sci., sér. sci. chim., 19, 305-308 (1971).
26. P.W.Feit, J. Med. Chem. 7, 14-17 (1964).
27. S.Kano, Y.Yuasa and S.Shiboya, Chem. Commun. 796 (1979).
28. O.Achmatowicz Jr. and A.Sadownik, unpublished results.
29. D.F.Elliot, J.Chem. Soc. 62-68 (1950).
30. K.Pihlaja and L.Luoma, Acta Chem. Scand. 22, 2401-2414 (1968); K.Pihlaja, G.M. Kellie and F.G.Riddell, J. Chem. Soc., Perkin II, 252-256 (1972); U.Burkert, Tetrahedron 35, 691-695 (1979).
31. B.Szechner, unpublished results.
32. N.Miyashita, A.Yoshikoshi and P.A.Grieco, J. Org. Chem. 42, 3772-3774 (1977).
33. Y.Lefebvre, Tetrahedron Lett. 133-136 (1972); R.Laliberté, G.Medawar and Y.Lefebvre, J. Med. Chem. 16, 1084-1089 (1973).
34. M.Černy and J.Staněk Jr., Adv. Carbohydr. Chem. Biochem. 34, 23-177 (1977).
35. O.Achmatowicz Jr. and P.Bukowski, Rocz. Chem. 47, 99-114 (1973).
36. J.G.Buchanan and H.Z.Sable, in B.S.Thyagarajan (Ed.) Selective Organic Transformations, Vol.2, p. 1-95, Wiley, New York (1972).
37. G.Grynkiewicz and H.M.Burzyńska, Tetrahedron, 32, 2109-2111 (1976).
38. O.Achmatowicz Jr. and B.Szechner, Carbohydr. Res. 50, 22-33 (1976).

CARBOHYDRATE DERIVATIVES IN THE ASYMMETRIC SYNTHESIS OF NATURAL PRODUCTS

B. Fraser-Reid*, Tim Fat Tam and King Mo Sun

*Guelph-Waterloo Centre for Graduate Work in Chemistry,
University of Waterloo, Waterloo, Ontario, Canada N2L 3G1*

In 1975[1] we suggested that carbohydrate derivatives possessed four attributes which made them particularly appealing as chiral synthons: (a) enantiomeric purity, (b) conformational bias, (c) ready proof of structure, and (d) versatile latent functionalities. The large number and variety of substances whose syntheses have been accomplished[2] in the interim exemplify these attributes fully, and demonstrate the surprising versatility of this hitherto unappreciated source of starting materials[3,4]. Of course any optically active substance is a chiral synthon (of sorts!), but we have suggested that *for the synthetic organic chemist the primary appeal might very well be the high degree of stereoselectivity observed in carbohydrate reactions*[2]. This results directly from their conformational properties which offer attractive opportunities for stereo−electronic control of reactions.

Three striking examples of this are shown in Scheme I where compounds 1[5], 3[6] and 7[1] were all formed stereospecifically in the reactions indicated. Compounds 1 and 3 may be described as *annulated pyranosides*[2], and the potential of such synthons was demonstrated by the transformation of 1 into *either* the (+) or the (−) enantiomer of *trans* chrysanthemic dicarboxylic acid (2)[5].

SCHEME I

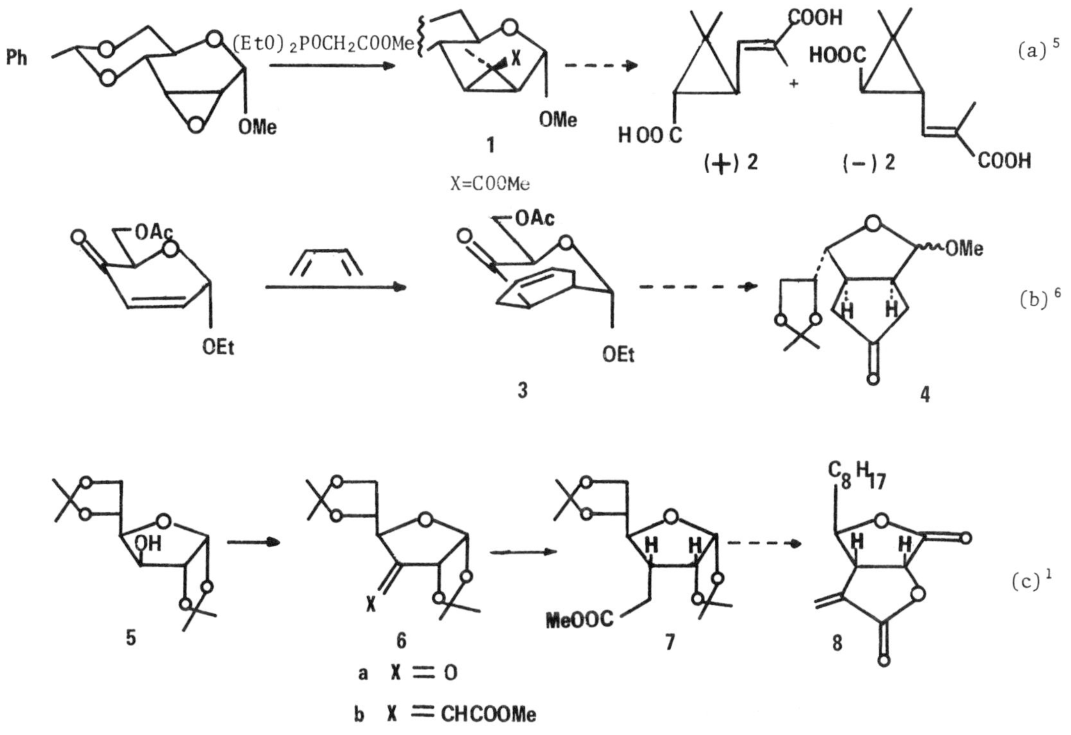

X=COOMe

(a)[5]

(b)[6]

(c)[1]

a   X = O

b   X = CHCOOMe

---

*Please address correspondence to this author at Chemistry Department, University of Maryland, College Park, Md., U.S.A. 20742.

SCHEME II

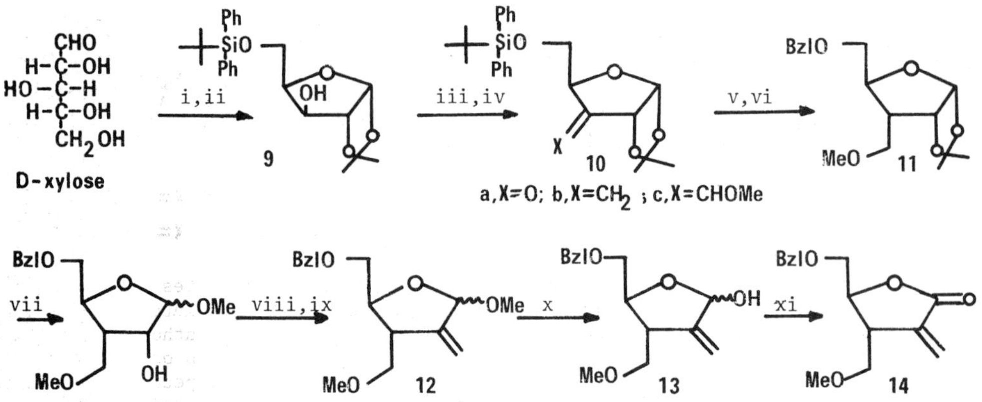

$$a, X=O; \quad b, X=CH_2 ; \quad c, X=CHOMe$$

(i) Me_2CO/H^+ ; (ii) t-BDPSiCl ; (iii) Collin's oxidation ; (iv) $Ph_3P=CHOMe$;
(v) H_2 ; (vi) F^-, $PhCH_2Br$; (vii) $MeOH/HCl$; (viii) PCC ; (ix) $Ph_3P=CH_2$;
(x) $DME/H_2O/reflux$; (xi) Fetizon's oxidation

Furanose derivatives have been used less frequently[2,4] although Woodward[7], and Yoshimura[8] have used diacetone glucose 5 for their approaches to tetrodotoxin. Furanose derivatives are frequently more accessible than pyranosides and are very versatile as may be judged by the conversion of 5 into natural avenaciolide 8[1] as well as its enantiomer.

Avenaciolide 8 contains an α-methylene-γ-butyrolactone, a structural feature which occurs in a wide variety of sesquiterpenes; but, as is obvious from Scheme I, this unit was not developed from the carbohydrate nucleus. However such a transformation ought to be possible and so we initiated the experiments shown in Scheme II[18].

SCHEME III

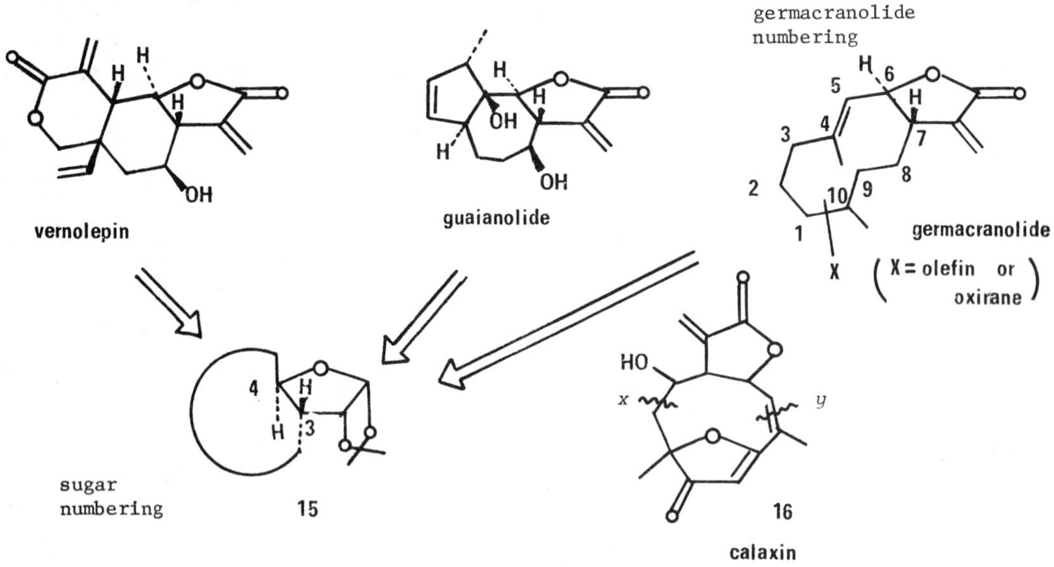

Xylose reacts with acetone to give the monoketal 9 from which the alkene 10b is obtained by standard reactions. However, hydroboration attacked the acetonide ring, so our route to 11 was via the vinyl ether 10c. Compound 12 was obtained readily but it proved to be exceedingly sensitive even to the most dilute acetic acid solutions. Fortunately we had found[5] that highly activated glycosides can be hydrolysed by boiling with aqueous glyme. The conversion 12→13→14 then proceeded smoothly[10].

The procedure in Scheme II shows that 1,2-O-isopropylidene furanoses may be considered synthons for α-methylene-γ-butyrolactones. Thus structural types as varied as the germacranolides, guaianolides and elemanolides (Scheme III) could conceivably be obtained from appropriate "annulated furanoses" such as 15.

SCHEME IV

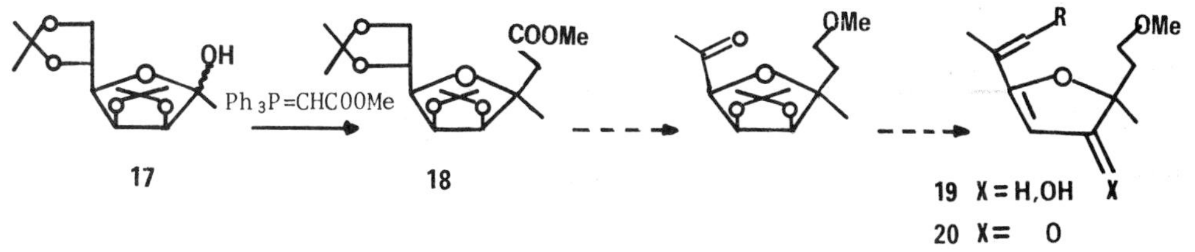

17   18   19 X = H,OH X
                                                                                20 X= O

Perusal of the excellent survey by Fischer, Oliver and Fischer[11] indicates that in the majority of sesquiterpenes, the lactone ring is trans-fused and has the absolute stereochemistry
shown in Scheme III. Thus the stereochemistries at C3 and C4 of the synthon 15 are cognate
with those at C7 and C6, respectively, of the sesquiterpene. Inspection of structures 7 and
11 show that the C3/C4 arrangement required in 15 is obtainable stereospecifically. *Thus the
readily prepared acetonides of glucose (5) and xylose (9) may be considered plausible synthons for the naturally occurring enantiomers of the majority of sesquiterpene lactones!*

Prior to discussing our current work, we will note some of our earlier studies relating to
the germacranolides. Structure 16 is representative of several germacranolides which may be
considered as derivable from two furanose species linked through the bonds x and y. A synthesis of the "upper" portion of 16 could be modelled after the work in Scheme II. Studies
designed to familiarise ourselves with the idiosyncracies of the highly functionalised
"lower" portion were undertaken. One of these proved to be the extreme lability of the alcohol 19 towards oxidation; only Fetizon's reagent afforded ketone 20. The full study has been
published[12] and so only fragments are shown in Scheme IV.

The success of the Wittig reaction giving 18 suggested that the two "furanoid" rings of 16
could be obtained from the condensation product (23) of isopropylidene ribose, 21, and the
Wittig reagent 22, Scheme V. (The numbering in 16 and 23 are comparable.) Accordingly 22
has been prepared as indicated in equation (b) from the readily obtained precursor 24[13]. Of
cautionary note is the oxidation of benzyl to benzoate. In view of the stringent requirements[12] for the Wittig reaction of 17, we have first tested 22 with the simple aldehyde 27,
and find that the condensation proceeds smoothly giving the enone 28[14].

SCHEME V

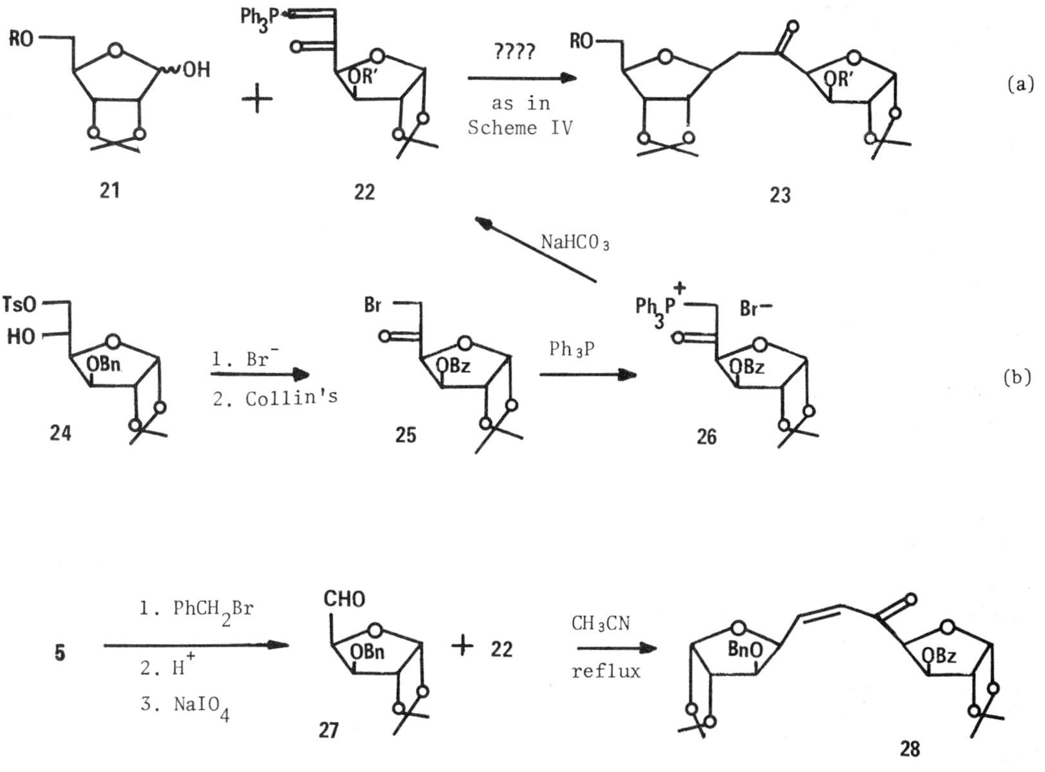

21   22   23   (a)

24   25   26   (b)

27   + 22   28

SCHEME VI

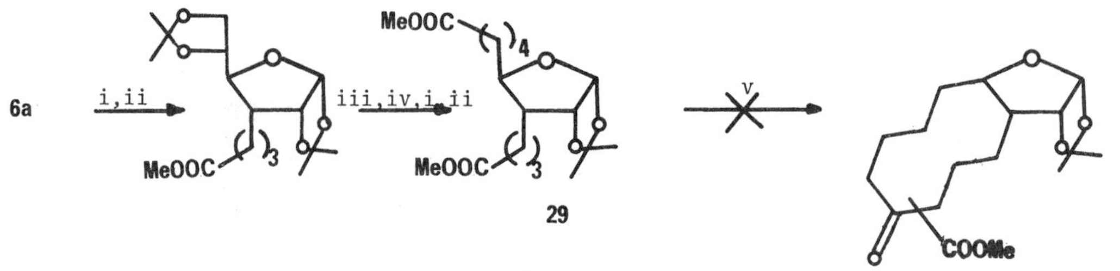

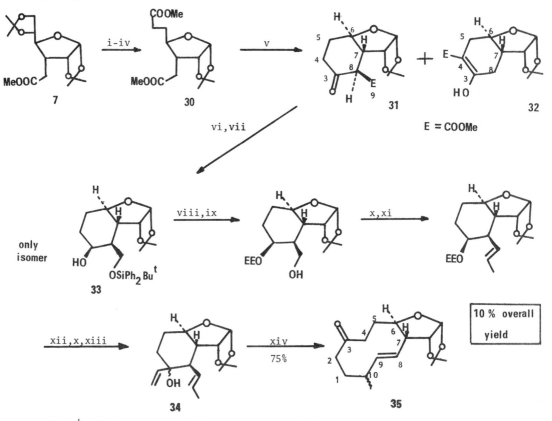

(i) Ph$_3$PCH-CH=CH-CO$_2$Me ; (ii) H$_2$; (iii) H$^+$; (iv) NaIO$_4$; (v) t-BuOK

But the tricyclic germacranolides such as 16 could conceivably be derived by intramolecular ring closure of the bicyclic analogues (Scheme III)[11]. In this case there is the daunting challenge of making a 10-membered ring. The Dieckmann cyclisation is usually futile here, although the presence of a double bond or a ring sometimes helps[15]. However with the diester 29 (Scheme VI), no cyclisation whatever was observed after seven days with potassium t-butoxide in refluxing benzene.

The Cope rearrangement of a suitable divinyl cyclohexane is a well-known route to 10-membered rings and the Evans modification[16] has been featured recently in the work of Still[17]. We therefore set out to see whether this procedure could be carried out with annulated furanoses.

The diester 30 was prepared readily from 7, and treatment with potassium t-butoxide gave an excellent yield of 31 and 32 in the ratio 10:1. Surprisingly the preferred site of deprotonation was C8 rather than C4 in spite of the proximity to the bulky acetonide residue. Notably the minor component, 32, exists entirely in the enolic form.

SCHEME VII

(i) H$^+$; (ii) NaIO$_4$; (iii) Ph$_3$P-CHCO$_2$Me ; (iv) H$_2$; (v) t-BuOK; (vi) LAH ; (vii) t-BDPSiCl ; (viii) EtOCH=CH$_2$; (ix) F$^-$; (x) Collin's ; (xi) Ph$_3$P=CHCH$_3$; (xii) PPTS ; (xiii) Ph$_3$P=CH$_2$; (xiv) KH/THF/Bu$_4$NI/reflux/12 h.

Compound 31 which is a crystalline material has the methoxycarbonyl group in equatorial orientation, judging from the coupling constant J_{78} = 13.5 Hz. This substance was subjected to some transformations (Scheme VII) which are seen to go in very good to excellent yields, so that the entire sequence goes in ten percent overall yield. Notably, lithium aluminum hydride reduction gives the C3 alcohol of 33 exclusively in the *endo* (or β) epimer. With regard to the final step, rearrangement did not occur in the absence of tetra-N-butyl-ammonium iodide; and the use of 18-crown -6[16] led to a messy product.

The formation of 35 was satisfying since it showed that annulated furanoses might be feasible synthons for germacranolides. However, as indicated in Scheme III methyl, and oxirane or olefinic groups are usually present at C4 and C1 respectively. Hence, some capability for such structural types must be built into the synthons. With regard to the C4 methyl group, compound 32 would be a better prospect than 31; however, since it proved to be the minor product of the Dieckmann reaction an alternative approach was required.

SCHEME   VIII

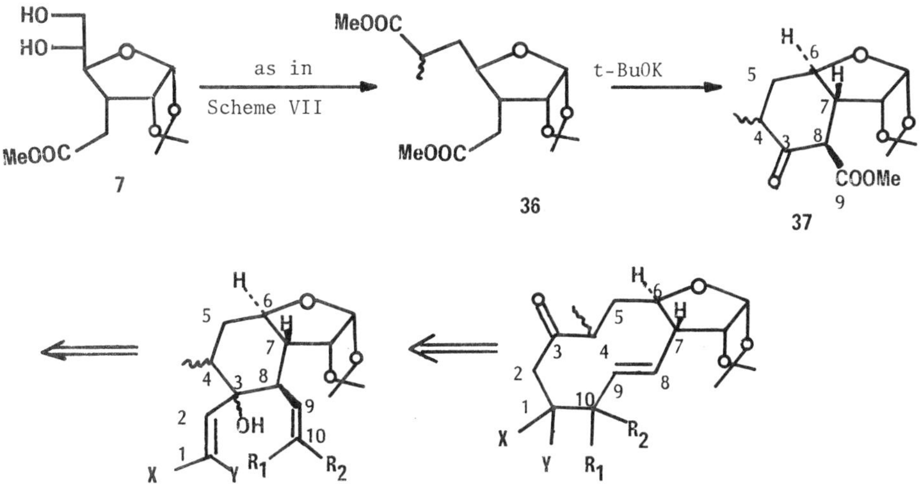

The α-methyl diester 36 (Scheme VIII) underwent cyclisation smoothly, giving 37 which has the C4 methyl in place. We are currently working at preparing 38 in which the stereochemistries of X, Y, R' and R" are being introduced specifically so as to be able to control the stereochemistries of X, Y, R' and R" in 39.

The annulated furanoses considered so far are synthons for sesquiterpenes in which the lactone is *trans*-fused (Scheme III). However, as seen in Scheme IX, the lactone ring may also be *cis*-fused and a suitable synthon is therefore required. A Diels-Alder condensation was an interesting option, particularly since it had served us well in the synthesis of 3. Now, however, the roles would be reversed, with the diene being the chiral partner.

SCHEME   IX

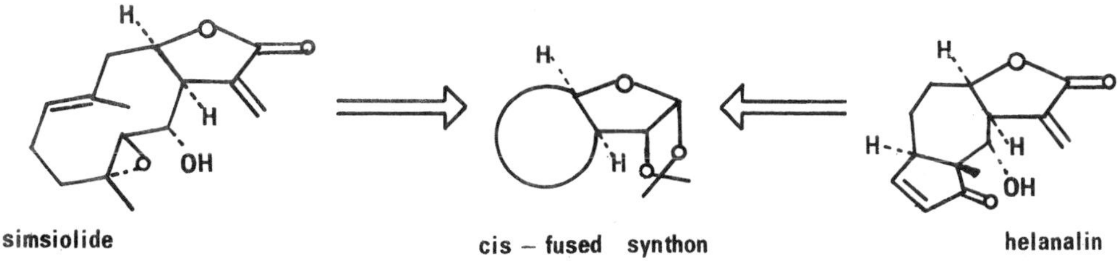

simsiolide   cis – fused   synthon   helanalin

The aldehyde-mesylate 40 (Scheme X) which is readily obtained by standard procedures[2] undergoes β-elimination quantitatively upon refluxing with pyridine; and a Wittig reaction leads to diene 42. Condensation with maleic anhydride is facile and the product 43 is formed in 80 percent yield as the single diastereomer shown. Predictably, the bulky acetonide in 42 forced addition from the β-face. Dreiding models of 43 confirm that the projected angle

SCHEME   X

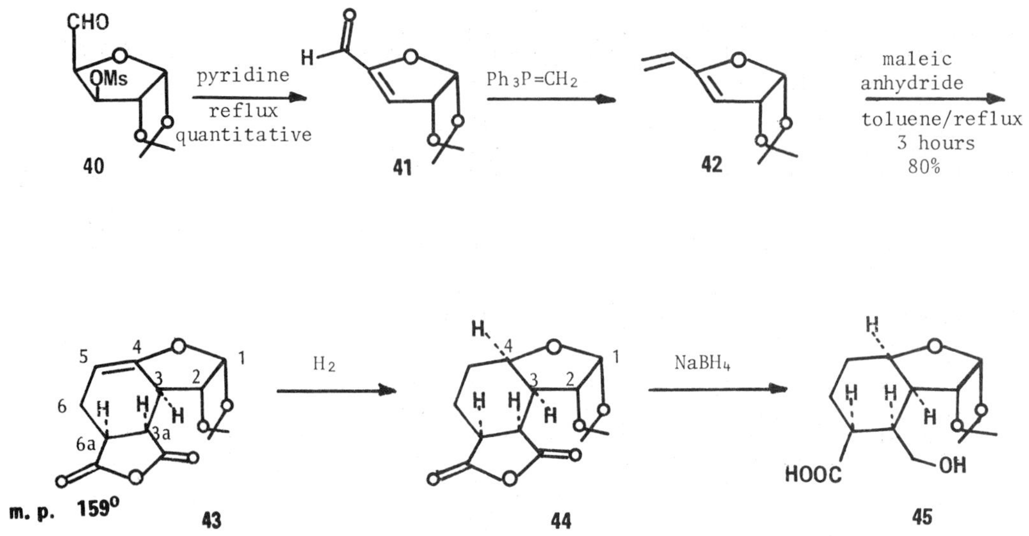

between H2 and H3 is ~90° as bourne out by the value $J_{23} = 0$. The observed *exclusive endo*-addition was unexpected; but the reality is apparent from a comparison of the coupling constants for H3a of 43 and H8 of 31 ($J_{3,3a} = 5.8$ and $J_{7,8} = 13.5$ respectively) indicating that H3 and H3a (and hence H6a) are *cis*-related in 43.

The next step involved reduction of the double bond in 43 which we hoped would occur from the α-face. Models showed that the α-approach from the convex exterior of 43 is favored, the bulky acetonide notwithstanding. Indeed, hydrogenation was quantitative and stereospecific giving 44, as the coupling constant ($J_{3,4} = 4.1$ Hz) reveals.

Compound 43 is a crystalline substance although it decomposes (retro-Diels Alder?) on heating. However, 44 melts cleanly as indicated. The potential of 44 is enhanced by the finding that reduction with sodium borohydride occurs at the more hindered site leading to the hydroxy-acid 45. This substance lactonises very slowly owing, undoubtedly, to the location of the functionalities on the concave face of the assembly.

SCHEME   XI

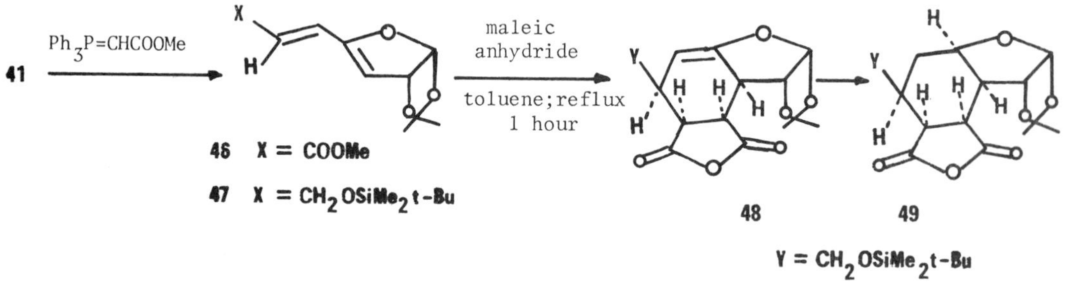

The synthon 44 could be furnished with an even greater selection of functional implements by taking advantage of stereochemical modes of addition to the aldehyde 41, Scheme XI. Thus the Wittig addition indicated gave the dienic ester 46 as the *exclusive* product, and careful reduction with lithium aluminium hydride led to the alcohol which was protected, as 47. Diels-Alder addition to the latter was complete in one-third the time of 42, again affording a single substance assigned as 48 in keeping with the previously encountered *endo*-addition. As before hydrogenation is stereospecific leading to the cis ring junction in 49.

The knowledge gained in the preparation of the foregoing structures is being utilised for a variety of synthetic objectives which will be published in due course.

ACKNOWLEDGEMENT

We express enormous gratitude to the following agencies for financial support of our work: the National Science and Engineering Research Council of Canada, and in the U.S.A., the National Institutes of Health, Merck, Sharp and Dohme and the Upjohn Company.

REFERENCES

1. R.C. Anderson and B. Fraser-Reid, J. Am. Chem. Soc., 97, 3870 (1975).
2. B. Fraser-Reid and R.C. Anderson, Progress Chem. Org. Natural Products, 39, 1 (1980).
3. B. Fraser-Reid, Accts. Chem. Res., 8, 195 (1975).
4. S. Hanessian, ibid., 12, 159 (1979).
5. B.J. Fitzsimmons, J. Am. Chem. Soc., 101, 6123 (1979).
6. A. Rosenthal and L.B. Nguyen, Tetrahedron Lett., 2393 (1967).
7. Harvard Theses of work supervised by R.B. Woodward: R.D. Sitrin, 1972, and J. Upeslacis, (1975).
8. J.K. Yoshimura, K. Kobayashi, K. Sato and M. Funabashi, Bull. Soc. Chem. Japan, 45, 1806 (1972); M. Funabashi, J. Wakai, K. Sato and J. Yoshimura, Abs. 9th Intl. Symp. Carbohydrate Chem. SD-5, 1976.
9. H. Ohrui and S. Emoto, Tetrahedron Lett., 3657 (1975).
10. T.F. Tam and B. Fraser-Reid, Chem. Commun., in press.
11. N.H. Fischer, E.J. Oliver and H.D. Fischer, Progress Chem. Org. Natural Products, 38, 47 (1979).
12. T.F. Tam and B. Fraser-Reid, J. Org. Chem., 45, 1344 (1980).
13. O.T. Schmidt, Methods in Carbohyd. Chem., 2, 318 (1963).
14. For some other examples of reactions of carbohydrate-derived Wittig reagents with aldehydo-sugars, see: Yu. A. Zhdanov and V.A. Polenov, Carbohydrate Res., 16, 465 (1971); J.A. Secrist III and S-R. Wu, J. Org. Chem., 44, 1434 (1979).
15. C. Illuminate, L. Mandolini and B. Masci, J. Am. Chem. Soc., 97, 4960 (1975).
16. D.A. Evans and A.M. Goleb, ibid., 97, 4765 (1975).
17. W.C. Still, ibid., 99, 4186 (1977); 2493 (1979).
18. After our work had been completed a publication in this area appeared: V. Nair and A.K. Sinhababu, J. Org. Chem., 45, 1893 (1980).

A PRACTICABLE SYNTHESIS
OF THIENAMYCIN

S. H. Pines

Merck, Sharp & Dohme Research Laboratories Division of Merck & Co., Inc.,
Rahway, New Jersey 07065, USA

Abstract - The development of a practicable, stereoselective process to
thienamycin from methyl acetonedicarboxylate will be described. Chirality
is introduced by a unique resolution at an early stage.

Thienamycin is a metabolite of *Streptomyces cattleya* with high antibacterial activity, both
gram positive and gram negative. Since its disclosure in 1976 (1,2), it has been the target
of considerable interest in several laboratories. It possesses no amide function at C_6, but
rather an hydroxyethyl group bound directly to the lactam ring. The structure and absolute
configuration (Fig. 1) were published in 1978 (3). In that paper, one can see some of the
many decompositions this tender molecule suffers when mistreated. Several of its mass
spectral fragmentations have been duplicated in solution.

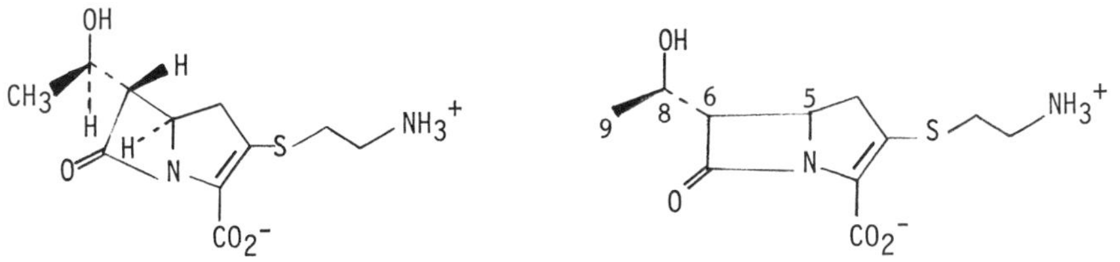

Fig. 1 Thienamycin
(5R,6S,8R)

When this work started, the first route to dl-thienamycin had been published (4), and a
chiral synthesis was on the drawing boards in our Basic Research Laboratories (5). Indeed,
both paths were destined to be useful for preparing analogs, the usual reason for synthesizing a natural product. Fermentation is almost invariably used for the manufacture of
complex natural products when it is applicable; nevertheless, it was deemed important to
develop a synthesis for thienamycin which might ultimately serve that purpose.

Our actual assignment was to develop a practical synthesis not of the parent, necessarily,
but the N-formimide of thienamycin which had been chosen as a clinical candidate (6).
There were a number of self-imposed, pragmatic constraints. Ideally, the chemistry should
be readily translatable to our pilot plant facilities; the starting materials should be
available in quantity; toxic and noxious reagents should be avoided if at all possible; and,
paramount, the processes must be safe.

The proposed strategy was guided by two features; namely, the three contiguous chiral
carbons, C_5, C_6, and C_8, and the known limited stability of the bicyclic system. The
initial goal was to produce stereoselectively an hydroxyethyl azetidinone acetic acid which
could be elaborated to the α-diazo-β-keto ester 7 (Fig. 2). At that point, the efficient

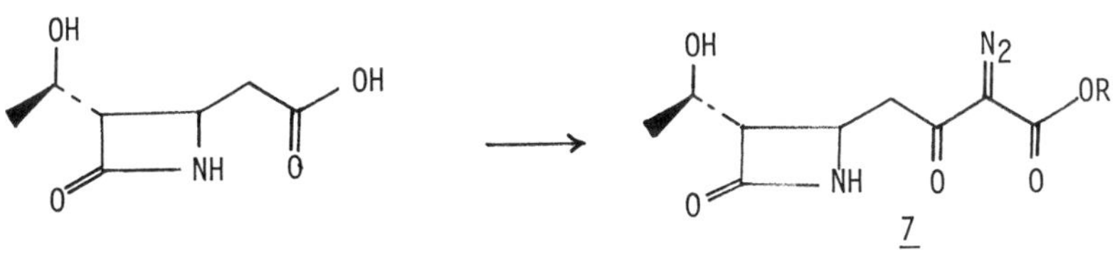

Fig. 2

annelation procedure developed by Salzmann, Ratcliffe, and Christensen, first reported for
the fused [4,6] bicyclic system (7) (Fig. 3), then used for thienamycin (5), could be
applied. (*Vide infra*, Fig. 4, 7 → 8.)

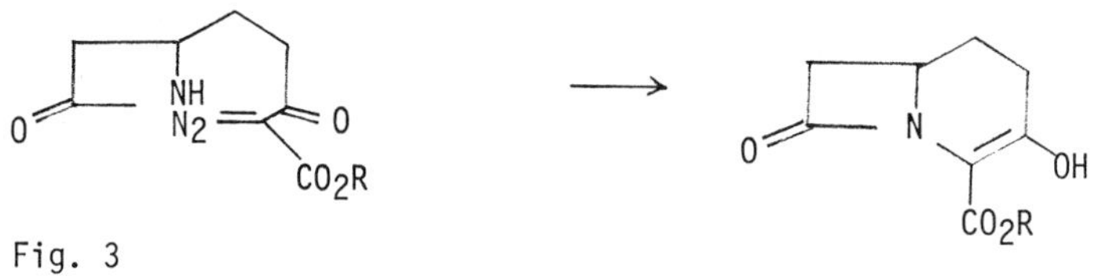

Fig. 3

Use of this rhodium acetate-catalyzed carbenoid insertion would allow most of the structural - and all of the stereo - chemistry to be achieved prior to ring fusion. In such a way, problems of molecular strain would be minimized until the last few steps. The proper attachment of a protected cysteamine moiety to a similar five-membered cyclic β-keto ester had been accomplished (9); the current task would, however, include developing that chemistry to a high yield, smoothly operable sequence.

A short review of the chiral route to thienamycin which was under study at this time (and has since been completed) is appropriate (5) (Fig. 4).

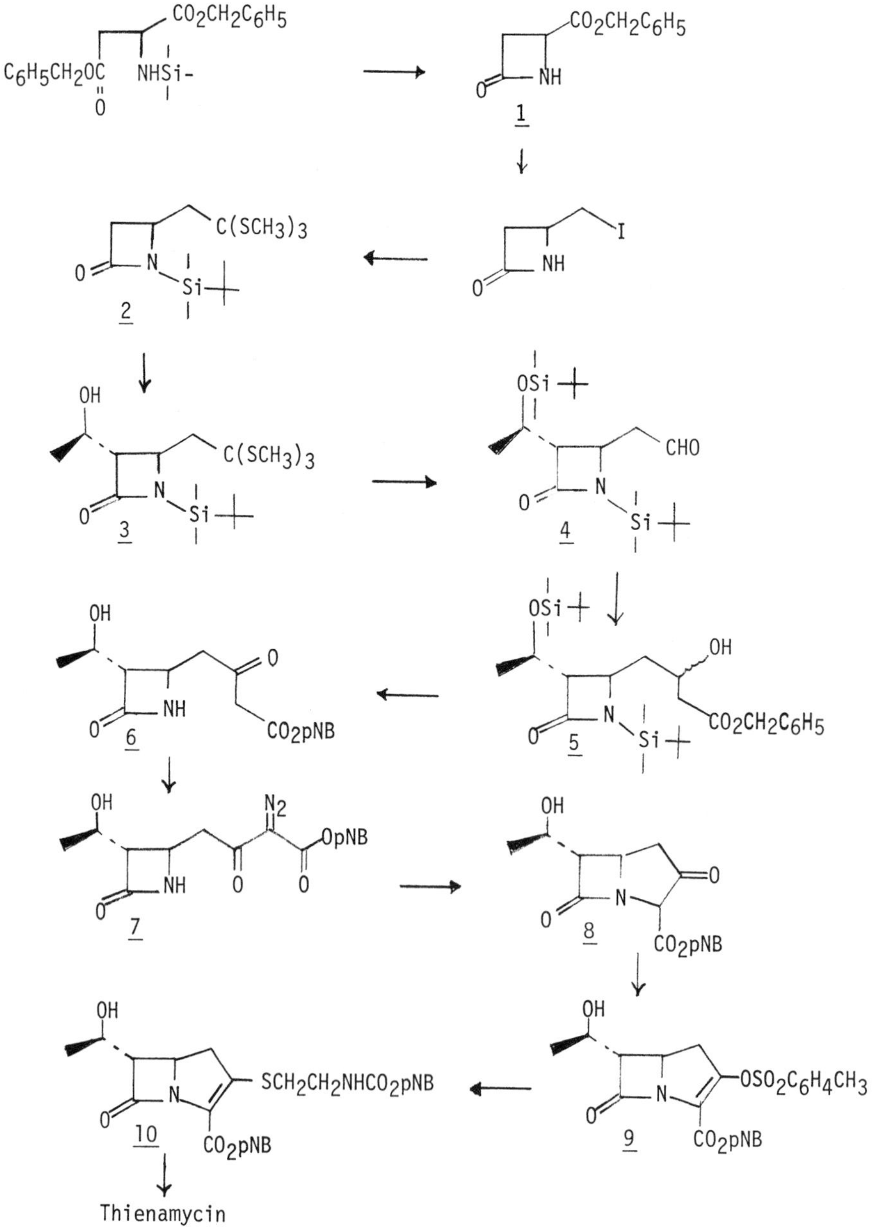

Fig. 4

N-silylated dibenzyl aspartate was treated with t-butyl magnesium chloride, a method which owes its origin to Breckpot in 1923 (10), to give chiral lactam 1. Borohydride reduction, mesylation, conversion to the iodomethyl lactam followed by silylation of the amide nitrogen, and then alkylation of trimethyl orthothioformate (11) gave 2. The hydroxyethyl group could be introduced by aldolization of the enolate at C-3, or with better stereoselectivity by the two-step method: acylation with acetylimidazole followed by potassium Selectride® reduction. Intermediate 3 was converted to the methyl ester by treatment with mercuric chloride in methanol containing p-toluene sulfonic acid, silylated at the hydroxyl, then reduced at -78° with diisobutyl aluminum hydride to form the aldehyde 4, which was condensed with lithiobenzyl acetate to form hydroxy ester 5. Since the p-nitrobenzyl ester was known to be more easily removed later, hydrogenolysis and alkylation with p-nitrobenzyl bromide preceded oxidation of the carbinol to the keto ester, then removal of silyl groups gave 6, an intermediate where both processes would converge. The steps common to both routes are discussed at their appropriate point in the sequence.

Simple, symmetrical acetonedicarboxylic acid ester was chosen as starting material for this work (Fig. 5). It was converted to its N-benzyl enamine, 11, quantitatively by stirring

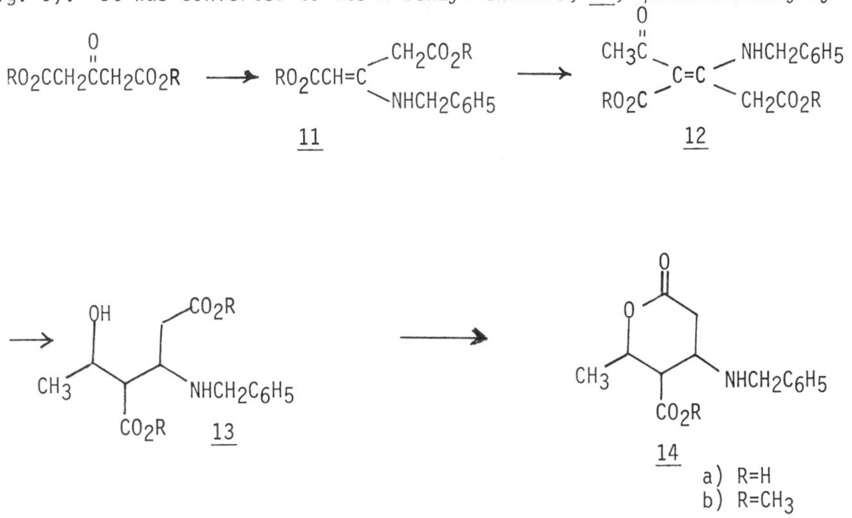

Fig. 5

the two reactants in toluene with sodium sulfate or molecular sieves. Treatment of 11 with ketene or, more simply, heating with acetic anhydride gave the C-acyl enamine, 12, which could be crystallized from aqueous alcohol. Borohydride reduction gave, mainly, one isomer which was crystallized pure from the mixture as its lactone 14. Proper choice of cyclization conditions permitted isolation of the lactone as its ester or its carboxylic acid (hydrochloride).

The expected stereochemistry of 14 could not be immediately confirmed by its NMR. Without a firm knowledge of its conformation, reliance upon the coupling constants of the ring protons was risky at best. The coupling constants, in fact, were not in accord with the most reasonable predictions, based on standard conformations. The best approach to structure proof was to convert the lactone to a β-lactam; a move that paralleled the ultimate process.

When the lactone acid 14a was allowed to stand in methanol, solvolysis proceeded giving acyclic ester 15, a β-amino acid derivative, which was cyclized with dicyclohexylcarbodiimide to provide the N-benzyl β-lactam, 15a (Fig. 6).

There are several methods which have been used to cyclize β-amino acids and their derivatives. As mentioned earlier, a Grignard reagent was useful for the aspartic acid route to thienamycin. Woodward, in his approach to cephalosporins, used triisobutylaluminum (12). Activating reagents such as thionyl chloride, phosphorous oxychloride, and acetic anhydride, etc., have a long history of application (13).

John Sheehan first used carbodiimides to form β-lactams in 1959 (14), and obtained a 25% yield of the 6-tritylaminopenicillin (methyl ester). That precedent seems to have inspired few researchers, and those reagents are rarely used for such syntheses. In fact, a 1978 review states that . . ."the effectiveness of carbodiimides and other peptide-forming reagents, has not been explored for the construction of monocyclic β-lactam rings" (15). Isaacs ignored the reagent completely (16); however, he noted the limitations of acetic anhydride, thionyl chloride, alanes, etc., in the presence of functional groups such as hydroxyl, carbonyl, and the like, which are vital to most complex syntheses.

Continuing with the case at hand, the ^{1}H NMR spectra of β-lactam 15a could be compared with those of analogs available in our laboratories (17). The low coupling constant, $J_{5,6}$ for 15a (Fig. 6) strongly supported a trans configuration, especially since it remained low

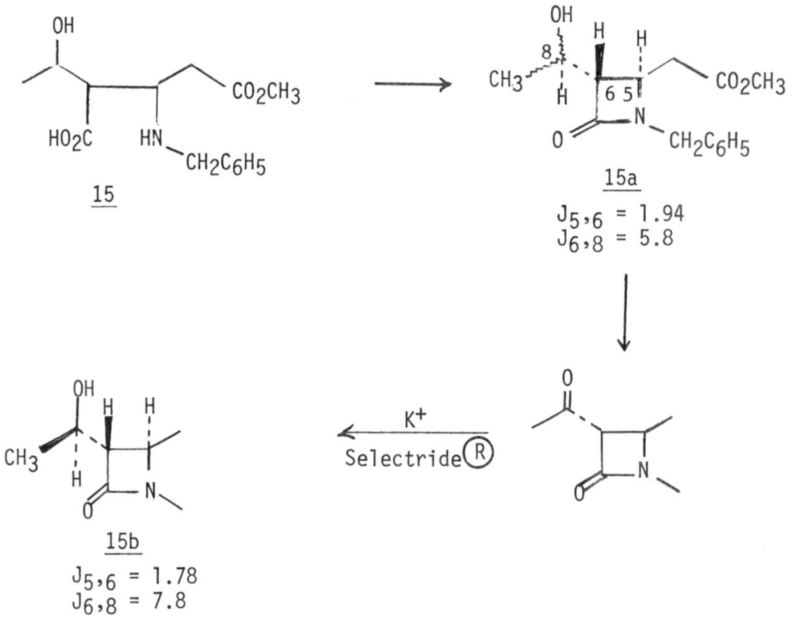

Fig. 6*   *Position numbers as in thienamycin.

after the oxidation-reduction sequence, 15a → 15b, which was known to give the trans lactam. Cis lactams show couplings of ca. 5 Hz for $J_{5,6}$. The increase in $J_{6,8}$ to 7.8 from 5.8 for the major product of oxidation and K-Selectride® reduction paralleled an earlier Merck study, and required the conclusion that in 15a, C-8 had the wrong relative configuration. Thus, inverted chirality prevailed in lactone 14 at position 6 (18). The solution to that problem will be described, but proper strategy demands an answer to the question of resolution first. If a yield loss must be taken, it should be done as quickly as possible.

The lactone amino acid 14a whose relative stereochemistry was now in hand (Fig. 7), gave nicely crystallizable salts with 10-camphorsulfonic acid. Their optical activity was high, allowing facile process control. One minor technical problem was the small, but measurable, amount of lactone opening which occurred in protic solvents. This was minimized by attention to detail.

The final stereochemical question was answered by single crystal x-ray analysis of the hydrobromide from (+)-14a (19). By chance, that isomer which formed the least soluble salt with the d(+)-10-camphorsulfonic acid possessed the R,R,S configuration, opposite to what was needed ultimately. That enantiomer could, however, be used for continuing chemical studies, and its crude antipode, which could be purified with ℓ-(-)-10-camphorsulfonic acid, could be reserved for conversion to the desired final product.

It is worth noting at this point that while much chemistry can be accomplished studying racemates, their use frequently introduces pitfalls in process research that are not easily predictable. In particular, crystallization of a pure product from a reaction mixture can be a vastly different problem with a racemate than with an enantiomer. Examples of this were apparent in this work and, therefore, optically active material was used as soon as it was available.

It should be noted also that physical chemical examination demonstrated conclusively that the lactone racemate was a mixture, and not a compound; thus, it would be possible to design a resolution which would separate the enantiomers of 14 without resorting to diastereomer formation.

Before continuing with the route of choice, an alternative which would have utilized the chiral lactone methyl ester 14b deserves mention (Fig. 7). Careful hydrolysis gave its acyclic form which, when treated with triphenylphosphine and an azodicarboxylic ester, gave the inverted R,S,R lactone 16a (20).

An equilibrium exists between these lactones and their acyclic counterparts. A result of that equilibrium in this process is the presence, sometimes as low as 5%, of uninverted (S,S,R) lactone 14b in the R,S,R product 16a. The yield, too, was not in the 90% range

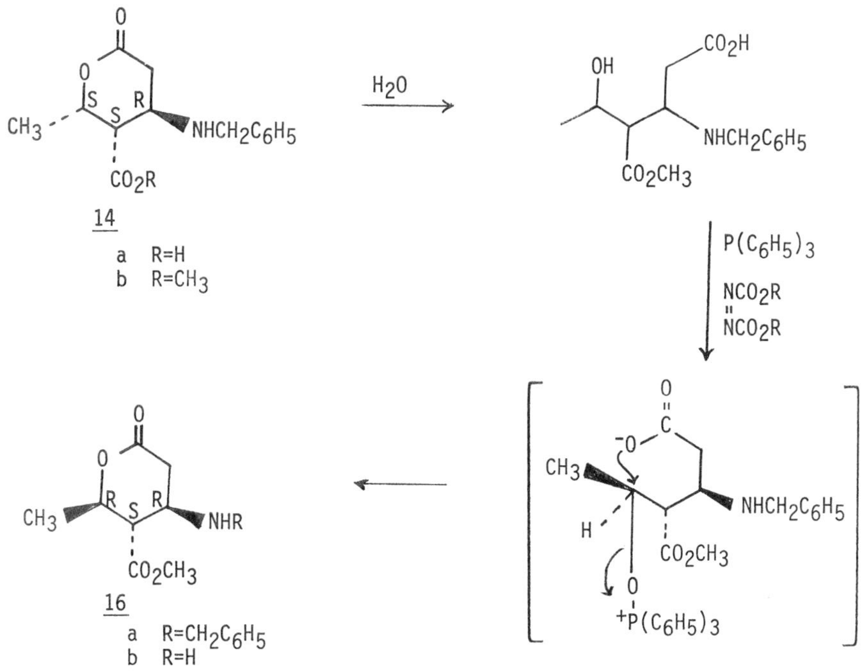

Fig. 7

expected from literature models. Further observation showed the presence of benzaldehyde, a by-product of an oxidative debenzylation (21). When, in the same process, an extra equiva- lent of azodicarboxylate was used, up to 50% overall yield of the debenzylated lactone 16b could be isolated after hydrolysis of its Schiff base. This sequence, nevertheless, could not be brought to desired yields; thus, conventional hydrogenolysis was employed and the inversion step was postponed until it could be achieved without interfering side reactions.

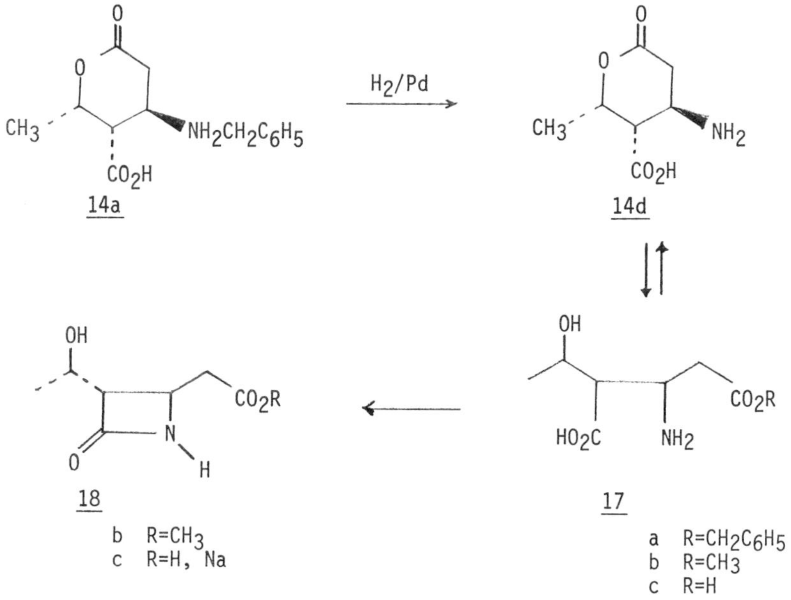

Fig. 8

Catalytic debenzylation (Fig. 8) of 14a over palladium gave the primary amine 14d ready for solvolysis. At this point, questions arose concerning subsequent steps and the necessary manipulations and work-up procedures between those steps. The intended strategy presupposed the worst case, the use of blocking groups at the hydroxyl and the lactam nitrogen before elaboration to the sensitive bicyclic β-keto ester. For such reasons, benzyl alcohol was elected for lactone opening, reserving the option to remove that group selectively by hydrogenolysis. This later caused considerable difficulty in working out simple, efficient separations of 17a, unreacted 14d and excess benzyl alcohol. Fortunately, concurrent studies uncovered methods of operation without *any* protecting groups, and methanolysis of 14d was substituted after a selective saponification of 18b was developed.

Methyl ester, 18b, has unusual properties. It is quite water-soluble; its distribution coefficient between water and methylene chloride exceeds 10:1, favoring water. The sodium salt of its acid can be crystallized allowing conventional isolation and storage if need be. The sodium salt is stable; it can be dried at temperatures above 100°C. As further testimony to its stability, it has been shown that ethyl acetate, for example, reacts more readily with hydroxylamine at pH 7 than does this β-lactam (ir: 1745 cm^{-1}; KBr pellet). In water at room temperature, however, the half-life of 18c is only about 3 hours at pH 2.3, whereupon it reverts to 14d without change in chirality. The net result of that treatment is to erase three efficient steps in short order!

Summarizing then, the preferred route for this segment of the synthesis is as follows: the chiral lactone, 14d, is solvolyzed in methanol and the hexanoate methyl ester, 17b, is crystallized from methylene chloride as its zwitterion. The β-lactam is formed with dicyclohexylcarbodiimide in aqueous acetonitrile and, after removal of the co-formed dicyclohexylurea, the product is saponified in water and its sodium salt crystallized. The hydroxyl inversion can be added to this sequence prior to the saponification, but for ease of purification of the intermediates, it is best postponed again.

The lactam acetic acid may be converted to the β-keto ester 8-epi 6 by two chemically equivalent methods (Fig. 9). Both utilize imidazolide activation of the lactam acid. In the former (22a), Meldrum's acid is acylated and the product converted to 8-epi 6 by reaction with p-nitrobenzyl alcohol. Alternatively, p-nitrobenzyl alcohol and Meldrum's acid are allowed to react to form a half malonate ester, the magnesium salt of which is then acylated directly (22b). Side chain extension can also be achieved via the lactam acid chloride; however, that sequence still requires hydroxyl protection.

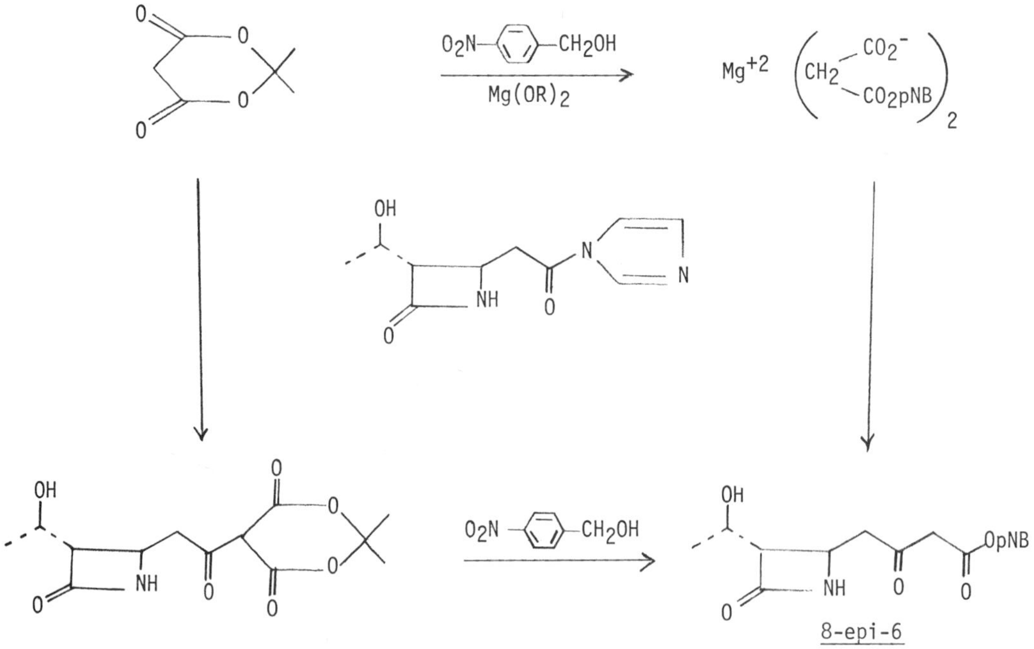

Fig. 9

The much delayed inversion of the hydroxyl group can now be performed conveniently using the azodicarboxylate-phosphine method mentioned earlier, with formic acid providing the nucleophile. The formate group is readily removed with dilute acid, giving 6 (Fig. 4).

The safety of the introduction of the diazo functionality (6 → 7) is cause for concern. The usual reagent for this reaction is tosyl azide or p-carboxybenzene sulfonyl azide which are both shock sensitive and of limited thermal stability. Polymer-bound tosyl azide (23) is reported to be safe; however, its insolubility would hamper purification of 7. An alternative reagent, para-dodecylbenzenesulfonyl azide, was developed. It is much less hazardous than the usual reagents, and the salts of its corresponding sulfonamide, which remain after reaction, are soluble and readily separated from 7. Exothermic decomposition of this new reagent begins at about 135°C; drop weight shock tests show its lesser sensitivity, also (24).

The remaining steps of the synthetic sequence are common to the chiral, stereocontrolled total synthesis (5) of Fig. 4. The cyclization of 7 to 8 behaved as expected (8), giving high yields of crystalline product. A variety of activating groups were explored to form its enol and, of course, provide the leaving group in the conversion of 8 to 10. The criteria for the choice reagent included selectivity for the ketone oxygen, rapid reactivity and, in turn, rapid displacement. The enol phosphate allowed both steps to be carried out without intervening work-up, giving the crystalline bis-protected thienamycin 10, in greater than 80% overall yield.

Efficient deprotection via catalytic hydrogenation is unpredictable with β-lactam antibiotics. Benzyl esters of penicillin respond nicely whereas the corresponding cephalosporins behave poorly. In this case, the p-nitrobenzyl groups were smoothly removed over platinum oxide catalyst in a medium buffered at pH 7. After hydrogenolysis, the aqueous solution was treated with excess benzyl formimidate, forming thienamycin formimidate, which was crystallized from aqueous ethanol, completing the process.

One final, serendipitous point deserves comment. It may be recalled that the resolution of the lactone 14a with d(+)-camphorsulfonic acid (CSA) gave the R,R,S isomer directly, which was to be used for development of the process. The S,S,R enantiomer in the mother liquors was to be purified with ℓ-(-)-CSA for making the natural product. d(+)-CSA is relatively inexpensive and commercially available; its enantiomer is not. Before becoming heavily committed to the more expensive CSA, a remarkable observation was made: *the enantiomer of the debenzylated lactone 14d which crystallized most readily with the (+)-CSA, turned out to possess S,S,R absolute configuration.* Thus, it was possible to hydrogenate the reserved mother liquor residues and obtain pure, crystalline S,S,R salt in more than 70% yield. This is the only resolution of diastereomers thus far encountered that gives, effectively, both antipodes with the same crystallizing partner!

Acknowledgement - This paper reports the achievements of many dedicated chemists in the Merck Process Research Department. It is *their* paper. I am honored to be associated with them and pleased to be their spokesman.

Drs. D.G. Melillo, I. Shinkai, M. Sletzinger, Messrs. T.M.H. Liu and K.M. Ryan (25) gave birth to this research and carried it into its development phases. Drs. and Messrs. H.R. Barkemeyer, F.W. Bollinger, G.C. Dezeny, G. Gal, E.J.J. Grabowski, V.J. Grenda, G.G. Hazen, W.H. Jones, M.A. Kozlowski, R.B. Purick, D.F. Reinhold, F.E. Roberts, W.K. Russ, E.F. Schoenewaldt, W.F. Shukis, G.B. Smith, R.P. Volante, L.M. Weinstock, and A.J. Zambito studied many of the steps in-depth and ably translated them into reproducible, large-scale processes. Drs. Gal and Reinhold deserve mention, in addition, for the resolution. Dr. A.W. Douglas and Mr. R.A. Reamer contributed numberous critical NMR studies.

The cooperation of the Merck Basic Research chemists whose work is cited, and our Analytical Research Department is gratefully acknowledged.

REFERENCES

1. At the 16th Interscience Conference on Antibacterial Agents and Chemotherapy, Chicago, Illinois [1976]; published form: Ref. 2.
2. J.S. Kahan, F.M. Kahan, R. Goegelman, S.A. Currie, M. Jackson, E.O. Stapley, T.W. Miller, A.K. Miller, D. Hendlin, S. Mochales, S. Hernandez, H.B. Woodruff, J. Birnbaum, Antibiotics, 32, 1-2 (1979).
3. G. Albers-Schönberg, B.H. Arison, O.D. Hensens, J. Hirshfield, K. Hoogsteen, E.A. Kaczka, R.E. Roades, J.S. Kahan, F.M. Kahan, R.W. Ratcliffe, E. Walton, L.J. Ruswinkle, R.B. Morin, B.G. Christensen, J. Am. Chem. Soc., 100, 6491-6499 (1978).
4. D.B.R. Johnston, S.M. Schmitt, F.A. Bouffard, B.G. Christensen, J. Am. Chem. Soc., 100, 313-315 (1978).
5. (a) Reported at "Penicillin--50 Years After Fleming", London, England, May 2, 1979.
   (b) T.N. Salzmann, R.W. Ratcliffe, F.A. Bouffard, B.G. Christensen, Proc. Royal Soc. (London), *in press.*
6. K.J. Wildonger, W.J. Leanza, T.W. Miller, B.G. Christensen, Interscience Conference on Antimicrobial Agents and Chemotherapy, Boston, Massachusetts, October, 1979.
7. T.N. Salzmann, R.W. Ratcliffe, B.G. Christensen, 176th ACS National Meeting, Miami, Florida, September, 1978. See also Ref. 8 for annelation of 5-membered rings.
8. R.W. Ratcliffe, T.N. Salzmann, B.G. Christensen, Tetrahedron Letters, 21, 31-34 (1980). An oxapenam has also been cyclized by this reaction: L.D. Cama, B.G. Christensen, Tetrahedron Letters, 4233 (1978).
9. European Patent Application Publication Number 0010312; publication date April 30, 1980.
10. R. Breckpot, Bull. Soc. Chim. Belg., 32, 412 (1923).
11. Alternatively, 2-trimethylsilyl-1,3-dithane. This sequence makes use of a carboxylic acid intermediate instead of the aldehyde. T.N. Salzmann, R.W. Ratcliffe, B.G. Christensen, and F.A. Bouffard, submitted for publication.
12. R.B. Woodward, K. Heusler, J. Gosteli, P. Nalgeli, W. Oppolzer, R. Ramage, S. Ranganathan, and H. Vorbrüggen, J. Am. Chem. Soc., 88, 852 (1966).

13. For early background see J.C. Sheehan, E.J. Corey, Organic Reactions, 9, 338 (1957).
14. J.C. Sheehan and K.R. Henery-Logan, J. Am. Chem. Soc., 81, 5838 (1959).
15. A.K. Mukerjee and A.K. Singh, Tetrahedron, 34, 1731-1767 (1978). Kametani has reported a β-amino acid cyclization with DCC in 22.5%; however, not without epimerization. T. Kamatani, S.-P. Huang, and M. Ihara, Heterocycles, 12, 1183-1187 (1979).
16. N.S. Isaacs, Chem. Soc. Revs., 5, 181 (1976).
17. cf. Ref. 4. More detail may be found in F.A. Bouffard, D.B.R. Johnston, and B.G. Christensen, J. Org. Chem., 45, 1130-1135 (1980).
18. These results were not unanticipated, and they reflect a cis attack by hydride to 12. Hydrogen bonding of NH to acetyl oxygen is observed with 12 to give a pseudo cyclic structure, and an analogous 1,2,6-boroxazine appears to be implicated during reduction.
19. We thank J.M. Hirshfield of these laboratories for this study.
20. H. Loibner and E. Zbiral, Helv. Chim. Acta, 59, 2100 (1976) and references found therein. The exact nature of the bracketed intermediate in Fig. 7 is not known.
21. G.W. Kenner and R.J. Stedman, J. Chem. Soc., 2089-2094 (1952).
22. (a) Acylation of Meldrum's acid: Y. Oikawa, K. Sugano, and O. Yonemitsu, J. Org. Chem., 43, 2087-2088 (1978).
    (b) For the malonate half ester approach, see D.W. Brooks, L.D.-L. Lu, and S. Masamune, Angew. Chem., Int'l Ed., English, 18, 72-74 (1979).
23. W.R. Roush, D. Feitler, and J. Rebek, Tetrahedron Letters, 1391 (1974).
24. We thank R.C. Connell of these laboratories for these studies.
25. D.G. Melillo, I. Shinkai, K.M. Ryan, T.M.H. Liu, and M. Sletzinger have submitted a manuscript covering much of the chemistry described herein.

APPROACHES TO THE SYNTHESIS OF TOPOLOGICALLY SPHERICAL MOLECULES. RECENT ADVANCES IN THE CHEMISTRY OF DODECAHEDRANE

L. A. Paquette

Department of Chemistry, The Ohio State University, Columbus, Ohio 43210, USA

Abstract - In this lecture, an account is presented of the high points in our progress towards the synthesis of the pentagonal dodecahedrane. The point of furthest advance is represented by the obtention of monosecododecahedrane 50, a molecule which possesses eighteen cis-syn locked methine groups and lacks only a final carbon-carbon bond.

The pentagonal dodecahedrane (1) is the parent of a fascinating and important group of molecular entities which has successfully defied synthesis to the present time. Conceived initially in abstract terms in 500-400 B.C. by Greek mathematicians, the dodecahedrane molecule ($C_{20}H_{20}$) has in recent years become a focal point of much synthetic and theoretical

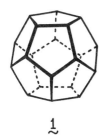

1

chemical research (1). Time will not allow me to deal here with the many significant recent achievements which have been made by several research groups. However, the attention of the audience is directed to an excellent timely review by Professor P. E. Eaton (2).

Although dodecahedrane has yet to be adopted as a symbol for an International Symposium on Organic Synthesis, it is hoped that such a compliment to this most symmetric of molecules (point group I_h) might be forthcoming. A lesser honor has befallen the structure in another way. Although no dodecahedrane-like molecule is known to occur in nature, St. Pyrek has viewed 1 to be a diterpene hypothetically derivable by proper biogenetic-like fusion of four isoprene units (3)! In more realistic terms, we see that our ultimate objective is constructed of 20 methine units, 30 carbon-carbon bonds, and 12 multiply fused cyclopentane rings. Furthermore, the topological requirements are such that all ring fusions must be cis-syn, a feature which requires passage through a large number of intermediates which possess unusually high levels of steric strain. As discussed in our earlier treatment of this subject (4), the difficulties associated with the surmounting of the adverse entropic and steric demands of such chemical transformations have caused the convergent approaches devised by Woodward (5), Eaton (6), and Paquette (7) to seriously falter. For this reason, the more recent efforts in this laboratory have been concentrated on synthetic pathways which deploy all of the constituent 20 carbon atoms in proper relative spatial relationship at the outset.

ACCESSIBILITY OF CLOSED DILACTONE 4

A simple method for arriving at diacid 2 in a single laboratory manipulation by domino Diels-Alder addition of dimethyl acetylenedicarboxylate to 9,10-dihydrofulvalene has been detailed previously (8,9). This reaction simultaneously generates a molecule having six cis-locked methine hydrogens and useful C_{2v} symmetry. By means of an efficient (80% overall) four-step sequence, 2 can be transformed into the C_2-symmetric diketo diester 3. Additional carbons can now be conveniently introduced with retention of axial symmetry to give the

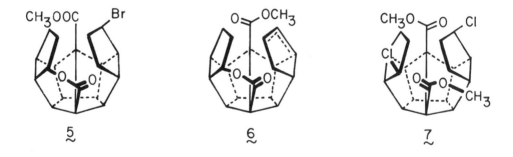

$$2 \qquad\qquad\qquad 3 \qquad\qquad\qquad 4$$

closed dilactone 4 (8b,10). Our reasoning for proceeding in this fashion is entirely apparent. An additional six methine hydrogens have taken their proper place on the exterior of the developing sphere. It may be noted also that 4 possesses the right number of carbon atoms rather rigidly held in proper spatial juxtaposition. Furthermore, the sphericality of the molecule is now such that entry of solvent molecules into its cavity will be seriously impeded.

We have studied various means for cleavage of the two lactone rings in 4 with somewhat unanticipated results (10). Chemistry does not, of course, occur concurrently at both carbonyl groups. Following attack at the first ester linkage, there is ruptured a ring vital to retention of spherical topology. The seco intermediate, once formed, drastically modifies its conformation in such a way as to increase the distance between the two central (opposed) methylene groups as much as possible. This alteration in geometry so modifies the chemical characteristics of the second carbonyl group that its reactivity falls off very appreciably. Thus, while 5 is formed rapidly upon exposure of 4 to hydrogen bromide in methanol at 0°C, no further chemical change occurs upon heating the reaction mixture at the reflux temperature for three weeks. In like fashion, trimethyloxonium fluoroborate in dichloromethane solution successfully converts 4 to 6 without disturbing the second

$$5 \qquad\qquad\qquad 6 \qquad\qquad\qquad 7$$

lactone functionality. A lone exception to this allosteric behavior has been uncovered. When 4 is treated with hydrogen chloride in methanol solution at room temperature, 7 is obtained in 62% yield. Although we can offer no explanation for this singular phenomenon, the successful preparation of 7 has provided handsome dividends (see below).

PREPARATION OF OPEN DILACTONES AND THE COMPLICATIONS OF TRANSANNULAR REACTIVITY

Closed dilactone 4 is a most attractive intermediate because it carries a minimum of functionality. More significantly, its two carbonyl groups are disposed in a 1,4-relationship with stereoelectronic features suitable for reductive cleavage of the unnecessary internal bond. In the presence of sodium and trimethylsilyl chloride in refluxing toluene, 4 is converted readily and in high yield to bistrimethylsilyl enol ester 8, hydrolysis of which in methanol delivers the open dilactone 9 (10). The exact structure of 9 has subsequently been confirmed through an x-ray crystal analysis by Professor W. Nowacki and Dr. P. Engel (11).

At this stage of our projected synthesis, 14 of the methine hydrogens have been properly positioned on the framework. Disappointingly, however, 9 proved to be extremely susceptible

$\underset{\sim}{4}$ $\xrightarrow[\text{(CH}_3)_3\text{SiCl}]{\overset{\text{Na}}{\text{toluene}}}$

(CH$_3$)$_3$SiO OSi(CH$_3$)$_3$

$\underset{\sim}{8}$

$\xrightarrow{\text{CH}_3\text{OH}}$

$\underset{\sim}{9}$

1. Na, NH$_3$
2. CH$_3$I or CH$_3$SSCH$_3$

R

R

$\underset{\sim}{10a}$, R = CH$_3$
    $\underset{\sim}{b}$, R = SCH$_3$

$\xrightarrow[\text{2. H}_3\text{O}^+]{\text{1. LiAlH}_4}$

CH$_3$ HO

OH

CH$_3$

$\underset{\sim}{11}$

$\xrightarrow[\text{CH}_3\text{OH}]{\text{[HCl]}}$

CH$_3$

H$_2$

CH$_3$

$\underset{\sim}{12}$

to symmetry-destroying transannular cyclization under both alkaline and acidic conditions (10). This most undesirable feature appears to arise from the absence of solvation in the molecular interior and the reasonable proximity of transannular centers. A reactive functional group, once generated, will understandably acquire hyperreactive qualities under these circumstances.

At this stage, we were led to block the enolizable centers in 9. Soon it was discovered that the dianion of 9, prepared by sodium-liquid ammonia reduction of 4, was adequately stable to be doubly functionalized as in 10. It will be easily imagined that the ^{1}H NMR spectra of our molecules are characterized by enormously complex upfield regions. However, the singlet absorption due to the pair of methyl groups in 10a stands out as a beacon signaling the C$_2$ symmetric nature of the species. Because this phenomenon subsequently allowed us to perform many diagnostic experiments on a few milligrams, this substitution pattern was gainfully deployed in much of our ensuing work.

Whereas 10a can be reduced to dilactol 11 (thermodynamically preferred isomer illustrated), this molecule has shown no tendency to adopt hydroxy aldehyde tautomeric forms. The interest in 11 resides in its potential for hypothetical two-fold dehydration (see 13) that could result in the formation of 14. Should appropriate Prins reaction conditions transform 14 to the triseco dienedione 15, we would be well on our way. The actual situation is remarkable and unfortunately less glamorous. Experiments directed toward the acid-catalyzed dehydration of 11 uniformly resulted in smooth conversion to 12 (12). The outcome of these reactions was reduction in one portion of the molecule and oxidation in the other. All attempts to modify conditions had the invariable result of driving the system inexorably to this unsymmetrical stable product. This finding hinted at the involvement of a highly efficient transannular hydride shift which was energetically profitable because of the transient formation of a dioxonium ion from a simpler oxonium species. This hypothesis was substantiated by LiAlD$_4$ reduction of 10a to 16 and its acid-catalyzed rearrangement to 17. The isotopic labels which were originally introduced on opposite sides of the hemisphere ultimately become positioned on the same carbon atom. Regrettably, no transformations involving 11 and its derivatives, interesting as they might be in them-

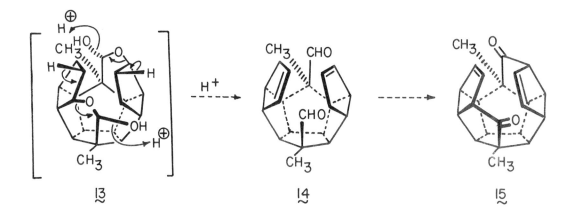

13   14   15

selves, led in the desired direction (12). Unquestionably, success in the construction of
additional peripheral bonds was attendant upon new approaches which would effectively by-

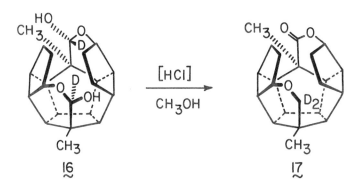

16   17

pass the transannular behavior so characteristic of the spherical molecules at this stage
of elaboration.

UNUSUAL PHOTOISOMERIZATION OF A BRIDGED α-DIKETONE

Our first attempt to control the regioselectivity of carbon-carbon bond formation was designed to take advantage of the well known ability of α-diketones for photoinduced hydrogen transfer and cyclization (13-15). In unconstrained systems, γ-hydrogen abstraction is
overwhelmingly preferred and 2-hydroxycyclobutanones are formed exclusively. In 18, only
δ-hydrogens are available in close spatial proximity to the n → π* excited carbonyl groups.
Our expectation was that 18 would follow the example of numerous monocarbonyl compounds

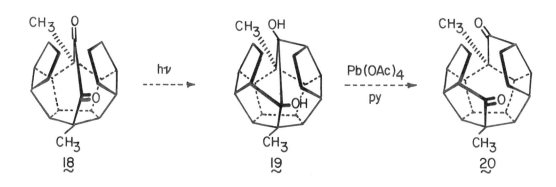

18   19   20

which yield cyclopentanols upon irradiation (16) and experience two-fold cyclization to 19. Although 18 does contain an interconnective bond improperly predisposed for arrival at 1, it was anticipated that this bond would become subject to cleavage with lead tetraacetate in 19.   An expedient route to triseco diketone 20 would then be in hand.

Having envisaged this wishful formalism, we proceeded to accomplish a very direct synthesis of 18.   As shown below, dichloro diester 7 can be reduced with tri-n-butyltin hydride and

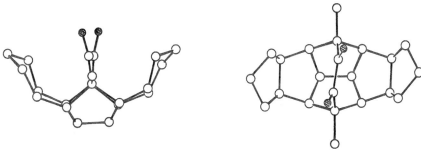

7

$(\underline{n}-Bu)_3SnCl$
────────────→
$NaBH_4$, AIBN
glyme, Δ

21
1. Na, NH_3
2. CH_3I

18

1. Na, Me_3SiCl
toluene
←────────────
2. $FeCl_3$,
HCl, ether

22

the resulting diester (21) reductively methylated in the usual way. Under acyloin conditions, 22 experiences facile cyclization to an α-hydroxy ketone which, because of excessive steric compression, is oxidized to 18 with exceptional ease (17). Thanks to the x-ray contributions of Dr. Peter Engel in Bern, the three dimensional feature of this beautifully crystalline yellow solid were unveiled. As seen in Fig. 1, the molecule is C_2 symmetric and projects an O=C-C=O dihedral angle of 19.7°. On this basis, the carbonyl groups in 18 were expected to be highly responsive to photoexcitation (18).

Fig. 1

When 18 was irradiated in benzene-acetone-tert-butyl alcohol solution, there was produced
a diol whose ^{13}C (11 lines) and ^{1}H NMR spectra revealed that twofold axial symmetry had
been maintained.   Particularly exciting was the willingness of this substance to undergo
cleavage to an equally symmetric diketone upon reaction with lead tetraacetate in pyridine
(15).   Before long, however, suspicion arose that this substance was not 20, especially
when H/D exchange could not be effected despite the deployment of forcing conditions.
Furthermore, an extraordinarily facile reconversion of the diketone to the diol was observed in the presence of a slight excess of lithium diisopropylamide in tetrahydrofuran

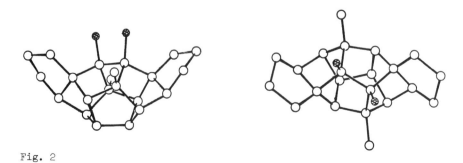

solution at room temperature.   While the ability of LDA to reduce certain types of carbonyl
groups has been previously noted (19), no examples of intramolecular pinacolization have
been reported.   In the spherical contour adopted by 20, the pair of carbonyl groups are
not in close proximity and consequently do not appear predisposed for reductive coupling
under such unusually mild conditions.

Due consideration of these observations led us to the conclusion that 26 was the diol
photoproduct.   This structural assignment was subsequently confirmed by an x-ray analysis
once again performed in Dr. Engel's laboratory.   The twisted shape of this unusual deca-
quinane is shown in Fig. 2.

Fig. 2

A reasonable pathway for the photocyclization of 18 to 26 involves 1,6-hydrogen transfer to give 23. Perhaps because of conformational factors, this biradical does not cyclize to generate an added peripheral cyclopentane ring. Rather, rearrangement to isomeric radical 24 appears to be kinetically favored. Closure of 24 gives 25, a product which is isolable at shorter reaction times. Similar intermediates presumably intervene in the conversion of 25 to 26 upon continued irradiation. Models indicate that the oxidative cleavage of 26 does not substantively relieve the contorted features of the carbon framework. Consequently, the carbonyl groups remain locked in close proximity.

Although the preceding efforts were not crowned with success, they were crucial to a fuller appreciation of the potential which resides in ketone photochemistry as a tool for carbon-carbon bond formation. Armed with a reasonable knowledge of the structural dimensions of our intermediates (x-ray data and molecular model studies) and an appreciation of the distance and geometric requirements necessary for excited-state intramolecular hydrogen abstraction reactions(20), we next opted literally to shine light on the problem.

TRISECO DERIVATIVES VIA DICHLORO DIESTER REDUCTION (AND METHYLATION)

Ester carbonyl groups are rarely photoactive and dilactones 4, 9, and 10, as well as diesters 21 and 22, proved not to be exceptions to this general rule. The task, then, was to transform 7 into one or more highly condensed ketonic intermediates by means of an appropriate intramolecular reductive coupling scheme. Although a few examples of metal-promoted ω-halo ketone cyclizations have been reported (21), 7 could not be encouraged to conduct itself analogously under a wide range of conditions, despite rigorous adherence to the highest standards of experimental precision. Since formation of carbanionic centers at the originally chlorinated carbon atoms in 7 led most frequently to simple reduction with formation of 21, the alternate electronic process in which the ester functionalities are made to experience reduction first was examined.

The dissolving metal reduction of 7 in liquid ammonia was viewed as a viable means of generating radical anion 27 where proximity considerations were expected to favor ready (hypothetical) S_N2 displacement of chloride ion and generation of 28. In our thinking, 29 would inevitably evolve upon continued reduction of the latter intermediate. In initial investigations of this process, quenching of the reduction mixture was brought about with an excess of methyl iodide. These circumstances led to the isolation of 30 and 31 in a combined yield in excess of 65%. The mechanistic considerations just presented are seen to deny ring closure from materializing a second time on the opposite face of the sphere. Nonetheless, the transformation has the undeniable value of catapulting the synthetic scheme (presently only 11 steps removed from starting material 2) to the tetraseco stage of construction.

There is no question that 31 is formed during solvent evaporation as a consequence of the alkalinity of the medium and the presence of residual alkylating agent. In fact, conditions can be controlled such that 30 predominates by a wide margin. Nonetheless, many of the exploratory studies were carried out on 31. When irradiated in benzene-tert-butyl alcohol (4:1) solution, this keto ester underwent smooth conversion to 32 whose dehydration to 33

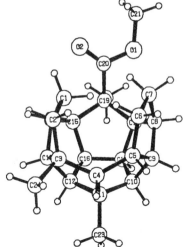

Fig. 3

could be routinely achieved under acidic conditions without evidence of skeletal rearrangement or Wagner-Meerwein methyl migration. With ensuing diimide reduction, the formation of triseco ester 34 was achieved. At this point, Dr. John Blount of Hoffmann-LaRoche expeditiously substantiated the structural assignment by x-ray methods (Fig. 3).

It will be noticed that the opposed methylene groups in 34 are unfunctionalized. In a perhaps too brief effort to correct this situation, epoxide 35 was prepared. On exposure to boron trifluoride etherate, remarkably clean isomerization to 36 was observed. The tertiary allylic alcohol does not respond in prototypical fashion to chromium-based oxidants (22). Rather than transmute into the 1,3-transposed enone, 36 gives rise to epoxy ketone 37. While reactions of this type have seen precedent (23), we deemed structural confirma-

tion a necessity. To this end, 36 was epoxidized and caused to undergo an interesting and unusual epoxy alcohol-epoxy alcohol rearrangement. The subsequent oxidation of 39 also gave 37.

Our detailed studies of the reduction of 7 led to the discovery that treatment with 14 equivalents of lithium in liquid ammonia followed by an ethanol quench provides a mixture of 40 and 41 in addition to a hydroxy ester formed by transannular bonding. Unoptimized yields of 31 and 26%, respectively, have been realized at this time. With extensive experience having been gained in the di- and trimethyl series, our next step was to apply this expertise to 40. Fortunately, these proved to be realistic prototypes and the conversion of 40 through 42 and 43 to 44 was found to be delightfully uneventful and efficient. The endo stereochemical disposition of the hydroxymethyl groups in 40 was deduced by spectroscopic techniques.

THE MONOSECO LEVEL OF ELABORATION

The next tantalizing problem which we faced was to engage the oxygenated carbon atom of the triseco derivatives into dual cyclopentane ring formation. This facet of the work was pioneered with the C_2 symmetric dimethyl ester 45 which was available to us in three precedented steps from 30. Because it was already obvious that the carbomethoxy group in 45 could not be enticed into chemical reaction by photochemical means, we made the decision to proceed with aldehyde 46. The conversion of 45 to 46 was an excellent one, but possible complications from the fact that the -CHO unit was bonded to a fully substituted carbon now had to be contended with. The literature dealing with the photochemical cyclization of aldehydes to cyclobutanols (24) leaves no doubt that these structural features are most conducive to decarbonylation. We nevertheless squarely faced the prospect of lowered yields and concerned ourselves with low-temperature conditions and the like. While 46 did prove highly prone to carbon monoxide extrusion, a 29% yield of the epimeric "homo-Norrish" cyclopentanols could be realized. With subsequent pyridinium chlorochromate oxidation, the desired diseco diketone 47 was obtained. Needless to say, we were delighted to find that 47 could be photocyclized reproducibly in high yield and that removal of the tertiary hydroxyl group in 48 and saturation of the double bond in 49 were encouragingly simple and efficient steps.

With arrival at the beautifully crystalline secododecahedrane 50, a return to C_{2v} symmetry status materializes, a phenomenon reflected in the appearance of only eight lines in its ^{13}C NMR spectrum. By every standard, the non-bonded interactions in 50 should provide a heightened level of steric congestion between the methylene groups. Just to what extent the system is inflexible in deterring the interfering endo hydrogens from moving apart must await the result of x-ray studies.

The situation at this point was such as to make pursuit of the unsubstituted secododecahedrane irresistible. However, initial attempts to oxidize 44 with chromium reagents invariably promoted unwanted dehydration and overoxidation, undoubtedly because considerable steric congestion is released during the conversion of 51 and 52. Analogous complications materialize with 43. These difficulties now appear to have been satisfactorily resolved

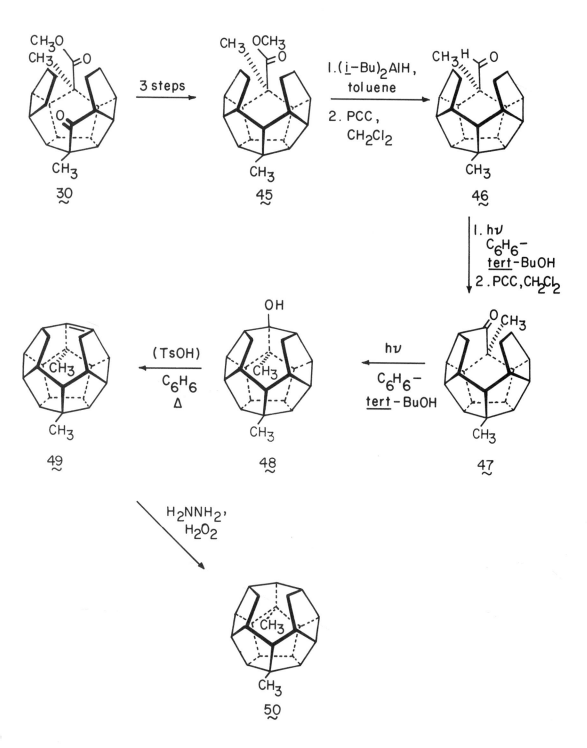

through the agency of activated manganese dioxide, a reagent which converts 44 predominantly to 53. In contrast to the behavior of 46, it is our belief that 53 will not decarbony-
late readily and will undergo photocyclization with striking efficiency. This hypothesis
awaits imminent experimental substantiation.

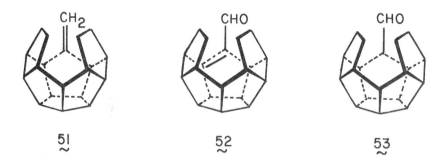

51   52   53

INSTALLATION OF THE FINAL BOND

With construction of four new peripheral bonds, eighteen of the framework atoms are seen to possess the proper syn,cis stereochemistry. The remaining two dodecahedryl carbons must now be conjoined to arrive at our objective. It is worthy of mention here that the preparation of 49 requires but 21 steps from cyclopentadiene. Although we would have been delighted to complete our self-appointed task in 20 laboratory stages (one per ring carbon!), we next concerned ourselves with the direct cyclization of 49. Despite the extreme congestion about the "unmade" bond in this hydrocarbon, our efforts to date have not been rewarded. Nor has its fully saturated analogue 50 yielded to catalytic dehydrogenation.

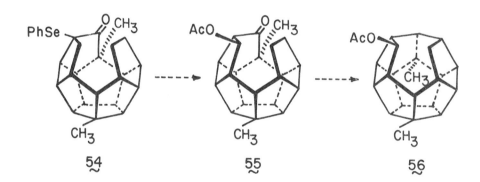

54   55   56

Although investigations along the preceding lines continue, other workable solutions to this final maneuver are currently being examined. One such possibility hinges on the phenyl-seleno ketone 54 which can be prepared from 47. The virtue of this intermediate lies in its potential for elimination (via the selenoxide) and subsequent 1,4 capture of acetic acid by the resultant α,β-unsaturated ketone. The conversion of 55 to 46 is to follow earlier precedent. At this point, a sound basis for further advancement will have been established.

ACKNOWLEDGEMENTS

It is a pleasure to pay a richly deserved tribute to the three men (Douglas Balogh, William Begley, and Martin Banwell) whose skill and zeal have enabled us to advance this far on an often difficult and frustrating venture. My good fortune in having such gifted collaborators summons forth a deep sense of appreciation. We should also like to thank the National Institutes of Health for the financial support which permitted the reduction of our ideas to practice.

REFERENCES

1. T. Heath, A History of Greek Mathematics, Vols. I and II, Oxford University Press (1921).
2. P. E. Eaton, Tetrahedron, 35, 2189 (1979).
3. J. St. Pyrek, Polish J. Chem., 53, 1557 (1979).
4. L. A. Paquette, Pure Appl. Chem., 50, 1291 (1978).
5. (a) R. B. Woodward, T. Fukunaga, R. C. Kelly, J. Am. Chem. Soc., 86, 3162 (1964); (b) O. Repic, Ph.D. Dissertation, Harvard University (1976).
6. (a) P. E. Eaton, R. H. Mueller, J. Am. Chem. Soc., 94, 1014 (1972); (b) P. E. Eaton, R. H. Mueller, G. R. Carlson, D. A. Cullison, G. F. Cooper, T.-C. Chou, E.-P. Krebs, ibid., 99, 2751 (1977); (c) P. E. Eaton, G. D. Andrews, E.-P. Krebs, A. Kunai, J. Org. Chem., 44, 2824 (1979).
7. (a) L. A. Paquette, W. B. Farnham, S. V. Ley, J. Am. Chem. Soc., 97, 7273 (1975); (b) L. A. Paquette, I. Itoh, W. B. Farnham, ibid., 97, 7280 (1975); (c) L. A. Paquette, I. Itoh, K. Kipkowitz, J. Org. Chem., 41, 3524 (1976).
8. (a) L. A. Paquette, M. J. Wyvratt, J. Am. Chem. Soc., 96, 4671 (1974); (b) L. A. Paquette, M. J. Wyvratt, O. Schallner, D. F. Schneider, W. J. Begley, R. M. Blankenship, ibid., 98, 6744 (1976).
9. D. McNeil, B. R. Vogt, J. J. Sudol, S. Theodoropoulos, E. Hedaya, J. Am. Chem. Soc., 96, 4673 (1974).
10. L. A. Paquette, M. J. Wyvratt, O. Schallner, J. L. Muthard, W. J. Begley, R. M. Blankenship, D. Balogh, J. Org. Chem., 44, 3616 (1979).
11. W. Nowacki, P. Engel, Z. Kristallogr., Kristallgeom., Kristallphys., Kristallchem., 4 (1977).
12. D. W. Balogh, L. A. Paquette, J. Org. Chem. in press.
13. Reviews: (a) M. Rubin, Fortschr. Chem. Forsch., 13, 251 (1969); (b) B. M. Monroe, Adv. Photochem., 8, 77 (1970).
14. (a) W. H. Urry, D. J. Trecker, J. Am. Chem. Soc., 84, 713 (1962); (b) W. H. Urry, D. J. Trecker, D. A. Winey, Tetrahedron Lett., 609 (1962); (c) N. J. Turro, T. Lee, J. Am. Chem. Soc., 91, 5651 (1969); (d) T. Burkoth, E. Ullman, Tetrahedron Lett., 145 (1970).
15. P. J. Wagner, R. G. Zepp, K.-C. Liu, M. Thomas, T.-J. Lee, N. J. Turro, J. Am. Chem. Soc., 98, 8125 (1976).
16. (a) P. J. Wagner, A. E. Kemppainen, J. Am. Chem. Soc., 94, 7495 (1972); (b) P. J. Wagner, P. A. Kelso, A. E. Kemppainen, R. G. Zepp, ibid., 94, 7500 (1972); (c) E. J. O'Connell, ibid., 90, 6550 (1968); N. C. Yang, M. J. Jorgenson, Tetrahedron Lett., 1203 (1964).
17. D. W. Balogh, L. A. Paquette, P. Engel, submitted for publication.
18. α-Diketones have been intensively studied by photoelectron spectroscopy and the effect of CO/CO dihedral angle on ground and excited states discussed: (a) J. L. Meeks, H. J. Maria, P. Brint, S. P. McGlynn, Chem. Revs., 75, 603 (1975); (b) D. Dougherty, J. J. Bloomfield, G. R. Newkome, J. F. Arnett, S. P. McGlynn, J. Phys. Chem., 80, 2212 (1976); (c) R. Gleiter, R. Bartetzko, P. Hofmann, H.-D. Scharf, Angew. Chem., Int. Ed. Engl., 16, 400 (1977); (d) H.-D. Martin, H. J. Schiwek, J. Spanget-Larsen, R. Gleiter, Chem. Ber., 111, 2557 (1978); (e) R. Bartetzko, R. Gleiter, J. L. Muthard, L. A. Paquette, J. Am. Chem. Soc., 100, 5589 (1978); (f) H.-D. Martin, B. Albert, H.-J. Schiwek, Tetrahedron Lett., 2347 (1979).
19. (a) L. T. Scott, K. J. Carlin, T. H. Schultz, Tetrahedron Lett., 4637 (1978); (b) C. Kowalski, X. Creary, A. J. Rollin, M. C. Burke, J. Org. Chem., 43, 2601 (1978).
20. J. R. Scheffer, A. A. Dzakpasu, J. Am. Chem. Soc., 100, 2163 (1978).
21. (a) E. J. Corey, I. Kuwajima, J. Am. Chem. Soc., 92, 395 (1970); S. Danishefsky, D. Dumas, JCS Chem. Commun., 1287 (1968).
22. W. G. Dauben, D. M. Michno, J. Org. Chem., 42, 682 (1977).
23. (a) O. Rosenheim, H. King, Nature (London), 139 1015 (1937); (b) O. Rosenheim, W. W. Starling, J. Chem. Soc., 377 (1937); (c) V. A. Petrow, W. W. Starling, ibid., 60,(1940); (d) S. Lieberman, D. K. Fukushima, J. Am. Chem. Soc., 72, 5211 (1960); (e) E. Glotter, S. Greenfield, D. Lavie, J. Chem. Soc. C, 1646 (1968) and references cited therein; (f) E. Glotter, Y. Rabinsohn, Y. Ozan, J. Chem. Soc., Perkin Trans 1, 2104 (1975); (g) P. Sundararaman, W. Herz, J. Org. Chem., 42, 813 (1977); (h) R. Ellison, Ph.D. Dissertation, Northwestern University (1976); (i) J.-H. Shau, W. Reusch, J. Org. Chem., 45, 2013 (1980).
24. (a) I. Orban, K. Schaffner, O. Jeger, J. Am. Chem. Soc., 85, 3033 (1963); (b) K. Schaffner, Chimia, 19, 575 (1965); (c) W. C. Agosta, D. K. Herron, J. Am. Chem. Soc., 90, 7025 (1968); (d) J. D. Coyle, J. Chem. Soc. B, 2254 (1971); (c) M. Lischewski, G. Adam, E. P. Serebryakov, Tetrahedron Lett., 45 (1980).

INDEX